TQM Engineering Handbook

QUALITY AND RELIABILITY

1. Designing for Minimal Maintenance Expense: The Practical Application of Reliability and Maintainability, *Marvin A. Moss*
2. Quality Control for Profit: Second Edition, Revised and Expanded, *Ronald H. Lester, Norbert L. Enrick, and Harry E. Mottley, Jr.*
3. QCPAC: Statistical Quality Control on the IBM PC, *Steven M. Zimmerman and Leo M. Conrad*
4. Quality by Experimental Design, *Thomas B. Barker*
5. Applications of Quality Control in the Service Industry, *A. C. Rosander*
6. Integrated Product Testing and Evaluating: A Systems Approach to Improve Reliability and Quality, Revised Edition, *Harold L. Gilmore and Herbert C. Schwartz*
7. Quality Management Handbook, *edited by Loren Walsh, Ralph Wurster, and Raymond J. Kimber*

8. Statistical Process Control: A Guide for Implementation, *Roger W. Berger and Thomas Hart*
9. Quality Circles: Selected Readings, *edited by Roger W. Berger and David L. Shores*
10. Quality and Productivity for Bankers and Financial Managers, *William J. Latzko*
11. Poor-Quality Cost, *H. James Harrington*
12. Human Resources Management, *edited by Jill P. Kern, John J. Riley, and Louis N. Jones*
13. The Good and the Bad News About Quality, *Edward M. Schrock and Henry L. Lefevre*
14. Engineering Design for Producibility and Reliability, *John W. Priest*
15. Statistical Process Control in Automated Manufacturing, *J. Bert Keats and Norma Faris Hubele*
16. Automated Inspection and Quality Assurance, *Stanley L. Robinson and Richard K. Miller*
17. Defect Prevention: Use of Simple Statistical Tools, *Victor E. Kane*
18. Defect Prevention: Use of Simple Statistical Tools, Solutions Manual, *Victor E. Kane*
19. Purchasing and Quality, *Max McRobb*
20. Specification Writing and Management, *Max McRobb*
21. Quality Function Deployment: A Practitioner's Approach, *James L. Bossert*
22. The Quality Promise, *Lester Jay Wollschlaeger*
23. Statistical Process Control in Manufacturing, *edited by J. Bert Keats and Douglas C. Montgomery*
24. Total Manufacturing Assurance, *Douglas C. Brauer and John Cesarone*
25. Deming's 14 Points Applied to Services, *A. C. Rosander*
26. Evaluation and Control of Measurements, *John Mandel*
27. Achieving Excellence in Business: A Practical Guide to the Total Quality Transformation Process, *Kenneth E. Ebel*
28. Statistical Methods for the Process Industries, *William H. McNeese and Robert A. Klein*
29. Quality Engineering Handbook, *edited by Thomas Pyzdek and Roger W. Berger*
30. Managing for World-Class Quality: A Primer for Executives and Managers, *Edwin S. Shecter*
31. A Leader's Journey to Quality, *Dana M. Cound*
32. ISO 9000: Preparing for Registration, *James L. Lamprecht*
33. Statistical Problem Solving, *Wendell E. Carr*
34. Quality Control for Profit: Gaining the Competitive Edge. Third Edition, Revised and Expanded, *Ronald H. Lester, Norbert L. Enrick, and Harry E. Mottley, Jr.*
35. Probability and Its Applications for Engineers, *David H. Evans*
36. An Introduction to Quality Control for the Apparel Industry, *Pradip V. Mehta*
37. Total Engineering Quality Management, *Ronald J. Cottman*
38. Ensuring Software Reliability, *Ann Marie Neufelder*

39. Guidelines for Laboratory Quality Auditing, *Donald C. Singer and Ronald P. Upton*
40. Implementing the ISO 9000 Series, *James L. Lamprecht*
41. Reliability Improvement with Design of Experiments, *Lloyd W. Condra*
42. The Next Phase of Total Quality Management: TQM II and the Focus on Profitability, *Robert E. Stein*
43. Quality by Experimental Design: Second Edition, Revised and Expanded, *Thomas B. Barker*
44. Quality Planning, Control, and Improvement in Research and Development, *edited by George W. Roberts*
45. Understanding ISO 9000 and Implementing the Basics to Quality, *D. H. Stamatis*
46. Applying TQM to Product Design and Development, *Marvin A. Moss*
47. Statistical Applications in Process Control, *edited by J. Bert Keats and Douglas C. Montgomery*
48. How to Achieve ISO 9000 Registration Economically and Efficiently, *Gurmeet Naroola and Robert Mac Connell*
49. QS-9000 Implementation and Registration, *Gurmeet Naroola*
50. The Theory of Constraints: Applications in Quality and Manufacturing: Second Edition, Revised and Expanded, *Robert E. Stein*
51. Guide to Preparing the Corporate Quality Manual, *Bernard Froman*
52. TQM Engineering Handbook, *D. H. Stamatis*
53. Quality Management Handbook: Second Edition, Revised and Expanded, *edited by Raymond Kimber, Robert Grenier, and John Heldt*

ADDITIONAL VOLUMES IN PREPARATION

TQM Engineering Handbook

D.H. Stamatis
Central Michigan University
South Pleasant, and
Contemporary Consultants Co.
Southgate, Michigan

CRC Press
Taylor & Francis Group
Boca Raton London New York

CRC Press is an imprint of the
Taylor & Francis Group, an **informa** business

First published 1997 by Marcel Dekker, Inc.

Published 2018 by CRC Press
Taylor & Francis Group
6000 Broken Sound Parkway NW, Suite 300
Boca Raton, FL 33487-2742

First issued in paperback 2019

ISBN 13: 978-0-367-44820-2 (pbk)
ISBN 13: 978-0-8247-0083-6 (hbk)

Visit the Taylor & Francis Web site at
http://www.taylorandfrancis.com

and the CRC Press Web site at
http://www.crcpress.com

Stamatis, D. H.
TQM engineering handbook / D.H. Stamatis.
p. cm.
Includes bibliographical references and index.
ISBN 0-8247-0083-X (hc : alk. paper)
1. Quality control. 2. Total quality management. I. Title
TS156.2.S73 1997 97-13117
658.5'62--dc CIP

Ειζ τουζ αγαπητουζ
Γιαννη, Χαριαννα,
Γιωργω, Τασσο και Φωτη

To my dear
John, Harianna,
George, Tasso and Foti

About the Series

The genesis of modern methods of quality and reliability will be found in a sample memo dated May 16, 1924, in which Walter A. Shewhart proposed the control chart for the analysis of inspection data. This led to a broadening of the concept of inspection from emphasis on detection and correction of defective material to control of quality through analysis and prevention of quality problems. Subsequent concern for product performance in the hands of the user stimulated development of the systems and techniques of reliability. Emphasis on the consumer as the ultimate judge of quality serves as the catalyst to bring about the integration of the methodology of quality with that of reliability. Thus, the innovations that came out of the control chart spawned a philosophy of control of quality and reliability that has come to include not only the methodology of the statistical sciences and engineering, but also the use of appropriate management methods together with various motivational procedures in a concerted effort dedicated to quality improvement.

This series is intended to provide a vehicle to foster interaction of the elements of the modern approach to quality, including statistical applications, quality and reliability engineering, management, and motivational aspects. It is a forum in which the subject matter of these various areas can be brought together to allow for effective integration of appropriate techniques. This will promote the true benefit of each, which can be achieved only through their interaction. In this sense, the whole of quality and reliability is greater than the sum of its parts, as each element augments the others.

The contributors to this series have been encouraged to discuss fundamental concepts as well as methodology, technology, and procedures at the leading edge of the discipline. Thus, new concepts are placed in proper perspective in these evolving disciplines. The series is intended for those in manufacturing, engineering, and marketing and management, as well as the consuming public, all of whom have an interest and stake in the products and services that are the lifeblood of the economic system.

The modern approach to quality and reliability concerns excellence: excellence when the product is designed, excellence when the product is made, excellence as the product is used, and excellence throughout its lifetime. But excellence does not result without effort, and products and services of superior quality and reliability require an appropriate combination of statistical, engineering, management, and motivational effort. This effort can be directed for maximum benefit only in light of timely knowledge of approaches and methods that have been developed and are available in these areas of expertise. Within the volumes of this series, the reader will find the means to create, control, correct, and improve quality and reliability in ways that are cost effective, that enhance productivity, and that create a motivational atmosphere that is harmonious and constructive. It is dedicated to that end and to the readers whose study of quality and reliability will lead to greater understanding of their products, their processes, their workplaces, and themselves.

Edward G. Schilling

Preface

In a famous cartoon Pogo said long ago: "We have found the enemy, and it is us." How profound that statement is and how relevant to quality. In the past 50 years or so, quality has been the Gideon's trumpet for a variety of ailments. Programs, in the name of quality, have come and gone, but the problems persist. Is quality a bad thing, or are we going about it the wrong way? Is it corporate greed that is causing our problems? Is it that technology has increased in complexity? Is it that the desire for "return on investment" is causing a frenzy for better prices? Is it the competition? Is it that we don't use the appropriate tools in pursuing our goals? Obviously, we can keep on asking questions. However, what is the answer?

In this book we address some of these issues and we provide an answer. The answer, in simple terms, is Quality—not just in name, but internalized throughout the organization, for the benefit of the customers (including ultimate consumers and stockholders), suppliers, society at large, and even the employees. The quality philosophy and program we cover here is what some people have called Total Quality Management, Total Improvement Management, Total Quality Improvement, and even a Breakthrough Philosophy to Improvement. The name is not as important as what is presented.

One may ask at this early juncture, why another book on quality, or even on Total Quality Management? A fair question, which needs to be addressed. To begin with, there is no doubt that corporations not only in the United States but all over the world proclaim that quality is a way of doing business, and furthermore they aim to please the customer. We suggest that that claim is only partially true. To be sure, improvements have been made because of quality awareness and implementation practices emphasizing quality. However, that is not good enough. We still produce and market for price! Why? Because we have been conditioned, all of us, to buy on price. This is a serious charge. But what happens if we dig a little deeper to identify the enemy and find it is us? There are, for instance, more than 20 million American families who own stock. It is safe to assume they are not all investing in environmentally aware companies. Rather, they are putting their money in companies with the expectation that they will get even more money out.

Many millions of individuals are counting on a comfortable retirement thanks to the strong financial performance of the companies in which their pensions are invested. When corporations fail to offer a good return, a responsible pension fund manager gets rid of the stock and buys something else. A company whose profits fall below expectations quickly comes under pressure from managers of pension funds across the country. Such managers have a responsibility to provide the best return possible to their clients, many of whom are teachers, state police, janitors, and so on. Institutional investors are much more demanding than they were 30 years ago, and much more willing to do things to bring about change, because they are under pressure themselves. Pressure is everywhere. Corporate managers, individual investors, and institutional investors all are looking for a return on their investment, to the point where quality often takes a second seat to profits. Let us look at what has been happening in the name of quality. In the semiconductor industry, the engineers push their hard drives on the basis of price and quality. But if you look at the actual product the megabytes may not be correctly identified. In some cases, they are off by 5%.[1] Why? In the software industry, it is not unusual to ship a new version of a program even though the beta site testing is not complete or "bugs" in the program are still present.[2] This is done in the name of beating the competition to the market. What happened to quality. Why? In the food industry, if you look at labels, you may notice that what the picture shows and the ingredient list presents are not the same.[3] Why? In the automotive industry, if you examine the end product, a car, you may find that when consumers have a problem, they have a rough time fixing it—that is, if they are lucky enough to find the appropriate representative. On the other hand, even when the company knows about a problem, it may still try to dismiss it as "misleading" and not important.[4] Why? In health care, politics, education,[5] and many industries the same tune is being played, that is; stretching the truth, outright lying, misrepresentation, and so on. Why? If quality is as important as everyone claims it to be, especially in the automotive industry, why are none of the American automotive companies in the top ten for perceived quality cars in the United States,[6] even though they all have practiced quality control for a very long time? Nevertheless, in all industries quality is proclaimed as the only way to true improvement.

The pressure is high for everyone to perform, but performance alone will not do it. An organization must produce effectively and efficiently in order for it to survive. One way of surviving is through quality. This quality, however, has to be totally in the minds of employees, corporate management, and the public at large. We must as a society try to do our best and as organizations try to be good corporate citizens. If that means that we have to tell the truth, educate our employees and our stockholders, and focus on long-term survival rather than short-term gains, so be it. If we practice quality, then the integrity of everything we do is the primary issue. We must recognize that cutting corners is not the way to practice quality. Quality is practiced by having a vision, goals, and appropriate action plans. In this book we present the philosophy of Total Quality Management, an implementation strategy, basic and advanced tools, and the leading issues every organization should embrace in order to be successful in the years to come.

Why this book? Many books have been written about quality including Total Quality Management. This book is quite different in many ways. Not only does it present the basic concepts of quality and the traditional basic tools, but it goes much further to address issues that in the 1990s and the next century will be of paramount importance in any organization for survival. Some of the topics that will be addressed and the reader will find unique are the chapters on a design of experiments, issues of reliability, advanced topics of quality—QED, FMEA, benchmarking, meetings, teams, quality awards, training, international standards, and much more.

How can this book be used? The book can be used as both a reference and text on quality. To facilitate these objectives each of the chapters is designed independently of the others and is self-contained. Also, additional references are provided either at the end of each chapter or in the selected bibliography.

Who can use this book? This book is directed to several groups. First, quality practitioners will find the book refreshing because it provides many tools with which quality can be defined, monitored, and evaluated. It goes beyond the content of other quality books, as it expands the discussion of traditional concepts and tools and, furthermore, provides the reader with the crucial issues of quality for the years to come. Second, anyone interested in quality of the future will find the book an excellent reference for both traditional concepts and tools, and advanced tools and approaches to quality. The academic group will find the book contains a variety of topics to be used in a classroom format, since the topics are geared for both graduate and undergraduate work. In all cases the book serves as a springboard for further research in any of the topics covered or as a textbook in quality-related classes.

D. H. Stamatis

NOTES

These notes are not an exhaustive study on the issue of "price" versus "quality." Rather they are a cross section of some specific examples that show the power of price.

1. A discussion on how the software industry is able to manipulate the disk space in a computer is covered in the November 1995 issue of *Multimedia World*, p. 133, in the article "Q & A: Working with Multimedia" by Bronwyn Fryer.
2. All of us who have used any type of computer software have come across "bugs" in the software system itself, in spite of the fact that all software packages go through a "beta" testing process.
3. During a *20/20* television program on March 1, 1996, a program on "food and apple juice" was aired that showed how large companies were manipulating data to sell watered-down maple syrup as real "maple syrup."
4. During the television program of *Prime Time* on February 28, 1996, a program on "auto" lemons was aired that showed how the dealers of some of the largest automotive companies were passing "lemon" cars from state to state, and in the process they were making thousands of dollars.

   Fix, J. (May 3, 1996). "Memo implies Ford recall left out some fire-prone cars." *Detroit Free Press*—Business, p. E1. This article reports that an internal document suggests the automaker did not recall vehicles that may be prone to ignition switch fires other than the 8.7 million in last weeks recall. The best that a Ford representative had to say was that "we are not hiding anything. Fires happen for many different reasons."
5. (September 13, 1993). "End the phony 'asbestos panic.'" *USA Today*, p. 11A. The article notes that millions of dollars were wasted, where in fact there was no danger of asbestos.

   (April 15, 1996). "Truth-in-savings law may be gutted." *Detroit Free Press*—Business Monday, p. 4F. The article discusses the greed the banking community is displaying by its efforts to repeal the Truth-in-Savings Act. Although the customer is better off knowing the actual annual percentage yield (APY), the banks do not want to disclose that figure. If the act is repealed, the result will be an unfair comparison between financial institutions, since the APY is calculated by using different methods. The customer is the loser.

Khanuja, G. H. (February 14, 1996). "Is service on par with the price?" *Mid-Day.* (New Delhi, India), p. 17. The article points out that while price tags are getting higher, service is deteriorating to the point of becoming nonexistent.

Pepper, J. (April 14, 1996). "GM quietly takes care of business the old fashioned way, one step at a time." *The Detroit News*—Business Section, p. 1-D. In the article, Mr. Pepper quotes a University of Michigan study in which price was the overwhelming criteria over quality in the selection of a family car.

Gottlier, M. (March 1992). "Hospital's ethical quandaries in crunch to pass inspection." *New York Times*, pp. A1, B4. The author reports how a particular hospital falsified records to ensure certification. Never mind the safety of the patients. What was important was that the hospital continue to do its business.

(April 25, 1996). "Power acknowledges Japanese excellence." *Business Free Press*, p. 1G. The article points out that in the category of reported problems per 100 cars by nameplate, none of them were American.

Brennan, M. (May 8, 1996). "Japanese automakers top poll." *Detroit Free Press*—Business, pp. E1–2E. The article notes that American automakers still lag behind the Japanese. The Japanese are also winning the battle of light trucks, as they won top honors in three of the five categories.

Gardner, G. (April 25, 1996). "Luxury models riding high in quality survey." *Business Free Press*, pp. 1G-2G. The author points out that even in the luxury category American automobiles are behind. They are catching up, but they are not in the top group.

Acknowledgments

William Blake defined gratitude in one of his poems as "heaven itself." In this spirit, it is a pleasure to record the many archival and intellectual debts that I have accumulated while writing this book. Many people and organizations have contributed. To identify them all by name is an impossible task, but I would like to name some to whom I am deeply indebted. They are:

The SAS Institute for granting me permission to use information from Littel, Ramon C., Freund, Rundolf J., Spector, Philip C., *SAS System for Linear Models*, Third Edition, Cary, NC, 1991, pp. 57–84. Copyright SAS Institute, Inc. The material is found in the regression discussion.

Quirk's Marketing Research Review for granting me permission to summarize some information regarding the various regression approaches. The material may be found in the chapter on regression.

The American Management Association for granting me permission to use the comparison tables of Classical, Taguchi, and Shainin. They were exerpted by permission of the publisher, from World Class Quality © 1991 AMACOM, a division of the American Management Association. All rights reserved.

Marketing News for granting me permission to use the creative case analysis method figure from their issue of July 23, 1990, p. 23.

Cutting Tool Engineering and Mr. B. Dovich for granting me permission to summarize the data on C_{pk} in reference to its variability.

Jossey-Bass for granting me permission to reproduce and modify the six-stage model cycle of needs assessment from Brinkerhoff, Robert O., *Achieving results from training: How to evaluate human resource development to strengthen programs and increase impact.* Figure 2, p. 27 and Table 3, pp. 28–29. Copyright 1987 by Jossey-Bass, Inc., Publishers.

ESD Technology for granting me permission to adopt and modify Figure 5 from G. Allmendinger's and L. Ribeeiro's article on performance measurement, the December 1990 issue.

Mr. J. Barone, a paint supervisor from the old Clark plant in Detroit, Michigan, for the stories about Mr. Kettering.

Mr. P. Grias for the poem "It couldn't be done."

Mr. G. Pallis for the story of processionary caterpillars.

Mr. H. Jamal for his comments and suggestions throughout the project.

Mrs. C. Stamatis for her enthusiastic support and editing earlier drafts.

Miss C. Stamatis for typing several sections of the manuscript without complaining.

Mr. S. Stamatis for helping with the computer work—both hardware and software—throughout the project.

Dr. R. Roy for his thoughtful comments on the Taguchi section of the book.

Mr. T. A. Bylsma for his thoughtful suggestions as well as the summary of the DISC leadership attributes.

Mr. R. Munro for his thoughtful comments and excellent suggestions throughout the project.

Ms. E. Rice for her helpful suggestions on teams and problem solving.

Ms. D. Fletcher for typing several earlier drafts.

The editors and reviewers of the book for their support and helpful suggestions to make this book a better product.

Finally, to all my seminar participants—over the years—for their suggestions and insights.

Contents

Introduction

Much has been written about quality and more specifically, Total Quality Management (TQM). The benefits, the advantages, and disadvantages have been discussed to the point where lately we have seen articles that TQM is dead and not appropriate for the rest of the 1990s and the twenty-first century. How wrong can they be! TQM is alive and kicking and will be with us for a very long time. We may change the name, but its essence will be with us. To believe that the TQM will no longer exist means that organizations will not care about continual improvement and all of its limitations, customer satisfaction, measurement, and efficiency. Concurrently, without TQM the customer will not care about quality, value, and satisfaction of their purchase.

Obviously these statements are wrong. Why? First, because as time passes, organizations become much more cognizant about productivity, efficiency, and costs, and second, customers become ever more conscious about value. How can TQM help? To see what is TQM and how it can affect both the customer and the organization as well as the supplier, let us focus on the improvement process of TQM.

TQM is a philosophy rather than a program; this means that it is always evolving. It is a formalized, yet "common sense" approach to the elimination of wasted time, energy, and materials. By using time, energy, and materials in their simplest form, we are making something simpler or easier to do, and in so doing, have created an easier task, conserved time, energy, or material, and accomplished a better result in less time and with a minimum of waste. TQM can be used not only in factories and offices, but also in our homes and in any phase of human activity. Remember it is a philosophy—a way of thinking—rather than something concrete and definite. There is nothing difficult about the application of the fundamental principles of the techniques used to implement the TQM philosophy. It simply involves thinking in an orderly manner about the work under consideration, and then selecting the best available method that we know of at the present time.

Fundamentally, when we speak of wasted effort and wasted activity, we refer to the unnecessary, nonproductive, tiring actions that are present in anything we do. We are concerned with the elements of work which tire us but contribute nothing what-so-ever to the product.

TQM is also a way of thinking as well as a relatively simple analytic approach (although it can be very complicated to implement) with which it is possible to eliminate most of the useless movements and to adapt more productive and useful actions. TQM will make it possible to produce more work with less expenditure of effort. Sutermeister (1969) said it best when he discussed the issue of productivity. He said that the only way tually to increase the average income and purchasing power of the worker is to find the means to increase his productivity. This is a rather straightforward observation although few people seem willing to understand it. Whenever we increase a worker's productivity, nothing has ever prevented the worker from benefiting in proportion to the increase. Neither depression nor boom, politicians nor profiteers have prevented the average income per worker from remaining proportional to productivity.

I. ADVANTAGES TO THE ORGANIZATION AND TO THE EMPLOYEE

Educating and training in the use of the principles and tools of TQM has numerous advantages to help anyone in their daily work. Through the application of the tools provided and the resultant simplification of a few of the primary jobs in the work environment, the following advantages should begin to develop.

Training of new workers should be quicker and easier on the simplified jobs, thus making available to you additional time for your other duties as well as additional education. (There is a distinction between education and training in that education answers the "why" and the training answers the "how." Both are necessary. Education presents the "whole" picture, whereas training presents the details needed for the task.) A truly simplified job will require less attention on your part to see that production and quality standards are met and maintained.

As more and more jobs are simplified and waste is eliminated, production has a tendency to increase. The additional production will undoubtedly be accomplished with less physical effort and fatigue to the operator. Increased production will make it easier to meet promised delivery dates to the customers.

A simplified job is a safer job and reduces hazards to the operator. Simplified machines, equipment, and maintenance procedures save much trouble and loss of production time in the work environment; this just another way of helping meet production goals.

The benefits from consistent application of the principles and tools of TQM will make more of your time and energy available for more complete and efficient work, and for further experimentation and development of ideas which are essential to progress. These ideas are equally applicable to both manufacturing and service industries.

II. OBJECTIVE OF TQM

Before starting any job, we must have an objective. The objective of TQM is to make a better product/service at a lower cost in shorter time with less effort, every time. This philosophy is designed to make individuals think creatively toward the better use of time, energy, and materials. It employs a common sense system for measuring and analyzing a given process in order to find the berrer way of doing work. TQM should help do a better job by:

1. Bringing people together for discussion and exchange of ideas
2. Developing an alert, open-minded, questioning attitude on the part of all employees
3. Uncovering and eliminating wasteful and time consuming-methods, thereby leaving more time for the development of effective procedures
4. Eliminating fear in all levels of the organization

III. ATTAINING THE OBJECTIVES

In order to attain the objectives an organization must:

1. Produce the best product and give the best service of which the organization is capable
2. Operate at the lowest cost by the most direct method and with proper consideration for the interest of all groups necessary to obtain the desired result
3. Deliver the product/service at the right time

These points should be the objectives of everyone in the organization, regardless of title or position, if the organization is expected to make continual progress. Above all, it should be realized that attaining these objectives involves a human problem, perhaps even more than a material one. TQM, as applied in industry or any phase of human activity, deals with human beings. In most organizations, we are fascinated by what we see in the way of machines, methods, and materials. These things are perhaps more intriguing to the average person than the men and women whose efforts are expended toward putting life into lifeless machinery. However, we must not forget that the machines, methods, and materials themselves never invented anything; they never presented a new idea to the world, and are only as good as the people who use them.

IV. HUMAN ENGINEERING

To attain our objectives, the problem of human engineering is of the utmost importance. It has been said that Thomas Edison once said, "Problems in human engineering will in the future receive the same thought that the last century gave to engineering in more material forms." Can we apply the engineering approach that is used in more material forms in trying to solve our problems of getting the human energy and force to work correctly? To answer this question intelligently, we must have an understanding of what is meant by the engineering approach. It is:

1. Collection of all the facts possible
2. Analyzing the facts at hand as a basis for judgement
3. Making decisions based upon that judgment of the facts
4. Acting on those decisions and checking for improvement confirmation through a systematic follow-up

To say it in more simple terms, it is: find out what is wrong with something, correct it, and then confirm the results. However, whereas this approach is a workable model and has served industry for a long time, it is not enough. In the 1990s and beyond we must look at something and even though nothing is wrong with it, we should try to improve it. It is that improvement and vision that the TQM encourages throughout the organization.

The engineer is concerned with more material things such as the generation, transportation, and application of power. The effectiveness of the application of power is such that we are almost certain that throwing a switch will immediately turn on the power. If the right switch is turned on and no mistakes are made, it is not necessary to go back and check. Management actions, however, are quite different from engineering actions. Do we find the same effectiveness in translating a decision made by management into action at the bench or machine as we find in transmitting steam or electric power from the coal pile or waterfall to the machine? By checking back on your own experience, you will find that the answer is quite obvious—no!

The probable reason is that no two of us are exactly alike: we do things differently, we do not see things the same way, and we react differently. Knowing that this is so, we realize that beyond a certain point it is necessary to consider each human being as an individual. First of all then, we must consider each human being and the things that affect them. Then we can discuss the applied power which is commonly called "work."

Remember that the engineer's method of approach is finding out what is wrong with a thing and then correcting it. Applying this approach to our human problems, we find that there are three common concerns with people that must be recognized and changed appropriately, if we are going to do a good job in human engineering. The three concerns are:

1. It is human nature to resist changes.
2. It is human nature to resist the new.
3. It is human nature to resent criticism.

A. We Resist Change

It is a universal trait in human nature to resist change. Why is it so? Probably the reason is that changes require thinking and considerable effort to get out of the old groove. The easiest way is the familiar way. Complacency—the feeling that all is right with the world, especially with us—is a comfortable feeling. When we feel that this comfortable state of affairs may be upset, we go into action; or when it has been upset, we act to restore it. There is grave doubt that we ever do anything at any time except to prevent our complacency from being disrupted or to recover it if it had been disrupted. This has nothing to do with reasoning. It is an emotional reaction. A Greek story will help explain this:

> A myth of the ancient world states that the Wind and the Sun one day wager who would be the first to cause a pedestrian to remove his hat. The Wind, of course, was very certain of his win and he chose to go first. So, with the first blow, the pedestrian raised his hand and pushed the hat deeper onto his head. The Wind blew stronger and the pedestrian repeated the action. The Wind tried again, to no avail; the pedestrian just as stubborn kept placing his hat deeper onto his head. Finally, the Wind blew with all his might, thinking that the pedestrian not only will loose his hat, but will bounce him all around as well. The pedestrian, however, put both his arms on top of his hat and took cover in the trunk of a nearby

> tree. When the Wind saw his reaction, he told the Sun to take over.
>
> The Sun slowly and continuously warmed the ambient temperature, as well as the earth. Before too long, the pedestrian felt warmer and before he knew it, he took his hat off to wipe his sweat from his head. The Sun won.

So it is with us. Nothing gets accomplished by force, anger, and shouting, except that the other person gets more defensive, argumentative, withdrawn, and so on. Just like the pedestrian we button up and run for cover when there is a force that we do not like or that is too sudden. On the other hand, just like the pedestrian when the change is slowly and consistent we get used to it and we learn to leave with it. The moral of the story of course is that when you make a change slowly, very few will object to or resist it.

B. We Resist the New

Like a parachute, the mind functions only when it is open. Check your own reaction to this first challenge. When somebody presents you with a new idea, what do you do about it? I believe it is true that one of the first things you say is: "It can't be done." Perhaps you immediately start searching for reasons why the idea will not work. What was your first reaction, for example, to Quality Circles, Participative Management, Management By Objective, Streamlining, and so on? Compare your first reaction to man flying across the ocean in the beginning of aviation with the present day service which we now enjoy. What would you have said years ago if anyone had outlined our present air transport service? What is your reaction to pictures without film? And so on. Your reaction to these ideas is probably no different from the reaction of the people who laughed at Columbus, Fulton, and Whitney for their impossible and unimaginable ideas. History is full of examples of this resistance to the new.

Some years ago, it took thirty seven days to paint a Cadillac car and twenty one days for a Buick. As automobile production increased, it was obvious that something had to be done to bring about an expansion. There was not room in the General Motors plants to store all the cars made in thirty seven days, so a conference of plant executives was called.

Again, C. F. Kettering suggested that the paint-drying time might be cut down, possibly to one hour. This statement broke up the meeting and as the men left, they pitied Kettering for making himself so ridiculous and said: "The man is out of his mind." Not long after this meeting, Mr. Kettering was walking down Fifth Avenue in New York City. In a window, he saw a lacquered pin tray and purchased it. Inquiry revealed that the tray was made in Newark, NJ. He went and requested some of the lacquer to "paint an automobile door" and he was told by the company, "You can't paint an automobile with that; it dries too damn fast." From that day on, Kettering spent all his time experimenting with various paints and lacquers.

At a later date, one of the executives who was very positive in his statement that it was impossible to paint a car in an hour, drove to New York to attend a meeting with Mr. Kettering. On the way to lunch, Mr. Kettering commented on how shabby the car looked, but the executive stated that he could not spare the car long enough to get it painted. On the way in, Kettering made a telephone call.

After lunch, they returned to where the car had been parked, but it could not be found. Finally, Kettering asked his associate if the new looking blue car parked there could possibly be the right one. It was. An hour to paint a car—not so crazy after all!

But the paint story is not unique. Throughout human history there have been many examples of "It can't be done." For example, consider the following.

1. The first cast iron plow invented in the United States in 1797 was rejected by New Jersey farmers under the theory that the cast iron poisoned the land and stimulated the growth of weeds.
2. Commodore Vanderbilt dismissed Westinghouse and his new air brakes for trains, stating that he had no time to waste on fools.
3. The people who loaned Robert Fulton money for his steamboat stipulated that their names be withheld for fear of ridicule were it known that they had supported anything so "foolhardy."
4. In 1881, when the New York Y.M.C.A. announced typing lessons for women, vigorous protests were made on the grounds that the female constitution would break down under the strain.
5. Men insisted that iron ships would not float, that they would damage more easily than wooden ships, that it would be difficult to preserve the iron bottoms from rust, and that the iron would deflect the compass.
6. Joshua Coppersmith was arrested in Boston, MA, for attempting to sell stock in the telephone. "All well-informed people know that it is impossible to transmit the human voice over a wire."
7. The first reaction to bathtubs was that the new device was a menace to health and morals. The city of Canton, OH, at one time had an ordinance which prohibited the ownership of a bathtub.

If we are to get along with people, consideration must be given to the human failing of resistance to the new. The moment we forget this, we are wasting our time and the results will be disappointing.

C. We Resent Criticism

Do you believe that people like to be told when they are wrong? I believe that we will all admit as we look back over our own experience that those who helped us the most were those who had been able to show us where we were wrong. But how many are ever willing to take advice? We do not refer to "destructive criticism," but deal exclusively with "constructive criticism." But if criticism is so good for us, why do we constantly resent it? Is not the answer that we would much rather be told that we are right? When we ask for advice, most of us merely want to have our own good opinions of ourselves verified. If we do not get this, we will very likely dismiss the advice. The best way to avoid the "malady of self-delusion" is to be known as a person who welcomes criticism—a person who wants the facts as people see them no matter how much they may deflate their ego or upset their present method or future plans.

There is no truer test of a man's qualities for permanent success than the way in which he takes criticism. The little man cannot stand it. It breaks his ego and he makes excuses. Then when he finds that excuses will not take the place of results, he sulks and pouts. It never occurs to him that he might profit from the experience.

V. ARE YOU TOO OLD TO LEARN? ARE YOU TOO OLD TO CHANGE?

How old are you? The answer to this question depends upon what you call old. Age is only a state of mind and a point of view. We know old people who are "young" and young people who are "old." In business if you are at the point where you think you know all about something, you are definitely old. And if you believe you are doing anything as well as it can be done, you are very old. You are old enough to die. In fact, you are dead and waiting for your competitors to hold your funeral. But if you are glad to admit that you know very little about anything, you are young; *and if you are sure that everything can be done much better than it is now being done or ever will be done*, then you are young and growing, and prosperous years are ahead of you.

Another story may help in understanding the issue of age: In ancient Athens, the advanced in age Plato, was making the rounds for his lecture in the town square. It was a hot day and he was getting very tired from walking and carrying his books and his personal items on his shoulder. As he was passing by a water spring, he decided to take a breather and drink some water. He unloaded his things from his shoulder and as he was preparing his "mug" to get some water, he noticed a child drinking water from the spring by putting the two hands together forming a cup. At that point Plato threw away his mug saying his now famous words "As I get older, I learn new things." Plato was indeed a young man because he recognized the opportunity to improve. His improvement was to carry one less item around which meant less weight, and of course the application of a new tool formed with his own hand. You see Plato was willing to learn from a child. He was indeed a young man.

VI. PROGRESS IN INDUSTRY

Progress always involves risks—you cannot steal second base and keep your foot on first. Even though we claim that we know many things, actually if you think it, you will find that we do not know much about many basic things. For instance, what does sunlight do to make people well, what is magnetism or electric current, and so on. We know very little about some chemical reactions, why we transmit voices over a wire, and various other phenomena common to industry today. (It must be emphasized that just because we do not know or understand something, that does not mean that we cannot use it. In fact, the examples just given are used in industry and personal life all the time.)

As far back as 1886, a Federal Commissioner of Labor allegedly has stated "while new processes of manufacturing and new discoveries will undoubtedly continue, these will no longer be on the scale that will permit us to make such remarkable advances as have been witnessed during the last fifty years from 1836 to 1886." If this commissioner was alive today, he would see in full operation many, many industries that have started since he made this statement; and these industries supply jobs for well over one-third of all the people who work in the United States.

If you really want progress and the changes that result from that progress, you may find yourself in the minority, for only a few are willing to pay the price of progress. As has been already discussed, most people resist the new and resent criticism. If this is true, how do we explain the marvelous progress that we have made, especially in the last fifty years? There are three reasons for this:

1. Occasionally, there is an individual intensely interested in trying to make something better through inventions.
2. Competition
3. Quality

History is full of stories of the early inventors, such as Marconi, Edison, Westinghouse, and many others who were more or less individualists. These men planned and completed their own jobs although some of the first inventions were quite obvious. We needed machines to improve the quality and reduce the cost of clothing; we needed better communication methods; we needed improved office machinery to take care of increased business. These inventions, although obviously needed, were not readily accepted when first developed. The cry went up that we would save so much labor that people would be out of work. This did not happen; these new inventions actually created entirely new industries that resulted in more jobs than had ever been thought possible. In comparison to the time of these early inventors who were more or less individuals and who planned and completed their own jobs, new inventions today are usually the product of the cooperative effort of both engineers and practical manufacturing people. Problems in industry today are too vast for one person to have all the answers, and specialization is the order of the day.

This brings us to our problem of finding and developing better ways of processing, conditioning, testing, handling, and fabricating materials in organizations throughout the United States and the world at large. The reason why we are now engaged in TQM is because many of the problems which we encounter today are too involved for one person, or even a small group of people, to solve and to make the necessary improvements required to keep pace with the demands made on us today.

Under present conditions, it is necessary to change production schedules more often and more rapidly than was necessary even a short time ago. This time must be shortened if we are to get material where it is needed and when it is needed. The best way of doing this is by eliminating waste, unnecessary steps, and finding simpler methods of handling.

Now let us consider the issue of competition. Competition in business has been called "the great universal supervisor," for it is the force that puts us to work or out of work. There are rare cases where companies or individuals who have no competition make continual improvements in products, quality, or service or voluntarily reduce their prices. Most companies and individuals take the line of least resistance and require competition to force them to change or improve.

The only place where you can sit down and rest is immediately in front of the undertaker's establishment, for the moment you are satisfied, the concrete has began to set in your head.

What if there were no competition? As we know, industry today is a very strenuous undertaking. Things move fast and it is a struggle to stay in the game. Many times when pressure is high, people think wishfully how fine it would be if there were no competition and how easy their jobs would be.

This is not sound thinking, however. The elimination of competition would not make jobs easy, it would simply allow people to take it easy. The stiffer the competition, the more necessary are the services of men who can help the organization better its position. One way to better our position is to attain the objective of TQM. In order to obtain this objective, we cannot follow old habits, good or bad, but we must develop new methods and new ideas. We must not be discouraged because people think new things will not work. We must challenge our paradigms and encourage innovations.

The final concern is that of quality. Quality is something that we all know, we all have experienced, yet it is an elusive concept that depends on the customer's expectation. In today's business environment, however, Quality is the name of the game, and the way that the organization will pursue it, communicate it, and internalize it will be the difference between a successful and a not so successful organization.

In the following chapters we will explain the issues of quality, the implementation strategy and some of the new ideas and approaches to improve the efficiency of an organization for the years to come.

REFERENCE

Sutermeister, R. A. (1969). *People and Productivity*. McGraw-Hill. New York.

1

The Quality Revolution

As Juran has pointed out in his most recent book on the history of quality (Juran 1995d), quality is nothing new. In fact, there has been quality in both the ancient world and the modern world. What makes it different in the modern world, however, is our perspective of quality. Whereas in the past the emphasis was on appraising quality through inspection techniques, sorting, and so on, the modern world is focusing on prevention methods, namely, to design quality into the product. That change of attitude in the world of quality obviously was not overnight. It took time and it was slowly developed over a period of years. There are many people who have contributed to this development. This chapter covers this paradigm change and the individuals who were instrumental not only in the theory, tools, and methodologies of pursuing quality, but also gave us the inspiration and rationale for aiming our goals to nothing but excellence. In our view these individuals are the true pioneers who indeed revolutionized the concept and the practice of quality.

I. QUALITY OVERVIEW

What is quality? Many definitions are given to the word quality. From the ancient world to the modern world we all talk about quality as an abstract. However, the basis for that abstract has been in the notion that if we produce something that does not meet customer requirements, we need to find the deficiency before giving it to the customer. Hence, our efforts have been in defining sampling plans, operating curves, average quality levels, and so on.

To be sure, that mentality was acceptable and worth pursuing for a long time. It made sense. However, in the 1990s and beyond, the old methods are not acceptable. The modern approach to quality is to base its notion on the old adage "an ounce of prevention is worth a pound of cure." Quality then, must be viewed as designing processes that prevent errors as opposed to simply finding them. It is a customer perception issue as well as a value added to the product and/or service for the customer.

II. THE PARADIGM CHANGE

Quality for most American companies—at least up to the mid-1980s—was an issue of "wishful thinking." This wishful thinking allowed many problems to be swept downstream. In fact, a large number ended up in the hands of the customer. Since the mid-1980s to about 1990 we saw some aggressive companies address their problems based on a Total Quality Control system. This system of quality was based on the notion of finding and correcting the root causes of the problem. The information is fed upstream so that the same problem is not introduced back in the process.

Both approaches worked in an acceptable manner for a long time and with good result for many organizations. However, to be successful in international competition today and in the future, it is absolutely necessary to achieve a level of quality that is based on prevention. This philosophy of prevention is based on the notion that potential problems and their root causes are identified before they occur. Optimization positions the design of quality as far as possible from all potential problems. The information is fed downstream to ensure that the problem is not introduced. In other words, *quality must be designed into the product, not inspected out of the product.*

We must change our paradigms of quality or, as Deming (1986) pointed out, we will pay some series consequences in the world of competition. If we do not change our thinking about quality, we are going to end up like the caterpillars in the following story.

> Processionary caterpillars feed upon needles. They move through the trees in a long procession, one leading and the other following, each

> with his eyes closed and his head snugly fitted against the rear extremity of his predecessor. Jean Henri Fabre, the French naturalist, after experimenting with a group of these caterpillars, finally enticed them to the rim of a large flower pot where he succeeded in getting the first one connected with the last one, thus forming a complete circle which started moving around in a procession which had neither a beginning nor an ending. The naturalist expected that they would soon get tired of their useless march and start off in a new direction, but not so.
>
> Through sheer force of habit, the living, creeping circle kept moving around the rim of the pot, around and around, keeping the same relentless pace for seven days and seven nights, and would have continued longer had it not been for sheer exhaustion and ultimate starvation.

Incidentally, an ample supply of food was close at hand and plainly visible, but outside the range of the circle, so they continued along the beaten path. They were following instinct, habit, custom, tradition, precedent, past experience, "standard practice," or whatever you may choose to call it, but they were following it blindly. They mistook activity for accomplishment. They meant well, but they got no place. There must be a better way.

The inspection approach and technology to achieve quality was the "new and better way" of the 1850s. It is now widely recognized that inspection in any environment has many problems. Everyone who is involved in achieving quality in any organization now recognizes that some form of online control is needed to react to problem tendencies before the problem actually occur.

To achieve improved quality and reliability, lower cost, and shorter development time, performance must be elevated to a new level. The old approaches and thinking have to be changed in order to accommodate the new wave of quality which is:

- prevention of problems
- reduction of unreliability
- elimination of waste

III. THE GURUS OF QUALITY

A. W. Edwards Deming

One of the most inspirational and influential gurus of quality in the twentieth century, is W. Edwards Deming. Much has been written about both him and his philosophy, so this section simply summarizes his philosophy and his 14 points for transformation of management. The reader is encouraged to see other sources about Deming (such as Deming, 1986; Scherkenbach, 1991; Dobyns and Crawford-Mason, 1994; and Latzko, Saunders, and Saunders, 1995.

Deming and Quality

- Real profits are generated by loyal customers, not just satisfied customers.
- The company that develops loyal customers has much higher earnings than the company that just pushes the product out the door.
- Merit reviews, by whatever name (e.g., management by objectives), are the single most destructive force in American management today.
- The belief that the worker is responsible for the poor quality and low productivity of American firms is wrong.
- Workers cannot change the system, only management can.

Summary of Deming's Philosophy

- Quality and cost are not opposites or trade-offs, with one being improved at the expense of the other. Instead both can be constantly improved.
- Quality is best understood from the point of view of the customer, but one important component of quality is improving uniformity.
- Variation is a naturally occurring phenomenon.
- Cooperation is a fundamental ingredient that leads to improvement. In conventional thinking competition is always preferred over cooperation.

The Typical Deming Company

The following characteristics, beliefs, and philosophy are necessary for the typical Deming company.

- Quality leads to lower costs.
- Inspection is too late. If workers can produce defect-free goods, eliminate inspectors.
- Quality is made in the board room.
- Most defects are caused by the system.
- Process never optimized; it can always be improved.
- Elimination of all work standards and quotas is necessary.
- Fear leads to disaster.
- People should be made to feel secure in their jobs.
- Most variations are caused by the system. Review systems that judge, punish, and reward above or below average performance destroy teamwork and the company.

- Buy from suppliers committed to quality.
- Work with suppliers.
- Invest time and knowledge to help suppliers improve quality and costs. Develop long-term relationships with suppliers.
- Profits are generated by loyal customers.
- Running a company by profit alone is like driving a car by looking in the rearview mirror. It tells you where you have been, not where you are going.
- Workers are really paying for the errors of management.
- Quality must come first. As quality increases, costs are decreased.
- If quality is sacrificed in trying to increase profitability, the actions will backfire.
- We bought European and Japanese products because we got stung by the quality of domestic goods. Our problem has been quality.
- The financial statements are not reality. They are a financial description of the past, a one-dimensional picture of a multidimensional world.
- Our competitive problems are not due to the workers in the system. They are due to poor management. More accurately, they are due to our managing under a set of ideas that is outmoded and incorrect.
- Quality has to be considered from the point of view of the user. One definition of quality is anything that enhances the product from the viewpoint of the customer.
- One important ingredient of quality is uniformity.
- The benefit of loyal customers and happy employees is unknown and unknowable. The improvement in costs because of better quality is also rarely measurable in full.
- As long as your attitude toward quality is that we just need to meet the competition or obtain a curtain level of quality, we are going to be in trouble.
- By allowing and even urging workers to experience the intrinsic rewards that come from doing something well and by using their innate and acquired abilities, productivity improves, quality improves, and customer satisfaction improves.
- Knowledge is the key ingredient of quality.
- Profound knowledge is knowledge universal to all businesses, large or small, in service or manufacturing, profit-making or not-for-profit. Clarity in the definition of quality and knowledge of the principle that increasing quality leads to increase productivity and higher profits are two elements. Other aspects include knowledge of variation, some psychology, and knowledge of the need for cooperation.
- Quality is everyone's responsibility.

- The people who are most in need of profound knowledge are the managers, particularly top managers.
- Quality defined from the user's point of view is anything that enhances satisfaction.
- No business can operate for long unless a certain amount of stability has been achieved. Deming estimates that in most business situations 94% (originally 85%) of the problems are problems of the system while only 6% (originally 15%) are special in nature.
- Giving workers higher pay will not improve quality but will cause resentment among those who do not receive incentive pay.
- Quotas double the cost of production.
- It is not unusual to find workers stopping an hour before the whistle blows. Great peer pressure is expected to keep production down so that all can meet the quota. No one makes suggestions that may improve production. The workers in such an environment are unhappy, but from management's point of view the quota is being met and that is all that counts.
- Neither the workers nor the foremen are capable of making the changes necessary to improve the quality of the incoming materials are changes in policy that only management can authorize.
- Management is responsible for making sure the employees are properly trained for their new job.
- When top management blames every accident on the lax behavior of the workers, they are admitting their ignorance and abdicating their responsibility.
- Whenever management uses fear, they will get incorrect numbers and misleading information.
- The causes of common problems cannot be attacked directly. It is a mistake to believe that one is improving quality when an inspector recognized and rejects a defective product or when a major quality flow is found. That is just recognizing a defect being produced by the system; it is not improving quality, it is not improving the system.
- A company that operates using fear, positioning top management against workers and middle management against the top, cannot produce the continual improvement in quality necessary to compete in the market place.
- One part of profound knowledge is obvious to children but not to some college presidents. In any group, half, will be below average.
- Cooperation is one of the key ingredients of improvement. Quality cannot be obtained and improvement is impossible without cooperation. Competition that most benefits the consumer occurs in a framework of cooperation.

- Quality Control Circles are nothing more than formalized meetings of workers, foremen, and engineers from various departments. By pooling their knowledge of a process, they are better able to tackle problems of quality.
- Perhaps 10% of the hourly workers and 2% of middle management really enjoy their work. By this measure, at least, American management has failed.
- It is the leader's job to lead the charge toward better quality.
- There is one job that belongs to the leader alone, and that is making sure all the parts and all the people work together.
- A leader's job is to see that everyone in his/her group works together and that his/her group works with the rest of the organization harmoniously to achieve the aims of the organization.
- Management is people.
- It is the leader's job to foster joy in work, harmony, and teamwork.
- Leaders have to be the primary agents for improvement.

Deming's Cycle of Continuous Improvement

1. Plan to change whatever you are trying to improve.
2. Carry out the change on a small scale.
3. Study the results.
4. Decide on an action, based on what you have learned from the change.
5. Repeat the cycle over and over.

This cycle can be seen in Figure 1.1.

Deming's 14 points for Transformation of Management

1. Create constancy of purpose toward improvement of product and service with the aim to become competitive, to stay in business, and to provide jobs.
2. Adopt a new philosophy.
3. Cease reliance on mass inspection to achieve quality. Eliminate the need for inspection on a mass basis by building quality into the product in the first place.
4. End the practice of awarding business on the basis of price alone.
5. Constantly improve the system of production and service; this will improve quality and productivity and constantly decrease costs.
6. Institute training on the job.
7. Institute leadership; the aim of supervision should be to help people and machines and gadgets to do a better job.
8. Drive out fear so that everyone may work effectively for the company.

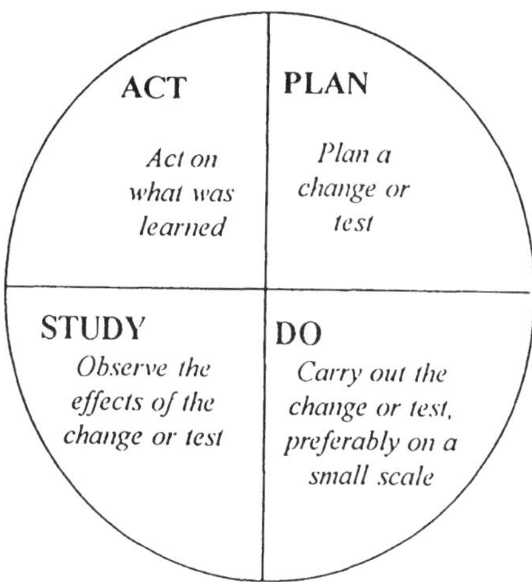

Figure 1.1 The Plan-Do-Study-Act cycle.

9. Break down barriers between departments.
10. Eliminate slogans.
11. Substitute leadership for work standards and management by objectives.
12. Remove barriers that rob the hourly workers and management of their right to pride of workmanship. The responsibility of supervisors must be changed from mere numbers to quality. End annual reviews or merit rating and management by objectives.
13. Institute a rigorous program of education and self-improvement.
14. Put everybody in the company to work to accomplish the transformation—it is everybody's job.

Barriers to Pride of Workmanship

1. Lack of direction
2. Goals without the tools to achieve them; time, resources
3. Arbitrary decisions by supervisors
4. Lack of clear goals and objectives
5. Lack of clarity as to how contribution is valued
6. Lack of expectations setting up criteria
7. Insufficient information available
8. Different organizational goals withing the company
9. Too much group management

10. Deadline anxiety
11. Lack of product definition: purpose and product arbitrarily changed by consumer and/or customer within company
12. Organization not valued by line organization
13. Hierarchy tries to run a technology that it does not understand
14. Lack of communication
    - conflicting and unclear objectives
    - lack of advance information
    - inadequate information flow
    - inadequate feedback
    - lack of authority to do what needs to be done
15. Lack of resources, time, and proper tools and equipment
16. Short-term objectives conflict with long-term ones
17. Nonuniform application of policy
18. Poor training
19. Specifications constrain creativity and procurement and manufacturing
20. Fear
21. Company and union adversarial relationship
22. Red tape/bureaucracy.
23. Unrealistic goals and objectives

The Seven Deadly Sins

In his wisdom, Deming, in addition to crystallizing his philosophy in the 14 points, also identified seven deadly diseases that, if unattended, cause harm in the organization and prevent it from becoming a "world-class" organization. The seven deadly sins are:

1. A lack of constancy of purpose to plan products and services that will have a market and keep the company in business and provide jobs.
2. An emphasis on short-term profits: short-term thinking (just the opposite from constancy of purpose to stay in business), fed by fear of unfriendly takeover, and by the push from bankers and owners for dividends.
3. The evaluation of performance by merit rating or annual review. It nourishes short-term performance, annihilates long-term planning, builds fear, demolishes teamwork, nourishes rivalry and politics. It leaves people, bitter, crushed, bruised, battered, desolate, despondent, dejected, feeling inferior, some even depressed, unfit for work for weeks after receipt of rating, unable to comprehend why they are inferior. It is unfair, as it ascribes to the people in a group differences that may be caused totally by the system (management)

that they work in, and not by anything that they themselves could control.

4. A constant change in management due to job hopping or continuous replacement or exchange of people in leadership positions. This leaves everyone wondering about stability, having to deal with new styles of leadership and changes in direction.
5. Management by the use of only visible figures, with little or no consideration of the figures that are unknown or knowable. This is peculiar to industry in the United Sstates. (He that would run his company on visible figures alone will in time have neither company nor figures.) What are the figures for failing to satisfy the customer, improving quality and productivity, poor leadership, inept design of a product, and the failure to improve processes?
6. The presence of excessive medical costs, unsafe products, unsafe processes, unsafe work place, and job stress.
7. Excessive costs of liability, swelled by lawyers that work only on contingency fees.

B. J. M. Juran

Juran, like Deming, is credited with part of the quality success story of Japan, where he went in 1954 to lecture on how to manage for quality. He is the author of numerous books on quality, leadership, and management as well as delivering speeches and seminars all over the world.

Juran was the first to deal with the broad management aspects of quality, which distinguishes him from those who espouse specific techniques, statistical or otherwise. As far back as 1940s, he pointed out that the technical aspects of quality control had been well covered, but that organizations did not know how to manage for quality. He identified some of the problems as organization, communication, and coordination of functions—in other words, the human element.

According to Juran, there are two kinds of quality: "fitness for use" and "conformance to specifications." Furthermore, there are three steps to progress: structured annual improvements combined with devotion and a sense of urgency, massive training programs, and upper management leadership. In his view less than 20% of quality problems are due to workers, with the remainder being caused by management. Because of this, Juran believes that all managers should have training in quality in order to oversee and participate in quality improvement projects.

Juran is also a strong believer of avoiding campaigns to motivate the workforce to solve the company's quality problems by doing perfect work. He claims that these exhortations are only shallow slogans that fail to set specific goals, establish specific plans to meet these goals, or provide the needed resources. However, he also points out that upper managers like these programs because they do not detract from their time.

Other important issues that Juran brought to the quality field are:

- The notion that quality circles improve communications between management and labor.
- The use of statistical process control, but he warns that it can lead to a tool-oriented approach.
- The notion that quality is not free (because of the law of diminishing returns). There is an optimum point of quality, beyond which conformance is more costly than the value of the quality obtained.
- The recognition of purchasing's role in quality and control of the suppliers, because they are part of the quality chain. He is a strong supporter of supplier qualification and surveys to insure that the supplier can consistently manufacture to specifications.
- The notion that single sourcing is counterproductive for an organization, since a single source can more easily neglect to sharpen its competitive edge in quality, cost, and service.

Juran's philosophy can be summarized by his 10 steps to quality improvement.

1. Build awareness of the need and opportunity for improvement
2. Set goals for improvement
3. Organize to reach the goals (establish a quality council, identify problems, select projects, appoint teams, designate facilitators)
4. Provide training
5. Carry out projects to solve problems
6. Report progress
7. Give recognition
8. Communicate results
9. Keep score
10. Maintain momentum by making annual improvement part of the regular systems and processes of the company

For more detailed information on Juran's philosophy, see Juran and Gryna (1988, 1993) and Juran (1989, 1992, 1995a,b).

C. Armand V. Feigenbaum

Armand V. Feigenbaum has been credited with the approach to quality and productivity that has profoundly influenced the competition for world markets. That approach is the Total Quality Control concept. Feigenbaum's philosophy is articulated in a series of principles which define the quality of products and services as a primary business strategy and a fundamental determinant for business health, growth, and economic viability. The main principles of the theory are:

1. Total quality control for Feigenbaum may be defined as: An effective system for integrating the quality-development, quality maintenance, and quality improvement efforts of the various groups in an organization so as to enable marketing, engineering, production, and service at the most economical levels which allow for full customer satisfaction.
2. In the phrase "quality control," the word "quality" does not have the popular meaning of "best" in any absolute sense. It means "best for certain customer requirements." These requirements are the actual use and selling price of the product.
3. In the phrase "quality control," the word "control" represents a management tool with four steps: a) setting quality standards, b) appraising conformance to these standards, c) acting when the standards are exceeded, and d) planning for improvements in the standards.
4. Quality has to be integrated in the entire organization for any measurable impact and improvement.
5. The factors affecting quality are: a) technological and b) human. Of these, the human factor is of greater importance by far.
6. Effective control ovèr the factors affecting quality demands controls at all important stages. These stages fall in the four natural classifications: a) new design control, b) incoming material control, c) product control, and d) special process control.
7. The details for a quality control program must be tailored to fit the needs of individual organizations.
8. Quality costs are a means for measuring and optimizing total quality control activities. Quality costs are divided into four types: a) prevention costs, b) appraisal costs, c) internal failure costs, and d) external failure costs.
9. From the human relations point of view, quality control organization is both a channel of communication for product quality information among all concerned employees/groups, and a means of participation in the overall quality control program by these employees and groups.
10. Necessary to the success of the quality program in an organization is the spirit of quality mindedness, extending from top management to the employees at the work floor.
11. Statistics may be used in an overall quality control program, but statistics are only one part of the total quality control pattern; they are not the pattern itself. The five most basic statistical tools are:

a) frequency distributions, b) control charts, c) sampling tables, d) special methods, and e) product reliability.

12. The total quality program provides the discipline methodology and techniques to assure consistently high product quality throughout the organization. Furthermore, it coordinates the efforts of the people, the machines, and the information which are basic to total quality control to provide high customer quality satisfaction which brings competitive advantage to the company.

See Feigenbaum (1991) for his detailed theory.

D. Philip B. Crosby

Philip B. Crosby is best known for introducing the concept of zero defects in the 1960s. According to Crosby the definition of quality is conformance to requirements and it can only be measured by the cost of nonconformance. His approach to quality means that the only standard of performance is zero defects. To be successful at this, the focus is on prevention which he defines as perfection. There is no place in his philosophy for statically acceptable levels of quality.

Crosby's contention is that people go to great elaborate things to develop statistical levels of compliance. He claims that we have learned to believe that error is inevitable, and therefore we plan for it. He goes on to say that we must change our way of thinking to the point where we accept that there is absolutely no reason for having errors or defects in any product.

Crosby talks about a quality "vaccine" that organizations can use to prevent nonconformances. The three ingredients of this vaccine are determination, education, and implementation. He points out that quality improvement is a process not a program. His philosophy towards quality may be summarized with his 14 steps to quality improvement. They are:

1. Make it clear that management is committed to quality.
2. Form quality improvement teams with representatives from each department.
3. Determine where current and potential quality problems lie.
4. Evaluate the quality awareness and personal concern of all employees.
5. Raise the quality awareness and personal concern of all employees.
6. Take actions to correct problems identified through previous steps.
7. Establish a committee for the zero defects programs.
8. Train supervisors to actively carry out their part of the quality improvement program.

9. Hold a “zero defects day” to let all employees realize that there has been a change.
10. Encourage individuals to establish improvement goals for themselves and their groups.
11. Encourage employees to communicate to management the obstacles they face in attaining their improvement goals.
12. Recognize and appreciate those who participate.
13. Establish quality councils to communicate on a regular basis.
14. Do it all over again to emphasize that the quality improvement program never ends.

See Crosby (1979) and Philip Crosby Associates (1985) for more detailed information.

E. G. Taguchi

Taguchi’s contribution to the field of quality is in the area of parameter design. It is Taguchi’s contention that in order to improve quality and reduce cost, one must plan prevention methods to avoid these costs. In addition, Taguchi suggests that all quality problems have an associated cost attached to them and somebody always pays. (The entity that ends up paying may or may not be the producing agent of the problem.) This notion has been quantified through the concept of the “loss function,” which measures the cost, as the quality varies about the target (Figure 1.2). As in the case of all the gurus of quality much has been written about Taguchi (see Barker and Clausing, 1984; Taguchi, 1986, 1987; Taguchi and Konishi, 1987; Ross, 1988; Roy, 1990; Peace, 1993). Therefore, we will not elaborate the details of his philosophy; however, because his philosophy is a very potent one and is being utilized in many industries with much success we offer the following summary. (Readers interested in Taguchi’s approach to quality are encouraged to see Chapter 9 for more details and/or references.)

1. Problem prevention in the design phase is better than inspection of the design.
   - Comparing designs one by one and selecting the best one is a very inefficient way of optimizing design quality.
   - Do not wait to detect failures of the design. rather identify faulty designs before they fail and/or need improvement.
2. Use a quality loss function to put a dollar value on quality.
   - The loss function allows for early prediction of cost of quality and guides improvements before problems are detected.
   - The loss function helps in improving customer satisfaction, and not simply meeting the customer’s specification.

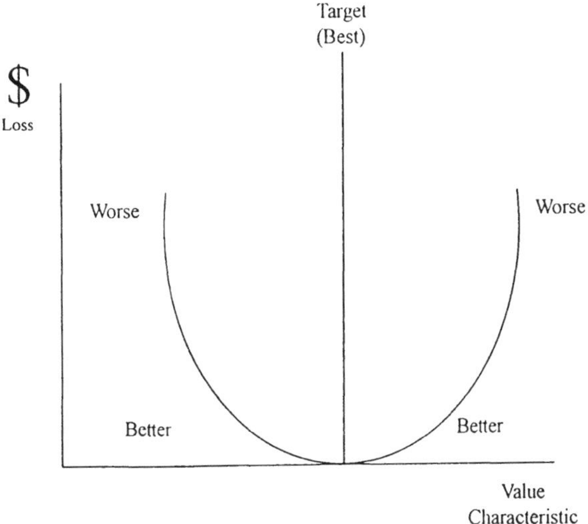

Figure 1.2 The loss function.

3. Use orthogonal arrays, signal-to-noise (S/N) ratios, and quality loss function to rapidly:
   - reduce low-cost production variances
   - enlarge customer-acceptable variances (production, environmental, time)
   - achieve the best balance between resources and quality
   - achieve mature designs with zero defects at launch
   - provide quantitative comparison with competitive benchmark
   - achieve high quality and low reject rates.
4. Common methods for product design and manufacturing:
   - break down traditional barriers
   - provide a common language and approach
   - help integrate suppliers

REFERENCES

Barker, T. B. and Clausing, D. P. (1984). *Quality engineering by design*. Paper presented at the 40th annual Reliability Society Quality Control conference.

Crosby, P. B. (1979). *Quality is free*. A Mentor Book: New American Library. New York.

Deming, W. E. (1986). *Out of crisis*. Massachusetts Institute of Technology. Center for advanced Engineering Studies. Cambridge, MA.

Dobyns, L. and Crawford-Mason, C. (1994). *Thinking about quality: Progress, wisdom and the Deming philosophy.* Quality Press. Milwaukee, WI.

Feigenbaum, A. V. (1991). *Total quality control.* 3rd ed. rev. McGraw-Hill. New York.

Juran, J. M. (1992). *Juran on quality design: The new steps for planning quality into goods and services.* Quality Press. Milwaukee, WI.

Juran, J. M. (1989). *Juran on leadership for quality.* The Free Press. New York.

Juran, J. M. (Ed.) (1995a). *A history of managing for quality.* Quality Press. Milwaukee, WI.

Juran, J. M. (1995b). *Managerial breakthrough.* rev. ed. Quality Press. Milwaukee, WI.

Juran, J. M. and Gryna, F. M. (Eds.) (1988). *Juran's Quality Control Handbook.* 4th ed. McGraw-Hill. New York.

Juran, J. M. and Gryna, F. M. (1993). *Quality Planning and analysis.* 3rd ed. McGraw-Hill. New York.

Latzko, W. J. and Saunders, D. M. (1995). *Four days with Dr. Deming.* Quality Press. Milwaukee, WI.

Peace, G. S. (1993). *Taguchi methods: A hands-on approach.* Addison-Wesley. Reading, MA.

Philip Crosby Associates (1985). *Quality improvement through defect prevention.* Philip Crosby Associates. Winter Park, FL.

Ross, P. J. (1988). *Taguchi techniques for quality engineering.* McGraw-Hill. New York.

Roy, R. (1990). *A primer on the Taguchi method.* Van Nostrand Reinhold. New York.

Scherkenback, W. W. (1991). *The Deming route to quality and productivity: Roadmaps and roadblocks.* Quality Press. Milwaukee, WI.

Taguchi, G. (1986). *Introduction to quality engineering.* Kraus International. White Plains, NY.

Taguchi, G. and Konishi, S. (1987). *Taguchi methods: Orthogonal arrays and linear graphs.* American Supplier Institute. Dearborn, MI.

Taguchi, G. (1987). *System of experimental design.* Vols. 1, 2. UNIPUB: Kraus International. White Plains, NY.

2

Traditional Quality Concepts

To understand and implement TQM with positive results, one must understand some of the key concepts that exist within the scope of quality. Without these concepts, TQM is only wishful thinking. This chapter discusses some of the old concepts in quality. None of them, of course, are given an exhaustive discussion. The intent of this chapter is to review these concepts because the foundations of quality are indeed built upon them. The references at the end of the chapter will help the reader to find the details necessary for a more comprehensive discussion and understanding.

I. DATA

In any plan, any single observation about a specified characteristic of interest is called a datum point. It is the basic unit of the quality person's raw material. Any collection of observations about one or more characteristics, of interest, for one or more elementary units, is called a data set. A data set is said to be univariate, bivariate, or multivariate depending on whether it contains information on one, two, or more than two variables (Becker and Harnett, 1987).

The set of all possible observations about a specific characteristic of interest is called a population or universe. It is important to note that when we speak of population or universe, by definition, we speak of all possible observations about a variable. When we speak of any subset (portion) of the population or universe we call it a sample. Conventionally, we represent this dichotomy of population and sample with Greek letters and English letters respectively. For example: A population standard deviation is represented by σ and the sample standard deviation is s.

Caution should be exercised here, however, because a population and/or sample depends entirely on how the question is raised. A sample may be a population and a population may be a sample. For example, Company XYZ had a recall of 1000 parts. The industry that his company belongs to had a recall of 1,000,000 parts. If Company XYZ wants to analyze the 1000 parts only, then the 1000 parts become the population of this study. If on the other hand, Company XYZ wants to analyze and compare the 1000 parts with the 1,000,000 parts then the 1000 parts are in fact a sample of all the recalled parts.

In addition to the sample and population concerns, any given characteristic of interest can differ in kind or in degree among various elementary units. A variable that is normally expressed numerically because it differs in kind rather than degree, is called a qualitative variable. Qualitative variables can be dichotomous or multinominal. Dichotomous qualitative variables are also called attribute or categorical variables because they are of two categories (e.g., go/no go, male/female, good/bad, on/off, good/reject.) Multinominal qualitative variables can be placed in more than two categories (e.g., job titles, types of business, colors; Daniel and Terrell, 1989; Becker and Harnett, 1987; Freund and Williams, 1972).

On the other hand, a variable that is normally expressed numerically, because it differs in degree rather than kind, is called a quantitative variable. Quantitative variables can in turn be discrete or continuous. Observations about discrete quantitative variables can assume values only at specific points on a scale of values, with gaps between them (e.g., cars in stock, rooms in houses).

Observations about continuous quantitative variables can, in contrast, assume values at all points on a scale of values, with no breaks or gaps between them (e.g., weight, time, temperature.) No matter how close two values are to each other, it is always possible for a more precise device to find another value between them. The accuracy of the measurement depends on the ability of the measurement instrument.

The distinction between qualitative and quantitative variables is visually obvious. The observations about one type of variable are recorded in words; those about the other type in numbers. Yet the distinction can be blurred: quantitative variables can be converted into seemingly qualitative ones and the opposite is also true. For instance, when we speak of high or low temperature/pressure we have in fact taken a quantitative (measurable) variable and transform it to an attribute (subjective) variable. When this coding takes place information is lost in the process. Conversely, it is not uncommon to code attribute variables with numerical values such as, for example: "good" as 1 and "bad" as 0.

A. Types of Data

The assignment of numbers and/or values to characteristics that are being observed—measurement—can yield four types of data of increasing sophistication. It can produce nominal, ordinal, interval, or ratio data, and different statistical concepts and techniques are appropriately applied to each type. For the actual selection of statistical techniques see Appendix A.

The weakest level of measurement produces nominal data which are numbers that merely name or label differences in kind and thus can serve the purpose of classifying observations about qualitative variables into mutually exclusive groups. No mathematical operations are possible except counting.

The next level of measurement produces ordinal data which are numbers that by their size order or rank observations on the basis of importance, while intervals between those numbers are meaningless. Again, no arithmetic operations are possible.

In the third level of sophistication are the interval data that permits at least addition and subtraction. Interval data are numbers that by their size, rank observations in order of importance and between which intervals or distances are comparable, while their ratios are meaningless. This kind of data possesses no meaningful origin and thus by a given definition establishes the zero point as well as equally arbitrary intervals between numbers.

The most useful types of data are the ratio data which are numbers that rank observations by their in the order of importance, and between which intervals as well as ratios that are meaningful. All types of arithmetic operations are possible because these types of numbers have a natural or a "true" zero point that denotes the complete absence of the characteristic they measure and makes the ratio of any two such numbers independent of the unit measurement.

B. Data Collection

Data can be gathered in different forms. First, however, the purpose of the data must be determined. Only then can a decision be made as to what kind of data would best serve the purpose. There can be many purposes for collecting data in the world of quality. Some common purposes are:

1. Understand the actual situation. Data is collected to check the extent of the current process.

2. Analysis. Data are collected to perform statistical analysis for a historical perspective of the process and/or future behavior of the same process.
3. Process control. After investigating product quality, this data can be used to determine whether or not the process is normal. In this case, control charts are used in this evaluation and action is taken on the basis of these data.
4. Regulation. This data is used as the basis for regulating the process based on a previously set goal.
5. Acceptance or rejection. This form of data is used to approve or reject parts and/or products after inspection. There are two methods total inspection and sampling. On the basis of the information obtained, a decision of what to do with the parts or products can be made.

It is imperative that the purpose of data collecting is not to put everything into neat figures but to provide a basis for action. The data may be in any form, but is generally divided into the measurement data and the countable (attribute) data. Once the purpose of the data has been defined, data can be generated by conducting a survey or by performing an experiment. A survey or observational study, is the collection of data from basic units without exercising any particular control over factors that may make these units different from one another and therefore affect the characteristic of interest being observed. An experiment on the other hand, involves the collection of data from basic units while exercising control over some or all factors that may make these units different from one another and therefore affect the characteristic of interest being observed.

II. SAMPLING

Data collection is the first step in working on problems; this provides information to assure that everything is within allowable tolerances and under control. Data can also be analyzed when things go out of control to find out what went wrong. Ideally, 100% of the given output would be checked for problems. However, this is not usually feasible because of time and cost considerations. Instead, what experimenters do is to design a plan that will be a portion of the 100% data, yet representative of the original without any bias in the selection process. This portion of the original data is called a sample. Representative data can be acquired by closely sampling the given output. This allows the use of smaller numbers, to be used for generalizations to the entire population (output). There are three types of samples used in all forms of statistical analysis; they are:

1. *Convenience sample.* When expediency is of primary concern, a-not-so-representative sample may be selected.

2. *Judgement sample.* A personal judgement is used for the selection of the sample based on "some" previous experience. Although used in the field of quality, caution should be exercised because of inherent biases by the selector of the sample.
3. *Random sample or probability sample.* Nonrepresentativeness is not one of the characteristics of this sample; rather, this sample is a subset of a population, chosen by a random process that gives each unit of that population a known positive (but not necessarily equal) chance to be selected. If properly executed the random selection process allows no discretion to the experimenter as to which particular units in the population enter the sample. This form of sampling maximizes the chance of making valid inferences about the totality from which it is drawn. There are four types of random sampling and they are:

Simple random sample. This is a subset of a population chosen in such a fashion that every possible subset of like size has an equal chance of being selected. Here be cautioned in that the implication of each individual unit of the population has an equal chance of being selected, but the converse is not true (i.e., giving each individual unit an equal chance of being selected does not assure that every possible subset of like size has an equal chance of selection).

Systematic random sample. This is a subset of a population chosen by randomly selecting one of the first elements and then inducing every *ith* element thereafter. In this procedure the i is determined by dividing the population size, N, by desired sample size, n.

Stratified random sample. This is a subset of a population chosen by taking separate (simple or systematic) random samples from every stratum in the population, often in such a way that the sizes of the separate samples vary with the importance of the different strata.

Clustered random sample. This is a subset of a population, chosen by taking separate censuses in a randomly-chosen subset of geographically distinct clusters.

A. Errors in Sampling Data

The process of gathering information is one of collecting and filtering data. The ultimate solution is a combination of good information at all levels, structures that size the decision-making process, and personal reconnaissance on the part of the decision maker at each level.

Statistics, even when most accurate, can never be the complete substitute for an in-depth knowledge of the situation of collecting, filtering, and analyzing data. Deming (1986) pointed out that there is a difference between visible and invisible numbers. He comments that some managers look only at the visible numbers, "but the visible numbers tell them so little. They know nothing of the invisible numbers. Who can put a price on a satisfied customer, and who can figure out the cost of a dissatisfied customer?"

Every sample from a large population is subject to a random error or chance error or sampling error. This error is the difference between the value of a variable obtained by taking a single random sample and the value obtained from the entire population. The second error is the systematic error or bias or nonsampling and is the difference between the value of a variable obtained by taking the population and the true value. Another way of explaining these errors is by thinking of sampling error as the reliability of data and nonsampling as the validity of the data.

Reliability is a concept like repeatability. If you keep repeating in all executionary details, then there is a probability that your sample will have an operating range. Furthermore, it will have a degree of stability that is based on that operating range. Note that this has nothing to do whatsoever with how accurate your sample was. This is a trap that most practitioners fall into and they are justified with what is called "confidence statement" or "statistical significance." To speak of 90%, 95%, or even 99% confidence is an issue of very limited value in management knowing that findings of a particular sample would probably be very similar to those of a second and/or third sample, if they were identically conducted. Such knowledge begs the issue of whether the sample methodology was any good in the first place. Statements of statistical significance beg the issue of data validity and hence its usefulness.

One can hardly list all possible types of nonsampling error, all the ways that a sample can yield misleading data, and all sources of invalid information about a target process. Only selected few are listed here:

1. In the planning stage
   a. Selection bias is a systematic tendency to favor the inclusion in a sample of selected basic units with particular characteristics, while excluding other units with other characteristics.
   b. Response bias is a tendency for selection of a sample to be wrong in some systematic way.
2. In the collection stage
   a. Selection bias is apt to enter the sample when experimenters are instructed to select within broad guidelines or the particular characteristics that they will sample.
   b. Response bias can arise for a number of reasons during data collection. Both the experimenter and the process may be at fault.
   c. Nonresponse bias may arise when no data (legitimate) is available from the sample.

3. In the processing stage. The emergence of bias during data collection can conceivably be minimized by the careful design of the sample. Nevertheless, bias can enter even at the data processing stage. People who code, edit, keypunch, tabulate, print, and otherwise manipulate data have many opportunities for making noncanceling errors. One of the areas that is a major concern in the quality area is the issue of data "outliers" or "wild values." We have a tendency to eliminate unbelievable data (high, low or just different from the majority), and/or to substitute zero for a "no value" and vice versa.

B. Controlling the Sample Data

To optimize the results of your sampling data the following may be considered.

1. Weighting sample data. This technique involves the multiplication of sample observations by one or more factors to increase or decrease the emphasis that will be given to the observation. The troublesome aspect of weighting is related to the selection or calculation of the weighting factors. The specifications of the weighting scheme must be defined in terms of our overall objective: What is the purpose of weighting? In most cases, the obvious answer is that we would like our sample data to be representative of the population. The immediate follow-up to the first question is another: In what ways are the data to be representative of the population? The answer to this question should lead us to select an appropriate technique.
2. Beware of the homing pigeon syndrome. This is where you become completely dependent for data on the incoming paper flow; you lose the interactive process and find out only what the sender wants you to know.
3. Reports/data required from the bottom up must be balanced by data interchange from the top down. Asking the same old questions gets the same old answers. If the system does not allow for an interchange in the data flow process, you will soon find yourself asking the wrong questions, at which point the answers do not matter.
4. Have appropriate sample for the specific project. It is imperative that we determine the correct sample plan before we begin experimenting.

III. VARIATION

Variation is so important and fundamental to all quality issues that everyone in the organization must understand it. Variation, in its simplest definition, may be defined as "change." In a more formal definition, variation is the difference between the target measurement and the actual measurement.

No two products or characteristics are exactly alike because any production process contains many sources of variability. The differences among products may be large or they may be almost unmeasurably small, but they are always present. The diameter of a cylinder, for instance, would be susceptible to potential variation from the machine (e.g., clearances, bearing wear), tool (e.g., strength, rate of wear), material (e.g., hardness, grade), operator (e.g., part feed, accuracy of centering), maintenance (e.g., lubrication, replacement of worn parts), and environment (e.g., temperature, constancy of power supply).

Some sources of variation in the process cause very short-run piece-to-piece differences such as backlash and clearances within the machine and fixuring, or the accuracy of the operator's work. Other sources of variation tend to cause changes in the product only over a long period of time, either gradually as with tool or machine wear, step-wise as with changes from one raw material lot to the next, or irregularly, as with environmental changes such as power surges. Therefore, the time period and conditions over which measurements are made will affect the amount of the total variation that will be measured.

From a specification standpoint, the only concern is with the total variation, regardless of source. Parts within specification tolerances are acceptable; parts beyond specification tolerances are not acceptable. However, to manage a manufacturing process the total variation must be traced back to its sources. The first step is to make the distinction between common (i.e., natural, inherent) and special (i.e., assignable) causes of variation.

Common causes refer to the many sources of variation within a process that is in statistical control. They behave like a constant system of chance causes. While individual measured values are all different, as a group they can be described as a distribution. This distribution can be characterized by its:

- location (typical value)
- spread (amount by which the smaller values differ from the larger ones)

Figure 2.1 Piece-to-piece variation.

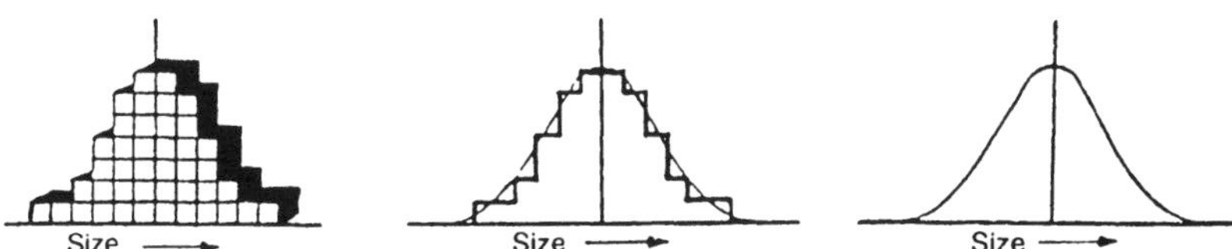

Figure 2.2 Pieces that form a distribution.

- shape (the pattern of variation—whether it is symmetrical, peaked, and so on.)

The illustrations in Figures 2.1, 2.2, and 2.3 reflect the nature of distributions as affected by these factors. Specifically, Figure 2.1 shows pieces vary from each other; Figure 2.2 shows pieces that form a pattern that, if stable, is called a distribution; Figure 2.3 shows that distributions can differ in location, spread, and shape. If only common causes of variation are present, the output of a process forms a distribution that is stable over time; and that is predictable. This is shown in Figure 2.4.

Special causes refer to any factors causing variation which cannot be adequately explained by any single distribution of the process output, as would be the case if the process were in statistical control. Unless all the special causes of variation are identified and corrected, they will continue to affect the process output in unpredictable ways. This is because if special causes are present, the process output will not be stable over time and therefore will be unpredictable. This is shown in Figure 2.5. In general then, we can view the area of process control as one in which special (assignable) causes of variation have been eliminated. This is shown in Figure 2.6. Table 2.1 displays some of the characteristics of variation. For a more detailed discussion see Wheeler (1993), Grant and Leavenworth (1988), and Chrysler, Ford, and General Motors (1995).

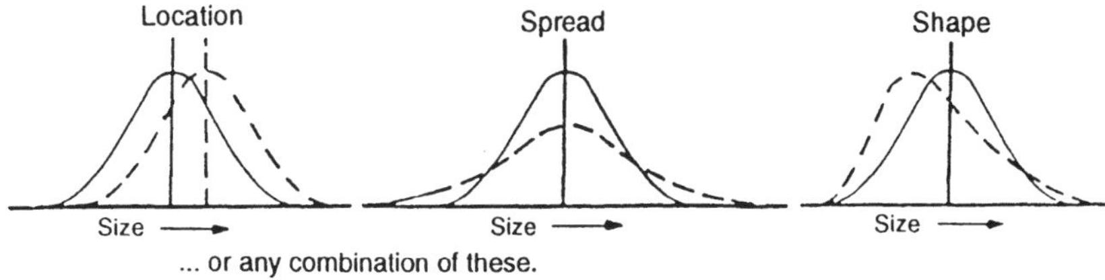

Figure 2.3 Distributions with different location, spread and shape.

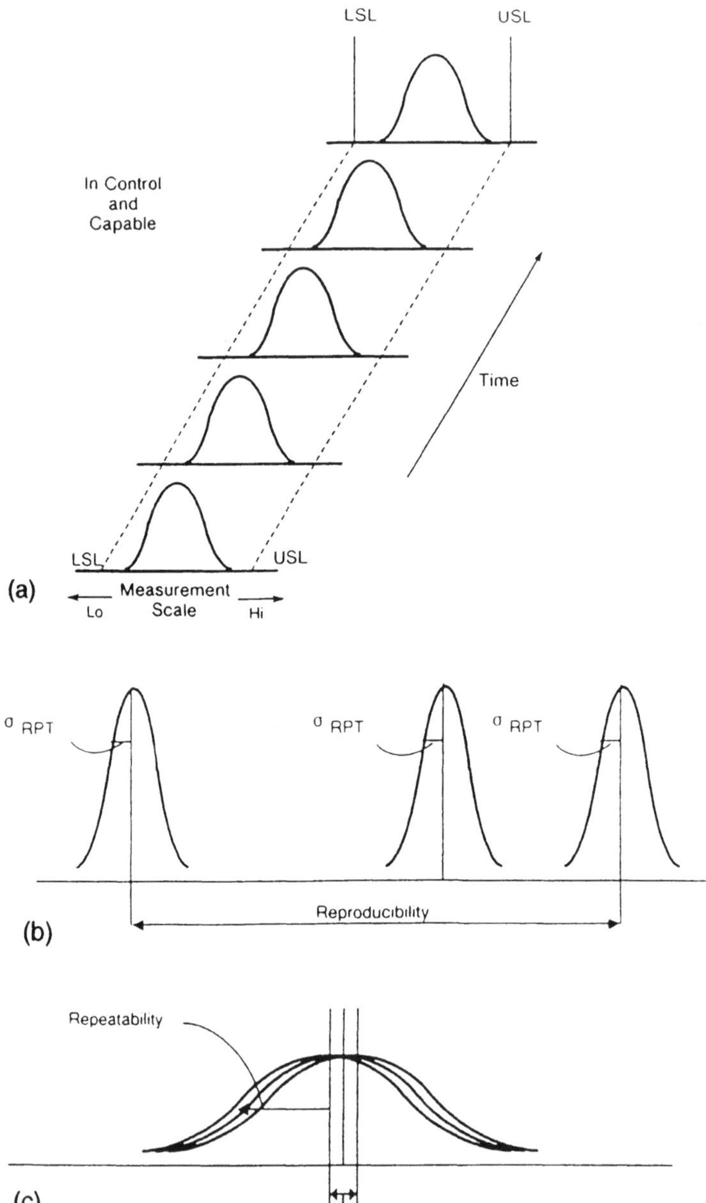

Figure 2.4 (a) A distribution with common variation. (b) Repeatability distributions. (c) Distributions with averages that are almost equal.

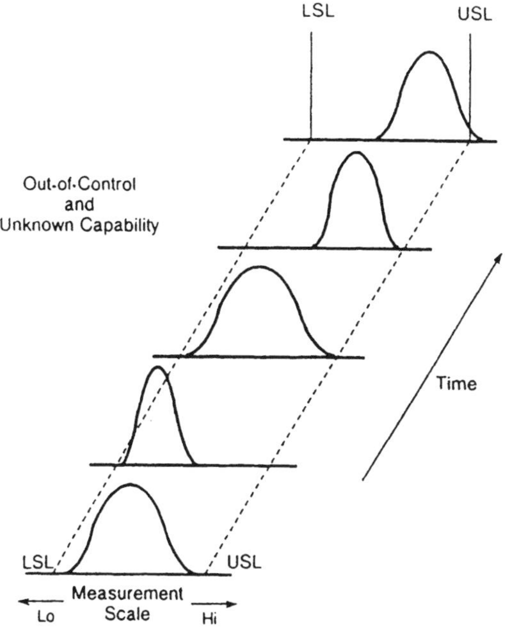

Figure 2.5 An unpredictable and unstable process.

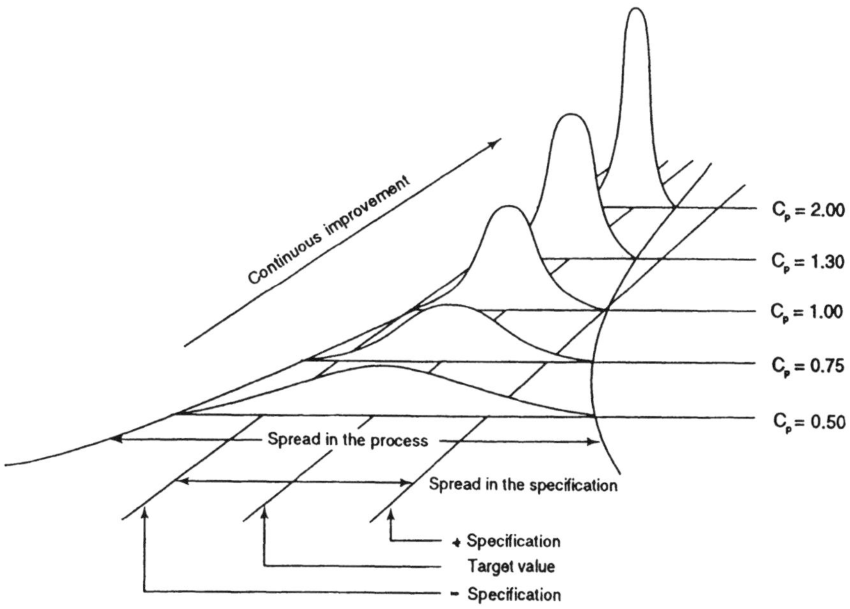

Figure 2.6 A process in which assignable causes have been eliminated.

Table 2.1 Characteristics of Variation

	Common cause	Special cause
Scope of influence	All data in similar manner	Some (or all) data in dissimilar manner
Typical identity	Many small sources	One or a few major sources
Nature	Stable: relatively predictable; inherent in the process	Irregular: unpredictable, appears and disappears
Improvement action	Remove common cause(s)	Eliminate special cause
Improvement action responsibility	System fault: management	Local fault: operator, supervisor

IV. MEASUREMENT

People measure product characteristics or process parameters so they can assess the performance of the system of interest. The measured values provide feedback of the process so that people may adjust settings, replace tools, redesign fixtures, or to allow the operation to continue on its current course. The measurements are indeed the data that allow people to make decisions critical to improvement efforts.

As critical these measurements are, no measurement process is a set of perfect activities. Sometimes different numbers or readings result when the same part or sample is measured a second time. Different readings may be made by different people and gauges, or by the same person using the same gauge. The difference in successive measurements of the same item is called measurement error. This source of variation must be analyzed because the validity of the data directly affects the validity of process improvement decisions.

The measurement system is a major component of the process. In fact, studying variation within the parameters of the measurement system is of paramount importance because:

- Measurement error contributes to process variation and has a negative influence upon the process capability level.
- Measurement error is present whenever measurements are made.

The effects of measurement error influence the assessment of all other items of the process.

In addition to being a part of the process or system, measurement activities also form a process. A typical measurement process is shown in Figure 2.7. There are five components which combine together as the oper-

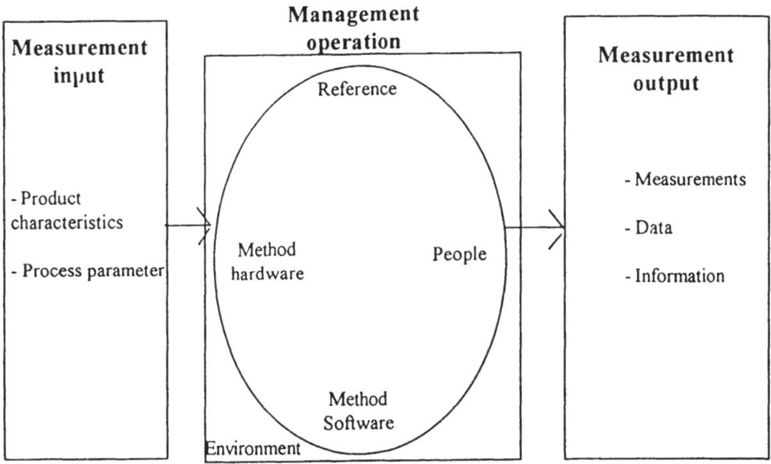

Figure 2.7 Measurement process.

ation or activity of the measurement system to result in the output. The output is the data, readings, or measurements from the process.

It must be emphasized that it is totally inappropriate to view measurement error as merely a function of measurement hardware or instruments. Other components of the measurement process are equally important to measurement error or validity. For example, people contribute to measurement error by having different levels of tactile, auditory, or visual perception. These characteristics account for calibration and/or interpretation differences. Another example in measurement error is the contribution of a "method" change. This kind of error is one of the largest sources of variation in the measurement process. The significance of this is compounded when different people or instruments are used to evaluate the same item or process. Obviously, a standard procedure is needed for every measurement activity. Only this procedure should be used by all people who operate test equipment. Measurement errors that are sometime attributed to differences in people are actually due to differences in methodology. People are usually able to produce similar readings when they use the same methods for operating the measurement equipment. Other examples where measurement error may be introduced include: changes in environment, test equipment, standards, and so on.

In dealing with measurement error, one must be familiar with:

1. *True value.* The true value is a theoretical number that is absolutely "correct" description of the measured characteristic. This is the value

that is estimated each time a measurement is made. The true value is never known in practice because of the measurement scale resolution, or the divisions that are used for the instrument's output. For example, one person may be satisfied with a dimension of up to tenth of an inch (0.1 in.) while another person may define that dimension with a different instrument up to the ten thousandth of an inch (0.1043 in.), which of course is closer to the true value. The appropriate level of measurement resolution is determined by the characteristic specifications and economic considerations. A common practice and "rule of thumb" calls for tester resolution that is equal to or less than one tenth of the characteristic tolerance (i.e., the upper specification limit minus the lower specification limit). The true value is considered as part of tester calibration and discussions of measurement accuracy.

2. *Accuracy.* The accuracy of a measurement system is the relationship between the average of a number of successive measurements of a single part and that of the true value. When the measurement process yields a mean of successive measurements that is different than the true value, the instrument is not calibrated properly. For more information see Ellis (1995).

3. *Precision.* The precision of a measurement system is quantified by the spread of readings that result from successive measurements of the same part or sample. The standard deviation of measurement error is used to quantify the spread of the precision distribution. The common variation that makes the precision distribution comes from two different sources. These are typically referred to as *repeatablility* (RPT) and *reproducibility* (RPD).

Repeatability is a measurement system variation that is due to a specific situation or set of conditions. The differences in successive measured readings for one item that are made by one person using one instrument in one setting/environment, and using one calibration of reference is error due to repeatability. This is the variation within a situation. Repeatability is present in every measurement system.

Figure 2.2 shows a distribution of repeated measurements that were made on one item by one person and with one tester. The average of the curve is not located near the true value for the item. This indicates the inaccuracy of the measurement system and the instrument should be recalibrated. The spread of the curve illustrates the degree of error due to repeatability. The calculated standard deviation for this distribution quantities the level of repeatability error.

Reproducibility describes the difference in successive measurements for the same item that are due to differences between hardware, people, methods, or environments. This source of variability is quantified as the spread (range or standard deviation) between the means of several repeatability distribution (Figure 2.3). Reproducibility only exists when there is more than one measurement situation.

The study that quantifies repeatability and reproducibility contains much diagnostic information. This information should be used to focus measurement system improvement efforts. There are two basic ways of determining the repeatability and reproducibility. One is the graphical way (Figure 2.4a) and another is the calculation method via the standard deviation of measurement error and the precision/tolerance (P/T) ratio.

Figure 2.4b shows three repeatability distributions. Error due to repeatability is smaller than the error due to reproducibility. Although each situation contains relatively small amounts or repeatability error, there is a large difference between each of the measurement activities. This is typical when people use different calibration or measurement methods. Figure 2.4c has averages that are almost equal. This indicates a very small degree of reproducibility error. The error that is inherent in each measurement situation is very large. In this case, all of the measurement activities are using the same calibration and measurement procedures. In spite of these similarities, all of the distributions experience a high degree of repeatability error. This signals a source of error common to each of the measurement situations. The measurement hardware or anything else common to each of the distributions should be investigated as a major source of measurement error.

A. The Standard Deviation of Measurement Error

Because precision is separated into repeatability and reproducibility, the spread of the precision distribution is calculated as a composite. The standard deviation for measurement error is calculated as

$$\sigma_e = (\sigma_{RPT} + \sigma_{RPD})^{1/2}$$

Six standard deviations (6σ) of measurement error describe the spread of the precision distribution. The magnitude of this spread is evaluated by the precision/tolerance ratio.

B. Precision/Tolerance Ratio

A measurement system is declared adequate when the magnitude of measurement error is not too large. One way to evaluate the spread of the precision distribution is to compare it to the product tolerance. This is an absolute index of measurement error because the product specifications should not change. A measurement system is acceptable if it is stable and does not consume a major portion of the product tolerance. The ratio between the precision distribution (six standard deviations of measurement error) and the product tolerance (i.e., upper specification limit and lower specification limit) is called the precision/tolerance (P/T) ratio and quantifies this relationship.

$$\text{P/T ratio} = \frac{6\sigma_{error}}{\text{USL} - \text{LSL}}$$

where

σ_{error} = standard deviation of measurement error
USL = upper specification limit
LSL = lower specification limit

The following general criteria are used to evaluate the size of the precision distribution:

P/T Ratio	Level of measurement error
0.0–0.10	Excellent
0.11–0.20	Good
0.21–0.30	Marginal
0.31	Unacceptable

C. Stability

Repeatability and reproducibility are indices of measurement error based on relatively short periods of time. Stability describes the consistency of the measurement system over a long period. The additional time period allows additional opportunity for the sources of repeatability and reproducibility error to change and add error to the measurement system. All measuring systems should be able to demonstrate stability over time. A control chart made from repeated measurements of the same items documents the level of a measurement system's stability. For more detailed information see Pennela (1992); Morris 1991; Farago (1994); Griffith (1994); National Tooling and Machining Association (1981); Chrysler, Ford, General Motors (1995); ISO 5725–1 (1994); ISO 5725–2 (1994); ISO 5725–3 (1994).

D. Methods of Calculating the Repeatability and Reproducibility (R&R Study)

To calculate the R&R using the *short method*, the following are necessary:

1. Accuracy. Have one operator measure the same part ten times and a standard ten times. Average the readings from each. The difference between the averages is the inaccuracy or bias.
2. Repeatability and reproducibility (R&R). To find the R&R the following steps are necessary:
   Step 1. Randomly select and number 5 parts.

Table 2.2 Form for a Short R&R Study

Parts	Operator A	Operator B	Range (R)
1			
2			
3			
4			
5			
			Sum of ranges =

Step 2. Two operators each measure the parts and record them in the row corresponding to the part number in Table 2.2.

Step 3. The difference in readings is recorded as the range (R) in the appropriate column. This column should have only positive numbers

Step 4. Calculate the average range, $\overline{R}$

$$\overline{R} = \frac{SR}{5}$$

Step 5. $RR_{error} = \frac{\overline{R}}{d_2} \times 5.15.$

This represents 99% of the normal distribution. The d_2 values for distribution of $\overline{R}$ are shown in Table 2.3.

Step 6. Divide RR_{error} by the tolerance and multiply by 100%

$$\frac{RR_{error}}{tolerance} \times 100\%$$

To calculate the R&R using the *long method* see Table 2.4.

Table 2.3 d_2 Values for a Short R&R Study

Parts	2 Operators	3 Operators
1	1.41	1.91
2	1.28	1.81
3	1.23	1.77
4	1.21	1.75
5	1.19	1.74

Table 2.4 Form and Calculations for a Long R&R Study

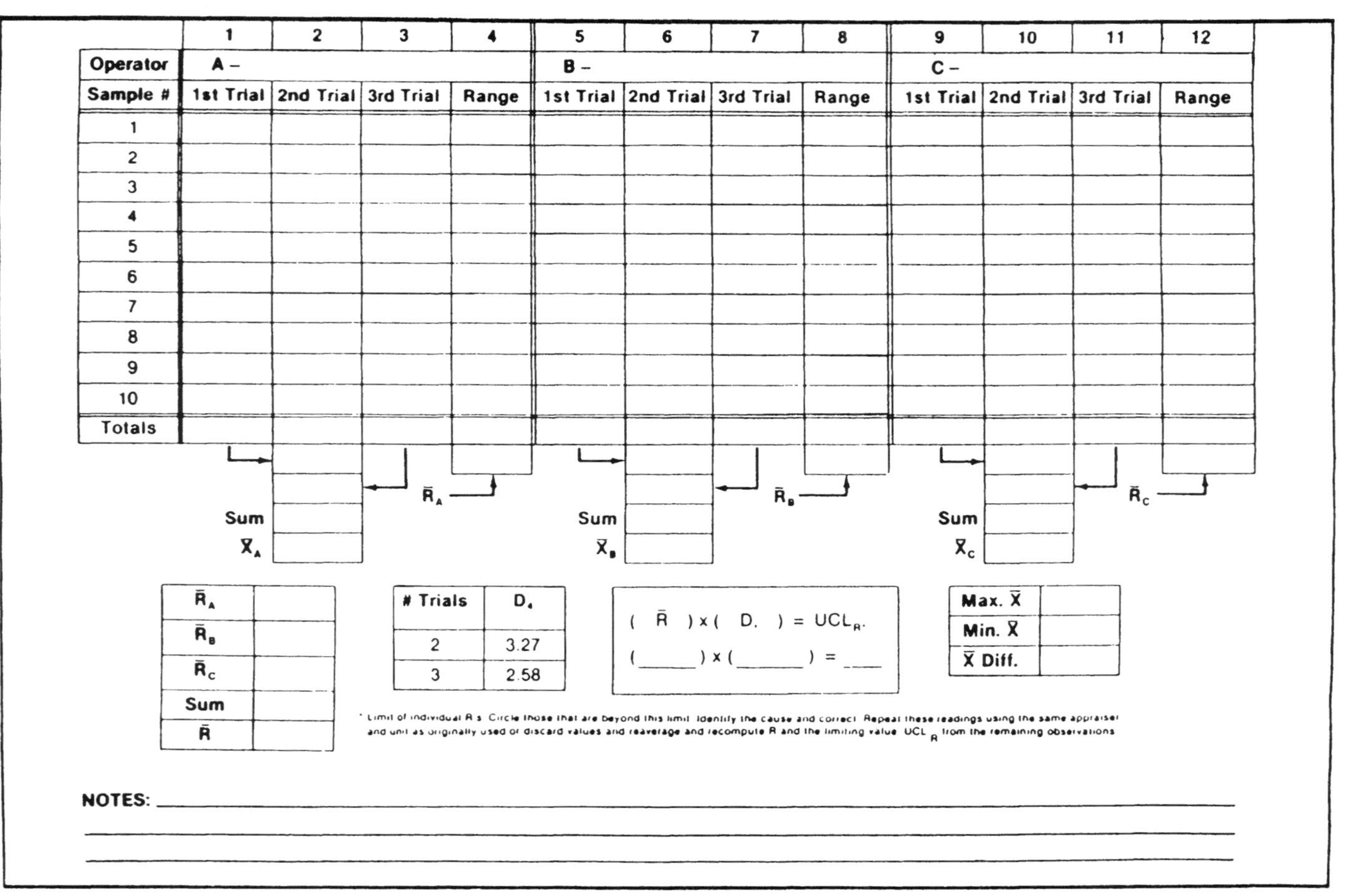

	1	2	3	4	5	6	7	8	9	10	11	12
Operator	**A –**				**B –**				**C –**			
Sample #	**1st Trial**	**2nd Trial**	**3rd Trial**	**Range**	**1st Trial**	**2nd Trial**	**3rd Trial**	**Range**	**1st Trial**	**2nd Trial**	**3rd Trial**	**Range**
1												
2												
3												
4												
5												
6												
7												
8												
9												
10												
Totals												

		$\bar{R}_A$			$\bar{R}_B$
Sum		**Sum**		**Sum**	
$\bar{X}_A$		$\bar{X}_B$		$\bar{X}_C$	

$\bar{R}_C$

$\bar{R}_A$	
$\bar{R}_B$	
$\bar{R}_C$	
Sum	
$\bar{R}$	

# Trials	D_4
2	3.27
3	2.58

$(\ \bar{R}\ ) \times (\ D_4\ ) = UCL_R$*

$(\_\_\_\_\_\_) \times (\_\_\_\_\_\_) = \_\_\_\_$

Max. $\bar{X}$	
Min. $\bar{X}$	
$\bar{X}$ **Diff.**	

* Limit of individual R's. Circle those that are beyond this limit. Identify the cause and correct. Repeat these readings using the same appraiser and unit as originally used or discard values and reaverage and recompute $\bar{R}$ and the limiting value UCL_R from the remaining observations.

NOTES: ____________________________

Part No. & Name ______ Gage Name ______ Date ______

Characteristic ______ Gage No. ______ Performed By: ______

Specification ______ Gage Type ______ ______

From Data Sheet: $\bar{R}$ = [ ] $\bar{X}_{Diff}$ = [ ]

MEASUREMENT UNIT ANALYSIS

REPEATABILITY · EQUIPMENT VARIATION (E.V.)

TRIALS	2	3
K_1	4.56	3.05

E.V. = ($\bar{R}$) x (K_1)

= (____) x (____) = [ ]

n = number of parts
r = number of trials

REPRODUCIBILITY — APPRAISER VARIATION (A.V.)

OPERATORS	2	3
K_2	3.65	2.70

$$A.V. = \sqrt{[(\bar{X}_{diff}) \times (k_2)]^2 - [(E.V.)^2 / (n \times r)]}$$

$$= \sqrt{[(\_\_\_\_) \times (\_\_\_\_)]^2 - [(\_\_\_\_)^2 / (\_\_ \times \_\_)]}$$

= [ ]

REPEATABILITY AND REPRODUCIBILITY (R & R)

$$R \& R = \sqrt{(E.V.)^2 + (A.V.)^2}$$

$$= \sqrt{(\_\_\_\_)^2 + (\_\_\_\_)^2} = [\ ]$$

NOTE: All calculations are based upon predicting 5.15 σ (99% of the area under normal curve)

% TOLERANCE ANALYSIS

% E.V. = 100 [(E.V.) / (TOLERANCE)]

= 100 [(____) / (____)]

= [%]

% A.V. = 100 [(A.V.) / (TOLERANCE)]

= 100 [(____) / (____)]

= [%]

$$\% R \& R = \sqrt{(\% E.V.)^2 + (\% A.V.)^2}$$

$$= \sqrt{(\_\_\_\_)^2 + (\_\_\_\_)^2}$$

= [%]

A negative value under the square root sign causes the appraiser variation to default to zero

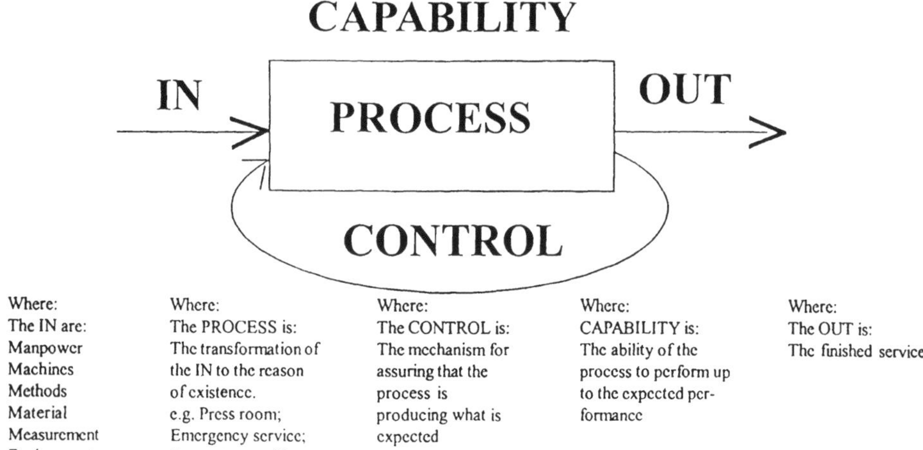

Figure 2.8 A typical process.

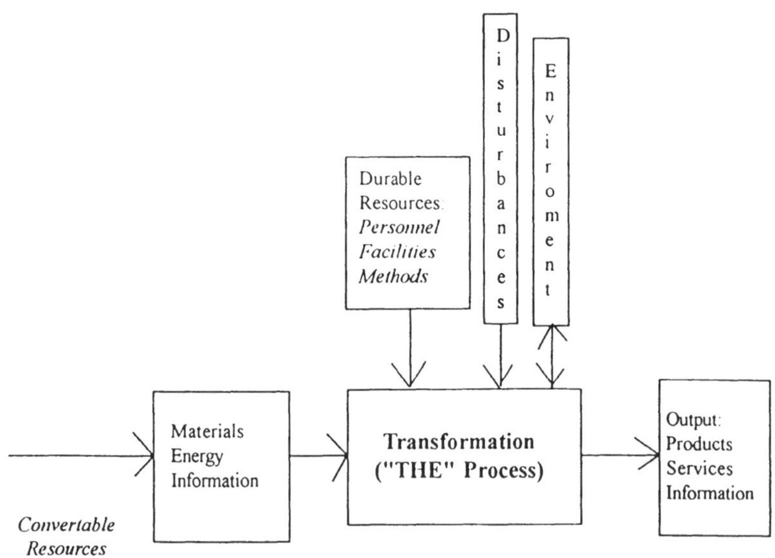

Figure 2.9 A process with its typical components.

V. PROCESS

A process may be defined as a combination of inputs, both durable and convertible resources, for the purpose of obtaining desired quality outputs. The transformation of the inputs into the output(s) is the process. A typical process is shown in Figure 2.8; Figure 2.9 show a typical process with its typical components.

Using this definition, a process may be thought of in global terms as all the operations of a business collectively or in a much more narrow sense as a particular operation of a specific machine. Both views are appropriate. Typically, opportunities for improvement are illuminated by removing as much of the noise as possible by narrowing the focus to smaller elements of the total process.

The effort to control a process and its output consists of essentially four elements (IBM 1984; Ford 1985). They are:

1. *The Process.* By the process, we mean the whole combination of people, machines, materials, methods, measurements, and environments that work together to produce output. The total performance of the process—the quality of its output and its productive efficiency—depends on the way the process has been designed and built, and on the way it is operated. The rest of the process control system is useful only if it contributes to improved performance of the process.

2. *Information About the Process.* Much information about the actual performance of the process can be learned by studying the process output. In a broad sense, process output includes not only the products that are produced, but also any intermediate outputs that describe the operating state of the process, such as temperature, cycle times, and so on. If this information is gathered and interpreted correctly, it can show whether action is necessary to correct the process or the just-produced output. If timely and appropriate actions are not taken, however, any information gathering effort is wasted.

3. *Action on the Process.* Action on the process is future oriented, as it is taken when necessary to prevent the production of out-of-specification products. This action might consist of changes in the operations (e.g., operator training, changes to the incoming material, etc.), or the more basic elements of the process itself as a whole (e.g., changes in shop temperature or humidity). The effect of actions should be monitored, and further analysis and action should be taken if necessary.

4. *Action on the Product.* Action on the output is postoriented because it involves detecting out-of-specification products already produced. Unfortunately, if current output does not consistently meet specification, it may be necessary to sort all products and to scrap or rework any nonconforming items.

This must continue until the necessary corrective action on the process has been taken and verified, or until the product specifications have been changed.

Obviously, the action on the process is the most crucial element in this area of process control. The action on the process involves prevention, whereas the action on the output involves sorting and appraisal methods of identifying nonconformities.

A. Process Control Versus Product Control

The distinction between process control and product control can best be depicted through the Figures 2.10 and 2.11. Product control orients the classical control cycle in a feed-forward (in time) mode. On the other hand, the control cycle is oriented as a feed-back system with process control. Figure 2.12 shows the movement from product control to process control. In addition, Table 2.5 summarizes some of the most important characteristics of product and process control.

B. Capability

By definition capability is the total range of a stable process's inherent variation. There are many ways to measure capability; here, however, we address the most common.

First Run Capability

Perhaps one of the easiest and simplest ways to calculate the capability of a process. Many managers and quality professionals use first run capability

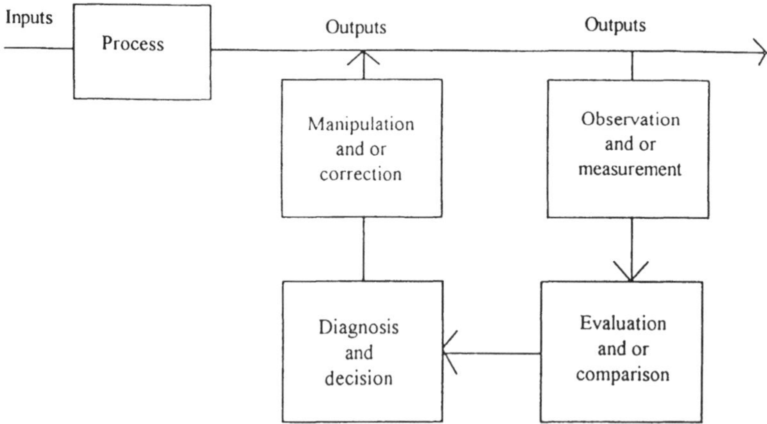

Figure 2.10 Process control.

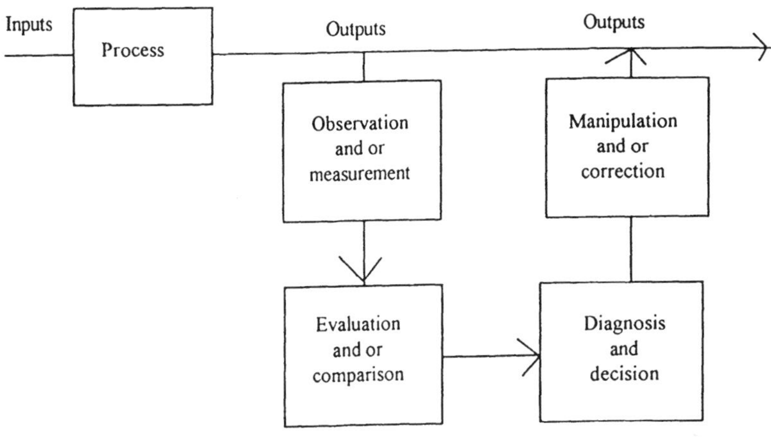

Figure 2.11 Product control.

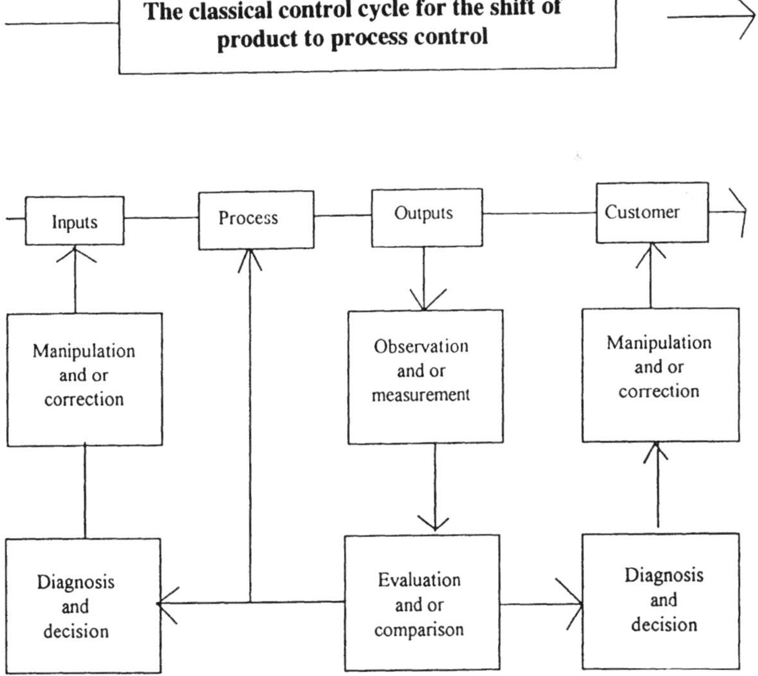

Figure 2.12 Movement from product to process control

Table 2.5 Characteristics of Product and Process Control

Characteristic	Product control	Process control
Focus	Product (appraisal)	Process (prevention)
Goal	Variability within specification limits	On target with smallest variation
Typical tools	Acceptance sampling plans	Control charts
Improvement nature	Outgoing quality only	Quality plus productivity
Philosophy	Detection and containment of problematic occurrences	Prevention of problematic occurrences

(FRC) data, as one measure of performance. Often, however, what is called FRC is really the percent of acceptable parts that make it through an inspection point (Percent OK) including those that required rework or repair. The difference could be significant. Although Percent OK and other like measures are useful to help identify trends, an important ingredient is masked, namely, the proportion of production that is scrapped, repaired, or reworked prior to reaching the inspection point.

First run capability as defined below considers only those products which are "done right the first time," thereby exposing all of the waste reduction opportunities in the process. The definition is intended to be generic in nature such that it can be applied to any process or system to foster continuous improvement by eliminating waste. Furthermore, it should be noted that FRC does not preclude using other methods to monitor process effectiveness. However, if the other methods presently in use do not comply with FRC as now defined, they should be renamed to reflect more closely what they are measuring.

A formal and functional definition for the FRC of a process or system (or a specified portion of a process or system) is the percentage of input which flows through the process or system without being scrapped or incurring rework or repair.

Given that N is the number of pieces, units, or transactions put into a process or system and W is the amount of given process/system input that was scrapped, repaired, or reworked (i.e., waste), then FRC can be calculated as:

$$\text{FRC} = \frac{N - W}{N} \times 100\%$$

An application of this definition of FRC to a hypothetical process or system is illustrated in Figure 2.13. The FRC of this total process/system is 79%. This result can be used to identify process improvement opportunities, and

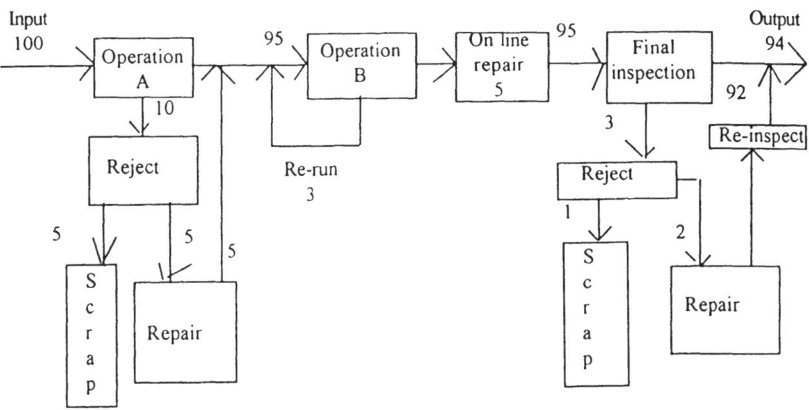

Where:

N = Number of input pieces, units, transactions = 100
W = Waste

Operation A scrap	5
Operation A repair	5
Operation B rerun	3
On line repair	5
Final inspection:	
Scrap	1
Repair	2
Total W	21

Therefore,

FRC = {(N-W)/N} X 100% = {(100-21)/100} X 100% = 79%

Figure 2.13 Example of first run capability.

to verify the effectiveness of changes made to the process/system. The FRC of a section of a process/system can be illustrated by calculating the FRC of Operation A. Operation A FRC would be 90% (i.e., $[(100-90)/100] \times 100\%$.)

First run capability as defined is generic and can be applied to any process or system. The process flow shown in Figure 2.13 could represent a typical manufacturing process or an administrative process (e.g., an accounts payable system where the units are paper transactions.)

First run capability studies should be completed using process/system input of a known quality level, so that the FRC of that process/system can be studied analytically utilizing statistical methods. Since the calculation of

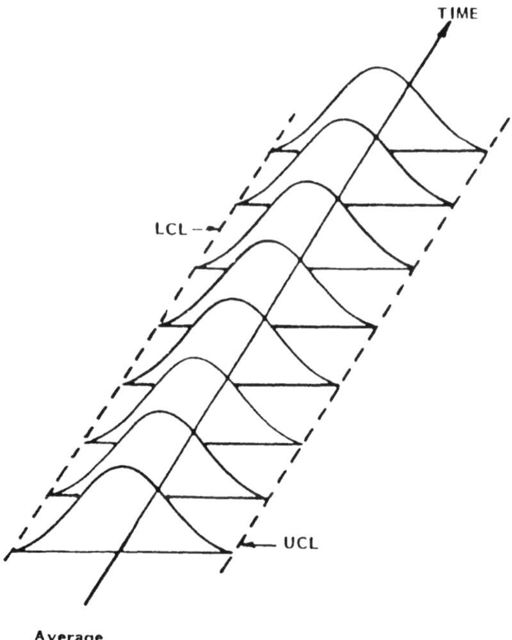

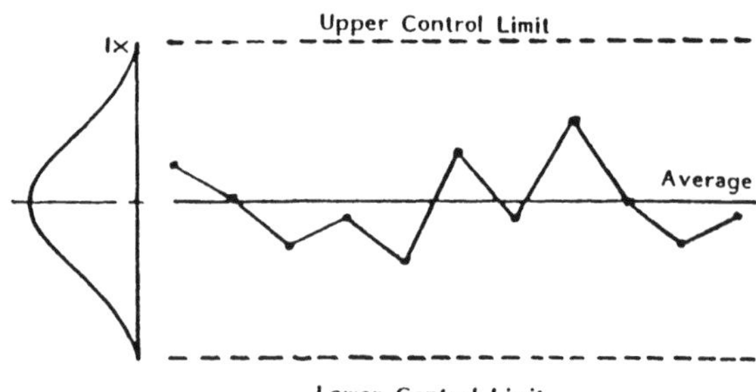

Figure 2.14 A capable process.

a process/system FRC may require additional resources, periodic FRC studies can be considered as one means of identifying continuous improvement opportunities. Also, other measures of process effectiveness (e.g., such as Percent OK) at a tally point or inspection point may be utilized if appropriate. It is also important to recognize the limitations of these other measures.

The C_p and C_{pm}

Interest in process capability is growing, due partially to the changing philosophy in quality. Process capability has been defined in many ways; as a result, several measures of process capability exist. One common definition describes process capability as the range over which the output of a process varies. This quantity is also referred to as the actual process spread. Measures in this group depend on the measuring units (meters, kilograms, and so on) and do not encourage comparisons among processes with different quality characteristics.

The process capability index (C_p) relates the allowable process spread (an engineering requirement, usually taken as the difference between the upper and lower specification limits) to the actual process spread as a ratio:

$$C_p = \frac{\text{allowable actual process spread}}{\text{allowable process spread}}$$

This index is unitless and thereby allows comparisons between processes with different quality variables; the index also promotes similar inferences regardless of the product or quality characteristic measured. For example, an index value of one indicates that the allowable process spread is equivalent to the actual process spread. A process capability index of two indicates that the allowable process spread is twice that of the actual process spread, suggesting that the process is quite capable of producing within specifications (Figure 2.14). Index values less than one indicate that the actual process spread is larger than the allowable process spread, suggesting that nonconforming product will result (Figure 2.15).

The actual process spread is generally taken to be 6s, which represents, in normal distribution theory, the width of the interval that contains 99.73% of the population. The difference in the specification limits is used to indicate allowable process spread. The allowable process spread is considered fixed, while the actual process spread must be estimated. For a more technical discussion see Spiring (1991).

A measure similar to the process capability index is the C_{pm} and falls into a group of second generation measures of process capability that consider both process variation and proximity to the target when assessing the ability of a process. The shift from the original index to the modified index in terms of calculation is subtle, but the inferences and interpretations are vastly different. For a detailed technical discussion see Spiring (1991).

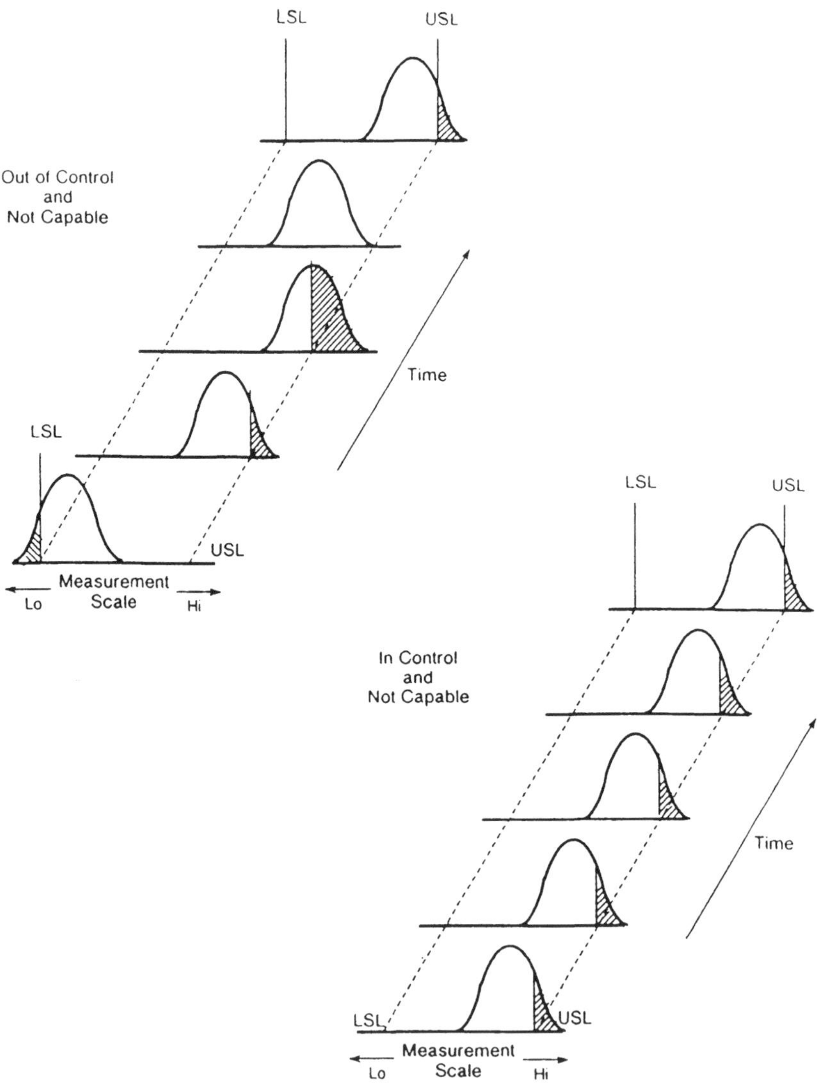

Figure 2.15 A process producing nonconforming parts.

Capability Ratio

The capability ratio (C_r) index is used to describe the comparison of the tolerance to the measurement spread (six standard deviations) without worrying about its location within the specification limits. The C_r is the opposite of C_p. Therefore, we represent the C_r index with:

$$C_r = \frac{6\sigma}{\text{tolerance}}$$

This ratio gives (as a percent) the tolerance of the measurement spread. An acceptable index is 75% or less for a potential capability study. This potential capability is always completed before the machine/process is put into production. Notice that the smaller the percent, the better the potential capability. However, this does not tell you whether or not the spread is within the specifications. The C_r percent index will always be smaller than the TRp percent index if the spread is not centered.

Target Ratio Percent

The target ratio percent (TRp) index will give a percent that compares the average of the machine or process with the nearest specification limit by using a measurement spread equal to three standard deviations. It will indicate what percent that the spread (three standard deviations) is using of the difference between the average and its nearest blueprint specification. The equation for TRp is:

$$\text{TRp} = \frac{3\sigma \times 100}{\text{the minimum of } (\text{USL} - \bar{X}) \text{ and } (\bar{X} - \text{LSL})}$$

where

USL = upper specification limit
LSL = lower specification limit
$\bar{X}$ = distribution mean

An acceptable index is 100% or less for a long-term capability study. Notice that the smaller the percent the better is the long term capability. The TRp index will always be larger than the C_r unless the average and the standard deviation are the same.

C. Normal Distribution

The normal distribution is an important concept in quality and especially when using statistical process control (SPC). Learning about it allows people to use many statistical tools and methods that can be used to help improve a production process.

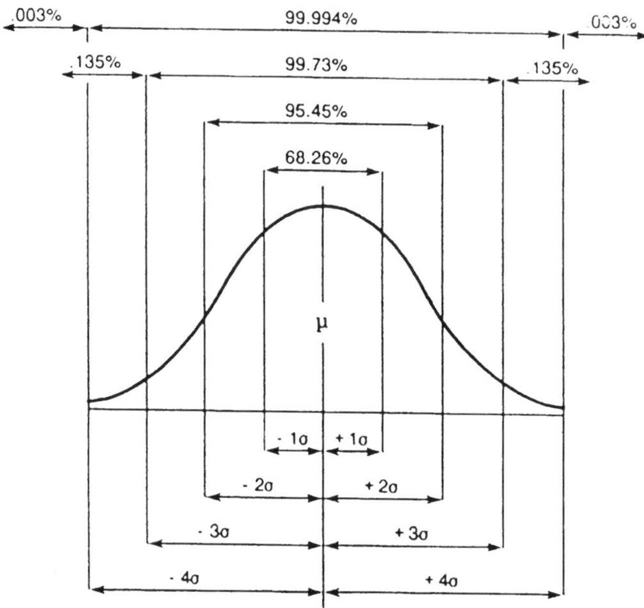

Figure 2.16 The normal distribution curve.

The normal distribution (curve) has a clear pattern and it is the result of many operations. Fundamentally, it is described by its mean and standard deviation. It has the following features (Figure 2.16):

- The point of greatest frequency is at the mean.
- The mean, median, and mode are the same.
- The distribution curves equally around the mean (i.e., each side mirrors the other). All normal distributions are symmetrical; however, not all symmetrical distributions are normal.
- The curve slopes down on both sides of the mean. The tails of the curve never touch the horizontal axis. They approach infinity.

Numbers can also describe the normal distribution, by dividing the curve into standard deviation zones. The area under the curve (probability) can be calculated with the zones (Figure 2.17). Given the specific zones, the following may be calculated:

- The mean plus or minus one standard deviation is 68.27% of the distribution.

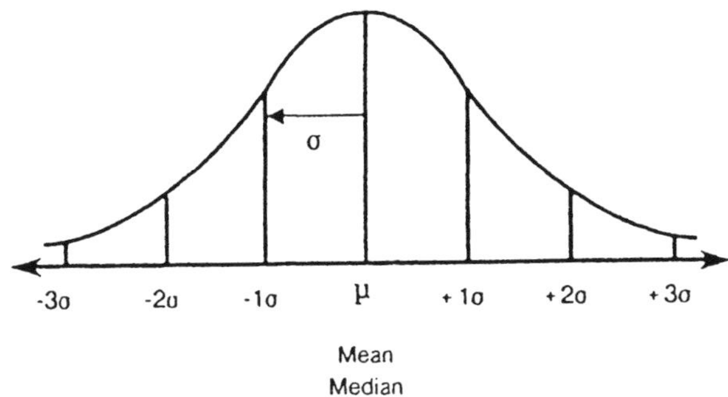

Figure 2.17 The standard deviation zones.

- The mean plus or minus two standard deviations is 95.45% of the distribution.
- The mean plus or minus three standard deviations is 99.73 of the distribution.

Obviously, the normal distribution continuous past three standard deviations. However, the area beyond three standard deviations is very small.

To show the applicability of the normal distribution let us look at a manufacturing process. 125 shaft diameters were measured. They are shown in a histogram and with descriptive statistics in Figure 2.18. The shape of the sample group matches the normal distribution. This means that the estimates for the population are good. (Here caution should be exercised, because the

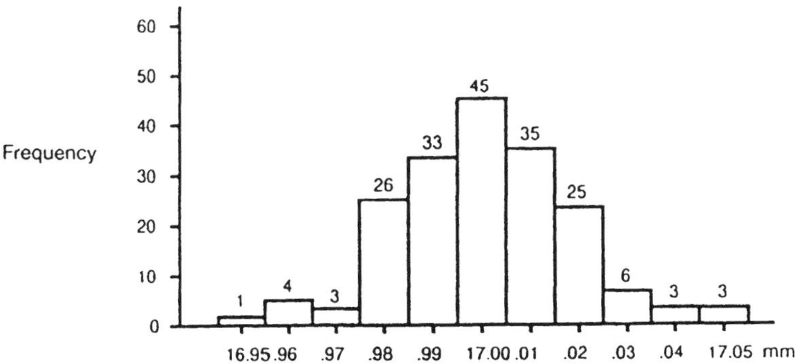

Figure 2.18 Histogram with descriptive statistics for shaft diameters.

normal distribution gives poor estimates for skewed or nonnormal histograms.) Using Figure 2.18 the following information may be of value:

- 68.27% of the diameters are between 16.991 and 17.025 mm.
- 95.45% of the diameters are between 16.974 and 17.042 mm.
- 99.73% of the diameters are between 16.957 and 17.059 mm.

(Note that the values do not always fall on whole number standard deviations. This is the case with measuring diameters.)

D. Standardized Values

Standardized values (Z) make the normal distribution useful especially in capability studies. Z values measure the area under the curve of the normal distribution. To use the Z values we do the following:

1. Define the zone of interest. Do this by standard deviations or fractions of a standard deviation.
2. Change the actual measurement (raw data) scale in our example (mm) to a standard scale Z.
3. Use Z for any distribution matching the normal curve. You may use any unit of measurement; however, make sure to standardize the measurement scale. The conversion formula from raw data to Z values is the following:

$$Z = \frac{x - \overline{X}}{\sigma}$$

where:

Z = standard value
x = value to be standardized
$\overline{X}$ = distribution mean
σ = distribution (sample) standard deviation

The Z formula yields a scale without units. The output values show the distance (how many standard deviations) of a point from the mean. Of course, the sign of the answer can be positive or negative. A positive value indicates that the point is above the mean, where a negative value indicates that the point is below the mean.

The following example to show how the Z value is calculated is shown below based on the example of the shaft diameters.

1. Gather the needed information. Draw the problem (Figure 2.19).
   a. x = value to be standardized (i.e., the largest part that will not freeze = 17.035 mm.)
   b. $\overline{X}$ = distribution mean = 17.001 mm.
   c. s = Distribution (sample) standard deviation = 0.018 mm.

2. Calculate the Z values:
   a. $Z = (x - \overline{X})/\sigma$
   b. $Z = (17.035 - 17.001)/0.018 = 1.8888889$
3. Round the Z value to two (2) decimal places. Z = 1.89.
4. Use Table 2.6 to find the area of probability.
   a. The left column of Table 2.6 is labeled Z. The numbers in the column are the units and tenths digits of the standard Z value. (For our example the unit is 1. The tenth digit is 8 and the hundredth digit is 9. The row is indicated by *x.x*0, *x.x*1, *x.x*2 ... *x.x*9, where the 0, 1, 2...9 indicates the actual hundredth digit.
   b. The answer is found where the row (1.8) and column (*x.x*9) cross. The answer from the table is highlighted and is shown as 0.02938 (we rounded to 0.294). The values in Table 2.6 are proportions or percentages and are always in decimal form. Multiply them by 100 (0.0294 × 100 = 2.94%) to turn them into whole number percentages. The meaning of this percentage is always dependant upon the shaded part of the normal distribution curve on the very top of the table. Without the shaded part the table is useless. In our example the shaded part indicates the reject region. Therefore, 2.94% of our shafts will freeze.

E. C_{pk}

The C_{pk} is yet another index of capability; however, because it is dependent on the normal distribution the user must be very knowledgeable about skewness and kurtosis of distribution. If the process is not normally distributed, then the C_{pk} is useless and it should not be used as a measure of capability. Therefore, the testing for normality is imperative before any C_{pk} analysis takes place.

A simple test for normality is that of skewness and kurtosis. With the proliferation of statistical packages this task is very simple (in fact, just about all statistical packages provide the values by default) and its usual interpretation is as follows. Skewness is an indication of whether the data in the histogram has a normal distribution. Specifically, it is a measure of symmetry. Values of skewness have the following meaning:

0 Symmetrical distribution.
> 0 Positive skewness. The distribution has a longer "tail" to the positive side.
< 0 Negative skewness. The distribution has a longer "tail to the negative side.

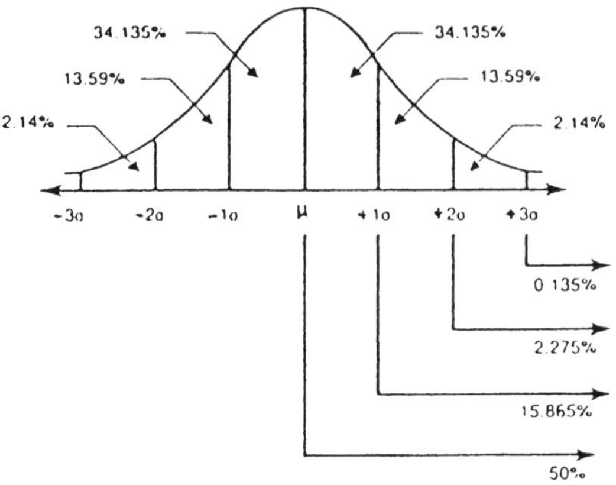

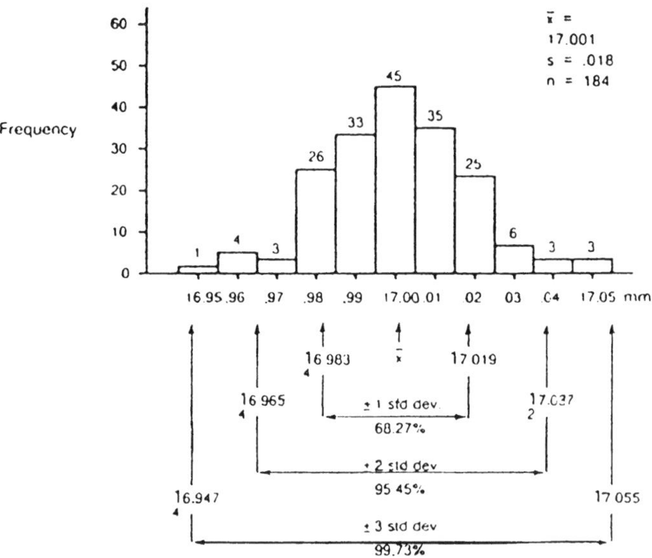

Figure 2.19 An example of using the normal curve.

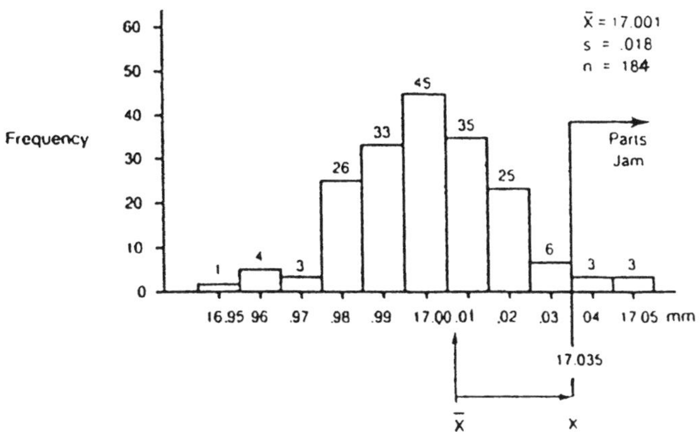

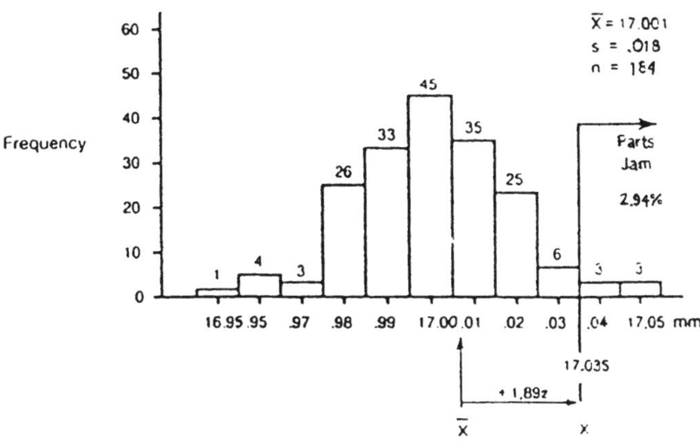

Figure 2.19 (Continued)

Table 2.6 Tail Probabilities for the Standard Normal Distribution. The table entry is $a = Pr(Z \geq z)$, the shaded area under the standard normal density function from z to infinity.

z	0.00	0.01	0.02	0.03	0.04	0.05	0.06	0.07	0.08	0.09
0.00	0.50000	0.49601	0.49202	0.48803	0.48405	0.48006	0.47608	0.47210	0.46812	0.46414
0.10	0.46017	0.45620	0.45224	0.44828	0.44433	0.44038	0.43644	0.43251	0.42858	0.42465
0.20	0.42074	0.41683	0.41294	0.40905	0.40517	0.40129	0.39743	0.39358	0.38974	0.38591
0.30	0.38209	0.37828	0.37448	0.37070	0.36693	0.36317	0.35942	0.35569	0.35197	0.34827
0.40	0.34458	0.34090	0.33724	0.33360	0.32997	0.32636	0.32276	0.31918	0.31561	0.31207
0.50	0.30854	0.30503	0.30153	0.29806	0.29460	0.29116	0.28774	0.28434	0.28096	0.27760
0.60	0.27425	0.27093	0.26763	0.26435	0.26109	0.25785	0.25463	0.25143	0.24825	0.24510
0.70	0.24196	0.23885	0.23576	0.23270	0.22965	0.22663	0.22363	0.22065	0.21770	0.21476
0.80	0.21186	0.20897	0.20611	0.20327	0.20045	0.19766	0.19489	0.19215	0.18943	0.18673
0.90	0.18406	0.18141	0.17879	0.17619	0.17361	0.17106	0.16853	0.16602	0.16354	0.16109
1.00	0.15866	0.15625	0.15386	0.15151	0.14917	0.14686	0.14457	0.14231	0.14007	0.13786
1.10	0.13567	0.13350	0.13136	0.12924	0.12714	0.12507	0.12302	0.12100	0.11900	0.11702
1.20	0.11507	0.11314	0.11123	0.10935	0.10749	0.10565	0.10383	0.10204	0.10027	0.09853
1.30	0.09680	0.09510	0.09342	0.09176	0.09012	0.08851	0.08691	0.08534	0.08379	0.08226
1.40	0.08076	0.07927	0.07780	0.07636	0.07493	0.07353	0.07215	0.07078	0.06944	0.06811

1.50	0.06681	0.06552	0.06426	0.06301	0.06178	0.06057	0.05938	0.05821	0.05705	0.05592
1.60	0.05480	0.05370	0.05262	0.05155	0.05050	0.04947	0.04846	0.04746	0.04648	0.04551
1.70	0.04457	0.04363	0.04272	0.04182	0.04093	0.04006	0.03920	0.03836	0.03754	0.03673
1.80	0.03593	0.03515	0.03438	0.03362	0.03288	0.03216	0.03144	0.03074	0.03005	0.02938
1.90	0.02872	0.02807	0.02743	0.02680	0.02619	0.02559	0.02500	0.02442	0.02385	0.02330
2.00	0.02275	0.02222	0.02169	0.02118	0.02068	0.02018	0.01970	0.01923	0.01876	0.01831
2.10	0.01786	0.01743	0.01700	0.01659	0.01618	0.01578	0.01539	0.01500	0.01463	0.01426
2.20	0.01390	0.01355	0.01321	0.01287	0.01255	0.01222	0.01191	0.01160	0.01130	0.01101
2.30	0.01072	0.01044	0.01017	0.00990	0.00964	0.00939	0.00914	0.00889	0.00866	0.00842
2.40	0.00820	0.00798	0.00776	0.00755	0.00734	0.00714	0.00695	0.00676	0.00657	0.00639
2.50	0.00621	0.00604	0.00587	0.00570	0.00554	0.00539	0.00523	0.00508	0.00494	0.00480
2.60	0.00466	0.00453	0.00440	0.00427	0.00415	0.00402	0.00391	0.00379	0.00368	0.00357
2.70	0.00347	0.00336	0.00326	0.00317	0.00307	0.00298	0.00289	0.00280	0.00272	0.00264
2.80	0.00256	0.00248	0.00240	0.00233	0.00226	0.00219	0.00212	0.00205	0.00199	0.00193
2.90	0.00187	0.00181	0.00175	0.00169	0.00164	0.00159	0.00154	0.00149	0.00144	0.00139
3.00	0.00135	0.00131	0.00126	0.00122	0.00118	0.00114	0.00111	0.00107	0.00104	0.00100
3.10	0.00097	0.00094	0.00090	0.00087	0.00084	0.00082	0.00079	0.00076	0.00074	0.00071
3.20	0.00069	0.00066	0.00064	0.00062	0.00060	0.00058	0.00056	0.00054	0.00052	0.00050
3.30	0.00048	0.00047	0.00045	0.00043	0.00042	0.00040	0.00039	0.00038	0.00036	0.00035
3.40	0.00034	0.00032	0.00031	0.00030	0.00029	0.00028	0.00027	0.00026	0.00025	0.00024
3.50	0.00023	0.00022	0.00022	0.00021	0.00020	0.00019	0.00019	0.00018	0.00017	0.00017
3.60	0.00016	0.00015	0.00015	0.00014	0.00014	0.00013	0.00013	0.00012	0.00012	0.00011
3.70	0.00011	0.00010	0.00010	0.00010	0.00009	0.00009	0.00008	0.00008	0.00008	0.00008
3.80	0.00007	0.00007	0.00007	0.00006	0.00006	0.00006	0.00006	0.00005	0.00005	0.00005
3.90	0.00005	0.00005	0.00004	0.00004	0.00004	0.00004	0.00004	0.00004	0.00003	0.00003
4.00	0.00003	0.00003	0.00003	0.00003	0.00003	0.00003	0.00002	0.00002	0.00002	0.00002

On the other hand, kurtosis is an indication of whether the data in the histogram has a normal distribution. Specifically, it is a measure of the flatness or peakedness of a curve. Values for kurtosis have the following meaning:

3	Normal distribution.
< 3	Neptokurtic curve. The curve has a high peak.
> 3	Platykurtic curve. The curve has a low peak.

Once the determination of normality has been examined, then the experimenter is ready for the C_{pk} analysis. Very simply, C_{pk} is equal to the distance from the process mean to the nearest specification limit divided by three times the standard deviation.

$$C_{pk} = \frac{\text{process mean} - \text{nearest specification limit}}{3\sigma}$$

When the C_{pk} is larger than 1, you are in good shape. Just make sure the process average does not start drifting toward a specification limit. If C_{pk} is equal to 1, things are all right at the moment, but improvement is still needed in at least the following areas:

- centering the process average between the specification limits
- reducing process variation (σ)
- reviewing specification limits to see if they are where they need to be

When the C_{pk} is less than 1, somebody is going to lose money. The process is incapable and needs immediate attention. Even when C_{pk} is larger than 1 (meaning you are in good shape), it does not mean your work is completely done. Most major companies have already discovered that meeting specifications is not enough. In order to survive in competitive world markets, they have found it necessary to continuously shrink the limits of process variability. This produces a more uniform product. C_{pk} can be used to monitor the improvement as the variability of the process decreases, giving C_{pk} values of 2, 3, or even larger.

F. Capability and the Six Sigma (6σ) Relationship

Motorola's determination for quality excellence, introduced the concept of 6σ, as a statistical measure expressing how close a product comes to its quality goal. Sigma means 68% of products are acceptable, three sigma means 99.7%, and six sigma is 99.9997% perfect or 3.4 nonconformities per million parts. In fact, Motorola was so convinced that quality would be a ticket to their success, that in conjunction to their six sigma, they also introduced the concept of "zero defects" as an attitude (Gill 1990a,b), so that they could elim-

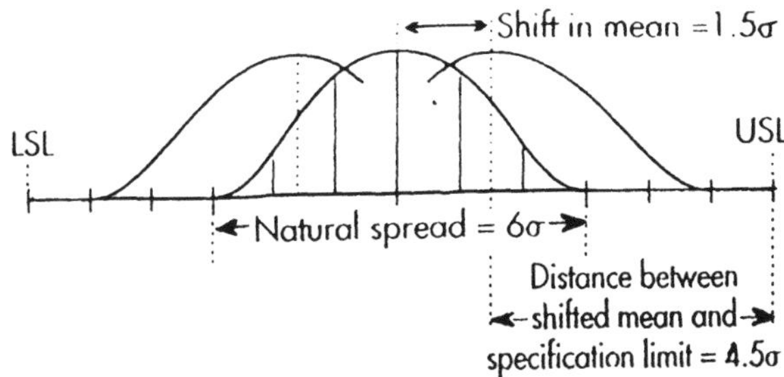

Figure 2.20 Normally distributed variable with $C_p = 2$ and shifts in the mean equal to 1.5σ.

inate defects throughout the entire organization. Their rationale for such an endeavor was that the finer your ability to measure defects, the easier it is to catch and eliminate them.

A second definition of six sigma accounts for shifts in the mean averaging 1.5 times the distribution's standard deviation to the right or left (Figure 2.20). Research and empirical data indicate this is consistent with shifts in means that occur in industrial processes. The process in the figure has a natural spread half as wide as the distance between specification limits and a process capability of 2. A process mean shift of 1.5 standard deviation to the right pushes the upper specification limit to 4.5 standard deviations above the mean. No values would be below the lower specification, and 3.4 values per million will exceed the upper specification limit (since the one-tailed area under the normal distribution beyond 4.5 standard deviation is 0.0000034). A similar process shift to the left of the mean produces similar results; on average, 3.4 values per million that do not conform to specifications. For a more technical and detailed discussion of the six-sigma approach see Dambolena and Rao (1994), Motorola (1992), and Harry and Stewart (1988).

Six sigma does not imply three defective units per million made. It signifies three defects per million opportunities. Some products may have tens of thousands of opportunities for nonconformities per finished item, so the proportion of defective items may actually be quite large. Nevertheless, using the six-sigma approach quite often results in a significant improvement.

G. Concerns about the C_{pk}

Many organizations have used the concept and the results of the C_{pk} to correct their processes. In many cases and to their amazement, they did nothing to improve the capability of a process or reduce production costs. Why or how did that happened? The answer is a very simple one, and it lies on the lack of understanding in the following three items.

1. The average C_{pk} value for each process that requires a capability study in a facility will be equal to, or greater than 1.5.
2. To demonstrate that the process meets requirements for C_{pk}, one must conduct a process-capability study using a minimum of 100 items to show that the value of C_{pk} is equal to or greater than 1.67.
3. Once an acceptable process capability has been demonstrated, the value of C_{pk} must be recalculated, generally monthly or quarterly, based on the number of items produced during that period. Whenever the values goes down, the concerned party will institute a cause-and-corrective-action study to determine the reason, implement corrective action, and ensure that the value of C_{pk} once again meets requirements.

These items are supposed to promulgate the philosophy of continuous improvement but, in many cases, use up valuable resources trying to fix processes that already meet customer requirements. Why? Because they are biased, deterministic, and ambiguous.

C_{pk} is a biased statistic that is ambiguous regarding the apparent process capability conveyed by analyzing the value. In practice, the C_{pk} is applied in a deterministic fashion by users. For example, "... our requirement is a C_{pk} of 1.67, however, the process shows that it has a capability of 1.63, so you fail." The problem is that the exact values obtained for C_{pk} do not, in most cases, accurately portray process capabilities.

When considering C_{pk}, first look at where this statistic is defined for use. To use C_{pk}, the process must be in a state of control over a period of time, and be repeatable and consistent. That means that the process output must follow a univariate, normal distribution with some parts falling outside quality requirements. If these conditions are not met, C_{pk} is of no use. Warning! Many believe that when a process is operating in a state of control, that this implies normality. The process may be consistently bad and, over time, still be in a state of control.

Dovich (1994) found through several simulation C_{pk} studies (of up to 10,000 sample size and 250,000 replications) that indeed the actual C_{pk} of a process is lower than the theoretical mathematical value. The results point out that the C_{pk} has a significant amount of variability as well as bias. In some cases 60% of the values estimated are below the true C_{pk}. To put this result in perspective, it means that when a requirement is 1.5—and you have met this requirement—the process when studied using 50 samples, will be deemed unacceptable 57% of the time.

So, is the C_{pk} a reliable statistic for the future? We believe that it is a gauge of performance; however, because of its variability and inherent bias, it should be used very sparingly. For example, a C_{pk} value of 1.33 can have anywhere from 32 to 63 nonconforming parts per million (NCPPM). A C_{pk} value of 1.278 could also have 63 NCPPM. So, you can see, the statistic is ambiguous. What should be used is a measurement of nonconforming parts per million? Because is is more reflective of the quality demands of the modern era.

Finally, what good are average C_{pk} values? A process could have an average C_{pk} of 2.00 on all critical characteristics and still be making nearly 100% nonconforming parts. This averaging could come from several characteristics having a C_{pk} of 4.0 or more, and other characteristics having C_{pk} values close to 0.0. The average looks good, but the product is bad.

VI. WARNING! NO ONE IS PERFECT

In many cases, our obsession for quality has pushed all of us to expect unrealistic perfection. We keep forgetting that nothing can be of greater quality than the creator of quality. We have come to believe and in fact associate our quality with high numbers of C_{pk}, not recognizing (or more appropriately, conveniently forgetting) that the two legs of the normal distribution never touch the horizontal axes. As long as that is true, we must recognize that quality is an issue of definition and perception from the customer's perspective, and no matter what the C_{pk} is, there is a possibility of nonconformance(s). Let me explain with three recent public relation nightmares.

Intel's Pentium chip was found to have a flaw, albeit one that would only effect a very small amount of users. My second example is the annual ridicule of the U.S. Postal Service around Christmas time. If you recall, the joke is that how many letters will be lost this year, even though they handle millions of letters daily? My third example is the disaster of Pepsi Cola. If you recall, someone claimed that they found a syringe in the can. In all three cases the jabs are not deserved; the point is that failures do occur, even though they are rare. Few organizations produce any product or service that is more reliable for less money than the companies just mentioned.

The Pentium chip, Intel's most powerful chip, is the brains of almost all of the top-of-the-line personal computers. It is a slice of silicon no bigger than a human thumb, embedded with 3.3 million transistors. The cost of the Pentium at the time of the problem was $587.00. Think about that: 3.3 million transistors, and a Pentium PC can do taxes, play CD-ROMs, or even design cars. It is indeed very tough to find a better value. By comparison, a 1,000 sheet roll of Charmin toilet paper costs 69 cents, so 3.3 million sheets would cost $2,277.00—four times the price of a pentium chip (at the time of the occurrence).

Table 2.7 Relationship Between C_{pk} and Percent Nonconforming

C_{pk}	Percent nonconforming	Percent yield	Number of parts out of specifications
0.10	76.4177	23.5823	764.2 parts per thousand
0.20	54.8506	45.1494	548.5 parts per thousand
0.30	36.8120	63.1880	368.1 parts per thousand
0.40	23.0139	76.9861	230.1 parts per thousand
0.50	13.3614	86.6386	133.6 parts per thousand
0.60	7.1861	92.8139	71.9 parts per thousand
0.70	3.5729	96.4271	35.7 parts per thousand
0.80	1.6395	98.3605	16.4 parts per thousand
0.90	0.6934	99.3066	6.9 parts per thousand
1.00	0.2700	99.7300	2.7 parts per thousand
1.10	9.6685×10^{-2}	99.9033	966.8 parts per million
1.20	3.1822×10^{-2}	99.9682	318.2 parts per million
1.30	9.6193×10^{-3}	99.9904	96.2 parts per million
1.33	6.3342×10^{-3}	99.9937	63.3 parts per million
1.40	2.6691×10^{-3}	99.9973	26.7 parts per million
1.50	6.7953×10^{-4}	99.9993	6.9 parts per million
1.60	1.5867×10^{-4}	99.9998	1.6 parts per million
1.67	5.7330×10^{-5}	99.9999	573.3 parts per billion
1.70	3.3965×10^{-5}	> 99.9999	339.7 parts per billion
1.80	6.6641×10^{-6}	> 99.9999	66.6 parts per billion
1.90	1.1981×10^{-6}	> 99.9999	12.0 parts per billion
2.00	1.9732×10^{-7}	> 99.9999	2.0 parts per billion
2.10	2.9765×10^{-8}	> 99.9999	297.6 parts per trillion
2.20	4.1116×10^{-9}	> 99.9999	41.1 parts per trillion
2.30	5.2003×10^{-10}	> 99.9999	5.2 parts per trillion
2.40	6.0213×10^{-11}	> 99.9999	0.6 parts per trillion
3.00	2.2572×10^{-17}	> 99.9999	0.0 parts per trillion
5.00	7.3419×10^{-49}	> 99.9999	0.0 parts per trillion

Assumptions used for this table:
1. The process is in control.
2. The characteristic being evaluated is normally distributed.
3. The characteristic being evaluated is on target at the midpoint of the specification limits.

How about the flaw? Well, the Pentium will make an error in 1 in 9 billion calculations. The odds of winning a state lottery jackpot are 1 in 4 million—2,000 times better than the odds of a Pentium gaffe. The odds of a high-school football player eventually leading an NFL team to a Super Bowl victory in the last 2 minutes of the game are 1 in 2 million. A Pentium PC running a personal tax program would make a mistake once every 2,000 years.

How about the U.S. Postal Service? Sure, it is clunky and very bureaucratic, it loses mail and it damages parcels, and it may be in debt. But it does deliver 177 billion pieces of mail, or 580 million pieces a day, to 123 million addresses. That is more mail in a day than the best of its competition does in a year. The service of the U.S. Post Office is second to none, especially when you consider the price of 32 cents for a first class letter. The Postal Service gets complaints for being slow. Yet 82% of letters get delivered on time.

Finally, the disaster with the Pepsi. The Chief executive of Pepsi was quoted as saying at the height of the controversy "I would assure you it is 99.99% assured that nothing is happening in the facilities. . . . It's physically impossible." (Strauss and Klemens, 1993). The problem with the statement, however, is that what the executive is saying is that it is a remote possibility, to the tune of 0.0001 (10000–9999) out of 20 million cans in 150 sites, that a syringe may in fact have contaminated the can(s).

The moral of the story? Intel, the Postal Service, Pepsi, and many other corporations are not perfect even though their capability is unquestionably high and they can prove it with their process C_{pk}.

The relationship of a C_{pk} and percent nonconforming is shown in Table 2.7. The reader will notice that as the C_{pk} increases the possibility of producing non conforming product is reduced, but never eliminated.

REFERENCES

Becker, W. E. and Harnett, D. L. (1987). *Business and economics statistics with computer applications*. Addison-Wesley Publishing Co. Reading, MA.

Chrysler, Ford and General Motors (1995). *Measurement systems analysis*. Chrysler, Ford and General Motors. Distributed by Automotive Industry Action Group. Southfield, MI.

Dambolena, I. and Rao, A. (1994). What is six sigma anyway? *Quality*. Nov.: 10.

Daniel, W. W. and Terrell D. (1989). *Business statistics*. 5th ed. Houghton Mifflin. Dallas, TX.

Deming, W. Edwards. (1986). *Out of the crisis*. Massachusetts Institute of Technology, Center for Advanced Engineering study. Cambridge, MA.

Dovich, B. (1994). Say no to C_{pk}. *Cutting Tool Engineering*. Apr.: 86–87.

Ellis, S. (1995). What is accuracy. *Sensors*. Sept.: 4–6.

Farago, F. T. (1994). *Handbook of dimensional measurement*. 3rd ed. Quality Press. Milwaukee, WI.

Ford (1985). *Continuing process control and process capability improvement*. Statistical Methods Office. Operations Support Staff. Ford Motor Company. Dearborn, MI.

Freund, J. E. and Williams, F. J. (1972). *Elementary business statistics: The modern approach*. 2nd ed. Prantice-Hall, Inc. Englewood Cliffs, NJ.

Gill, M. S. (1990a). Stalking six sigma.*Business Month*. Jan.: 42–46.

Gill, M. S. (1990b). The quality fanatics. *Business Month*. Jan.: 46.

Grant, E. L. and Leavenworth, R. S. (1988) *Statistical quality control*. 6th ed. McGraw-Hill. New York.

Griffith, G. K. (1994). *Measuring and gaging geometric tolerances*. Quality Press. Milwaukee, WI.

Harry, M. J. and Stewart, R. (1988). *Six sigma mechanical design Dimensioning and Tolerancing*. Motorola University Press. Shaumburg, IL.

IBM (1984). *Process control, capability and improvement*. The Quality Institute. International Business Machines. Thornwood, NY.

ISO 5725-1 (1994). *Accuracy (trueness and precision) of Measurement Methods and Results - Part 1: General Principles and Definitions*. American Society For Quality Control. Milwaukee, WI.

ISO 5725-2 (1994). *Accuracy (trueness and precision) of Measurement Methods and Results - Part 2: Basic Methods for the Determination of Repeatability and Reproducibility of a Standard Measurement Method*. American Society For Quality Control. Milwaukee, WI.

ISO 5725-3 (1994). *Accuracy (trueness and precision) of Measurement Methods and Results—Part 3: Intermediate Measures of the Precision of a Standard Measurement Method*. American Society For Quality Control. Milwaukee, WI.

Morris, A. S. (1991). *Measurement and calibration for quality assurance*. Quality Press. Milwaukee, WI.

Motorola. (1992). *Reliability and quality handbook*. Motorola Semiconductor Products Sector. Phoenix, AZ.

National Tooling and Machining Association. (1981). *Measuring and gaging in the machine shop*. Quality Press. Milwaukee, WI.

Pennela, C. R. (1992). *Managing the metrology system*. Quality Press. Milwaukee, WI.

Spiring, F. A. (1991). The C_{pm} index. *Quality progress*. Feb.: 57–61.

Strauss, G. and Klemens, M. (1993). Keeping the top on the PR disaster. USA Today, June 17, p. 1A.

Wheeler, D. J. (1993). *Understanding variation: The key to managing chaos*. SPC Press. Knoxville, TN.

3

Modern Concepts of Quality

To understand and implement TQM with positive results, one must understand some of the key concepts that exist within the scope of modern quality. Without these concepts, TQM is only wishful thinking. This chapter will discuss some of the new concepts in quality. None of them, of course, are given an exhaustive discussion. The intent of this chapter is to familiarize the reader with these concepts because the future will demand their understanding as well as implementation. The references at the end of the chapter will help the reader to find the details necessary for a more comprehensive discussion and understanding.

I. LEADERSHIP

A quality-focused leadership style is not a new management tool. It is just the perception of quality that's new. Today leaders perceive quality as a fundamental responsibility that cannot be delegated. Leaders communicate their operating style through actions, attitudes, and behavior. This is called the leadership cycle (Figure 3.1). They must clearly define and communicate what you are doing now and what the organization wants to be. (This is

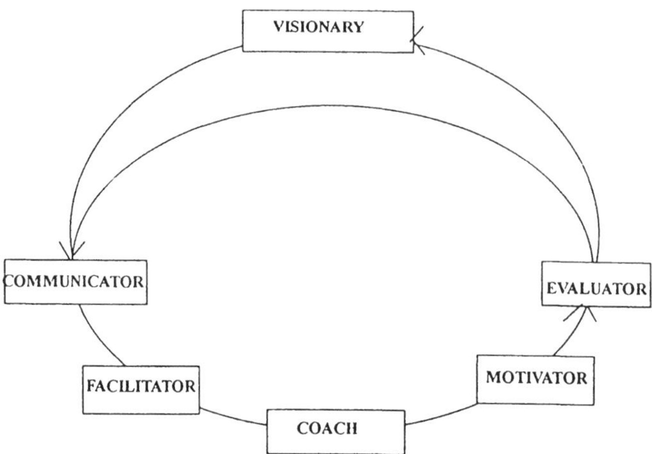

Figure 3.1 Leadership cycle.

articulated through the mission and vision.) Leaders must provide continued growth opportunities, as well as the tools and training needed to accomplish the mission.

For leadership to be effective, leaders must have certain ideals or core values. Minimum core values requirements are:

Integrity provides the foundation of trust. Demonstrated integrity helps in building a quality organization primarily because it recognizes the worth of every person's contribution to the team effort.

Character allows integrity to stand on its own. Demonstrated character is an inherent attribute of commitment to honesty.

Courage gives the moral strength to do the right thing in a situation where the outcome is uncertain. Long-term improvement calls for courage. Fight the temptation to apply a "quick fix" that gives fast results but offers only short-term gains. Courage is particularly important when you empower others. Empowerment places accountability, authority, and responsibility at the lowest possible level.

Competence is the watchword of master craftsmen. Competent members build their skills, knowledge, and experience in the classic tradition of experts who know competency is an ongoing experience. Competence in relationship to leadership is the characteristic that will allow the leader to strive to be the best, to be the expert. In fact, competence will also guide the leader to be selfconfident to delegate, as well as to ask and probe questions.

Tenacity drives individuals and teams to exercise determination and persistence. As you incorporate a cultural change, watch out for obstacles that threaten to slow your progress. Tenacity carries you over those barriers; it is the corporate value which allows you to stay the course.

Service leads to customer satisfaction. The organizational culture creates an environment in which you anticipate, meet and exceed the customers' needs.

Loyalty allows you to recognize the importance of accomplishing mission goals. It is often a sacrifice for the greater good. Do the best for the organization.

The core values of leadership must be supported by a set of principles, which provide a road map to help you reach the identified goals. Some of the most generic principles are:

Leadership involvement sets the pace for the quality journey. That means setting the vision, policies, priorities, and strategies. Next, leaders communicate these actions; they create an environment that supports trust, teamwork, risk taking, initiative, reward, and continuous improvement. Leaders help design quality into culture.

Dedication to mission is reflected in all you do as a team. No matter what the role is, every person is critical to the team that achieves "best in class."

Respect for the individual happens when you recognize everyone's skill and contribution. Value everyone as a professional. Do not let rank and level of responsibility be your only guides and respect. Success comes when you understand how each person contributes to your team.

Decentralized organizations help return decision-making authority to the individual departments within the organization.

Empowerment is one of the most misunderstood concepts in this culture. Some leaders think that empowerment means they must surrender their power to subordinates. This is not true! This is not about power; it is about giving people the tools they need to do their jobs. Empowerment at the point of contact gives people the opportunity, authority, and resources they need. Leaders who have learned to use empowerment find their role enhanced, not weakened. The goal is to create an environment in which properly trained subordinates can continually improve the organization; that encourages innovation and risk taking, which are important factors in the cultural change process.

Management by fact uses realistic measures to help indicate when, where, and how to improve your most important processes. Do not guess! Data-driven decisions can help you break through to a smarter, more productive way of doing business.

To be a leader in name only is not that big of a deal. What is a really big deal is to demonstrate that leadership. Unless the leadership is demonstrated and is part of the everyday work environment, the organization is going to experience a difficult time in changing its culture. The leadership behaviors help create a distinctive and easily recognized style. In Appendix B presents a very detailed summary of the different leadership styles based on the DISC (dominant, interesting, steady, curious) model. Here, however, we look at some elements from an operating style.

1. Creating a working environment that inspires trust, teamwork, and pride. Trust and teamwork instill pride and a sense of mission ownership. That creates quality professionals.
2. Delegate responsibility and authority to teams. This is the key to quality and innovation. Give people the training and resources they need and you will see them accept accountability for results.
3. Set attainable goals, measure progress, and reward performance. Develop and communicate goals that support your own organization's vision. You need to align your own and your organization's objectives from top to bottom. Evaluate progress and celebrate successes.
4. Give everyone a stake in the outcome. Empower people who own the processes and products.
5. Strive for continuous improvement. Challenge the concept of "business as usual." You need to understand your customers' needs and requirements. Learn new ways to do your job smarter and better, including your relationship with suppliers, for constant improvement.

For more information on leadership see Brake (1996), Johnson (1993), Juran (1989, 1995), Langdon (1993), Stamatis (1996), and Wickman and Doyle (1993).

II. TEAM LEADERSHIP FUNCTIONS AND ANTIFUNCTIONS

A. Task Functions

Task functions that help satisfy the problem-solving needs of the team include:

1. Initiating discussions. Suggesting new ideas or ways of looking at problem, or new procedures.
2. Information seeking. Seeking clarification of ideas presented or for facts and authority to support ideas presented.
3. Information giving. Offering examples, facts, or authority relative to problem or ideas presented.

4. Clarifying. Probing for meaning and understanding of problem or ideas presented. Restating the problem or ideas in different words.
5. Coordinating. Pointing out the relationships between the problem(s) and/or several ideas that have been presented.
6. Summarizing. Bringing the group up to date be reviewing content of discussion to that point.

B. Team Building

Team (as opposed to group building) and vitalizing functions that help satisfy the interpersonal needs in a team include:.

1. Encouraging. Being friendly and responsive to others, praising others and their ideas, and agreeing with the accepting contribution of others.
2. Mediating. Conciliating differences in points of view and being willing to compromise.
3. Gatekeeping. Encouraging and facilitating participation, making it possible for other members to contribute, suggesting limiting the time one person can talk.
4. Standard setting. Suggesting standards for selecting subject, determining procedures and evaluation, suggesting rules and ethical values.
5. Following. Going along with group, being a good listener, accepting ideas of others.
6. Relieving tension. Helping to drain off negative feelings by jesting or throwing oil on troubled waters, diverting attention from unpleasant to pleasant matters.

C. Antifunctions

1. Blocking. Going off on tangent, arguing too much on a point already settled by rest of group, rejecting ideas without proper consideration.
2. Aggression. Criticizing and blaming others, attacking the motives or deflating the ego or status of others, joking aggressively.
3. Seeking recognition. Calling attention to oneself by excessive talking, presenting extreme ideas, boasting, being boisterous.
4. Special pleading. Introducing and supporting ideas related to one's own "pet concerns" beyond reason.
5. Withdrawing. Being indifferent or passive, doodling, whispering to others, using excessive formality.
6. Dominating. Asserting authority to manipulate team or certain members, giving direction authoritatively, interrupting others.
7. Playboy. Displaying a lack of involvement in form of cynicism, nonchallenge, and/or horseplay.

8. Help-seeking. Trying to get sympathy response from others by self-depreciation beyond reason.

III. PARTICIPATION PHILOSOPHIES

Participative philosophy is a highly complex form of managing that requires a high level of insight into how people relate to each other and the tasks they perform. However, mastering this style of management can be highly rewarding to all concerned. In this section we look at the participation philosophies with a very cursory analysis and from the point of view of decision making, leadership, and quality.

Research by Peters (1992) and others has shown that companies that practice participative style of management possess:

Shared decision making	Participative leadership
High cooperation	Experimentation encouraged
Few layers of management	Team centered
Results oriented	Low degree of specialization
Changes welcomed	Innovation orientation
Great emphasis on employee development	Joint problem solving with union encouraged
Future direction of company to employees	Effective communications and information systems

When we talk about participation philosophy we talk about a set of concepts predicated on a philosophical conviction. This conviction is so important that Deming (1986) and Scherkenbach (1988) placed it on the very top of Deming's 14 Points of Management. This conviction is translated into the attitude and beliefs and actions on the part of management, which is: *That the majority of people can make significant contributions to business, human and social objectives, if they are provided the opportunity, knowledge, support, and reinforcement to do so.*

The notion of participation has considerable appeal even though in some cases it is totally misunderstood. Let us then look at some of the points that make the participative philosophy very appealing.

Participation assumes that most people want to work well, enjoy doing their work well, and are uniquely qualified to participate in establishing how the work should be done—because they are the ones doing it. The leader is saying to his fellow workers: "I want you to participate because I value your opinions, your experience, and your desire to do a good job." In essence then, we say that: "Two heads are better than one; participants in charge understand the change better" and that:

Participation clears the communication channels.
Participation reduces employee fear.
Participation increases morale.
Participation increases the chances of success.

To better understand the participative philosophy let us look at the levels of participation and compare them to the management style.

1. The leader makes the decision without consulting team members. In essence he/she tells someone to do something. The leader (boss) identifies a problem, considers alternative solutions, chooses one of them, and then reports this decision to his/her subordinates for implementation. He/she provides no opportunity for the subordinate to participate directly in the decision process.

2. The leader makes the decision with limited input from team members. In essence he/she sells the idea to someone. The leader (boss) takes the responsibility for identifying the problem and arriving at a decision. However, rather than simply announcing it, they try to persuade the subordinate to accept it. The leader in this case recognizes the possibility of some resistance among those who will be faced with the decision and seeks to reduce this resistance by offering incentives and/or other benefits.

3. The leader consults with team members before making the decision. In essence they take the time to consult with subordinates. The leader (boss) before finalizing the decision presents the ideas, invites questions, and presents a tentative decision, subject to change by the subordinates. The leader expects reaction from the subordinate and acts as though saying: "I would like to hear what you have to say about this plan that I have developed. I will appreciate your frank reactions, but will reserve for myself the final decision."

4. The leader encourages total involvement of all team members. The decision may be made by the team with the leader (consensus) or by the leader unilaterally. In essence he/she joins the team for the decision. The leader (boss) presents the problem, gets suggestions, and then makes the decision. The subordinate here plays a major role in the decision since his/her knowledge and experience is expected to increase the manager's choices of possible solutions.

5. The leader defines limits and permits the team to make the decision. In essence, he/she delegates. At this point the manager passes to the group the right to make decisions. Before doing so, however, the problem to be solved is defined and the boundaries within which the decision must be made is clear. Here the team undertakes the identification and diagnosis of the problem, and develops alternative solutions. The only limits directly imposed on the group by the organization are those specified by the superior. If the

Figure 3.2 The leadership continuum.

boss participates in the decision making process, he/she attempts to do so with no more authority than any other member of the group. They commit themselves in advance to assist in implementing whatever decision the team makes.

A pictorial view of these levels and the leadership continuum are shown in Figure 3.2. On the other hand, the continuum of development and authority of a leader is shown in Figure 3.3. One can see that participative management is not easy and it takes time to implement precisely because it is a major change from the way that the corporate culture has been doing business. Some may adopt to it very easy, many will find it very difficult. If applied, however, with a true commitment from management, this style of management can contribute to at least three major goals:

1. reduce costs
2. raise quality and productivity
3. enhance the work, so that all workers can derive feelings of accomplishment and well being

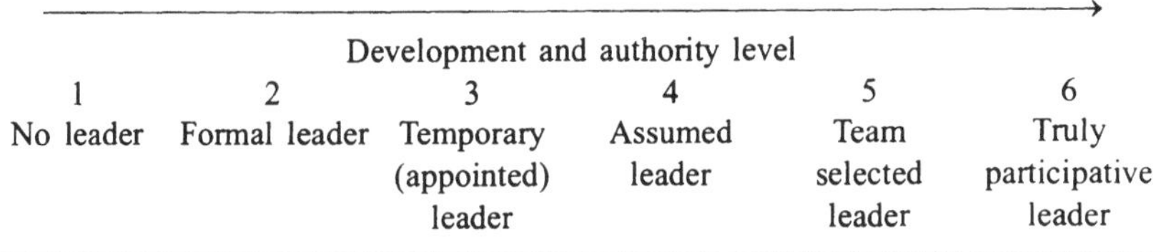

1, Group functions without direction and goal.
2, The leader has a formal position in the organization (management team).
3, A temporary leader is appointed by management (matrix team).
4, Leadership is assumed by a dominant member of a coalition of members.
5, The group selects a leader; leadership is rotated. (Selfmanaging team).
6, No leader is appointed or selected; no member or coalition assumes leadership.

Figure 3.3 Leadership development continuum.

Boss-centered leadership	Team-centered leadership

Use of authority by manager Area of freedom for subordinates
1 2 3 4
Directs Consults Joins Delegates

1. Managers only (Directs). The manager solves the problem or makes the decision without asking for information or opinion of others.
2. Manager consults (Consults). The manager asks for opinions and hears considerations of individuals from the team, and afterwards makes the decision on his/her own.
3. Subgroup decision (Joins). The manager discusses the problem with one or more of the team members and makes the decision together with them.
4. Team decision (Delegates). The manager brings the problem before the entire team. The team discusses it and makes a decision.

Figure 3.4 Characteristics of boss-centered versus team-centered leadership.

IV. WAYS OF MAKING A DECISION

There are many ways to make a decision; however, Figure 3.4 shows the typical development for a team-centered leadership decision making. In making a decision a leader must also recognize that influence and reaction are important. For example if the leader is influential using his/her knowledge, experience, and interpersonal skills in pursuing a particular goal, chances are that the leader will be successful. On the other hand, if the leader is seen as noncredible, then the success is diminished considerably. By the same token, if the reaction to a decision is not perceived as positive, then failure is near by. A list of positive and negative use of influences is shown in Table 3.1 and some typical reactions in Table 3.2.

V. CHANGE FORMULA PHILOSOPHY

When there is a new model for managing or organizing and there is a planned process for managing change and dissatisfaction, and when the cost of the status quo is greater than the cost of change to individuals and groups, then change will occur.

There are two kinds of changes. One is unplanned and the other one is planned. From a managerial and a quality point of view the planned change is important because we can control it. Because change takes time for people to accept, adapt to, and integrate into their lives, management has to allow the time and give the appropriate support as they go through each of these

Table 3.1 Use of Influence

Positive	Negative
Positive	Vindictive
Influence to accomplish organization's goals	Withholding power when you should exercise it
Mentoring or bringing others along	Personal gain at expense of others
Working to prevent problems	Control without choice—dictating
Motivating	Encourage passivity
Implement change	Destroy consensus
Stimulate innovation	Egocentric
Keep people on track	Discrimination
Coordinate many talents	Manipulation of people and issues
Putting self on line (take risks)	Close minded in negative fashion
Provide support	Opposite/undermine teamwork
Common good	Parochial
Positive role model	Retribution
To empower others	Sense of scarcity
Visionary	Imposing your will
Improve operations and service	Abuse of others
Reward/incentives	Devaluing others
Recognition	Taking credit for ideas or work that are not yours

stages. It must be also recognized that just like people, and to do so organizational cultures move at different rates through these stages.

It has been said that "change will always occur" and that "resistance to change is inevitable." Some of the barriers to change are:

1. lack of understanding of the purpose of the change
2. existing behavior patterns, such as habits, relationships, and expectations of self and others
3. unwilling to substitute the unfamiliar for the familiar
4. feeling of lack of control over something happening to one's life or person (having something done to you)
5. fear of the unknown
6. fear of failure at something new
7. lack of trust in the initiator or in the process

However, there are some things that management can do to minimize the effects of that change. They are:

Table 3.2 Reactions to Use of Influence

Positive	Negative
Recognition	May be intimidating or overbearing in getting our way
Achieve results	Focus of double standard
Acceptance as a team player	Perceived as critical and condescending
Feeling trusted by subordinates and peers	Perceived as self serving
Seen as knowledgeable	Conflict of interests
Sense of well being and pride	Unnecessary or additional burden
Objective	Potential danger
Sense you__________	Seen as enforcer
Reflect on reputation of department	Seen as target
More capacity to interact broadly seen as having integrity	Tension
Approachability positive to take on responsibility	Isolation
Confidence	Perceived as insensitive
Increased energy	Careful about what we say and do
Give credit where it belongs	Increased pressure to be right
Encourage risk taking	
Ability to communicate and persuade	
Ability to mentor	
Ability to share mutual respect	

1. forming a clear visual image of the desired future state
2. acknowledging that resistance to change exists and devising strategies for reducing or overcoming this resistance
3. recognizing the stages that individuals and teams in corporate cultures typically go through as they respond to change and giving them "permission" and support to go through each stage
4. keeping the appropriate expectations for each stage of change
5. managing conflict is also essential to effective change efforts

VI. STRATEGIC PLANNING

Strategic planning in the current and future world of TQM is very essential. The strategic plan includes the mission, vision, values, goals, objectives, and action plans. Without these basic items the thrust of TQM will not sustain itself, and as a consequence, TQM will fail.

How do we generate a strategic plan? We can choose to generate a strategic plan with either a macro or a micro focus. For a macro plan, the mission, vision, and values are the ingredients of a good strategic plan. On the other hand, the micro plan focuses on objectives and action plans, sometimes all the way down to the departmental level. To start a micro plan, chose a specific process within the organization (there are plenty of processes to chose from). However, whatever process you choose, make sure it contains the four generic phases of quality, which are:

1. Formulation. Identify the process, goals and expectations of improvement.
2. Deployment. Identify the plan of how you are going to get there.
3. Implementation. Identify the steps that you are going to use in the process of implementing the plans.
4. Review. Conduct a periodic review to measure improvements and identify areas of concern. Communicate the results to the decision makers.

Whereas the generic phases provide a clue of what has to be done, the following specific steps identify the action of each phase in a more detailed manner.

A. Planning to Plan

Leaders ask themselves if they are ready to begin the strategic planning process. Are they willing to commit the time and other resources necessary for an effective planning process? Is the entire organization ready to commit to strategic planning? Besides committing to the process, you need a strong level of trust and teamwork within the organization. Without strong trust and teamwork and commitment, the planning process is likely to fail.

B. Values Assessment

The values that exist within an organization define the organizational culture. This includes both personal and organizational beliefs. Some basic core values of all organizations should be: character, integrity, courage, competence, tenacity, service, and loyalty. (See Section I on leadership). The core values of your organization will define the boundaries of all planning and serve as a benchmark for decision making.

The goal of the value assessment is not to change anyone's values. Rather, it is to identify these values and understand how they affect others. Once the values are defined then you move on to analyzing the mission.

C. Analyze the Mission

Generally speaking, a mission statement expresses a reason for existence. As a consequence, that mission should be understood be everyone in the organization. It should be operationally defined. It should answer the questions of: What does the organization do? For whom does it do this? How does it do this? Why does the organization exist?

The clue of whether or not your organization has an effective mission is whether or not the statement itself is easily recalled. So how can you write a mission statement that is sure to be remembered? The following items may help.

1. gather information that impacts the mission
2. define customers, suppliers, and their requirements
3. define key result areas
4. define key processes

D. Envision the Future

Without a vision of the future, it is hard to plan for anything except sustainment. It is this envisioning of the future that makes strategic planning *strategic* in nature. The focus is to determine the direction of the organization, visualize its future state and describe what this looks like.

An effective vision is a clear, positive, forceful statement of what the organization wants to be in 5, or even in 10 years. It is expressed in simple, specific terms. The vision allows the organization to stretch and aim for a high target. The vision must be powerful enough to excite people and show them the way things can be. A well-crafted vision supported by action can be a powerful tool for focusing the organization towards a common goal.

Whatever form the vision takes, it is important that it is communicated throughout the organization frequently and with conviction. The timing and method of communication provide an opportunity for creativity by top management. It is important that the vision be followed soon by a concrete plan of action to avoid being dismissed as a hollow slogan.

E. Assess Current Capabilities

Evaluate your key processes. Do they meet customer requirements? One way to measure your current capabilities is to conduct a unit self-assessment. This assessment may be used as a baseline for future improvement as well as a gap analysis.

F. Gap Analysis

Identify the current capabilities of each key process result area to the future customer requirements spelled out in the most probable future scenario. Do you have gaps in capability? If you do, they are critical issues and need action. In fact, it is these gaps that will serve as the basis for near- and long-term goal setting.

The gap analysis also provides the organization with a reality check of the planning process up to this point. A careful study of the gap will help the organization to achieve their goals more in a more realistic manner. For example, if the gap between present capability and the future state is small and easily achievable, then the planning team probably did not stretch enough in visualizing the future. On the other hand, if the gap is large and overwhelming, maybe the team needs to readjust the future state.

A third possibility exists in which the gaps are all achievable, but there are too many to address them simultaneously. If that is the case, then the team may go back and prioritize key result areas, values, and environmental factors to set reasonable goals.

G. Develop Strategic Goals and Objectives

Perhaps the most important step in strategic planning is to tie the strategic goals and objectives to the vision. The goals and objectives must bridge the gap between current capability and the vision. The strategic goals and objectives form the basis for the functional plans. Because of their importance, a constant vigilance through a feedback system must be in place to assure that the goals and objectives are feasible, and attainable within the constraints of the organization.

H. Develop Functional Plans

At this point your planning team coordinates with middle managers and working groups to develop the action plans. These plans must address potential problem areas and they may also include "what if" situations in case the primary plans cannot be executed. To develop action plans, you need to define the subprocesses and tasks that align with and support key processes in your organization. Do you have indicators to measure the subprocesses? If not, develop them now. The indicators should align with and support the organizational metric system. It is important that before you implement any plan that the senior planning team has a chance to review the action plans. They will check for cross-functional integration, alignment, and system optimization.

I. Implement Plans

Get the action moving! Natural working groups and functional, process, action, and development teams all carry out the plans to move the organization closer to its future vision.

J. Periodic Review

Take time to review the functional plans. The time interval depends on the needs of the organization. However, a common practice of review is once every six months or once a quarter. The most important aspect of this review is to use metric (measurable) data to assess the situations. It is the metric data that will allow you to reset targets and to use the results as baseline observations for improvement, or even as input for the next planning cycle. The results of the review must be communicated both vertically and horizontally within the organization so that appropriate action may be taken.

For more details on strategic planning see Mintzberg and Quinn (1992), Abell (1980), and Stamatis (1997a).

VII. EMPOWERMENT

One of the most important inherent elements in TQM is the notion of empowerment. By definition, empowerment is the natural outcome of involving people in the decisions they must make in order for them to feel responsible and accountable for their work. By the same token, in order for empowerment to occur, the proper authority and responsibility must be delegated to those who are empowered. If appropriate and applicable authority and responsibility are not delegated, then the alleged empowered will not have clearly defined objectives of what must be done, and will not know what conditions that they should exercise their empowerment power and what are the limits of the empowerment. Another condition of a successful empowerment program is that the people asked to participate in empowerment situations must possess the information, education, and skill to do what is asked of them. To be sure, empowerment sounds easy and one may wonder why is empowerment not so prevalent; at the same time where it does exist, it is so misunderstood. Stamatis (1997a), Vogt and Murrell (1990), and Byham (1988) identified some clues. They are:

1. fear of failure
2. feeling "not ready" or "not qualified" to do the task or participate
3. fear of the unknown
4. unclear direction and sometimes conflicting direction
5. unclear and/or conflicting expectations

6. unaware of changed behaviors, skills and job requirements
7. lack of confidence or willingness in accepting responsibility
8. misconception of the decision process
9. misguided understanding of management's role in the empowerment process
10. refusal by both employees and management to change their old paradigms

On the other hand, Stamatis (1996a, 1997a) pointed out that whereas in theory many organizations try to unlock employees' creative potential through empowerment as well as team involvement, the reality is that there are employees who indeed find the opportunity exciting but there are also those employees who are threatened by it. Empowerment confuses many people and leaves others uncertain about how to achieve it.

To combat this dichotomy (the willingness to self-govern versus fear of the unknown), empowerment has to be viewed from two perspectives. The first is empowerment of the individual; this empowers people to do their work, to make decisions needed to satisfy their customers' requirements, and to operate with little or no supervision. This kind of empowerment enables the team members to do their work effectively by having the ability, means, and authority to resolve problems or delays that might occur. Management has empowered people to do what is necessary to complete the required task.

The second perspective of empowerment is empowerment of the team. This kind of empowerment expands the team's work to include the planning, controlling, coordinating, and improvement functions that were performed by superior and/or staff specialists. This new notion of managing and doing your own work is indeed frightening to some people, and unless appropriate training is given to these individuals, they will fail.

So, how does one go about implementing empowerment in the work place? Perhaps the most important step in empowerment is that we must recognize that empowerment must be an inside-out process. We cannot have empowerment by decree, where we impose empowerment on our organization. Nor can we "rewrite" the heads of our people or our managers so that they become and act empowered. We start empowerment by focusing on ourselves, evaluating and reforming our leadership paradigms, and building our trustworthiness. Then we develop the trust in our relationships with our people by looking at our management practices. Changing the way we think and the way we act are preconditions to lean manufacturing and employee empowerment. To be effective in empowerment, we must use a mechanism (sort of a bridge) to implement the concept of empowerment in our organizations and then to align the structure and systems to support it.

A typical bridge to implement this empowerment concept with success includes:

1. Recognize that the culture of the organization must change.
2. Establish and create a training program for teams.
3. Create an empowerment plan.
4. Establish and train coaches for the empowering process.
5. Select the individuals to participate.
6. Everyone involved in the empowerment process must be given authority, responsibility, education, and training.
7. Everyone must have the opportunity to be coached.
8. All employees must be rewarded.

VIII. MOTIVATION

For change to occur, there must be a motivation for that change. Whether that motivation is for any kind of improvement or competitive advantage, or whether it is based on organizational goals, the fact is that motivation plays a key role in the success of the project. For TQM to be successful, motivation must be present. There are two kinds of motivation. One, the intrinsic motivation which the single individual brings to the project because he or she has internalized the specific need and benefit from a pending behavior and/or activity. The second, the extrinsic motivation which is based on factors outside the "self."

Both kinds of motivation are important and relevant to quality. However, we know that some people, because of their high level of intrinsic motivation, do the best work they are capable of without any prompting from anyone (especially management). They are the people who will volunteer, go the extra mile, contribute beyond the expected contribution and so on.

On the other hand, there are also those individuals who need an occasional nudge, direction, very specific instructions, always proof of "What is in it for me," and so on. This extrinsic motivation to excel has to be imposed on them from above, sometimes subtly, sometimes obviously. For this group of individuals, it is the job of management to make them want to live up to their full potential. To improve management's ability to motivate, the following may prove helpful.

A. Keep the Standards Consistent

A person who knows that he/she is being judged by a single, fair standard has a target to aim for. He/she can modify his performance accordingly and try to meet that standard. No one wants to (or can) work without a goal. To be consistent to the point of inflexibility, however, is poor management. If you are going to modify your standards, communicate this in advance so the employees can remain flexible themselves.

B. Inform All Employees of Changes that May Affect Them

This includes all forthcoming changes in policies, organizational structure, promotions, benefits, programs affecting their responsibility, and so on. The idea is to demonstrate to your employees that they do indeed count and they are important. Every measure should be taken to avoid the indictment of ignoring the most valuable resource of your organization. This is true whether or not the employees are represented by a union.

C. Actively Encourage Feedback Regarding Changes

When a change is about to be introduced or it has been introduced, allow for active participation of everyone to ask questions. All questions should be answered as honestly as possible. If appropriate, an open meeting between management and employees should take place for the roll out program with the opportunity for everyone to ask specific questions.

D. Back People Up

If you want to motivate people, you must assume some responsibility for what they do. The employee who does not feel that he has the backing of his boss is not apt to stick his neck out with new ideas or bold solutions, unless he wants his head chopped off. But the employee who knows that he has the backing of his boss can employ his full energies to doing his job in the best way he knows.

E. Always Be Tactful

While tact probably cannot be taught, it includes such things as consideration for others, courtesy, and an appreciation for the views and feelings of others. A good way to learn how to be tactful is to observe those whom you yourself consider diplomatic. Listen to their conversations with others, observe their behavior, try to find out how they handle themselves in disagreements and uncomfortable situations.

F. Always Allow Freedom of Expression

Reassure people that there will not be any sort of retaliation for any difference of opinion and/or legitimate disagreement. Every employee should have the right to make his own job more interesting by doing it his own way. The requirement, however, for the licence to modify his job is that the results are what management seeks and that the employee is working within the framework of company policy.

G. Empower People

Give them the appropriate authority and responsibility to carry out the task(s). Delegate as much of your burden as possible. Pressure often motivates people, so let them bite off as much as they can chew. Then let them ride with their own decisions (as far as possible), let them learn through their own mistakes, and let them revel in their own successes.

H. Mistakes Happen

Do not humiliate employees in front of others. It is normal for humans to err; however, it is these errors that make us all learn how to avoid future problems. One of the worst thing for managers to do is to humiliate employees for failures, especially in front of others. If some one needs to be reprimanded, managers should do it in private. By the same token, managers should give praise whenever possible and in public.

I. Look for Hidden Talent

Before throwing in the towel, look at your employee's records. Do they possess any skill that have not been used in their current positions, but might be applied elsewhere within the organization? Ask them to tell you about their interests. These may provide a window of future opportunities for both the organization and the employee.

J. Reassure People

Always remind people of the importance of their jobs and their personal contribution. Everyone likes to be told that they are important, needed, and that their contribution is appreciated. Managers should take advantage of this knowledge and encourage all employees with positive feedback. However, this feedback must be for specific activities and timely. Otherwise it is meaningless and in some cases it may be interpreted as a negative response.

IX. REWARD SYSTEM

Traditional American management theory teaches all of us to optimize for the management of commonality. Managers assume that businesses, whatever their focus, share similar structures and strategies (e.g., organizational structures, labor practices, product development protocols, sales and marketing strategies, accounting requirements, quality standards, and so on.) Accordingly, they seek control over that which is known, prevalent and predictable (Allmendinger 1990). The measurement of that control, and the success it is intended to produce, is primarily driven by out-dated accounting and reward systems and/or practices. What is unmeasurable in the American management-by-the-numbers culture is generally considered to be unreal.

The allocation and the evaluation of resources, namely capital and people, is the unique province and responsibility of management. As organizations downsize and become increasingly reliant on knowledge specialists, success depends on making sure the right specialist link and focus on the right objective at the right time. Success depends on how well the creative energy of marketers and engineers are vectored to the needs of the marketplace and how well the production resources and service functions are harnessed to deliver the value of the product to the customer. Therefore, the reward system for such a success should be fair, appropriate, and across the board in the entire organization. Organizations cannot afford to have a reward system that was based only on financial indicators. Financial indicators, although easily quantified, are largely symptomatic. They are only indirectly linked to the variables that can be manipulated directly: people, technology, organization. This in fact may explain why traditional management by the numbers and award systems are not working very well.

A. Traditional Reward Systems

One of the oldest and perhaps the most used reward system is the annual merit raise. Regardless of the performance of the individual and the organization everyone in the organization receives a set percentage increase based on their current wage/salary. Obviously this system does not encourage improvement, efficiency, and accountability for one's performance. It is not based on any measurement and therefore it is one of the poorest ways to reward.

A second reward system is the one based on Management By Objectives (MBO) or a derivative of that system. The problem with this reward system is that more often than not, the MBO has been set by someone other than the person who is being evaluated. Therefore, it does not really measure the effectiveness and/or the improvement of the individual. At best, it is based on a system that is fundamentally bias against the individual who is being evaluated.

A third reward system is the one based on work incentives. It is the carrot method. The essence of this method is that when employees do what they are told, the manager gives them a bonus and says "Good job!" Typical incentives are strictly based on productivity and quite often are set against quality standards. So while employees make something, the product may not be good enough for the customer. In other words, the employee is encouraged to produce at the expense of quality. Incentives may take the form of stock options, sales commissions, trips, trophies, a set percentage wage increase over and above the standard wage/salary and so on. They are all variations of a single simple theory that people will be motivated if rewards are dangled in front of them.

Do all these traditional reward programs work? It depends on what is meant by work. If we are looking for temporary compliance, then rewards (and punishments) are extremely effective. But if we are more interested in innovation, effective problem solving, and quality over the long haul, then rewards and other tactics of control are not only ineffective but counterproductive to the organization as a whole.

Rewards based on the traditional approach are indeed undermining excellence and promote complacency through playing it safe. In addition, all these reward systems are geared to motivate people not for general improvement but rather to get the rewards, and that is quite often at the expense of excellence. Rewards, like punishments, are basically ways of doing things to people. But success requires that managers work with people, by letting them participate in making decisions about what they do every day. It requires collaboration in the work place, not only in one particular department but on the entire organization.

B. A New Approach to Reward Systems

To have an appropriate reward system that is meaningful and motivates everyone in the organization we must first develop performance measurements. These measurements will become the indicators and the barometer of success and includes: flexibility, cost, time, productivity, quality, innovation, and relationship. These indicators must be measured and compared with each other for viability, reliability, accuracy, and relevancy as well as efficiency. This comparison must be conducted both internally as well as externally. For an example see Table 3.3. [These indicators are based on the work of Allmendinger (1990). Where Allmendinger described these indicators as a way to measure more effectively process performance, we have extended their application as a basis for a reward system and have somewhat modified them as well.] Quite often, these indicators may be weighted for specific organizations as particular indicators may be more important than others.

In all organizations, the selected indicators must measure overall improvement before the reward is actually given. Ideally, the performance measurement system will serve a communication function and not solely a management control scheme. After all, the set of metrics typically used to measure

Table 3.3 Indicators, Dimensions, and Attributes for a Reward System

Indicators	Dimensions	Manufacturing	Engineering	Marketing
Flexibility	Product	Manufacture new products	Design and develop new products	Customer driven
		Manufacture current products to customer specifications; improve current design with new features	Offer a variety of products in relationship to competition	
	Volume	Capability	Design for manufacturability	Offer multiple sizes
		Capacity		Offer alternative timing
		New product lines		Recognition of manufacturing and engineering constraints
Cost	Direct	Product	Design and development	Cost of sale
	Indirect	Waste in process	Engineering training	Warranty
		Scrap	Engineering system support	Cost/Inventory
		Rework		
		Overhead		
Time	Internal	Throughput	Design lead time	Order lead time
		Value added per shift		
	External (supplier, customer)	Value added per investment dollar	Feature update rate	New product introduction rate
Productivity	People	Value added per person per shift	Design results per engineer per day	Sales per salesperson
		Total quality throughput		
	Capital	Value added per investment dollar	Design results per investment dollar	Marketing dollars per sale

Quality	Product	Reliability	Performance	Perceived versus design performance
		Quality programs implemented	Features	
	Support	Spares availability	Serviceability	Service responsiveness
		Quality training		
Innovation	Product	Unique process development	Technology developed in-house	Number and success of product concepts
		Process improvement		
		Current product improvement		
	Process	Automation implementation	Simultaneous engineering	Ability to leverage process strength
		Process improvement		
		Simultaneous engineering		
Relationship	Supplier	Effectiveness of relationship	Integration with supplier	Customer feedback
				Corrective action follow-up
	Customer	Complaints	Conceptualization and development, customer driven culture	
		Compliments		Effectiveness of relationship
		Customer satisfaction		
		Product meeting the customer specification		

and reward individual performance is only indirectly related to the individual's rewards and even more indirectly related to the organization's true performance. For a truly successful reward system to work, there needs to be an explicit and direct link between the performance of individuals with the performance of the company as a whole. We believe that the measurement of the above indicators not only measure the pulse of the organization as a whole, but more importantly they encourage single individuals across the organization to actively participate and to be willing to take charge for improvement. The result is ownership, motivation, and self-fulfilment, and is worth the reward. For more detailed information on reward systems see Wilson (1995) and Knouse (1995).

X. REENGINEERING

Reengineering in its most simplest meaning is the redesign of a business process. The assumption that makes this possible is the notion that the process has been improved to its maximum—through statistical process control, continual improvement, and/or the Kaizen approach—and because of diminishing returns it is time to revolutionize its design. Another way of looking at the reengineering concept is to think of it as a drastic change of the process.

Stamatis (1997b), Shores (1994), Lowenthal (1994), Roberts (1994), and Hammer and Champy (1993) addressed the issue of the need for reengineering under several conditions. They are:

1. when the performance of your organization is behind your competitors
2. when the organization is facing a major crisis, such as declining market share
3. when the organization is looking for improvements beyond the 50th percentile
4. when the market changes due to technological innovations
5. when your organization is determined to be the leader in the industry
6. when competition is changing due to advances in process innovation and your organization must respond in a timely fashion

Reengineering is not necessarily cutting the head count, although downsizing may occur. This is very important in that reengineering as practiced today by most companies has created a situation in which organizations are so "lean and mean" that they are not able to maintain manufacturing quality, effectively create new products, and promptly meet any demand for either increased product output and/or customer satisfaction.

Is reengineering real? Yes it is. However, it must be thought out very carefully and applied with a long-term perspective. Anything less than this long-term perspective will backfire for the organization. We believe that reengineering will play a major role in the future world of quality, but we also believe as Parr (1996) has noted that tomorrow's successful corporations will be those that quickly realize what went wrong with their reengineering programs and hire (or rehire) the people they need to fulfill long-term product development and improvement needs, but also have the good sense to constantly monitor and control growth in the future.

XI. AGILE MANUFACTURING

Whereas the old quality concepts were mutually exclusive with "mass production," the new and future quality is a concept that goes beyond the issues of reliability and satisfaction to customer gratification; it leads to products and/or services that appear to be custom-produced to individual customer specifications. This new approach is called agile manufacturing. Some of the elements of agile manufacturing were described by Womack, Jones, and Roos (1990); however, the essence of this concept was first described by Goldman and Preiss (1991) and conceptualized by Nagel and Dove (1991).

Agility is the ability to thrive in a competitive environment of continuous and unanticipated change, and to respond quickly to rapidly changing markets driven by customer-specified products and services. Therefore, agility entails the capability for:

- intense, sustained, interactive customer relationships
- rapid cost-effective, development of products and production facilities
- continuous improvement of product and/or service development, as well as production processes in ways that are easy to understand and easy to maintain
- rapid efficient, changes in production volume (up/down)
- low variability of volume/unit cost ratio with volume changes
- open electronic access to product data, change orders, enterprise performance and status reporting
- full, location-independent, interactive communications among all participants in the product/service design, development, production, marketing, and service process
- distributed authority, such that decisions are made at relevant functional knowledge points, not centralized in nodes of the managerial hierarchy

What is of major importance is the fact that with an agile organization, not one of the features of agility, taken by itself, is an innovation unique to it. Not flattening the organization. Not cross-functional work teams. Not workforce empowerment or even customer-focused organization. Not intensive collaboration with other companies, nor any other program that we have introduced in the last 20 years in the name of improvement. What is new and news about the agile organization is that by systematically integrating many existing technologies and techniques into a new organizational structure, it promises to be the twenty-first century analog of the turn-of-the-twentieth century crystallization of the mass production manufacturing system. Agility, too, promises to create new business strategies that will offer decisive competitive advantages to companies capable of implementing them.

An agile company can be identified by several characteristics (Goldman and Mellon 1993).

A. Organizational Structure

An agile company can be distinguished from others by the extent, and the intensity, of their networking relationships with other companies and by the interdependencies these relationships create. An agile company depends on current and correct information, as well as personnel with appropriate skills and knowledgeable enough to exploit information innovatively. To sustain that capability, the company's organizational structure must be as flat as possible in order to be dynamic, routinely reconfigurable in response to constantly changing circumstances, and effective in meeting constantly changing goals and performance objectives. Optimization of business processes drives organizational structure.

As a consequence, the company must be organized around the core competencies of its personnel and it must be rapidly reorganizable around changing core competencies as these evolve. Personnel must be organizable on their own initiative into customer opportunity teams whose agendas are driven by customer requirements.

B. Corporate Strategy and Focus

Agility generates profits by responding to the sale opportunity that each customer represents by providing each with customized products and/or services. Furthermore, the market of the agile company is global and its customers (and their individual requirements) are globally distributed.

In place of reliability, or even satisfaction with their purchase, quality for an agile company is measured by customer gratification and/or delight with the products or services they have purchased over the entire lifetime of the purchase (including ultimate disposal). For the agile company, the bottleneck for responding to customer opportunities is information flow within the company and between other companies. The limiting resources are the workforce skill and knowledge base.

C. Managerial Authority and Corporate Ethic

By definition, in the agile company, decision-making authority is distributed widely, right down to the operational level where the need for a decision to be made arises and where the information on which to base it must be available. Continuous workforce education must thus be considered as an investment, as an enhancement of a corporate asset. An ethic of trust must pervade an agile company in order to sustain such an open information environment. Personnel at all levels must be empowered to seek out the information they need in order to make the decisions they are best able to make. Furthermore, they must be empowered to make the appropriate and applicable decisions.

For such an organization, management is best effected by macromanagement, that is, by coaching, supporting, and leading, and by challenging the workforce with bold goals, providing the resources required, and leaving to them the precise means of accomplishing those goals. Agile managers are entrepreneurial and inspire entrepreneurial values in all personnel. They are positive about pursuing new opportunities and attempting new approaches to new challenges. All levels of the organization must be skilled in customer relations and management. Given the extent of networking in an agile company, trust must also pervade relationships with other companies by sharing information and joint decision-making (even with competitors).

D. Workforce Environment

Agile company personnel are required to think about their work and to make their own decisions about how it can best be performed as their work objectives change. They will function as individuals and in teams, and will be compensated for their individual skills and as well as their team's performance affects the company's bottom line.

The requirement of workforce initiative for the success of agile companies puts a premium on personal growth as a company asset and on a culturally diverse workforce as a peculiarly fertile seedbed for creativity.

E. Product Focus

Agile company products and/or services will have a high information content and will be designed by producers and customers working together to specify the products and services desired by the customer. Design and manufacture will involve peer-level interactions among companies, often functioning as a virtual organization. Partners and suppliers will be selected by an agile company based on the skills and the knowledge that they can bring to a product opportunity. Companies will be attractive partners in a given production effort because of their ability to complement one another's core competencies and because of their reputation for developing their skills and knowledge in the course of a collaborative relationship.

In an agile competitive environment, customers will purchase skills over time and will be more aware of the general cost; the manufacturing processes involved will be only a small part of the price they pay for products and services. For their part, producers in an agile competitive environment will create products that have their profit based on some fraction of the value that they provide rather than on their cost of manufacture. This implies short production runs of nearly unique products that are made to meet specific customers needs or opportunities. Some products, perhaps a growing number, will be information only rather than traditional hardware. For all customers, agile products will be configured at the time of order and reconfigurable over the product's life cycle to meet the customer's evolving requirements.

F. Information Environment

Current and accurate information is the central, critical, asset of an agile company. Information will be an increasing component of product content and information tools will become increasingly valuable company products. Collecting, evaluating, organizing, and distributing information becomes a decisive enabler and infrastructure requirement of agile product development and delivery.

Information will account for a significant fraction of the value customers pay for when they purchase agile goods and services, so the product potential of information must be highlighted within the company. Ironically, agility entails distributing information widely in order to collaborate effectively with partnering companies and in order to empower personnel, to nurture an environment of reciprocal responsibility and accountability, and to unlock initiative and creativity. This implies that an agile company must possess robust intellectual property models for protecting and for sharing information, internally as well as externally.

G. New Paradigms

The prevailing paradigms in an agile organization must include:

mass customization
customer value-based products/services
competition based on the company's core competencies and on the competencies of partner and supplier companies
customer gratification as the driving ideal of company activity

well-developed intellectual property protection models and sharing mechanisms that protect core information while sharing everything else
cooperation with competitors and peer interaction with suppliers
fluid organizational structure
virtual company formation
company infrastructure as evolving, active and dynamic, its development over time shared with other companies
investment in people as the most powerful means of affecting the company's bottom line

In essence then, an agile corporation is a virtual enterprise. A virtual enterprise is a means of responding to new market opportunities quickly and with a minimum of dedicated investment and risk. The virtual enterprise creates a new productive resource by aggregating into a single, customer-focused business entity a carefully selected set of existing resources—expertise, facilities, and functional capabilities distributed among the divisions or departments of a single, large corporation, or among a number of different enterprises. Instead of forming a new business unit by co-locating these resources, or by acquiring all of them and then beginning the process of developing a new product or service, the virtual enterprise leaves the resources in place and bases the development process on the synergy created by interconnecting them.

In the world of rapidly changing, and in fragmented, high value-added markets, as well as the twenty-first century quality demands, the ability to quickly bring together the kinds of expertise required to design, deliver, and service a new product can confer decisive competitive advantage. For a more detailed discussion on both theory and implementation strategies the reader is encouraged to see Womack and Jones (1996), Maskell (1994), Jackson and Dyer (1995), and Montgomery and Levine (1996).

XII. PARTNERING

Manufacturers and service organizations are demanding improved quality and efficiency and further price cuts. On the other hand, suppliers in general are complaining about dwindling profit margins. The evolution of quality has indeed affected the relationship between organization and supplier base. However, as it is currently observed, this relationship has changed for the better and is expected to improve even more.

One of the ways that this relationship has been improved and will continue to be so is partnering. Partnering demands trust, loyalty, cooperation and open communication. In fact, it is these ingredients that also help the improvement of quality in any organizational chain. The goals of partnering are win-win relationships for both the organization and the supplier. The benefits for such an agreement are quite varied and focus on controlling the price, waste, variation, scheduling, long-term contracts, and quality. For more information on this concept see Stamatis (1996c), Keenan (1995), Mieghem (1996), and Bell (1994).

XIII. TOTAL PREDICTIVE MAINTENANCE (TPM)

Of all the various and sundry operational improvement methods that can and probably should be implemented, perhaps one of the best (both in terms of do-ability and results) especially for people in places where equipment is used to make things, is TPM. The theory is simple. Equipment is used. The users of the equipment know the most about it. Chances are excellent that the equipment operation can be improved. That improvement can and should be expected if the performance of the equipment at its first day is used as a baseline from which improvement is measured.

Total predictive maintenance is not quantum physics. It is defined as activities aimed at: 1) eliminating breakdowns, defects, and all other equipment losses, 2) gradually increasing equipment effectiveness, 3) improving company profits, and 4) creating a satisfying workplace environments (Shirose 1995). TPM is a system that describes company-wide activities involving all managers and employees to obtain full and best use of existing equipment. Fundamentally the TPM approach is based on the following:

investigating and improving current machines, etc.
a constant search for ways to standardize methods for improvement
determining how to provide and guarantee reliable product quality through the use of existing equipment
learning how to improve efficiency of manufacturing operations
learning how to motivate and educate operators so they take an interest in their equipment and maintain it well.

The TPM program depends on several skills and includes: daily checking, equipment tuning, precision inspection, lubrication, adjustments, troubleshooting, and repair. The trick is, of course, figuring out how to implement a TPM program and then sustaining it. It is beyond the scope of this book to address the details of TPM; however, the reader is encouraged to see Shirose (1992), Robinson and Ginder (1995), Suzuki (1992), Steinbacher and Steinbachere (1993), Shirose, Kimura, and Kaneda (1995), Suzuki (1994), Campbell (1995), and Shimbum (1995).

A. Material Flow Management

As TQM moves towards a relentless improvement of the process, the practitioner finds himself/herself in such a position that it becomes mandatory to know what is the status of material management throughout the organization. Material flow management (MFM) through a manufacturing business is indeed a core process of the futuristic approach to quality. It involves not only manufacturing but also sales, finance, product management, and indeed the entire organization. But what if material management is poor? These characteristics indicate the case in point:

poor due date compliance
customer arrears
fire fighting driven
continually changing priorities
poor control of stock
benefits of manufacturing system redesign not realized
existing planning systems over complex
no definitive, single plan for the business/factory
factory overload due to limited consideration of capacity during planning

Material flow management is a structured system used to achieve effective planning and control of operations. It begins with customer demand and goes through to engineering, procurement, manufacturing, sales, and on to delivery to the customer. Business is divided into three levels for MFM: strategic planning (by product family), master production scheduling (by product number), and cell planning (by part number).

Leveled scheduling is a simple technique combining volume production with small-batch production and the classification of demand into runners, repeaters, and strangers. When designing materials flow control systems, it is important to distinguish these three different groups of material flows, as each requires a different type of control. Runners are repetitive elements that occur once every work cycle and are significant volume. Repeaters are elements of regular frequency, such as once every 10 work cycles. Strangers are occasional elements that occur irregularly and frequently in low volumes.

The supplies module is an organizational unit owning all resources, equipment, and systems required to obtain externally sourced material. This work cell has responsibility for all aspects of the quality, cost, and delivery associated with the acquisition of materials.

Sitting at the beginning of the operations process, the supplies module is the interface between suppliers and the manufacturing cells and engineering. The previously separate functions of Purchasing, Receiving, Receiving Inspection, Material Handling, and Supplier Development now are combined into one natural group under one manager, the Supplies Module.

Material flow management consists of five basic elements that encompass the development and implementation of a complete, effective planning and control system. All five elements work together and they are:

1. Classify products into runners, repeaters, and strangers to simplify the control system and aid the level scheduling process.
2. Create a simple three-level forecasting and scheduling system integrating cell, product unit, and business plans for each product group.
3. Create leveled scheduling system. Leveling the schedule and matching load to capacity at each planning level will simplify the planning task at lower levels. Simple tools are required to support the process of leveling.
4. Put organizational mechanisms such as multifunctional planning committees in place covering site, product unit, and cell levels. Cross-functional membership is mandatory and attendance nonnegotiable. Standard check lists, agendas, and supporting data and documentation should be agreed upon. Responsibilities for actions of members in the group should also be clearly defined.
5. Simplified support tools are essential. These might include visible planning boards, cell gateway, or PC spreadsheets linked to simplified site systems. Often only Material Resource Planning System I (MRPI) or Material Resource Planning System I (MRPII) plus support tools are needed to manage an MFM system.

These five elements provide the framework for many of the actions that are required to improve the material planning and control mechanisms and thus reduce stocks and arrears, and help increase output and eliminate firefighting.

There are several key differences between the way MFM works and the traditional approach:

- Business planning revolves around a process, not a computer.
- Factory plan/master production schedule (MPS) is owned by plant manager.
- Supplies module and front office are part of planning process.
- Order load is matched against factory capacity.
- Capacity is fixed (need to allow for a certain percentage of breakdowns, urgent jobs, and so on).
- Customer is told "no" if you cannot meet his requirement(s).
- Schedule is fixed to an agreed horizon.
- There is a weekly MPS meeting with fixed attendees.
- There is a team approach to the planning process versus planning by each functional group.

The effectiveness of the control of material flow at every stage, from order through suppliers through to manufacture and delivery, impacts very visibly on two main aspects of business competitiveness: profitability and customer satisfaction.

Profitability is influenced through the cost of stocks held and through high manufacturing costs due to complexity and variability. On the other hand, customer satisfaction is affected if products are not supplied exactly when customers want them, which leads to increased costs, a loss of business, and a dissatisfied customer.

The benefits of MFM are many some of which are the following:

Business performance. Improved customer due date compliance, greater customer responsiveness, realistic achievable demands upon manufacturing, consistent lead time, realization of anufacturing systems redesign (MSR) benefits.

Reduction of waste. Reduced nonvalue-added steps and elimination of redundant, overly complex systems.

Planning and control. Forward visibility, elimination of month-end panic, controlled work in progress (WIP), ability to plan around capacity fluctuations.

General benefits. Integration of cell plans, visible problems, forced resolution of difficult problems, improved company-wide communication.

XIV. OUTSOURCING

The history between suppliers and companies has not always been friendly. Most companies looked at suppliers as adversaries. In fact they were not suppliers at all. They were simply vendors. Communication and information was kept at a minimum. Very little information was shared about the product for fear of knowledge leaking out. There was no trust. There was an unwritten rule of multiple sources. The belief of boxing yourself in the corner, do business with the lowest bidder and having the freedom to move from vendor to vendor were quite prevalent. All these beliefs contributed to a false security of true competition and innovation among vendors.

There is a great distinction between vendors and suppliers. Vendors are usually the ones that deal in commodities since they sell their products and/or services based on price only. There is no loyalty between the members of the transaction team. The general philosophy is win-lose. On the other hand, supplier is someone who is working with a customer to improve the product. It is based on the general philosophy of win-win, mutual trust, loyalty, and improvement. Price is only part of the relationship. Other elements may enter into the relationship such as delivery, production schedules, capabilities, and so on. For outsourcing to work, a supplier relationship must be in place.

Today and in the future, companies are taking a different view of their relationships with their suppliers. Suppliers are becoming more and more important to a company and have many positive things to contribute. One may see the supplier's involvement in the product evolution process, from design cooperation to tracking warranty repairs. The idea is finally here—the company and supplier is viewed as a whole rather than two separate entities.

The holistic approach (the relationship between company and supplier) is becoming so prominent that companies are willing to reduce their supplier base. This strategy of single sourcing promotes mutual trust, cooperation, and shared information. This harmony eventually will lead to a better product through better communication.

Inputs to a production system, whether manufacturing or services, are important since the final output is a function of these inputs. Raw materials, knowledge, and services are a few of the varied forms that inputs may take. One of the more controversial topics in the area of inputs is outsourcing. Outsourcing is simply hiring an outside company to do what used to be done (or could be done) within the company itself. The advantages and disadvantages of outsourcing depend on the position and viewpoint of the individual making the decision and/or the organization and its goal.

Currently, it seems that there is a growing favoritism toward outsourcing by both customers and suppliers all across different industries and products. Among the advantages, this arrangement of hiring an outside company benefits both the supplier and the company being supplied. Outsourcing provides a way to extend the company's capacity and capabilities by providing additional resources with minimal risks. On the other hand, outsourcing allows the company to concentrate on the essential activities that produce profits. At the same time, the supplier gains business and support from the company it is supplying. When making a production and/or quality decision, outsourcing should be considered as a viable alternative.

At a time when people are considered to be the "greatest asset" of the organization, outsourcing is viewed by many as a great threat. This is perhaps because of a prevailing perception that individuals within a company feel they will be displaced once outsourcing has been implemented in their company. To eliminate such fears many proposals have been advanced including:

- better utilization of employees by the employer
- retraining by the employer for a new function within the company (one such position may be that of measuring the supplier performance and maintaining communications with the supplier)

A second threat to outsourcing is the fear of losing management control. The reference, of course, is to the flow of goods and overall quality. This may be minimized if not completely eliminated by a true commitment by both parties (customer and vendor) to build a relationship of trust. Some of the basic elements that should be considered before an outsourcing decision is made are: make-or-buy analysis, preappraisal activities, vendor appraisals, the actual selection decision, postdecision activities, and follow-up activities

A. Make-or-Buy Analysis

This is the most challenging and critical step of the process. It involves a decision whether to purchase an input from an outside supplier or to produce that input internally. One of the main considerations in this analysis is cost. As a general rule most companies have an objective to hold down cost and to increase profits. Other considerations may be labor costs, economics of volume (marginal costs), manufacturing time, lead time, whether or not the project is of temporary nature or not, scrap, rework, and others. Unfortunately, there is not one analysis that is right for all organizations. Even if there was standard method, every organization could reach a different decision because every individual and/or organization has different backgrounds, values expectations, goals, and priorities.

More and more we see that outsourcing has become a way to receive special expertise, experience, and technical knowledge as well as a second opinion in a very short time. To keep up with some of the most innovative ideas, suppliers have in fact formed alliances with specialized consulting firms and/or large corporations. For example, in the automotive industry we find that some of the most likely alliances are in the area of subassemblies primarily because of a very rapid changing technology that requires more research and development. By forming such alliances they are able to win larger market share while meeting the ever changing demands of business.

B. Preappraisal Activities

An appraisal committee should be formed that will make recommendations on the various available vendors. The members of the committee should represent the departments that have an interest in any or all aspects concerned with the product that will be supplied from the outside company. As a minimum representation, personnel from manufacturing, engineering, purchasing, and quality should make the nucleus of the team. Some of the concerns that the team should address at this stage are:

- What is the vendor going to supply?
- Who are alternative suppliers?
- How are suppliers evaluated?
- What are the responsibilities of the company to the vendor?

- Other issues of criteria include: security, innovation, efficiency, flexibility, reputation, technical competency, management techniques, financial soundness of vendor, processes, capability, credibility, integrity, commitment, etc.

C. Vendor Appraisal

Members of the committee meet with suppliers to see if they are able to supply the right product at the right time at the right cost. During these meetings, committee members should evaluate the vendors based on the criteria and characteristics outlines before the meeting. (There should be agreement on the evaluation criteria and no argument on the vendor's premises.) This process provides an excellent chance to check out potential supplier commitment, management attitude, and whatever the team deems important for the relationship to grow. The meeting at the vendor's place of work may in fact provide the added information the team needs for a sound decision. After all, a company is not only committing to buying a product, but the supplier's quality, capability, and process. The vendor should be a business that the company is proud to be associated with. To find all the necessary information and to make a sound decision, the committee must have access to the appropriate personnel, hardware, software, machinery, etc.

D. The Actual Decision

One (or perhaps more than one) supplier becomes the supplier. A single supplier is generally cheaper than multiple suppliers because of volume economics and there will be less of a personnel commitment on the part of the company. The committee at this stage needs to reach a consensus based upon the characteristics agreed upon initially as criteria. The consensus decision is presented to management for the final selection. Once the final selection is made, then the contract is written. At this stage, there are two forms that the contract may take. One is contract for hire where there are no exclusive rights on either party; the other is a dedicated contract where the parties will only deal with each other.

E. Postdecision Activities

A single person from the company should be named as the main focal point for dealing with the supplier. This person should also be on an in-house response team as one that would be established to deal with the problems arising from the supplied material or to assist the supplier with any problems they encounter. At the very minimum, there should be at least one person acting as an ombudsman for the company and the vendor to eliminate potential stalemates of problems.

In addition to the problems the ombudsman or the in-house response team have to handle, it is their responsibility to cultivate the communication lines between the company and the vendor. These lines should be maintained properly. A vendor at this point is an extension of the company. The smooth implementation of the desired results on both ends takes good communications, shared information, trust, and understanding. Time and money may be wasted if the lines of communication break down.

In this stage, because of the commitment and the high stakes of both parties, it is imperative that the involvement of both be at all levels (i.e., design, implementation, quality. In fact, if the supplier is having problems, the company should assist them in developing capabilities and solutions. The emphasis is always on the long term as opposed to the short term. A real commitment with a true relationship of mutual benefit is worth keeping healthy. In case of emergency, however, the company should have contingency plans to avoid complete shut downs.

F. Follow-Up with the Supplier

Continual improvement must be communicated. The desire for high and predictable quality must be stressed in the supplier relationship. This is especially important if the company is using "just in time inventory." The supplier's quality should correspond to the products cost and technology. This quality should be backed by either some form of certification and/or statistical quality data by the supplier.

A good starting point may be a vendor tracking program, so that the vendor(s) know exactly where they stand in relation to the criteria set by the company and in relation to the other vendors, if any. This tracking program should be the impetus for the company and vendor to get together to discuss the current status, future events, and/or communicate problems, and to ask for assistance.

A supplier education program should be instituted. It should include plant tours, statistical training, introduction to the company's philosophy and ways of doing business, and so on. The idea here is to emphasize a two way relationship and to use this education program as a feedback to the overall objective of continual improvement. In addition this will be the forum for creating action plans as well as deciding upon supplier motivation programs. Positive reinforcement could be part of the plan used to reward vendor commitment to quality goals, delivery, cost, and so on.

XV. JUST IN TIME

Just in time (JIT) is a discipline organized to improve quality and productivity through the elimination of all waste. Its goal is the same as it is with all production systems:

1. Maximize customer service levels through: delivery, quality, price, design, and flexibility
2. Minimize cost to the corporation through: overhead, material cost, scrap, inventory holding cost, productivity

The JIT system originated in Japan and its purpose is to eliminate waste and to improve the respect for individuals within the organization. Ohno (1988) of Toyota described waste as anything other than the minimum amount of materials, manpower, machines, or tools necessary to production. In order for JIT to be effective, a company must receive and produce the right number of quality parts. The parts must then be available at the right place and time and use a minimum of manpower, equipment, and facilities. Unlike many other systems, JIT does not require a software package.

It is difficult to decide which system and/or theory will work best in a given environment. However, many manufacturing experts believe that the various programs can work together as one. There is no such thing as a "prepackage" in manufacturing. There are many types of techniques that may be used, but the most popular are the Material Requirement Planning System (MRP) and Manufacturing Resource Planning (MRPII). Both these systems become more effective when used in conjunction with JIT. Many U.S. manufacturers have implemented MRP into their companies. It is a simple mathematical tool used as a planning and control device tailored for items with higher levels of demand. MRP advocates claim substantial benefits to their system including:, increase sales, reduced sale prices, reduced inventory, better customer service, better response to market demands, ability to change the master schedule, reduced idle times, and reduced set up and tear down costs,

On the other hand, MRPII is a complete management system. Its advocates say that MRPII is a simple, common-sense method of addressing the deeply rooted rescheduling problems existing in every manufacturing operation. MRPII formulates financial and operating plans and measures performance against those plans. It focuses on such fundamental manufacturing questions as:

- What are we going to make (production and master schedule)?
- What does it take to make it (bills of materials and routings)?
- What do we have in inventory (inventory status)?
- What do we need to make it (planned orders)?

Most MRP factories follow a production schedule that sends work from one location to another. This type of system is called a "push" system since it pushes materials through the factory. On the other hand, JIT is a "pull" system since the users send for the material as it is needed. The requirements for the pull system are:

1. The system must provide for pulling the material to adequately support downstream operations.
2. The system must provide constraints on the level of inventory within the system.
3. Each operation must know where to send for material (routings must be clear and standardized).
4. Ther must be a reasonably level workload.
5. There must be a close to repetitive flow of material.

The idea here is for the company to have what is needed to do the job and what products are in great demand. In other words, if a company's production is stacked at the point where it was made and no one comes for it, management can readily see that production of that part should stop. To avoid such problems, JIT only utilizes those parts that there is a demand for and if there is no demand then why waste the time and money producing those parts.

As we mentioned earlier, the MRP and JIT can work together effectively if implemented properly. Most companies using both systems used the planning concept of MRP and the execution concept of JIT while realizing that the MRP must be changed to accommodate the JIT (and never the other way). The reason for this is because JIT is such a strict philosophy.

To make the systems compatible, several changes in MRP are necessary. They are:

1. Managing product flow rather that product orders.
2. Managing product inventory by issuing to, and managing from, floor stock locations rather than on an order basis.
3. Replenishing orders at floor stock locations.
4. Enhancing materials usage variable accounting to avoid out-of-control inventory.
5. Improving materials routing through better product flow.
6. Acknowledging and managing overlapping operations.
7. Adapting manufacturing schedule capabilities to accommodate multiple flows and production lines.

These changes make the MRP and JIT systems compatible, especially on repetitive operations such as assembly lines. It must be emphasized, however, that there are some industries and some products that cannot be forced into the JIT philosophy, such as high technology products, custom orders, new technology, small volume orders, and items requiring special inspections of lots (pharmaceuticals).

XVI. IMPLEMENTING THE JIT PROCESS

In order for a company to begin, it must first detect the need for the JIT process. This will take the cooperation of everybody involved and the willingness for change. A plan must be drawn up stating where the company stands at this point in time. For example, how much inventory is at hand, the value of finished goods, the value of work in process inventory, vendor lead times, and so on. The company must be willing to respond to customer needs since JIT relies on the demand of a product. Smaller production lot sizes are essential if manufacturing systems are to function more responsively. Sometimes, this means that a change in plant layout is necessary for the implementation. This new layout involves the use of group technological work cells that will produce a similar sequence of operations. These work cells will be grouped together into smaller individual U-shaped factories within the factory. Each U-shaped factory will have similar machinery in order to act as a small assembly line type of process. The rationale for such layout change is to make each employee knowledgeable about the machinery and thereby contribute in the reduction of his boredom, absenteeism, and overall costs.

The company implementing JIT needs to have a well-established relationship with its suppliers, and vice versa. The supplier must be responsible enough to ship the required quantity at the required time. The user is responsible to maintain record integrity based on a fixed schedule.

Since employees are such an integral part of the JIT process, much education is necessary. The employees must realize the benefits reaped from using this system and be sold on it from early on. Management should use positive reinforcement as the solution to getting employees to cooperate. A team concept approach is highly encouraged, since the team concept encourages open communication and active participation. The JIT process is not simple to implement. Any new system that requires a lot of work, a lot of time, demands participation, and carries a certain amount of real or perceived risk will likely attract a great many excuses not to do it. For example:

- Employee and union opposition will be too strong.
- JIT, because it was developed by the Japanese, is too culture dependent for us to adopt.
- It will cost way too much to make changes such as: more flexible equipment, reduction in set up time, plant layout, material handling systems, and new information systems.

- We have too many production problems to be able to reduce inventory by any amount.
- We cannot operate in our marketing environment with the schedule and design limitations of JIT.

A specific negative JIT view by department may be of the type:

JIT image	Concerned party
Fewer jobs	Employees/unions
Fewer safety stocks	Purchasing
Few or no inspections	Quality
Fewer "opportunity" stocks	Sales
Uncertainties of new system	Management
Fewer internal controls	Accounting
Less machine utilization	Engineering
More transactions	Inventory control
More set ups	Manufacturing supervisors

When implementing JIT some rules to keep in mind are:

1. Whenever you feel comfortable with your inventory level, you have too much.
2. Poor systems breed more systems.
3. There is not end to what you can accomplish if you do not care who gets the credit.
4. When people do not understand fundamental principles, their solutions to basic problems will be grossly overly sophisticated.
5. Never split the responsibility for a limited resource.
6. Never run away. Always run to.
7. Anything you measure will improve.
8. Beware of the light at the end of the tunnel. It just might be the train coming the other way.
9. Always use the simplest system that works. Anything more complex must be justified with respect to people.
10. The ability of people to understand a system goes down exponentially as the complexity goes up.
11. Always make it easy to do it the right way, the first time.

XVII. KANBAN INVENTORY MANAGEMENT SYSTEM

Kanban is a special method of achieving JIT production and scheduling. It literally means "card" and is sometimes referred to as the "pull" system. It is a means of pulling parts and products through the manufacturing sequence as needed. The Toyota Motor Company of Japan pioneered the technique under Vice President Taiichi Ohno. Ohno's ideas about JIT and Kanban were inspired by the American supermarket, where shelves were replenished as emptied. Since space for each item was limited, more items were brought in only when there was a need for them (Esparrago, 1988).

Kanban is universally accepted in Japan as a way to maximize profits by minimizing costs and minimizing inventory. Suppliers ship parts directly to automobile assembly plants where side unloading trucks deliver parts directly to designated lines without counting or sorting. Even fabricating plants follow inventory reduction by receiving steel several times per week, and turning it over into a product in one week (Deming, 1988).

To understand the Kanban system one must know its primary components:

1. kanban authorization cards: move and production
2. standard containers
3. work stations (or work cell, work center): may be a machine, assembly line, or a single work table
4. withdrawal and production posts
5. input areas
6. output areas

Every Kanban system utilizes a marker such as a card, flag, light, hand signal, or voice command. In a two-card system, a move Kanban indicates a container can be passed on to the next work center. A production Kanban notifies the work center to produce one container of parts. Two areas, called withdrawal and production areas, exist side by side on the shop floor.

Every work center has a withdrawal and production area. The withdrawal system is basically responsible for obtaining more stock, raw material, or parts from the previous work cell or machine. When the material containers are empty, a move Kanban is placed on that container, allowing it to retrieve more stock. When refilled, its move card is taken out and replaced with a production card and the container is returned to its rightful work cell. The move card that was removed from it will be placed on the previous work cell's container, allowing that work cell to obtain more stock. Thus, each cell interacts with the preceding work station.

Suppose that at the production area a finished or full container waits for a move card. That container goes no further down the line unless authorized (or pulled) from the next downstream work center. When the downstream work center sends an up an empty rack it is signaling that more stock is needed for its operations. The full and empty container then switch containers and cards, thus authorizing the middle work center to make more parts, and the downstream work center to have more stock for its operations. Thus each cell receives raw material in the form of its upstream cell's finished parts, and, in turn, supplies its own finished parts as the raw materials for downstream cells.

A Kanban system may also be single-card, a combination of push and pull. In some domestic automobile fabricating plants, long die-change times and low uptime of high speed transfer presses, complex product designs, and an untrained work force make a dual card system difficult, because dies cannot be changed over quickly enough to justify small lot production runs. The customer assembly plants can pull subassemblies from stamping plants, but the metal assembly lines within the stamping plant have material pushed on them from their adjoining press rooms. Perhaps the Japanese and transplant automobile manufactures can utilize dual card system due to higher up time, speedy die changes, simpler products which require fewer dies, and a culture that accepts Kanban systems.

If an organization is producing a broad product line and/or has an established MRP, then the company may consider a Synchro MRP which is nothing more than a computerized Kanban system. Obviously, the computerized systems are more complicated. For more information on the JIT and Kanban systems, see Shores (1994), Ohno and Mito (1988), Shingo 1988, Ohno (1988), Barrett (1988), Belt (1987), Deleersnyder et al. (1989), Krajewski et al. (1987), and Landon (1988).

XVIII. QUALITY CONSIDERATIONS

Proper quality systems, at least of the statistical process control (SPC) level, are not only important for proper customer service and cost control, but also represent a critical consideration for JIT. Consider that JIT focuses on the uncertainties in the system. If rejects are high and quality is poor, we are uncertain as to the usability of any given part we select for our product. This uncertainty prompts us to build a buffer (inventory) against that uncertainty causing a disruption of the schedule.

Since the automotive industry is already involved in detailed implementation of SPC, and the topic is covered in Chapter 5, a detailed treatment of the concepts is not necessary here. However, it must be emphasized that SPC is totally compatible and critically important to JIT. A few points worth highlighting:

1. Product design should establish tolerances that are as realistic as possible to what will make the overall product meet the desired quality. Tight tolerances breed informal systems of "real" acceptance levels as people recognize the need to keep production going while still building quality product.

2. Quality and cost are both competitive weapons. While a typical argument talks about the trade-offs between the two, that is a short-term and narrow argument. You can have the best of both, especially if you do not constrain yourself to the short run.
3. Over the life of a long relationship, the process capabilities become more efficient and cheaper to produce. As a consequence, more manufacturers are relying on their suppliers to reduce costs. They are also achieving price reductions by single sourcing their suppliers. Costs are reduced by reducing the supply base and therefore, having a better control of quality.
4. The importance of SPC lends further support to keeping production processes as simple as possible. Simple technology is not necessarily old technology. It is, however, easily controlled technology.
5. The technology issue has developed to the point where it is not unusual for a company to rely exclusively on their supplier's expertise when designing a part for maximum value and quality. The suppliers provide information, insight, and knowledge that a company may lack. Furthermore, the suppliers are always on the leading edge of technology and continuous improvement to satisfy the never ending pursuit of continuous improvement. This relationship is kept and nurtured by mutual trust and understanding of the needs and goals of both company and vendor.
6. Management must constantly stress what the work force has accomplished, not what management has accomplished.
7. Keep problems and the state of product quality visible. Feedback needs to be immediate and constructive.
8. Build records to organize and classify problems in order to find and correct causes. The typical "cause and effect" diagram and the quality control circle approach to participative problem solving is appropriate for action and solution implementation. Keep in mind these methods are not merely for problem solutions, but also can be used for problem anticipation and prevention.
9. Make detailed comparisons of operations to strive for simplicity, efficiency, and uniformity.
10. The operator will become more important in the role of quality as automation and process control become more prevalent. Operators will become more involved as controllers of equipment, as problem detectives and solvers, and as contributors to reductions in variations.
11. The workers will become more important as an integral part of the quality system as well as a key toward correction and improvement in the overall production system.
12. Establish improvement goals and measure the results; publish them and stress the positive.

13. Delivery is a major concern and component because it affects all phases of manufacturing: production planning, service and receiving, purchasing, equipment set up, and quality.
14. The importance of the worker in JIT cannot be overly stressed. The employee approaches may be of the form of removing barriers, promoting communication, providing responsibility (with training), providing support, and promoting shared decision-making.

The expected changes to the organization are primarily in two areas: productivity gains and supervisor roles. The considerations concerning supervisor roles are:

1. Keep inventory low enough so that every one has to concentrate on the process and how to improve it.
2. Keep an equity of work. No one should have the luxury of relaxing while others struggle to improve the system.
3. Fine-tune proposed inventory levels to reflect current problems or improvement status within the department.
4. Work with industrial engineering to keep data and evaluate the quantity and type of people necessary to maintain different cycle rates.
5. Maintain rebalancing for standard schedule changes or process improvements.
6. If the card system is used, withdraw cards as necessary to keep the system tight.
7. As much as possible, be involved in planning, training, development, and process improvement.
8. Recognize the importance of the workforce in JIT and work to develop and improve workforce involvement.
9. Forget about the traditional "warm and fuzzy" security blanket of management and supervisors (i.e., plenty of "just in case" inventory to keep the workers busy and improve efficiency).

The differences between conventional and just in time attitudes are summarizecd in Table 3.4.

XIX. DESIGN FOR ASSEMBLY

On the face of it, design for assembly (DFA) and design for manufacturing (DFM) sound redundant, or at best, obvious. Logically, nothing should be designed that cannot be assembled or manufactured. Yet, the manufacturing process inevitably puts constraints on the product design process. Designers once were able to rely on the manufacturing people to figure out an assembly process for a given product. But as manufacturing has become increasingly

Table 3.4 The Differences Between Conventional and JIT Attitudes

Conventional	JIT
Large lots are efficient (more is better)	Ideal lot size is one unit (less is better)
Faster production is more efficient	Faster production than necessary is a waste; balanced production is more efficient
Scheduling and queues are necessary tradeoffs to maximize output from equipment and manpower	Tradeoffs are bad; they trade one waste for another and prevent the proper solution of problems
Inventory smooths production	Inventory is undesirable
Inventory provides safety	Safety stock is a waste

automated and product life cycles are increasingly shortened, the designer's prerogatives must accommodate these realities.

Designers are concerned about making more manufacturable designs and participating in cutting the time to market. In fact, DFA is a structured method and software system for evaluating the "global" efficiency of part designs and assembly systems. When a design is viewed as a system of interrelated parts and functions, it can be optimized by reducing the number of parts and assembly steps.

Design for assembly is often used interchangeably with design for manufacturing (DFM), but there are some differences. Design for assembly focuses on identifying the most appropriate assembly system, done in conjunction with a structural analysis of the design for its overall cost-efficiency, parts consolidation, and suitability for the chosen assembly method. Design for manufacturing is a broader concept centering on identifying appropriate materials and manufacturing processes for a product's component parts. Where DFA is concerned with parts and their orientation, DFM also considers surfaces, features, and textures. While DFM pays more attention to aesthetics and engineering, but you still have to be concerned that it is going to be a real product that is going to be assembled. On the other hand, when DFA and DFM are combined to form design for manufacture and assembly (DFMA), this combination requires the user to assess the contribution of each part as it relates to the structural and cost efficiency of the whole design. By definition, DFMA is a quantitative tool that allows designers to identify the ease-of-assembly and manufacturing-efficiency factors in their designs

Why do we want to use DFA or DFM? Because we want to maximize:

- simplicity of design

- economical materials, parts, and components
- economical tooling/fixtures, processes, and methods
- standardization
- ease of assembly
- ease of service
- ease of testing
- integrity of product features

and minimize:

- uniqueness of process
- critical processes
- material waste or scrap due to process
- energy consumption
- pollution, and liquid or solid waste
- scarce materials, components, and parts
- scarce, proprietary, or long lead time equipment
- degree of ongoing product and production support

For more detailed information the reader is encouraged to see Bralla (1986), White (1987), Dean and Susman (1989), and Drucker (1990).

XX. KAIZEN

Kaizen literally means gradual, continuous improvement, mainly in small matters, in all aspects of working life. Kaizen begins with the recognition that any corporation can have problems and improvements can be made as long as perfection is lacking. Kaizen solves problems by establishing a corporate culture in which everyone can freely admit to these problems.

In any organizational environment, the Kaizen approach can help towards the continuous improvement movement by contributing small changes throughout the organization. This can be accomplished through wide company programs, such as: TQM, suggestion system, quality circles, and so on. The limitation of the Kaizen approach is when it reaches the diminishing return point at which time it is appropriate to look at improvements using reengineering methods. For more detailed information on Kaizen see Imai (1986), Akiyama (1991), and Kobayashi (1995).

XXI. POKA-YOKE

This was developed and refined by Shigeo Shingo, poka-yoke literally means mistake-proofing or making a process fail-safe. The main idea behind it is to respect the intelligence of workers. By taking over repetitive tasks that depend on vigilance or memory, poka-yoke frees a worker's mind to pursue more creative and value-adding tasks. For more information see Shingo (1986, 1988, 1992), and Shimbum (1989).

XXII. CONCURRENT ENGINEERING

Concurrent engineering (CE) can be defined as a a multidisciplinary team approach is used to conduct product conception, design, and production planning at one time. Sometimes CE is also called simultaneous engineering. For more information see Turino (1992). The CE approach requires that each new-product start-up project be handled by a full-time, multidisciplinary team. This includes specialists from product design, engineering, manufacturing engineering, marketing, purchasing, and finance. Principal suppliers of process equipment, purchased parts, materials, and services may also participate at various times from the outset.

The key feature of CE is that major functions occur concurrently, not sequentially as in traditional engineering. Thus, various phases of product conception and design, test design, feasibility study, tool design, and process development may be going on at one time. The practice entails much discussion and give and take among team members.

Concurrent engineering is revolutionizing the way business is being done and is being credited with tremendous reductions in cycle time, process improvements, customer satisfaction, and positive employee morale.

XXIII. ZERO DEFECTS

The concept of zero defects was introduced by Shingo (1986) and is based on four elements. They are: successive checks, self-checks, source inspections, and poka-yoke devices. The idea of the zero defects is to sensitize the entire organization about waste. To facilitate the concept of zero defects the organization may use employee programs such as: TQM, Poka-Yoke, quality circles, cost of quality, and so on.

XXIV. TOTAL EMPLOYEE INVOLVEMENT

Total employee involvement (TEI) is the not-so-scientific art of getting everyone in the company to assume responsibility for making his or her job easier, more efficient, more productive, and safer. It is the cornerstone of the continuous improvement movement. For a frank discussion on employee involvement see Moe (1995).

REFERENCES

Abell, D. F. (1980). *Defining the business: The starting point of strategic planning.* Prentice Hall. Englewood Cliffs, NJ.

Akiyama, K. (1991). *Function analysis.* Productivity Press. Portland, OR.

Allmendinger, G. (1990). Performance measurement. *ESD Technology.* Dec.: 10–13.

Barrett, J. (1988). Case study: I. E.'s at Calcomp are integrating JIT, TQC and employee involvement for world class manufacturing. *Industrial Engineering.* Sept.: 26–32.

Bell, C. R. (1994). *Customers as partners: Building relationships that last.* Quality Press. Milwaukee, WI.

Belt, B. (1987). MRP and Kanban - a possible synergy? *Production and Inventory Management.* 1st Quarter: 71–80.

Brake, T. (1996). *The global leader.* Irwin Professional. Burr Ridge, IL.

Bralla, J. (1986). *Product design for manufacturing.* McGraw-Hill. New York.

Byham W. C. (1988). *Zapp! The lightning of empowerment.* A Fawcett Columbine Book. New York.

Campbell, J. D. (1995). *Uptime: Strategies for excellence in maintenance management.* Productivity Press. Portland, OR.

Dean, J. W. and Susman G. I. (1989). Organizing for manufacturable design. *Harvard Business Review.* Jan.-Feb.: 28–36.

Deming, W. Edwards. (1986). *Out of the crisis.* Massachusetts Institute of Technology. Center for Advanced Engineering study. Cambridge, MA. *also* (1988), 5th printing.

Deleersnyder, Jean-Luc, Hodgson, T. J, Muller, H, and O'Grady, P. J. (1989). Kanban controlled pull systems. *Management Science.* Sept.: 1079–91.

Drucker, P. (1990). The emerging theory of manufacturing. *Harvard Business Review.* May-June: 94–102.

Esparrago, R. A. (1988). Kanban. *Production and Inventory Management Journal.* 1st Quarter: 6–10.

Goldman, S. L. and Mellon, A. W. (1993). Agile manufacturing: Quality continuous to evolve. *Technology.* Aug.: 44–49.

Goldman, S. L. and Preiss, K. (1991). *21st century manufacturing enterprise strategy: An industry-led view.* Monogram published by Lehigh University's Iacocca Institute. Lehigh, PA.

Hammer, M. and Champy, J. (1993). *Reengineering the corporation.* Harper Business. New York.

Imai, M. (1986). *Kaizen.* Random House. New York.

Jackson, T. L. and Dyer, C. E. (1995). *Implementing a lean management system.* Productivity Press. Portland, OR.

Johnson, R. S. (1993). TQM: Leadership for the quality transformation. Quality Press. Milwaukee, WI.

Juran, J. M. (1995). *Managerial breakthrough.* Rev. ed. Quality Press. Milwaukee, WI.

Juran, J. M. (1989). *Juran on leadership for quality.* Free Press. New York.

Keenan, T. (1995). Suppliers and the '96s. *Ward's Auto World.* Oct.: 57–60.

Knouse, S. B. (1995). *The reward and recognition process in total quality management.* Quality Press. Milwaukee, WI.

Kobayashi, I. (1995). *20 keys to workplace improvement.* rev. Productivity Press. Portland, OR.

Krajewski, L. J., King, B. E., Ritzman, L. P., and Wong, D. (1987). Kanban, MRP and shaping the manufacturing environment. *Manufacturing Science*. Jan.: 39–57.

Landon, W. G. (1988). Kanban and Deming's 14 points. *Quality*. Sept.: 50–52.

Langdon, M. J. (1993). *Where leadership begins: Key skills of today's best managers.* Quality Press. Milwaukee, WI.

Lowenthal, J. N. (1994). *Reengineering the organization: A step by step approach to corporate revitalization*. Quality Press. Milwaukee, WI.

Maskell, B. H. (1994). *Software and the agile manufacturer: Computer systems and world class manufacturing*. Productivity Press. Portland, OR.

Mieghem, T. V. (1996). *Implementing supplier partnerships: How to lower costs and improve service*. Quality Press. Milwaukee, WI.

Moe, J. L. (1995). What does 'employee involvement' mean? *Quality Progress*. July: 67–71.

Mintzberg, H. and Quinn, J. (1992). *The strategy process: Concepts and contexts.* Prentice Hall. Englewood Cliffs, NJ.

Montgomery, J. C. and Levine, L. O. (Ed.) (1996). *The transition to agile manufacturing: Staying flexible for competitive advantage*. Quality Press. Milwaukee, WI.

Nagel, R. N. and Dove, R. (1991). *What it would take for U.S. industry to regain global manufacturing competitiveness by the early twenty first century*. Monogram published by the Lehigh University's Iacocca Institute. Lehigh, PA.

Ohno, T. (1988). *Toyota production system: Beyond large-scale production*. Productivity Press. Portland, OR.

Ohno, T. and Mito, S. (1988). *Just-in-time for today and tomorrow*. Productivity Press. Portland, OR.

Parr, G. L. (1996). Gladly fulfilling an editorial request. *Quality*. Feb.: 4.

Peters, T. (1992). *Liberation management.* Alfred A. Knopf. New York.

Roberts, L. (1994). *Process reengineering: The key to achieving breakthrough success.* Quality Press. Milwaukee, WI.

Robinson, C. J. and Ginder, A. P. (1995). *Implementing TPM: The North American experience*. Productivity Press. Portland, OR.

Scherkenbach, W. W. (1988). *The Deming route to quality and productivity*. Quality Press. Milwaukee, WI.

Shimbum N. K. (Ed.) (1995). *TPM case studies: Factory management series*. Productivity Press. Portland, OR.

Shimbum, N. K. (1989). *Poka-Yoke: Improving product quality by preventing defects*. Productivity Press. Portland, OR.

Shingo, S. (1988). *Non-stock production: The Shingo system for continuous improvement*. Productivity Press. Portland, OR.

Shingo, S. (1992). *The Shingo production management system: Improving process functions*. Productivity Press. Portland, OR.

Shingo, S. (1986). *Zero quality control: Source inspection and the Poka-yoke system*. Productivity Press. Portland, OR.

Shirose, K. (Ed.) (1995). *TPM team guide*. Productivity Press. Portland, OR.

Shirose, K. (1992). *TPM for workshop leaders*. Productivity Press. Portland, OR.
Shirose, K., Kimura, Y. and Kaneda, M. (1995). *P-M analysis: An advanced step in TPM implementation*. Productivity Press. Portland, OR.
Shores, A. R. (1994). *Reengineering the factory: A primer for world-class manufacturing*. Quality Press. Milwaukee, WI.
Stamatis, D. H. (1996a). *Total quality service*. St. Lucie Press. Delray Beach, FL.
Stamatis, D. H. (1996b). *Total quality management in healthcare*. Irwin. Burr Ridge, IL.
Stamatis, D. H. (1996c). *Q and A about teams*. Paton Press. Red Bluff, CA.
Stamatis, D. H. (1996d). *The nuts and bolts of reengineering*. Paton Press. Red Bluff, CA.
Steinbacher, H. R. and Steinbacher, N. L. (1993). *TPM for America: What it is and why you need it*. Productivity Press. Portland, OR.
Suzuki, T. (1992). *New directions for TPM*. Productivity Press. Portland, OR.
Suzuki, T. (Ed.) (1994). *TPM in process industries*. Productivity Press. Portland, OR.
Turino, J. (1992). *Managing concurrent engineering*. Van Nostrand Reinhold. New York.
Vogt, J. F. and Murrell, K. L. (1990). *Empowerment in organizations*. Pfeiffer. San Diego, CA.
White, J. (1987). *Production handbook*. John Wiley and Sons. New York.
Wickman, R. F. and Doyle, R. S. (1993). *Breakthrough quality improvement for leaders who want results*. Quality Press. Milwaukee, WI.
Wilson, T. B. (1995). *Innovative reward systems for the changing workplace*. Quality Press. Milwaukee, WI.
Womack, J. and Jones, D. (1996). *Lean thinking*. Simon and Schuster. New York.
Womack, J. and Jones, D. and Roos, D. (1990). *The machine that changed the world*. Harper Perennial. New York.

4

Implementation Strategy for Total Quality Management

I. OVERVIEW

Lately the literature on quality and more specifically on TQM has suggested that TQM is not successful (Keck, 1995; Dobbins, 1995; Bemowski, 1995; Cottrell, 1992), is losing ground (Whyte and Witcher 1992; Wilkinson, Marchington, and Dale, 1993; Witcher 1993), and in some cases it has been pronounced dead (Duff, 1995; Fellers, 1995). Whether one agrees with these assessments or not is not the issue in here. The issue is that if indeed TQM is on a downward slide, why is this so and can we reverse the direction to an upward direction?

Over the last 28 years working in the quality field there are at least a dozen observations that I have personally noted regarding quality in general and I submit that these observations may indeed be the reason for the negative press of TQM. These twelve items are being repeated over and over again in a variety of organizations regardless of product, service, geography and size. These observations are:

- The lack of significant results
- The lack of significant project selection
- Most organizations spend huge amounts of time educating staff on the TQM process and motivating teams, all the while concentrating on projects that do not matter.
- Most organizations are not taking the time up front to develop strategy; there is a mistaken belief that the TQM process alone will show the organization where it needs to be.
- Even those organizations with clear strategic direction are not allowing strategy to guide project selection, leaving the bottom-up TQM effort utterly unconnected to the top-down planning effort.
- Most organizations are not coordinating TQM activities, allowing teams to attack problems on multiple of fronts. The result is slight improvements in many different areas.
- Having selected projects that matter and aligned team efforts, next most important step is ignored or not given the necessary attention; this is setting goals that will spur teams to "breakthrough" results.
- Too many top managers (including the chief executive officer) stand back from their responsibility to lead TQM, and fail to provide a compelling vision of where organization must move and why.
- The four essential items for mobilizing support are missing: 1) Open communication which allows ideas, problems, and opportunities to be addressed by anyone in the organization. 2) Elimination of fear which holds back innovation, risk taking and entrepreneurship due to fear of failure, intimidation and the like. 3) True and honest leadership which gives meaning and purpose to TQM; the simple truth is that employees will work harder on project(s) whose importance they understand and believe in. Part of the honest leadership is to tell the employees what is in it for them. 4) Full-force commitment of key players; many organizations failing to bring on-board the essential power bases and/or middle management.
- The lack of middle management's cooperation, enthusiasm, and assistance means that many of most important improvement projects remain undoable. Without middle management commitment, support, and enthusiasm it is very hard to guarantee commitment from subordinates and general employees.
- Organizations misinterpret employee empowerment as an excuse to abdicate leadership.

Can we redefine the direction of TQM in such a way that all organizations may benefit from its philosophy? We believe so. To do that as an organization we must change our paradigm of what can and what can not be done. The following anonymous poem says an awful lot about the attitude of most organizations dealing with or about to deal with TQM.

II. IT COULDN'T BE DONE

Somebody said that it couldn't be done,
But he with a chuckle replied
That "maybe it couldn't but he would be one
Who wouldn't say so till he'd tried.
So he buckled right in with the trace of a grin
On his face. If he worried he hid it.
He started to sing as he tackled the thing
That couldn't be done, and he did it.
Somebody scoffed: "Oh, you'll never do that!
At least no one ever has done it."
But he took off his coat and he took off his hat,
And the first thing we knew he'd begun it.
With a lift of his chin and a bit of a grin,
Without any doubting or quiddit,
He started to sing as he tackled the thing
That couldn't be done, and he did it.
There are thousands to tell you it cannot be done.
There are thousands to prophesy failure;
There are thousands to point out to you one by one,
The dangers that wait to assail you.
But just buckle in with a bit of a grin,
Just take off your coat and go to it;
Just start in to sing as you tackle the thing
That "cannot be done," and you'll do it.

III. THE ROLE OF TQM

Traditionally, TQM has been identified as a philosophy of interdependent characteristics. These characteristics are:

1. *Customer Focus (Needs, Wants, and Expectations).* There is a tremendous need to identify the customer(s) (internal and external) and their needs. It is imperative that the practitioner of the TQM concept realizes that there is a customer and a supplier at every step in the process. To simplify the confusion about who is the customer or supplier think of the supplier as the one who passes on the product, service, information and/or material. On

the other hand, a customer is the one who receives the product, service, information and/or material.

2. *Commitment.* Only with a true and total involvement can positive results accrue to the organization. Management must demonstrate in no uncertain terms a commitment through a leadership style that fits the organization and is in agreement with the principles of TQM. Typical ways that management can demonstrate this commitment is to provide improvement opportunities for all employees, to delegate responsibility and authority for improving work processes to those who actually do the work and can do something about it (i.e., empower employees), and to create multidisciplinary, cross-functional work teams responsible for designing and improving products and services, processes, and systems.

3. *Measurement.* It is the responsibility of management to establish baseline measures with customers. However, in order for this to be done, management must be familiar with their own processes and/or systems. To better acquaint themselves with the process and/or the system, they must develop a system by which the processes and results are measured effectively, determine criteria for outputs of critical work, measure compliance with realistic and attainable customer requirements, and monitor corrective action and improvement.

4. *Systematic support.* For any idea, program, and/or philosophy to work appropriately, there must be a systematic supportive system in place (i.e., a good foundation). With TQM, the systematic support has to do with managing the quality process and building a quality infrastructure tied to the internal management structure. To be effective, the support system must be able to link TQM principles to existing management systems. Typical management systems are: strategic planning, performance management, reward system, communication, and so on.

5. *Continuous improvement.* Perhaps one of the most famous characteristics of the TQM philosophy is the notion of continuous improvement. It is based on the fundamental concept of measurement and more precisely the loop of the Shewhart/Deming Cycle of Plan-Do-Study-Act (Figure 4.1). To appreciate the concept of continuous improvement we all must understand that for improvement to occur we must:

- view all work as a process
- anticipate changing customer needs, wants and expectations
- make incremental improvements
- reduce cycle time
- encourage and receive gladly feedback

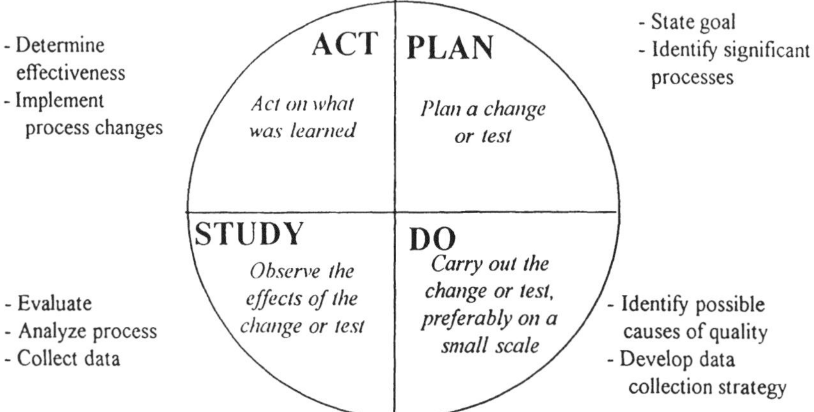

Figure 4.1 The Shewhart-Deming cycle.

IV. TRADITIONAL MODEL OF TQM

Oakland (1989) formalized the first model of quality (Figure 4.2). To be sure, the model was based on Feigenbaum's positions, which was founded on a quality system-based approach. In essence, Oakland's 1989 model placed the ISO 9000 quality system in the center and put teamwork, quality tools, and management commitment to the side. Later, this model was modified by Oakland (1993) because the emphasis of the first model was wrong. Oakland

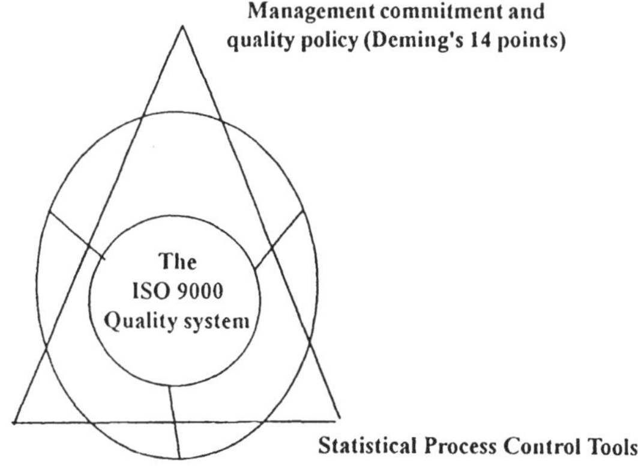

Figure 4.2 The Oakland TQM model.

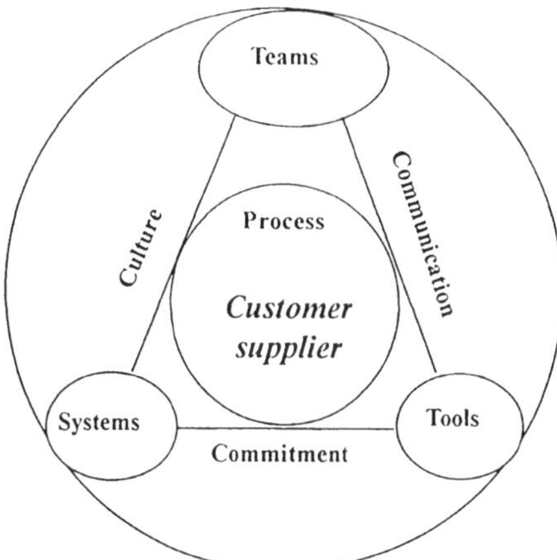

Figure 4.3 The modified Oakland model.

replaced the ISO 9000 center with "process-supplier-customer" (Figure 4.3). This change is quite significant as it mirrors a change in perceptions among quality practitioners. It also reflects a growing doubt about the role of systems in TQM. After all, the effectiveness of systems depends on how people keep to it in practice, and TQM favors formal means rather than market ends.

Witcher (1995) summarized the traditional model of TQM through eight propositions. They are:

1. TQM is a culture change program, separate from general management and organization, and associated with the attainment of excellence.
2. TQM is primarily a people-based way of working, requiring changes in the management of human resources, and where empowerment is task oriented.
3. TQM is based on the ability of employees to selfmanage quality, and total quality requires an organization-wide quality system.
4. TQM is based on customer requirements that are determined through specific process customer-supplier interactions, where continuous improvement is built into the routine of daily work.
5. TQM is a corporate philosophy that facilitates the independent use of an association of related management methodologies, particularly in the achievement of step change.
6. TQM is a way of producing a customer-focused organization, particularly where processes are reengineered around the needs of external customers and where TQM facilitates relationship marketing.

7. TQM is radically new form of managing organizations. It is changing the subjects of study for management academics, but, because of its holistic properties, it requires new ways of understanding from social science disciplines.
8. TQM, as a part of the fundamental socioeconomic change taking place in the organization of production and markets on a global basis, is an indicator of an emergent new international capitalism.

V. CONTEMPORARY MODEL OF TQM

The contemporary role of TQM focuses on three basic items: big results, fast results, and permanent results. This is shown in Table 4.1. Furthermore this contemporary view is based on the notion that quality is always evolving and moving forward. However, in order for this movement to take place, it is imperative that the organization is ready and willing to take the challenge. Figure 4.4 is a visual representation of this readiness.

As we already have seen the traditional ways of approaching contemporary quality have indeed approached their life expectancy. For all intended purposes they are indeed dying. Some of these approaches are shown in Table 4.2.

The fundamental idea of modern quality thinking is the notion of changing our paradigms from product orientation to process orientation. In fact, this notion is precisely why TQM is not dead nor will die in the future. No matter what improvements and no matter how we define TQM—even if we give it a different name—the switch is inevitable and the results can be quantifiable. Some of the benefits to the organization are shown in Table 4.3 (see also Appendix G).

VI. A GENERIC MODEL OF MODERN TQM

Total quality management is a philosophy and its fruits will materialize through major changes in the organization that will implement this philosophy. Some of the obvious changes are in:

- the way we think
- our corporate culture
- how we view and treat the customer(s)
- how we view and treat our employees
- how we view and treat our community
- how we view and treat share holder relationships

But what is TQM? In the simplest definition, when we breakdown or examine each of the words, the message is loud and clear.

Table 4.1 Contemporary Role of TQM

	Big results		Fast results	Permanent results
Focus the effort	Foster team creativity	Mobilize the organization	Control	Monitor and standardize
Strategic planning	Needs assessment	Executive champions	As needed specific training	Continuous monitoring
Focus on project selection	Benchmarking	Visionary leadership	Predict behavior of process	
Evaluate resources	Breakthrough targets Project ownership	Organizational convergent targets Efficient teams Aligning organizational incentives with organizational objectives Co-opting middle managers to actively participate	Optimization of trials and pilots	Manage by data

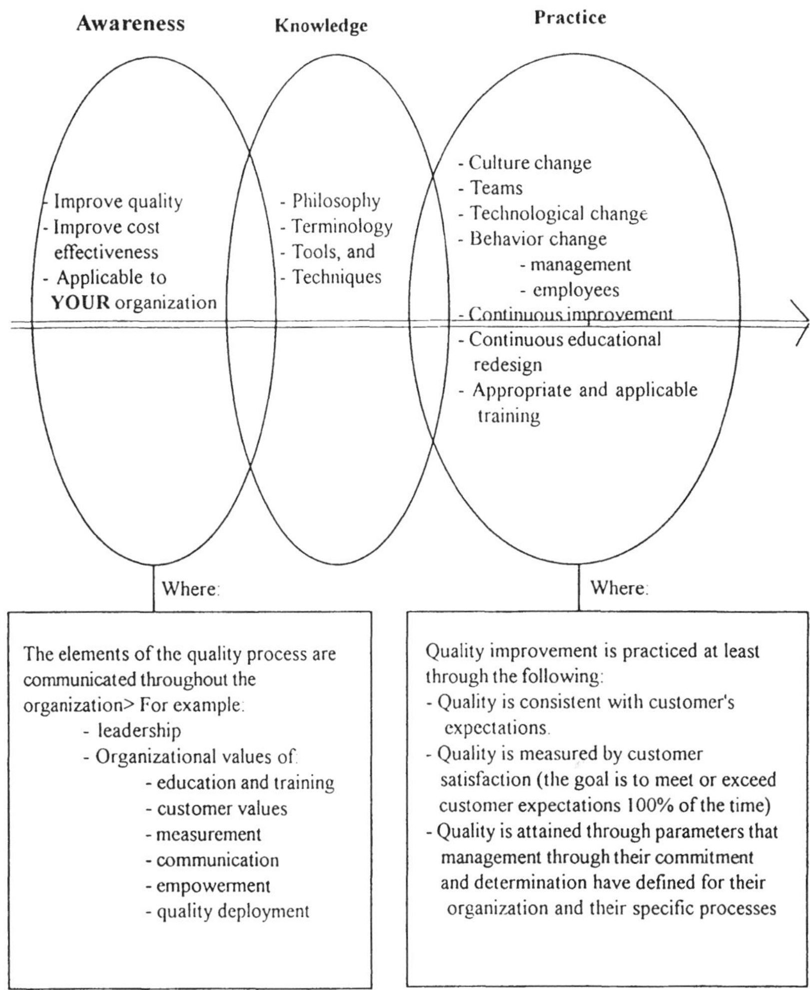

Figure 4.4 The readiness cycle of the organization for quality improvement.

Table 4.2 A Comparison Between Traditional and Modern Thinking About Quality

Traditional thinking	Total quality thinking
Quality improvements is management's responsibility	Quality improvements is everyone's responsibility
Customers are outsiders we sell to	Customers (internal & external) are vital components of our organization
"Good enough" is good enough	Nothing less than 100% effort will do
We need better people to improve our quality	We already have the best people for the job
Vendors & suppliers are our adversaries	Vendors and suppliers are important members of our team
Quality comes from inspection, rejection, and rework	Quality is built into products and services from the start
"If it ain't broke, don't fix it"	"If it ain't broke, improve it." Continuous improvement is the only way
Quality improvement is expensive and labor intensive	Quality improvement reduces cost and increases productivity
Target driven	Vision driven
Centralize	Decentralize
Control employees	Empower employees
Management	Leadership
Left brain	Whole brain
Big is always better	Small, adaptive, flexible systems
Efficiency centered	Effectiveness centered
Planned obsolescence	Recycling
People to fit jobs	Jobs to fit people
Specialization	Multidisciplinary
Management oriented	People oriented
Lose/win	Win/win
Stability at all costs	Harnessing change and risk taking
Advertisement driven	Quality driven
Keep your costs down	Take pride in quality
Imposed goals	Self-determination
Talking	Active listening
Fragmentation	Integration
Compartmentalization	Crossfertilization

Table 4.3 Benefits of an Improved Process

Drains	Traditional process	Improved process
Emphasis on push, without sufficient pull	Needs of customer not recognized Production at all costs was the objective	Focus on strategy of organization and appropriateness of current technology transfer for the product
Customer not taken into account	Engineer is the king Specialists within the organization are very important as consultants	Customer dependent!!!
Dependence on blind discovery as a success strategy	Emphasis on singular design Blind improvement without regard to effectiveness	Emphasis on team concept and consensus Focus on commitment, improvement and breakthrough
Very sensitive about product quality	Emphasis on aesthetics, good impressions and good demonstrations	Emphasis on optimization Legitimate "good" quality Emphasis on prevention of problems
Pretend designs	Design not production oriented, but engineering oriented Emphasis on "newness" rather than superior design	Emphasis on customer needs, wants and expectations Designs are customer oriented
Hardware gadgets	Focus on overlapped prototype iterations without regard to "real" improvement	Minimize prototype iterations with appropriate experimentation, so that maximum contribution to optimization of the process may occur

Table 4.3 (Continued)

Drains	Traditional process	Improved process
Product without appropriate facilities	Focus on product development and market penetration	Focus on "total" development process (with manufacturing and other department inputs)
	Manufacturing reacting to product development	Emphasize process capability
Focus on past practices and past solutions to current problems	Emphasize the status quo	Focus on process parameters with the intent of maximizing their use
		Design quality into the process
		Reduce cycle time
Appraisal	Focus on inspection	Focus on optimizing process to perform its best
	Emphasis on human error	Always emphasize prevention
Goal is based on targets	Lack of teamwork	Teamwork
	Focus on fear and intimidation	Focus on openness, communication and personal motivation
		Emphasis on managing problems

Total implies at every level in the company, each and every day, in every department and support group.

Quality continuous improvement in order to meet and/or exceed internal or external customer requirements and expectations.

Management by establishing systems and environments that support a continuous improvement culture.

TQM then is a strategic, integrated management system which involves all managers and employees and uses quantitative methods to continuously improve an organization's processes to meet and exceed customer needs, wants, and expectations. There are five assumptions required to implement TQM (Figure 4.5).

A. Customer Focus

The customer must be identified—whether that customer is internal or external—and their needs must be understood. To do this, customer requirements must be established and some kind of compliance must be in place. Also, there must be a partnering relationship with key customers and suppliers, so that appropriate and applicable operationalization of the requirements will take place.

Note that at every step in the process there is a customer and a supplier. A supplier is defined as the one who passes on the customer, patient, information, material, and so on. On the other hand, a customer is one who receives the customer, patient, information, material, and so on. Depending on the task and the level of the organization, is not unusual to see a customer treated as a supplier and a supplier treated as a customer. For more on this point see Stamatis (1996a).

B. Total Involvement (Commitment)

Management must demonstrate the commitment and leadership, through real opportunities for quality improvement for all employees, by appropriate delegation and empowerment for improving their own work environment through creation of multidiscipline and cross-functional teams and/or self-directed teams.

C. Measurement

Management must establish a baseline measure with customers and develop the appropriate process and results measure. To develop this, the proper input and output criteria must be identified, so that a relationship of congruence is established between customer requirements and process variables. If there is no congruence of the requirements and the process, correction and improvement are in order.

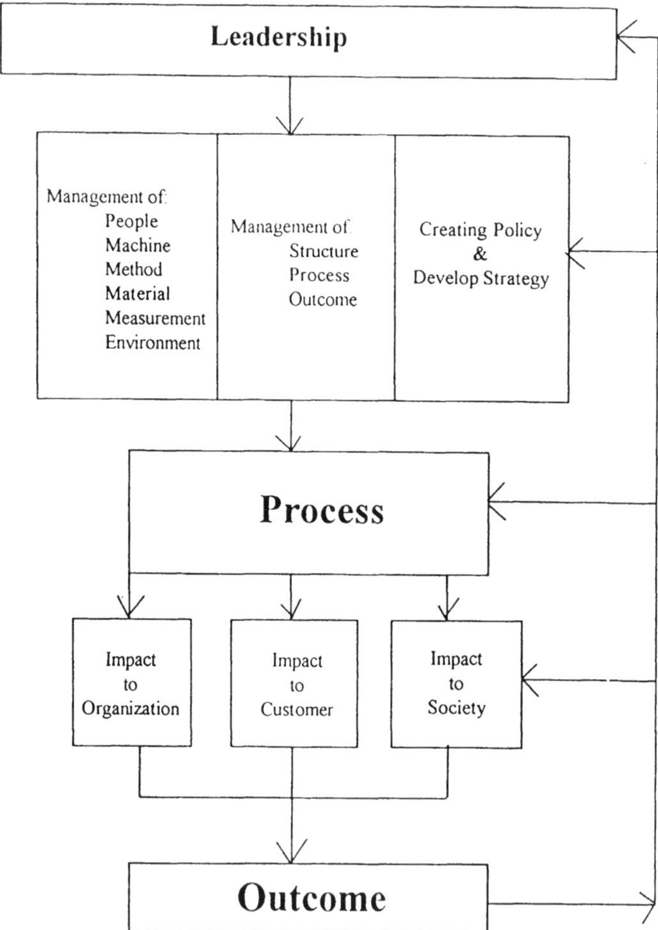

Figure 4.5 A generic modern TQM model.

D. Systematic Support

Management is responsible for managing the quality process. To do that, it must do two things. First, it must build a quality infrastructure tied to the internal management structure and second, it must link quality to existing management systems (unless proven unworthy of continuing the upkeep of such systems).

E. Continuous Improvement

Perhaps one of the most fundamental issues of TQM and management is the fact that management must recognize and preach that all work must be viewed as a process. As such, it is management's responsibility to anticipate changes as well as the ramifications to the customer, employee, and society. The changes should be either in a Kaizen (incremental) or reengineering (drastic) format, as found appropriate in the given organization. To be successful at this, the focus is on reducing cycle time (in all processes) and encouraging appropriate and timely feedback.

VII. LEADERSHIP

The foundation for TQM is indeed leadership and leadership has several functions.

A. Create a Vision and Establish a Mission for the Organization

The root of the improvement problem is the leadership skills. Most of us suffer from the "implementation-assumption trap." We assume that if we simply install new structures, policies, slogans, technology, management systems, products/services, or improvement programs (e.g., quality teams, statistical process control, etc.), they will improve performance. We confuse education with training, inspiration with habits, intention with behavior, knowledge with skills, and project results with culture change. The gap between assumptions and execution can be filled only by strong leadership skills. The two essential elements of leadership are developing a compelling vision and (more importantly) living it. It is the job of a leader to instill confidence and build self-esteem of every employee.

B. Set Realistic Quality Improvement Goals

The foundation for improvement is laid only when the people who do the work learn to solve their own problems instead of passing the buck. To "get" that ownership, however, the leader must set goals that are realistic (i.e., attainable, doable, and within the constraints of both the individuals involved and the organization).

C. Identify Customer Requirements

Just providing a product and/or service is not good enough anymore. What good leaders do for their organization is to find ways to surprise and delight the customer with their products and/or their service.

D. Lead the Culture Change to a Team Environment

For change to occur there must be a reason. That reason has to have a meaning that everyone in the organization must be familiar with and understand it. As a general rule that "meaning" is provided through the strategic vision. The key to motivate everyone in the organization for a culture change, is to ensure everyone's involvement in the process. How do we do this? By understanding the three most basic needs. The most compelling and empowering vision is the hope of a better tomorrow and even better, of being number one, the best in the class tomorrow, or working for a company that will be number one tomorrow. The second need is fulfilled by giving them a sense of ownership and empowering them to make decisions. Positive reinforcements, preferably nonmonetary, are required to fulfil the third need. Quality programs that do not ensure satisfaction of these three needs shall not have their employees on board and are therefore doomed from the start.

E. Active Participation On Strategic Planning and/or Quality Committees

To unleash the creativity of the organization, leaders must change the organizational culture and structure to encourage innovation and creativity. They must actively participate on everything that has to do with quality and/or improvement. It is the only way to get ahead of the competition. By being participative, leaders encourage others to do the same. As a consequence, the result of this active participation will be more diverse ideas, more discussion of ideas, more ownership of ideas, and ultimately more trust for exploring and daring new thoughts, ideas and so on.

F. Lead the Improvement Effort

Leaders must:

- provide the opportunity to form teams

- recognize the need for both education and training and provide both to the employees
- empower the employees to resolve issues affecting and effecting their own work
- encourage the employees to participate without fear
- listen and then listen some more
- provide timely feedback
- support, direct, coach the employees
- communicate with everyone in the organization (vertically and horizontally)
- apply the Plan-Do-Check (Study)-Act model as much as possible in everything within the organization
- reward and recognize the employees for teamwork, improvement, performance, and contribution to quality

Once the leadership has been defined, then policy, resources, and general management will guide the process; this, of course, is the real transformation of the input to the results based on the definitions of satisfaction of society, customer and the organization.

VIII. A WARNING ON THE ISSUE OF CONTINUOUS IMPROVEMENT

To be sure, there are some problems with the concept of continuous improvement. One of the most fundamental problems is the expectation of continuous improvement. Given today's shortened product life cycles, improving last year's model may hurt you. None of us are opposed to quality per se. We recognize that a good product is often the first requirement in creating a successful marketing campaign. However, to what extent will committing resources to producing the better version of an existing product lead to lasting marketplace gains? Sometimes it seems that we pursue improvement even though the product has no longer any life left. The pace of today's markets is so furious, companies often have more to loose by improving the current product line. Resources, both human and capital, are better spent creating entirely new technologies that produce wholly new products. For example, consider where the automotive industry would be today if we still used carburetors instead of fuel injection, or where the computer industry would have been today if the pioneers in that industry had waited for a perfect 286 PC, or the 386 PC, or the 486 PC, or the pentium and so on;. Consumers once marveled at the beauty of a 300 dpi laser printer until 600 dpi replaced it with remarkable gray-scale clarity. But before consumers could upgrade, color printers at 720 dpi replaced them. And so it goes. The changes in technology are so rapid that they call into question the idea that industries have time to improve existing product lines ad infinitum. In many cases, research and development of new technology should and must take precedence over improving existing models. For example, the translucent technology of LCD overhead projector panels has improved, but industry would be wiser to move forward on new mirror-based technology that renders the former version obsolete. Also, industry should discontinue improvements to the computer's mouse and move toward the next generation of computers that open files and applications with voice-activated commands.

The point to all this is the fact that life cycles have become so short that continuous improvement is increasingly irrelevant for many products and industries. Also, achieving bug-free products is always just out of reach, and you do not have time to look back; the competition is gaining on you.

Be that as it may, we still believe that Total Quality Management has a lot to offer even for the most progressive, most aggressive, and most technological advanced industries because they are process oriented (see Table 4.3). Just because a particular item may be better off with a totally new design, that does not mean the other variables of the TQM have to be abandoned. It will be truly a shame if TQM is cast out from the growth strategy of any organization just because is still has value.

IX. APPLICATION OF THE MODEL

As we already have mentioned, TQM is doing the right thing, doing it right the first time, on time and all the time; always striving for improvement, and always satisfying the customer. This requires a focus on customer needs, people, system, and processes, and a supportive cultural environment.

The model may be used for the following general as well as specific processes throughout the organization.

A. Quality Planning

- identifying target market segments
- determining specific customer needs, wants and expectations
- translating the customer needs into service and process requirements
- designing services and processes with the required characteristics. (Competitive benchmarking can assist in this part of the process.)

B. Quality Assurance

- planning for quality

C. Quality Control

- measuring actual quality performance versus the design goals
- diagnosing the cause of poor quality and initiating the required corrective steps
- establishing controls to maintain the gains

D. Quality Improvement

- establishing a benchmarking process
- organizing a quality function deployment project
- providing the necessary resources

If the model is to be used in an internal environment (i.e., a specific process), it is imperative that the selected process is:

- strategic in nature and very proactive
- competitively focused on meeting customer needs as opposed to techniques of analysis
- comprehensive in terms of level and functions
- managed in terms of quality, not defect reduction

X. IMPLEMENTATION STRATEGY

There is no one right way to implement TQM successfully in an organization. In fact, over the years I personally have been instrumental in implementing TQM in both manufacturing and service organizations using different methodologies (Stamatis, 1991, 1995, 1996, 1996a–c, 1997). Due to the variation of cultures each organization is quite unique and as such, there is no guaranteed recipe for success.

The implementation strategy proposed in this book is a synthesis of approaches used successfully by organizations all across industries. It is offered only as a guide in developing strategies and associated plans to carry out these strategies. The intent of this approach is to demonstrate a flexible method that an organization can use to mobilize its strong points and capitalize all available energy, so that it is focused on key improvement opportunities.

1. Recognize at the beginningthat your organization is unique. As a consequence, do not borrow someone else's experience and try to fit into it. It would not work even though the organization that you borrowed the plan from was very successful with it.
2. Conduct a needs assessment. Unless you know where you currently are, there is no way that you can measure your progress, or for that matter develop improvement plans. The needs assessment serves as

the springboard for your organization to develop a strategy for improvement. It will do that by identifying those vital processes to be targeted for change and it will provide a baseline measurement for judging progress.

3. Start small. The best plans are those that result in action—action which improves the processes of the organization and results in better services and products for the customer. A simple plan that generates action and gets results is better than an elaborate plan that collects dust. Some initial TQM actions might consist of specific projects designed to address system-wide problems which have potential for expanding to other processes of the organization; or, they might be efforts to implement TQM in one or more organizational components. Examples of such efforts might include:
   - conduct of customer identification, survey, and feedback efforts to be reflected in quality and timeliness indicators
   - designate quality teams to address specific operating problems
   - conduct of organizational assessment, leadership development, and group dynamics efforts
   - direct involvement of some line personnel (e.g., nonmanagers) in implementing some form of quality improvement effort reflected in the overall strategic plan
4. Identify your customers and their requirements; then review areas or quality indicators or both where TQM will be implemented.
5. Make a plan. The more specific your plans are, the more likely your organization will be successful.

A synthesized implementation process based on these five steps is: gain top management commitment, determine an organization's readiness, create a vision and guiding principles, set up a top management "quality council," and, if applicable, communicate the vision of TQM with union representatives.

XI. THE ROLE OF THE CONSULTANT

Some of us may have had and/or heard of horror stories about consultants, especially in recent years now that so many people have been downsized into "consulting" rather than choosing it voluntarily. This is a shame, because consultants receive a bum rap; they are indeed invaluable. Those consultants who do have the knowledge and experience to contribute in the organization that "extra" method, approach, knowledge, or whatever, can make a difference. By definition, consultants are hired to give advice in broad general terms rather than to strut their "stuff" with the instruments of their profession.

Most consultants enjoyed some measure of success early in their careers. The reputations they gained from those early triumphs enabled them to survive professionally. Times change, however, and so too do the formulae for achieving results. At least from my perspective, too many consultants tend to apply the same solution to every problem—they tend to rest on their laurels. That is most, not all, and that represents just my own experience. In fact, if I was to quantify it, I would say that 70% of the consultants are sorely lacking in both knowledge, experience, and/or integrity; however, the remaining 30% are worth their weight in gold (and charge that way, too).

How do you connect with someone who is good? You must be resourceful and do an enormous amount of homework. Get and check references of previous clients, particularly recent clients. Ask specific questions. You need to know exactly what services were performed and what results were achieved. Agreements prevent disagreements, so once you have made your choice, enter into a written contract with your consultant. Make sure that you can terminate it at any time, no questions asked, with a fixed price to exit. Also, make sure that all the work and development for a project that the consultant did under your contract, is your property.

Is it mandatory to use a consultant? This question is very difficult to answer because it depends on many variables. However, strictly speaking, the answer is a categorical "no." You do not need a consultant to implement a quality system. On the other hand, a consultant with knowledge and experience will help you get to your destination a lot faster and help you avoid mistakes. When you hire a consultant, make sure that their services are to coach and direct you in becoming self-sufficient in the task(s) that you need the help. Do not fall in the trap of allowing the consultant to do it for you. If they do it for you, not only is more expensive, but you did not learn anything from the experience.

REFERENCES

Bemowski, K. (1995). TQM: Flimsy footing or firm foundation? *Quality Progress*. July: 27–30.

Cottrell, J. (1992). Favorable recipe. *The TQM Magazine*. Feb.: 17–20.

Dobbins, R. D. (1995). A failure of methods, not philosophy. *Quality Progress*. July: 31–34.

Duff, R. (1995). Why TQM fails - and what companies can do about it. *Quality Digest*. Feb.: 50–54.

Fellers, G. (1995). *Why things go wrong*. Pelican. Gretna, LA.

Keck, P. (1995). Why quality fails. *Quality Digest*. Nov.: 53–55.

Oakland, J. S. (1989). *Total quality management*. Butterworth-Heinemann. Oxford.

Oakland, J. S. (1993). *Total quality management: The route to improving performance*. 2nd. ed. Butterworth-Heinemann. Oxford.

Stamatis, D. H. (1991). Total quality management implementation. In: *Quality Concepts: A national forum on total quality management.* Conference proceedings. The Engineering society, ASQC-Automotive Div. October 14–16, 1991. Detroit, MI.

Stamatis, D. H. (1995). *Integrating QS-9000 with your automotive quality system.* Quality Press. Milwaukee, WI.

Stamatis, D. H. (1996a). *Total quality service.* St. Lucie Press. Delray Beach, FL.

Stamatis, D. H. (1996b). *Understanding ISO and implementing the basics to quality.* Marcel Dekker. New York.

Stamatis, D. H. (1996c). *Integrating QS-9000 with your automotive quality system.* 2nd ed. Quality Press. Milwaukee, WI.

Stamatis, D. H. (1997). *The nuts and bolts of reengineering.* Paton Press. Red Bluff, CA.

Whyte, J. and Witcher, B. (1992). *The adoption of total quality management in Northern England.* Occasional Paper. University of Durham. Durham, U.K.

Wilkinson, A., Marchington, M., and Dale, B. (1993). Enhancing the contribution of the human resource function to quality improvement. *Quality Management Journal.* 1: 35–46.

Witcher, B. (1993). *The adoption of total quality management in Scotland.* University of Durham. Durham, U.K.

Witcher, B. (1995). The changing scale of total quality management. *Quality Management Journal.* Summer: 9–29.

5

Tools of Quality

In any endeavor, there are specific tools utilized to accomplish the defined tasks. Quality implementation is no different. In fact, the process and the tools of implementing quality have become quite generic. Much has been written in both areas (the process and the tools). However, our goal here is not so much to be redundant with other sources but to offer a cursory systematic review of the basic tools and their application. To accomplish this task, a review of each of the basic tools will be given, rather than a detailed discussion. Our focus is to give the basic understanding and approach of each of the tools discussed. The reader is encouraged to see Montgomery (1985), Gitlow et al. (1989), Ishikawa (1982), Grant and Leavenworth (1980), Mizunu (1988), Gulezian (1991), Brassard and Ritter (1988, 1994), and others for detailed explanations and descriptions.

I. APPLICATIONS OF QUALITY IMPROVEMENT TOOLS

To know how to apply the basic tools for quality improvement we must first understand the steps of problem solving. To be sure, there is no specific

Table 5.1 A Problem-Solving Approach

Category	Basic steps	Tools used
The problem	1. List and prioritize problem(s)	Data collection
	2. Define project and team	Process flow diagram
Research of problem	1. Analyze symptoms	Flow diagrams, data collection, pareto analysis, brainstorming
	2. Set hypothesis	Brainstorming, cause-and-effect diagram
	3. Test hypothesis	Brainstorming, data collection, graphs, flow diagrams, Pareto analysis, scatter diagrams, control charts
	4. Get to the root cause	Data collection, flow diagrams, graphs, Pareto analysis, scatter diagrams, control charts
Fixing the problem	1. Alternative solutions	Brainstorming, cause-and-effect diagram, flow diagram
	2. Define solutions and controls	Data collection. flow diagram, graphs, scatter diagrams, control charts
	3. Plan for resistance to change	Brainstorming
	4. Implement solutions and controls	Flow diagrams
Monitor the process	1. Monitor performance	Data collection, graphs and charts, Pareto analysis, histograms, control charts
	2. Monitor control system	Data collection, graphs and charts, control charts

number of steps. However, all problem solving approaches have some basic criteria as their core approach. Some have defined it as the scientific method: 1) set a hypothesis, 2) test the hypothesis, 3) analyze the results, 4) reject/accept the hypothesis); others have defined it as an eight-step approach (e.g., Stamatis, 1995; Ford Motor Co., 1987).

To demonstrate that a problem solving approach is not unique, we offer the following generic variation. It is based on four basic steps, each step with several substeps. The overall approach is a twelve-step methodology shown in Table 5.1 and in no way is it better or worse than any other method-it is simply different.

II. SUMMARY OF THE BASIC TOOLS

A. Brainstorming

This is a group decision-making technique designed to generate a large number of creative ideas through an interactive process. Brainstorming can be used to get the ideas of the team and or group organized into a process flow diagram or a cause-and-effect diagram.

Questions To Use During Brainstorming Sessions

Often the question regarding brainstorming is: "How do we begin?" The following basic four questions pose a good beginning.

1. What are the organization's three most important unsolved and or recurrent quality problems as you see them?
2. What kind of action plan is needed to solve these problems?
3. Which areas are most in need of such action? (Try to be as specific as possible.)
4. What are some major obstacles in the way of improving our quality?

Even though this is a creative session we must follow specific rules:

1. Clearly define the goal of the brainstorming session. This is very important. If the definition is not clear, then the participants will be discussing different problems.
2. Try to have *all* people give their ideas in three words or less.
3. Write down *all* ideas. Some may seem silly, but they may lead in an idea that could help solve the problem.
4. Generate a large number of ideas—quantity, not quality is important here. The greater the idea count the better! Combinations of ideas are alright, too.
5. Do not allow one or two individuals to dominate the session.
6. Make the ideas visible so that everyone can easily see them; use an overhead projector or a flip chart to list the ideas.

Brainstorming Procedures

In addition to the rules, there are appropriate procedures that must be followed in order for an optimum resolution. These procedures are divided into two categories which are:

1. Creating the ideas
   a. The leader asks each person to list his/her ideas on a piece of paper.
   b. The leader then solicits everyone's contribution. This is done by going from person to person in rotation. If a member does

not have an idea on a particular round, he/she simply says "I pass." However, he/she is asked on every round.

c. As ideas are given they should be listed on a transparency or a flip chart.
d. Omit only ideas that have been listed and that have already been contributed by someone else.
e. After each person has contributed their individual list, the leader asks for any additional ideas which may have been generated during the contribution process.
f. A final point. The human mind thinks at two different levels, the conscious and the unconscious. After a good brainstorming session everyone's mind will be spinning with all new, wild ideas. The unconscious mind will continue brainstorming even after you have stopped thinking about it at a conscious level. Thus the team should continue the brainstorming after an incubation period has passed.

2. Evaluating the ideas that were contributed
   a. Each of the ideas should be evaluated. Some ideas will be scrapped right away without a detailed discussion.
   b. The evaluation should be pointed towards the idea and never towards the person who suggested it.

B. Process Flow Diagram

One of the most often used and powerful tools in the quality improvement process. A process flow diagram is a road map of the process (Figure 5.1). Specifically, a process flow diagram illustrates and or clarifies events/tasks in a process and the events/tasks between them. It assists in highlighting: 1) the present situation, 2) the differences between what should/is thought to be happening and the actual situation, 3) the proposed situation, and 4) any potential problem areas. It can be used to facilitate effectiveness during brainstorming, constructing a cause-and-effect diagram, and in every other situation where there is an ambivalence about what the present state is. To construct a process flow chart, the following are important.

1. Assemble all the appropriate people.
2. Define the process and its boundaries.
3. Brainstorm the process.
4. Use the simplest symbols possible (i.e., o, operations; D, delay, ——→, flow; ∇, inventory; □, inspection). Decisions are usually shown as diamonds or o since they are activities (operations).
5. Draw the steps the process actually follows.
6. Make sure each feedback loop is accounted for.

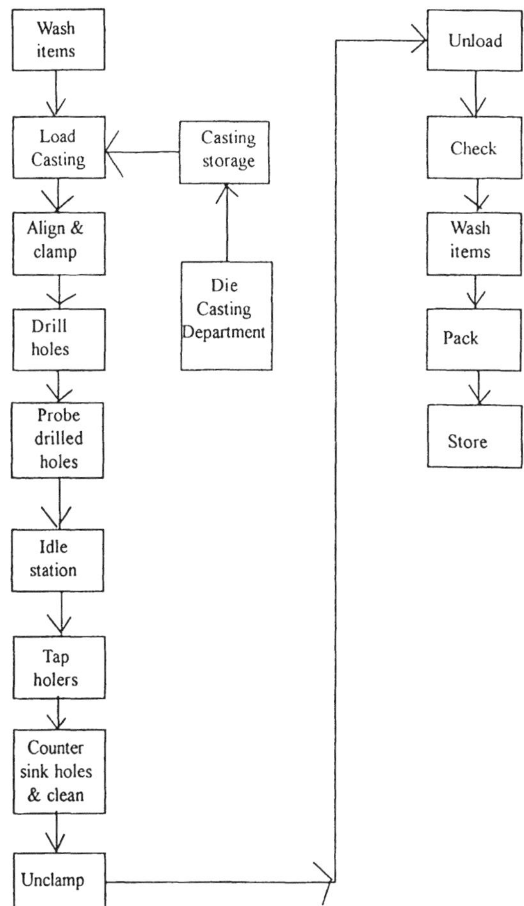

Figure 5.1 A typical process flow chart.

7. Draw the steps the process should follow.
8. Compare the two, and make the appropriate adjustments.

C. Check Sheet

Check sheets are forms that guide the experimenter in categorizing the data as they are being collected (Figure 5.2). Its construction is very simple and is as follows:

1. Agree on the item being observed.

Reason	Monday	Tuesday	Wednesday	Thursday	Friday	Total
Money	𝍸	||	|	𝍸	𝍸||	20
Sex	||	||	||	||	||	10
Children	𝍸	||	𝍸||	|	||||	19
Total	12	6	10	8	13	49

Figure 5.2 A typical check sheet.

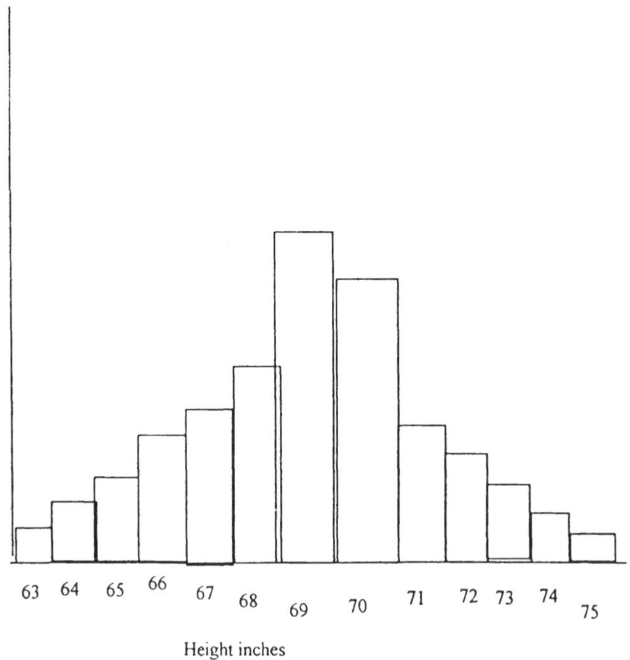

Figure 5.3 A typical histogram.

2. Decide on a time period to collect data.
3. Design a form that is clear and easy to understand.
4. Collect the data honestly and consistently.

D. Histogram

This is a bar graph that gives a historical and pictorial representation of a set of data (Figure 5.3). It is used to determine:

- the shape of a series of data values
- the readiness of a process to undergo a capability study
- the dispersion and central tendency of the data (quick analysis)
- the relationship of machines, customers, suppliers, etc. (quick analysis)

To construct a histogram,

1. Determine how many data values to use.
2. Determine the width of the data by computing the range.
3. Select the number of cells for the histogram.
4. Determine the width of each cell.
5. Determine the starting number for the first interval.
6. Calculate the intervals.
7. Assign data values to the appropriate intervals.
8. Construct the histogram by drawing bars to represent the cell frequencies.

E. Pareto Chart

The Pareto analysis is a chart based on the Pareto principle (Figure 5.4). This principle was named for an Italian Economist who, in the late 1800s, found that most of the wealth in Italy was owned by a few of the people. Today we find that 80% of our problems in the hospital or the office can be traced to 20% of the causes. The Pareto analysis is a process of ranking opportunities so we can determine which should be pursued first. In essence, the Pareto analysis allows us to examine the vital few from the trivial many.

Pareto analysis should be used at various stages in a quality improvement program to determine which step to take next. Pareto analysis is used to answer such questions as "What department should have the next team?" or "On what problem should we concentrate our efforts?" The chart itself is a graph that ranks factors in descending order of frequency or magnitude from left to right in the manner shown in Figure 5.4.

The following steps are necessary to construct a Pareto chart:

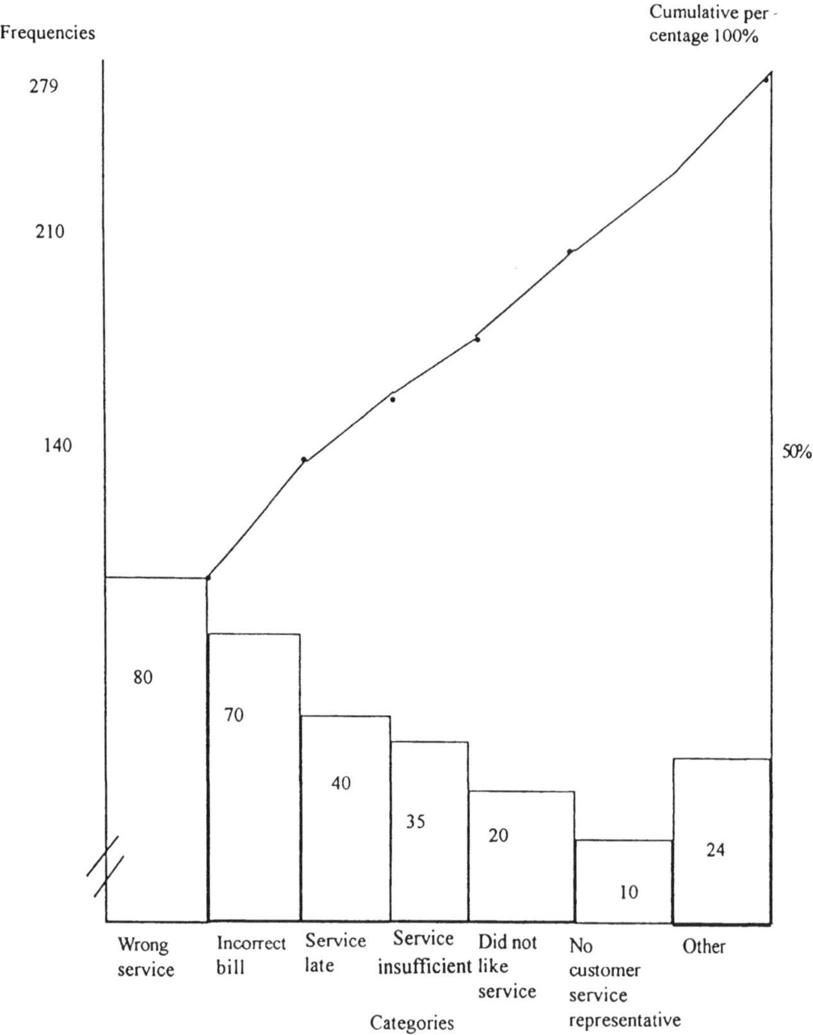

Figure 5.4 A typical Pareto diagram.

1. Decide the appropriate time interval.
2. Decide the number of classifications.
3. Decide the total number of occurrences for each category using primary or secondary data.
4. Calculate category percentage.
5. Rank categories in descending order.
6. Calculate the cumulative percentage.
7. Construct a Pareto chart for magnitudes and cumulative percentages.

F. Cause-and-Effect Diagram (Fishbone Diagram or the Ishikawa Diagram)

This diagram is a method of organizing information about a problem or a goal (Figure 5.5). It is a visual representation of what people think are the causes of a particular problem. In its simplest form it is a method used to help decide what to do to achieve a goal; it is based on the relationship among the variables of: manpower, machine, method, material, measurement, and environment. With the analysis of the relationships between these variables an isolated cause and effect resolution is identified.

There are three basic rules in constructing a cause-and-effect diagram:

1. Start with a clear definition of the problem or goal. Put this in a box on the right side of the paper.
2. Draw arrows using the “five Ms and E” (i.e., manpower, machine, method, material, measurement, and environment). Other labels may be used if appropriate and applicable.

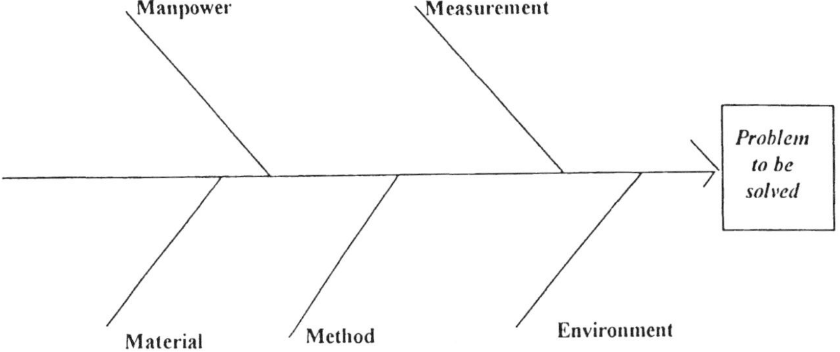

Figure 5.5 The skeleton of a cause-and-effect diagram.

3. Begin to fill in all the things that you think will cause the problem or help to achieve the goal.

A typical cause-and-effect diagram looks like Figure 5.5. Because this diagram is very popular in the quality improvement process, let us also address some issues in the construction process.

1. A cause-and-effect diagram may be constructed at the same time as the brainstorming is being conducted, or after a process flow diagram is completed, or even independently from either one of them.
2. Write down in the applicable category everything people suggest. Do not judge.
3. Remember, the information is important, *not* the form.
4. Where causes are put in the diagram is not important at first. They can go under any of the headings or even under multiple headings.
5. There is nothing wrong with placing the same cause in different places on the diagram, if people want it that way. If you find that a cause is recurring under different headings, it may be an indication of obvious trouble.
6. It is important for all persons to help in adding causes to the diagram.
7. If the diagram is not finished at the end of the meeting, you or others can add to it later.
8. Some companies post cause-and-effect diagrams so that their employees can add to them. They serve as displays rather than tools to identify problems and improvements.

G. Dispersion Analysis Cause-and-Effect Diagram

A major disadvantage of the cause-and-effect (C&E) diagram is that many major causes for a quality problem could appear on a single branch. This makes the diagram difficult to develop and use. A slightly different C&E diagram is a dispersion analysis diagram where all the major sources of variability are listed as branches (Figure 5.6). (Quite often the cause-and-effect diagram is used first and then a dispersion analysis diagram is used to identify and expand on the major causes.)

In choosing to use a C&E diagram versus other types of diagram, consider the following:

Advantages. 1) There is a clear grouping of potential causes which enables effective later analysis, and 2) the resulting diagram is not too complex.

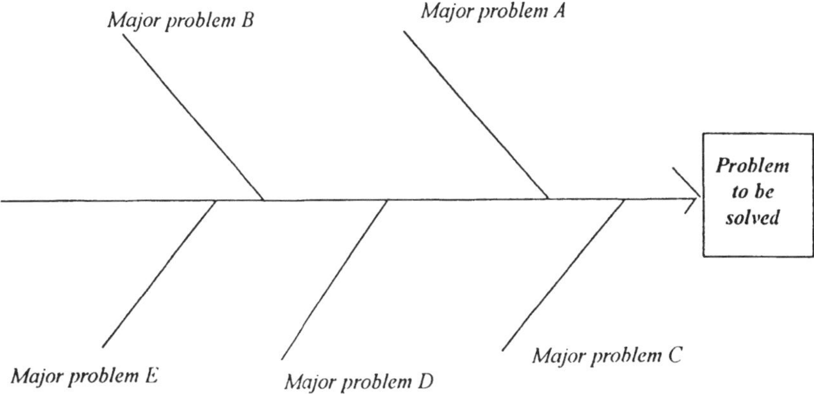

Figure 5.6 The skeleton of a dispersion analysis cause-and-effect diagram.

Disadvantages. 1) Major causes can be easily overlooked, if the 5M&E method is not used first; 2) it is sometimes difficult to classify major causes; and 3) the greatest knowledge of potential causes is required.

In essence then the construction question is: "Why is there variability in major cause (#) that could cause a quality problem (X)?"

There are many ways that the C&E diagram may be used. For an effective and efficient use in the workplace the following seven rules should be followed:

1. Develop a process flow diagram for the part of the process you want to improve.
2. Isolate and clearly define the problem to be solved.
3. Use the brainstorming technique to find all possible causes of the quality problem.
4. Organize the brainstorming results into logical categories.
5. Construct the appropriate cause-and-effect diagram that clearly and accurately displays the relationship of all the data in each category.
6. Explore and implement solutions to assess improvement.
7. Use statistical methods to assess improvement.

A C&E diagram is used to identify and categorize problems. Therefore, the person who is using it must be cognizant of some basic information as far as selecting a solution is concerned. The information is gained through a series of questions, and the following questions may serve as samples.

Will the solution solve our problem and is there information existing to improve this?

How long is it going to take to make the necessary changes and to see some signs of reducing or eliminating the problem?
What monetary investment will be needed?
What quality improvements will be achieved?
How many people/jobs are affected by the quality improvement solution?
How will the improvement affect people? Will they need to make changes? Will they choose to accept the changes?
What technical changes are needed? Will these changes have an impact on the 5M&E?
How much change can the company absorb at one time? How quickly can the changes be made and who is needed to help make the changes?

In addition to these questions, the team must use some rules for selecting a solution and then move to the approval process. The rules are indeed very basic, however, and are very important in the process. They provide consistency as well as a systematic approach to the C&E process. On the other hand, the approval process provides sort of a checklist of how the solution is approved and then implemented. To monitor the process, an action plan is used. A typical action plan is shown in Figure 5.7.

The Basic Rules

- Decide your goal (what you want to accomplish) and the method you are going to use to solve the problem.
- Be sure to let all the people who are affected by the problem be involved in selecting and planning its solution.
- Evaluate the costs, benefits, and time involved to implement the solution.
- Decide the steps that are necessary to implement the solution.
- Create a formal presentation for choosing the problem and your suggested solution(s).

The Approval Process

- Introducte what you are going to talk about.
- Clearly define the problem.
- Show the importance of the problem and its related costs (use evidence).
- Clearly state your proposed solution.
- Talk about the benefits of solving this problem (use evidence).
- State what is needed to solve the problem.
- Prepare a checklist to cover anticipated questions, objections, etc.

CONTROL PLAN *Proces:*____________ *Date:*____________				
Significant Characteristics	Evaluation Method	Frequency	Analysis of Result	Reaction and or Responsibility of Action

Figure 5.7 An example of an action plan form.

Implementing the Solution

- Clarify all tasks to be completed.
- Define the order of completion of the tasks.
- List the resources needed to complete each task. Who does what? How long will it take?
- Give the time frame for completion of task.
- Assign responsibility for each task.
- Assign how results will be measured and monitored.
- Set up procedures for measuring results of each task.
- Evaluate the effectiveness of solution.

Team Responsibilities for an Effective Solution

- Do not take anything for granted.
- Do not assume andything—ask questions.

- Tell each person what they are responsible for doing.
- Ensure that everyone works to complete and follow through with the action plan.
- Make sure completion of assignments are monitored.
- Remove any barriers to successfully implementing the solution.
- Be sure everyone affected by the solution is in the final process of implementation.

H. Scatter Plot

If one is interested in finding out if there is a relationship between variables, then the scatter plot is a useful tool for finding that relationship (Figure 5.8). Regardless of what the relationship is, the scatter plot *cannot* prove that one variable causes the other. What it shows in a graphical form is that a relationship exists and how strong that relationship is.

One of the shortcomings of the scatter plot is that it is not always possible to describe a relationship with a mathematical formula (i.e., it is a pictorial relationship); however, statistics can numerically describe the rela-

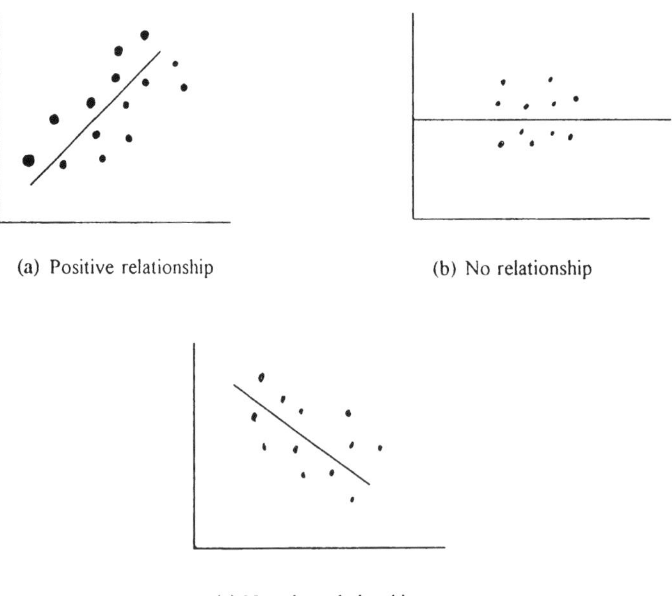

Figure 5.8 Scatter plot. (a) Positive relationship. (b) No relationship. (c) Negative relationship.

tionship between two sets of data. The methods used to do this are too difficult to present here, but we can gain an idea of the end result of the methods. A relationship between sets of data is called a correlation.

To construct a scatter plot:

1. Collect at least 50 paired samples of data.
2. Draw horizontal and vertical axes, making sure that the values on each axes become higher as you move away from the origin (as you move up and to the right in the vertical and horizontal axes respectively.) As a general rule the expected "cause" is the horizontal axes, where the "effect" is the vertical axes.
3. Plot the data. A typical scatter plot is shown in Figure 5.8.

I. Statistical Process Control

Statistical process control (SPC) is a bundle of techniques that identify random (common) causes versus identifiable (special) causes in a process. Both of these are potential sources for improvement. The amount of random variation effects the capability of a machine to produce within a desired range of dimensions. Hence, SPC could be performed to determine processing capabilities and how to achieve those levels. The determination and correction of recurring systematic changes is also a possibility.

The reduction of the random variation or the uncertainty of a process and the identification and correction of special causes are critical aspects of the total quality management process. Correction often requires a change in the total process and quite often multiple changes at the same time.

The first step of process improvement is to control the environment and the components of the system so that variations are within natural predictable limits. The second step is to reduce the underlying variation of the process. The undertaking of both of these are the issue of control charting.

Control Chart

A control chart is a pictorial representation of the process variation over time. A control chart identifies the changes in the process and unless the characteristic of the process is known, the change cannot be determined. All control charts are fundamentally based on the normal distribution, with statistical limits. The limits (upper and lower) are calculated and are drawn on either side of the process average. Do not confuse limits with specifications. Limits are calculated based on data from the process. Specifications are given by the customer and are the requirements we are expected to meet as a very minimum (Figure 5.9.)

In a sense, a control chart is a tool which sends a signal making possible the distinction between abnormal and normal variation. In addition, one of the very fundamental concepts in all control charts, is the notion of control.

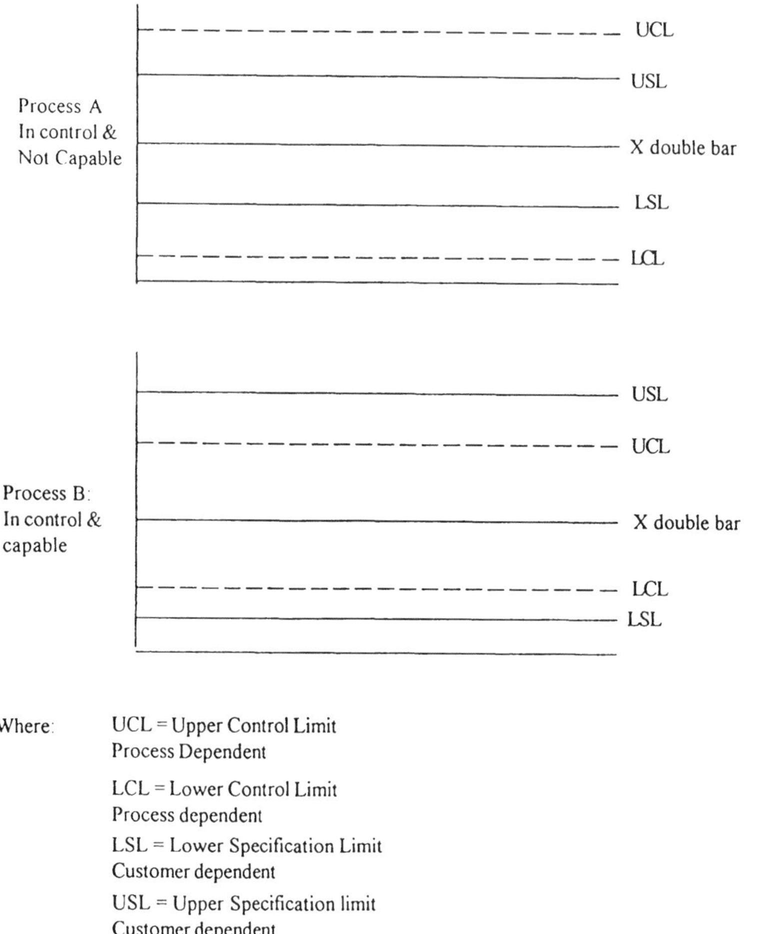

Where: UCL = Upper Control Limit
Process Dependent

LCL = Lower Control Limit
Process dependent

LSL = Lower Specification Limit
Customer dependent

USL = Upper Specification limit
Customer dependent

X double bar = overall average

Figure 5.9 The relationship between process limits and specifications. Note: control limits are always process dependent (calculated based on data from the process). On the other hand, specifications are customer dependent (given by the customer as part of the requirement).

Control does not necessarily mean that the product/service will meet your needs. It only means that the process is consistent and even then, it may be consistently bad or good. For a detailed discussion on control theory and application see Montgomery (1985) and Duncan (1986). For the constant values used in the construction phase of control charts see Appendix C. For the appropriate formulae used in the construction of control limits, capability, and basic descriptive statistics see Appendix D. In addition, Appendix E provides some general useful formulae used in the field of quality.

Types of Control Charts

There are many types of control charts.; however, their selection depends on the kind of data available (Figure 5.10). There are two kinds of data: variable and attribute. When variable data are available, a range of powerful charts are available. Variable control charts monitor or measure things that are actually measurable on a continuous scale, such as temperature, pressure, acidity, time, and so on. For example, typical charts that may be used in monitoring variable data are:

$\bar{X}$ and R chart. These charts require a number of consecutive units be taken n times per work period and analyzed for specific criteria; they display graphically the process stability and show data in terms of spread (piece-to-piece variability) and its location (process average). $\bar{X}$ charts cover averages of values in small subgroups (sample taken), which is known as the measure of location. R charts deal with range of values within each sample (highest minus lowest), which is known as measure of spread. To construct $\bar{X}$ and R charts:

1. Select the size, frequency, and number of subgroups.
2. Select and record the raw data.
3. Calculate the average of each of the subgroups as well as the range.
4. Calculate the overall average of the process as well as the average range.
5. Construct control chart scale. (A good rule of thumb in constructing the scale is as follows. For the range chart, take 2–3 times the largest range average. For the average chart, identify the largest and the smallest average of each of the subgroups, then take the difference of these two, multiply the difference by 2, and add it to the grand average.)
6. Plot the averages and ranges on the chart.
7. Calculate the control limits for the range chart, using appropriate formulae and constants.
8. Calculate the control limits for the average chart, using appropriate formulae and constants.
9. Interpret the range chart.
10. Interpret the average chart.

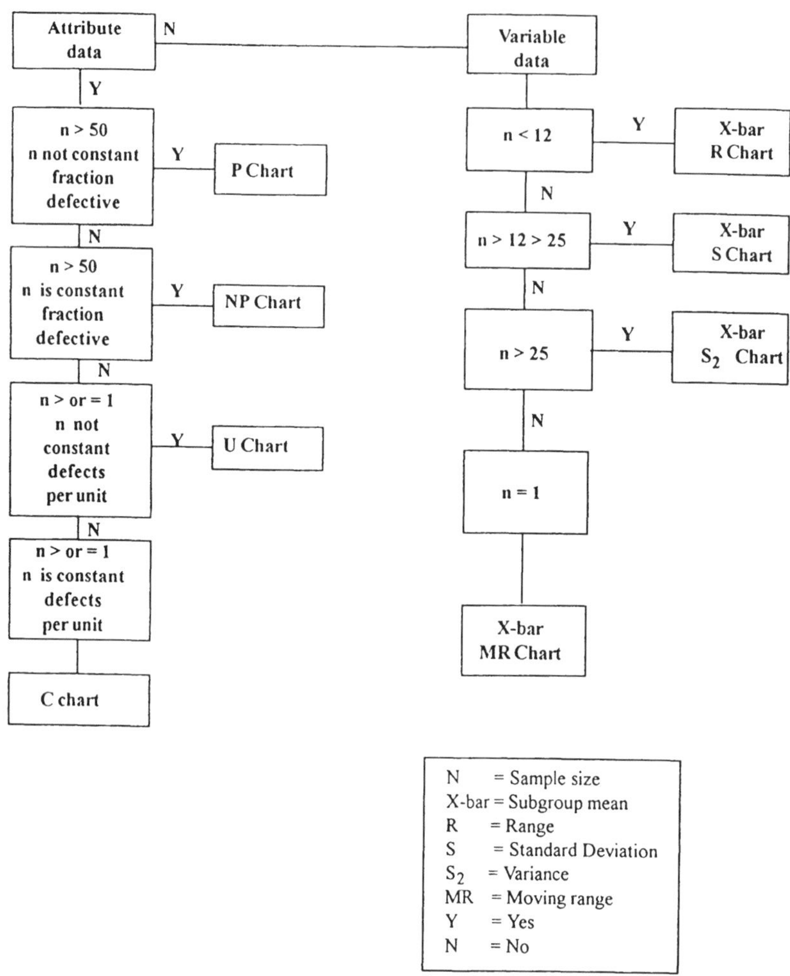

Figure 5.10 Selection process for an appropriate contol chart.

If process is consistent and repeatable then,

11. Calculate the process capability.
12. Continue to monitor the process.

If process is not consistent and repeatable then,

11a. Identify the cause(s) of inconsistency.
12a. Remove the cause(s).
13. Continue to monitor until the process is consistent and repeatable.
14. Move to step 11.

Individual and Moving Range Chart. These are standard control charts that require less samples than the X and/or R charts to establish stability. Even with one sample one can use this chart to establish trends and identify the variation between batches and or individual items. Other than the level of sampling, this chart is very similar to the $\bar{X}$ and R charts. To construct an individual and moving range chart:

1. Collect and record the raw data.
2. Select an appropriate interval of samples.
3. Calculate the range of each of the interval subgroups.
4. Calculate the overall average of the process and average range.
5. Construct the control chart scale.
6. Plot averages and ranges on the chart. (A good rule of thumb in constructing the scale is as follows. For the range chart, take 2–3 times the largest range average. For the individual chart, identify the largest and the smallest value, then take the difference of these two, multiply the difference by 2, and add it to the grand average.)
7. Calculate the control limits for the range chart, using appropriate formulae and constants.
8. Calculate the control limits for the average chart, using appropriate formulae and constants.
9. Interpret the range chart.
10. Interpret the average chart.

If process is consistent and repeatable then,

11. Calculate the process capability.
12. Continue to monitor the process.

If process is not consistent and repeatable then,

11a. Identify the cause(s) of inconsistency.
12a. Remove the cause(s).
13. Continue to monitor until the process is consistent and repeatable.
14. Move to step 11.

Median and R chart. This is a standard chart that is an alternative to the $\bar{X}$ and R charts for the control of processes. It is less sensitive to trends, however, and under some circumstances is considered to be more difficult to construct. It graphically displays the stability of a process, and yields similar information to $\bar{X}$ and R charts, but has several advantages.

1. It is easier to use since daily calculations are not required.
2. Individual values and medians are plotted, so the median chart shows the spread of process output and gives an ongoing view of process variation.
3. It shows where nonconformities are scattered through a more or less continuous flow of a function.
4. It shows where nonconformities from different areas may be evident.

To construct a median and range chart:

1. Select the size, frequency, and number of subgroups.
2. Select and record the raw data.
3. Select the median of each of the subgroups as well as the range.
4. Select the overall median of the process and the average range.
5. Construct control chart scale. (A good rule of thumb in constructing the scale is as follows. For the range chart, take 2–3 times the largest range average. For the median chart, identify the largest and the smallest median of each of the subgroups, then take the difference of these two, multiply the difference by 2, and add it to the grand median.)
6. Plot the median values and ranges on the chart.
7. Calculate the control limits for the range chart, using appropriate formulae and constants.
8. Calculate the control limits for the median chart, using appropriate formulae and constants.
9. Interpret the range chart.
10. Interpret the median chart.

If process is consistent and repeatable then,

11. Calculate the process capability.
12. Continue to monitor the process.

If the process is not consistent and repeatable then,

11a. Identify the cause(s) of inconsistency.
12a. Remove the cause(s).
13. Continue to monitor until the process is consistent and repeatable.
14. Move to step 11.

$\bar{X}$ and s charts. These are standard control chart that are similar to $\bar{X}$ and R charts, however, the s part of chart considers standard deviation (SD) and is more complicated to calculate. Because of the standard deviation, it is more sensitive to the process variation, especially with larger samples. On the other hand, it is less sensitive in detecting special causes of variation that produce only one value in a subgroup as unusual. To constructing an average and standard deviation chart:

1. Select the size, frequency, and number of subgroups.
2. Select and record the raw data.
3. Calculate the average of each of the subgroups as well as the range.

3a. Calculate the standard deviation for each subgroup.

4. Calculate the overall average of the process and the average range.

4a. Calculate the average standard deviation of the process.

5. Construct control chart scale. (A good rule of thumb in constructing the scale is as follows. For the standard deviation (SD) chart, take 2–3 times the largest SD average. For the average chart, identify the largest and the smallest average of each of the subgroups, then take the difference of these two, multiply the difference by 2, and add it to the grand average.)
6. Plot the averages and ranges on the chart.
7. Calculate the control limits for the standard deviation chart, using appropriate formulae and constants.
8. Calculate the control limits for the average chart, using appropriate formulae and constants.
9. Interpret the standard deviation chart.
10. Interpret the average chart.

If the process is consistent and repeatable then,

11. Calculate the process capability.
12. Continue to monitor the process.

If the process is not consistent and repeatable then,

11a. Identify the cause(s) of inconsistency.

12a. Remove the cause(s).

13. Continue to monitor until the process is consistent and repeatable.
14. Move to step 11.

Conversely, attribute control charts result from counting; they monitor or measure whether variables do or do not have certain characteristics or how often characteristics are present in a process. Examples are: delays, percent defects, percent of nonconformities, and so on. Basically, these control charts monitor pass/fail or go/no go situations. When the data is attribute in nature, then different charts are used.

p chart. THis is a standard control chart that requires a constant sample size, and charts either conforming or nonconforming items. Graphically, it displays stability of process, and it measures the actual number of conforming and nonconforming items rather than total number of faults. It expresses numbers in either fractional or percentile terms (whether conforming or nonconforming items are used) of total sample. (For example, total number of faulty forms in a batch irrespective of number of faults in any one form.) To construct a *p* chart:

1. Select the size, frequency, and number of subgroups.
2. Collect and record the raw data.
3. Calculate the proportion for each subgroup.
4. Calculate the average proportion as well as the average number of defectives.
5. Construct control chart scale. (A good rule of thumb in constructing the scale is as follows. For the *p* chart, identify the largest and the smallest proportion of each of the subgroups, then take the difference of these two, multiply the difference by 2, and add it to the grand average.)
6. Plot the individual proportions.
7. Calculate the control limits using appropriate formulae.
8. Interpret the *p* chart.

If the process is consistent and repeatable then,

10. Calculate the process capability.
11. Continue to monitor the process.

If process is not consistent and repeatable then,

10a. Identify the cause(s) of inconsistency.
11a. Remove the cause(s).
12. Continue to monitor and watch for process change.
13. Move to step 11.

np chart. This is a standard control chart that is similar to the *c* chart, but must be used if the sample sizes vary. Graphically, it displays stability of process, and it measures the number of nonconforming items rather than total number of faults (e.g., the total number of faulty forms in a batch irrespective of faults in any one form.) To construct an *np* chart:

1. Select the size, frequency, and number of subgroups.
2. Collect and record the raw data.

3. Calculate the process average number of nonconforming items.
4. Construct the control chart scale. (A good rule of thumb in constructing the scale is as follows. For the np chart, identify the largest and the smallest proportion of each of the subgroups, then take the difference of these two, multiply the difference by 2, and add it to the grand proportion.)
5. Calculate the control limits, using appropriate formulae.
6. Interpret the control chart.

If process is consistent then,

7. Calculate the process capability.

If the process is not consistent then,

7a. Identify the inconsistencies.
7b. Remove the inconsistencies.
8. Continue to monitor and watch for changes in the process.

c chart. This is a standard control chart for the total number of nonconformities, based on a constant sample size. Graphically, this chart displays stability of the process (e.g., the total number of errors in a batch of 100 forms rather than just the number of faulty forms.) To construct a *c* chart,

1. Select the size, frequency, and number of subgroups.
2. Collect and record the raw data.
3. Calculate the process average number nonconforming items.
4. Construct the control chart scale. (A good rule of thumb in constructing the scale is as follows. For the c chart, identify the largest and the smallest defectives of each of the subgroups, then take the difference of these two, multiply the difference by 2, and add it to the grand average defective.)
5. Calculate the control limits based on appropriate formulae.
6. Interpret the control chart.

If the process is consistent then,

7. Calculate the process capability.

If process is not consistent then,

7a. Identify the inconsistencies.
7b. Remove the inconsistencies.
8. Continue to monitor and watch for changes in the process.

u chart. This is a standard control chart that is similar to the *c* chart, but must be used if the sample sizes vary. Graphically, it displays the stability of process (e.g., the total number of errors in a batch of 100 forms rather than just the number of faulty forms.) To construct a *u* chart:

1. Select the size, frequency, and number of subgroups.
2. Collect and record the raw data.
3. Record and plot the nonconformities per unit in each subgroup.
4. Calculate the process average of nonconforming items per unit.
5. Construct the control chart scale. (A good rule of thumb in constructing the scale is as follows. For the u chart, identify the largest and the smallest nonconformities of each of the subgroups, then take the difference of these two, multiply the difference by 2, and add it to the grand average nonconformities.)
6. Calculate the control limits based on appropriate formulae.
7. Interpret the control chart.

If the process is consistent then,

8. Calculate the process capability.

If the process is not consistent then,

7a. Identify the inconsistencies.
7b. Remove the inconsistencies.
8. Continue to monitor and watch for changes in the process.

Control Chart Interpretation

Common Variation. A process with only common variation is predictable, consistent, and stable. The process under these conditions is said to be in control. An example of such a process is presented in Figure 5.11. As one looks at this control chart it is easy to spot that all points are within the control limits and they follow a random pattern. This is the ideal and very sought after condition in every control chart.

Special Variation. Unlike the common (inherent) variation a process may also have undue variation due to some special variation (i.e., due to assignable cause). That means that some part of the operation has changed in some degree and/or fashion. In other words, the present process is not part of the old one. The behavior of this change is shown in the control chart by the following signals:

1. *Points beyond control limits.* When this is shown in the chart, it means that the sample was collected from at least two different distributions of the process. Again, note that a change in the process may in fact be an improvement. The point is that this condition

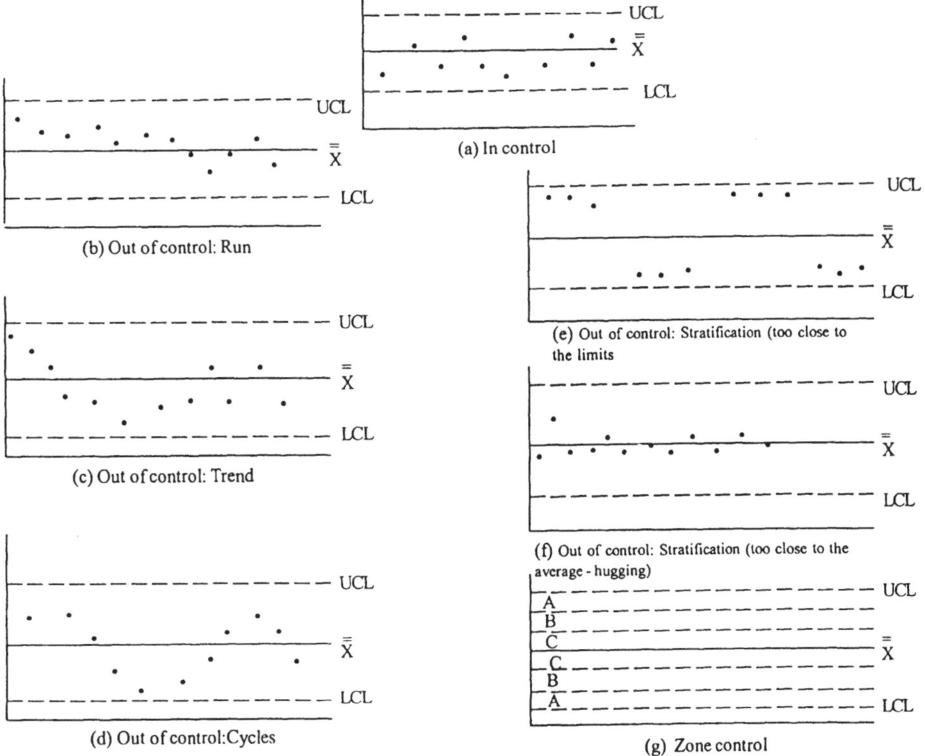

Figure 5.11 Process in control. (a) Process out of control, points out of control limits. (b) Process out of control, run. (c) Process out of control, trend. (d) Process out of control, cycle. (e) Process out of control, unusual. (f) Process out of control, stratification. (g) Zone control.

should be investigated and appropriately recorded in the process log. An example of this condition is shown in Figure 5.11a.

2. *Run of seven (7) points.* Even though all the points of the distribution are within the limits, this condition when present in the control chart shows that the process has a shift of the process average. That shift may be above or below the average and it must contain at least seven points in a row. When this happens, it is called a run. This is a special case and it should be highlighted in the control chart itself. It continues for a least twenty-five (25) subgroups, and means that the process has changed enough to recalculate the limits and the process average. An example of a run is shown in Figure 5.11b.

3. *Trend of seven points.* A trend is a special run of seven points that increase or decrease steadily. Think of it as drift in the process setting. Depending on the drift one may call an increasing trend a "run-up", while a decreasing trend one may call a "run down." An example of a trend is shown in Figure 5.11c. Note that in this special cause there are two signals of concern: the downward trend and the run below the process average. It is not uncommon to have more than one signal in a control chart.
4. *Cycle of points.* A cycle is a repetition of a pattern. It may have more than one cause and they are very easy to spot as well as correct. One of the easiest methods to solve a cycling problem may be to study the process through a process flow chart or a team effort approach. An example of a cycle is shown in Figure 5.11d.
5. *Unusual variation or stratification of hugging.* This is one of the most difficult signals to spot because it is easy to misinterpret as a good chart. The cause of this signal is not in the variation of the process, but rather it may be in the sampling itself or the scale of plotting. It is very important to note that as one looks at the chart the first impression is that it is "too good to be true" or "something is just not right." From a statistical point of view when this happens, it is referred to as a zone control problem. There are two main reasons for this unusual variation and they are: 1) basic changes have occurred in the process, and 2) process streams have had an effect on the overall process. An example of unusual variation is shown in Figure 5.11e. An example of stratification is shown in Figure 5.11f.

When one talks about zone control it is a reference to a control chart divided into three equal zones (each zone is one standard deviation; Figure 5.11g).

In analyzing a zone control chart, one should take note and examine what has changed and possibly make a process adjustment if:

1. Two out of three successive points are on the same side of the centerline in zone A or beyond.
2. Four out of five successive points on the same side of the centerline in zone B or beyond.
3. Seven successive points are on one side of the centerline.
4. Seven successive points are increasing or decreasing.
5. Fourteen points in a row are alternating up and down.
6. Fifteen points in a row are within zone C (above and below centerline).

A general algorithm to recalculate the control limits for any control chart is:

1. Does the process have enough points (usually 25) to generate a pattern?
   a. If yes, go to Step 2.
   b. If no, do not recalculate limits. The data is not sufficient.
2. Is there a pattern present on the control chart?
   a. If yes, go to step 3.
   b. If no, do not recalculate limits.
3. Do you know what is causing the pattern?
   a. If yes, go to step 4.
   b. If no, do not recalculate limits; focus the effort in identifying root cause(s) for patterns.
4. Do you like the pattern, as it appears?
   a. If yes, recalculate appropriate limits based on the formulae in Appendix D.
   b. If no, do not recalculate limits. Go to Step 3.

III. SUMMARY OF THE MANAGEMENT TOOLS OF QUALITY

As we already have mentioned, it is beyond the scope of this book to address in detail the basic tools as well as the management tools. However, because of their importance to the whole improvement process, their frequent use, and their simplicity, we will offer the following summary for each of the tools. For more detailed information see Mizuno (1988), Brassard and Ritter (1994), Kerzner (1995), and Tague (1995).

A. Affinity Diagram

An affinity diagram is the organized output from a team brainstorming session (Figure 5.12). It is a creative process and expresses data in the form of language without quantifying it. The purpose of an affinity diagram is to generate, organize, and consolidate information concerning a product, process, or complex issue or problem. The steps for constructing one are:

1. Choose a group leader.
2. State the issue or problem.
3. Brainstorm and record ideas.
4. Write each idea on a card.
5. Move the cards into like piles.
6. Name each pile with a title card.
7. Draw the affinity diagram. (Draw a circle or square around each group of like cards.)
8. Discuss the piles.

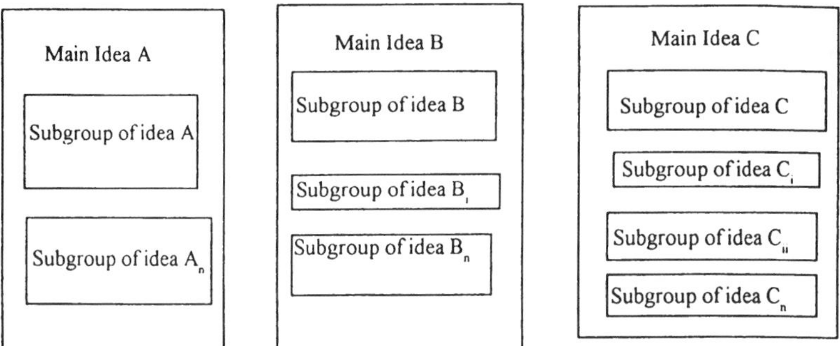

Figure 5.12 Affinity diagram.

B. Interrelationship Diagram

An interrelationship diagram is a pictorial representation of the cause-and-effect relationships among the elements of a problem or issue (Figure 5.13). The purpose of making an interrelationship diagram is to identify the root causes and root effects of a problem. Root causes are those factors or aspects of a problem which primarily influence other factors. Root effects are those factors of a problem which are primarily influenced by other factors. The steps of constructing an interrelationship diagram are:

1. Clearly define the issue or problem.

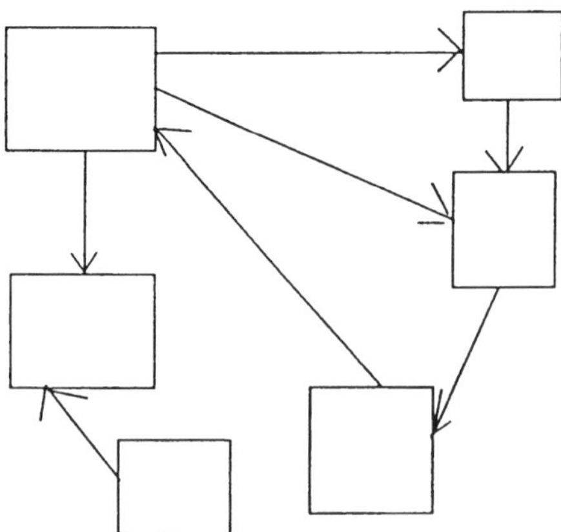

Figure 5.13 Interrelationship diagram.

2. Generate an affinity diagram. (If the affinity diagram has already been generated, go to Step 3.)
3. Construct the diagram layout. Write the problem statement at the center of a piece of paper and draw a circle around it. Then place the header cards from the affinity diagram in a circular pattern around the problem statement. Draw a circle around each header card.
4. Analyze the relations. If a relationship does not exist, do not draw a line between circles. If a relationship exists, draw a line between the two categories. Place an arrow at only one end of the line. The arrow indicates the category which is the effect and away from the category which is the cause.
5. Count the arrows. Once the analysis stage is complete, count the number of arrows going in and out of each category. Write the numbers above each category in the form number in/number out.
6. Identify the root causes and effects. The root causes are the categories with the greatest number of arrows going out and the root effects are the categories with the greatest number of arrows going in.
7. Study the final diagram. The focus the quality improvement efforts will be cased on the analysis of the root causes.

C. Tree Diagram

A tree diagram (sometimes called a systematic diagram) is a graphic representation of the different levels of actions used to accomplish a broad goal (Figure 5.14). The purpose of making a tree diagram is to generate the most specific level of action items that can be implemented to accomplish a broader goal. The steps to construct the tree diagram are:

1. Record the problem or goal statement.
2. Generate the first level of items by asking questions like: "What needs to happen to achieve this goal or solve this problem?" This

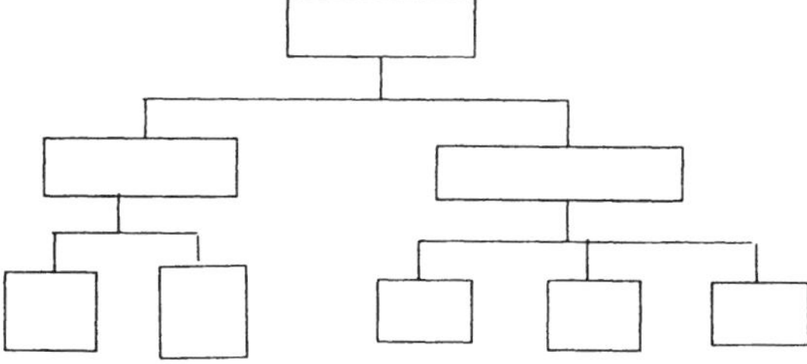

Figure 5.14 Tree diagram.

level of questioning is considered to be the means to achieve the end or main goal.

3. Complete the systematic diagram under each major path by asking the question "What needs to happen to achieve this goal or solve this problem?" Notice at this stage as you move from left to right in the diagram, the tasks are getting more specific. Of course, each level is the new means to accomplish the end (i.e., the prior level item).
4. Study the tree diagram. In reviewing the diagram, start with the last level on the far right and ask: "Will this set of items achieve the next higher level?" If the answer is yes for every level, then the diagram is complete. If not, then redo the diagram.

D. Prioritization Matrices

This approach is used when one wants to use a systematic approach to weigh the present options (Figure 5.15). Specifically, this approach helps at least in the following ways:

1. It quickly identifies basic disagreements.
2. It forces a team to focus on the best things to do, rather than everything they could do.
3. It limits private agendas since all the criteria identified are team generated.
4. It educes the chances of selecting someone's pet project.
5. It increases participation and ownership of both problems and solutions.
6. It increases morale and communication in the team and the organization.

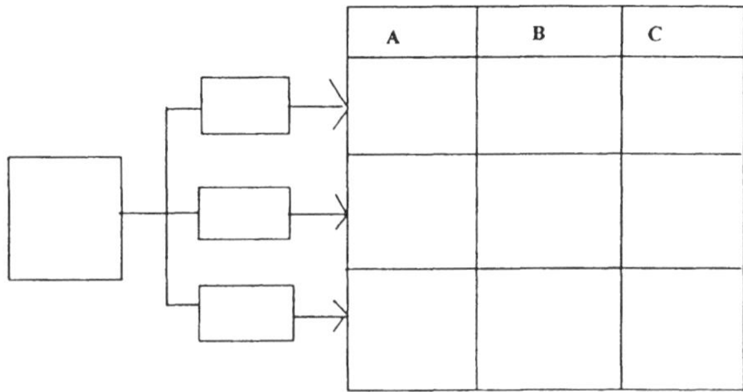

Figure 5.15 Prioritization matrices.

The approach to constructing the prioritization matrices is dependent upon the notion that the team is willing to decide based on a consensus. Unless consensus is understood and utilized in the decision process, this approach to prioritization will not be effective. Assuming that consensus is utilized, there are six required steps:

1. Agree on the goal to be achieved; the agreement must be clear, concise, and exact.
2. Create the list of criteria.
3. Using an L-shaped matrix that reflects the needs of the group. (See Section E, Matrix diagram.)
4. Compare all options relative to each criteria.
5. Using the L-shaped summary matrix, compare each option based on all criteria combined.
6. Choose the best option(s) across all criteria.

E. Matrix Diagram

The matrix diagram method is designed to clarify problematic spots through multidimensional thinking. Specifically, it is designed to seek out principal factors from many characteristics concerning a subject under study. The matrix diagram helps to expedite the process of problem solving by indicating the presence and degree of strength of a relationship between two sets of factors. By using the intersecting points of the factors, the experimenter may determine the strength of the relationship. There are many different types of matrix diagrams, however, they all focus on identifying the relationship(s) of the factors. Some of the most typical types are:

1. *The T-type matrix.* It is a matrix of A factors corresponding to B and C factors, respectively (Figure 5.16). This type may be used when there is a need for defect-reducing activities. Also, it may be utilized when exploring a new use of materials. In this case, the T-type matrix may be used to analyze ingredients and components of the material by characteristics and usage.
2. *The Y-type matrix.* This type of matrix is a combination of three L-type matrices: A factors and B factors, B factors and C factors, C factors and A factors (Figure 5.17). The intent of this matrix is to show how these factors correspond to each other. It can be used anytime there is a concern of optimizing three factors.
3. *The X-type matrix.* This type of matrix is a combination of four L-type matrices. This matrix shows the correspondence of four sets of factors: A and B, and AB and D; B and A, and BA and C; C and B, and CB and D; and D and A, and DA and C (Figure 5.18).

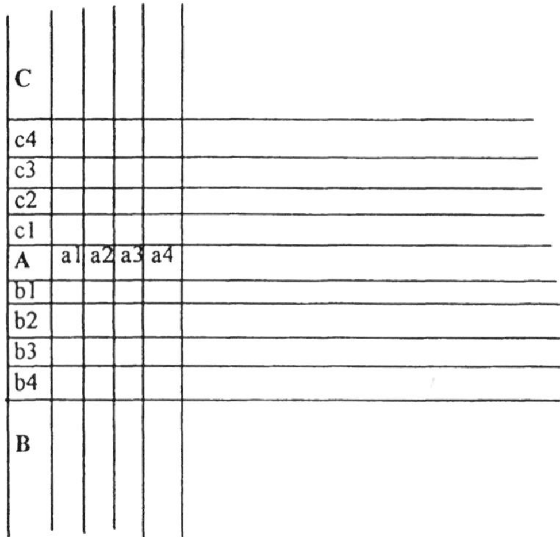

Figure 5.16 The *T*-type matrix.

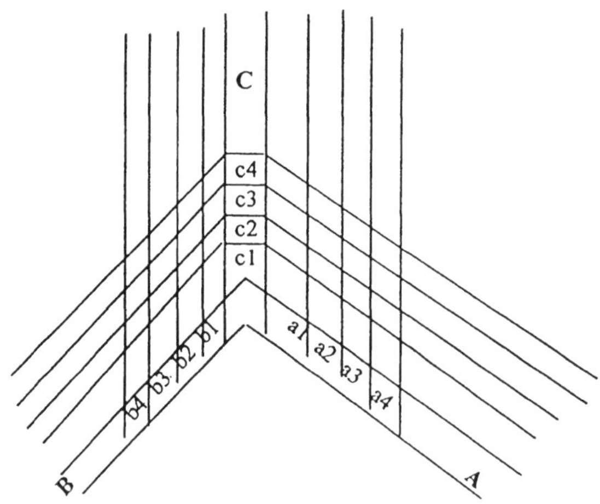

Figure 5.17 The *Y*-type matrix.

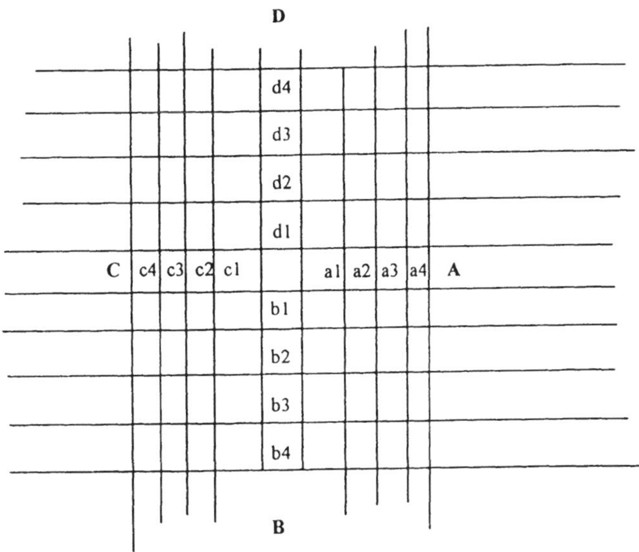

Figure 5.18 The *X*-type matrix.

Applications of this matrix are limited, but it can be used to consider the correspondence of management functions, management inputs, output data and input data.

4. *The C-type matrix*. This type of matrix is expressed in a rectangular cube whose sides are represented by three elements, A, B, and C (Figure 5.19). The main feature of this cubic type of matrix is the "point of conception of the idea," which is determined by three elements of A, B, and C in three-dimensional space.

F. Process and Decision Program Chart

The Process and Decision Program Chart (PDPC) is also a tool for improving implementation through contingency planning (Figure 5.20). It is fundamentally a tree diagram, but it is much simpler. To construct this chart:

1. Select a team closest to the implementation project.
2. Determine proposed implementation steps.
3. Identify as many as possible and likely problems of each step.
4. Identify as many as possible and reasonable responses of each likely problem.
5. Choose the most effective countermeasures and build them into a revised plan.

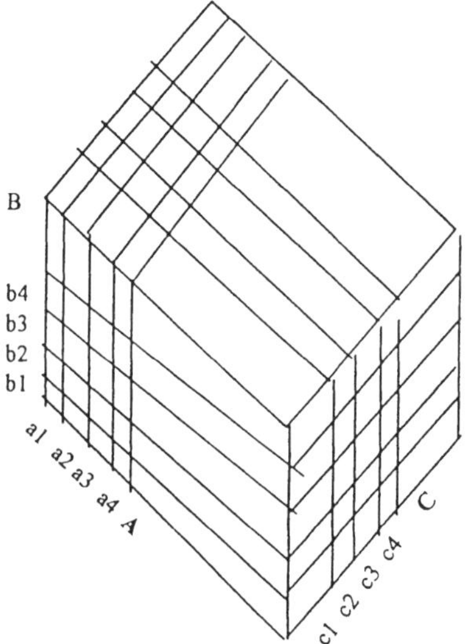

Figure 5.19 The *C*-type matrix.

G. Activity Network Diagram

The activity network diagram (AND) is also known as the Program Evaluation Review Technique (PERT). It is used to find the most efficient path—critical path—and a realistic schedule for the completion of any project by graphically showing the total completion time, the necessary sequence of tasks, those tasks that can be done simultaneously, the critical tasks to monitor, and the slack time of tasks within the project (Figure 5.21).

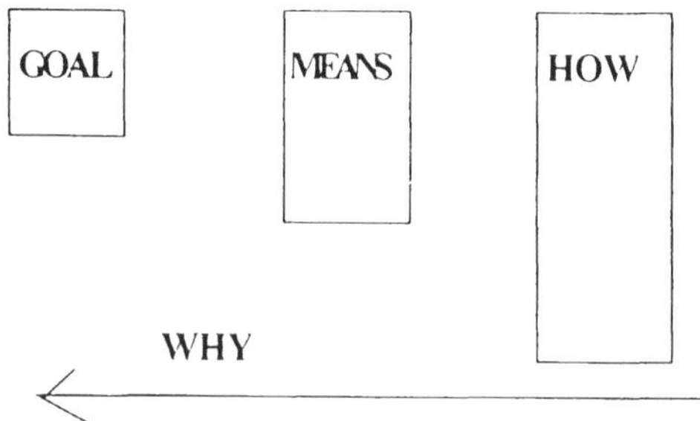

Figure 5.20 The PDPC diagram.

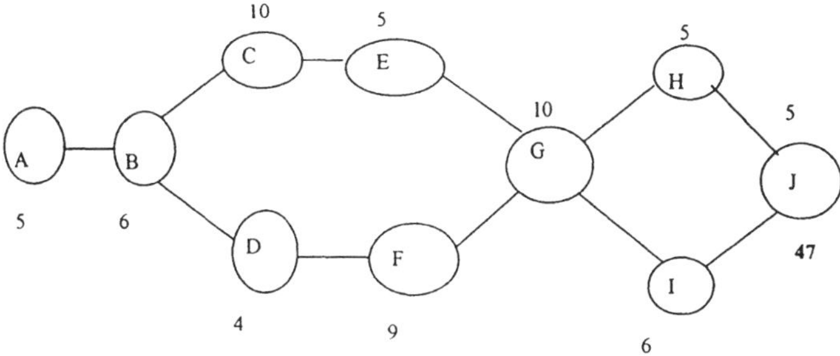

Figure 5.21 Activity network diagram.

Proper utilization of an activity network diagram presupposes an understanding of a specific vocabulary. The minimum vocabulary requirements are:

Activity Part of a project represented by an arrow, which has a defined beginning and end, and which requires time and resources.

Best (most optimistic) estimate (B) Earliest possible completion time.

Critical path (CP) The most stringent schedule with no slack time. Or, the most time consuming path through the network. Or, the sequence of activities imposing the most rigorous time constraint on attainment of the final event.

Earliest start time (EST) Earliest possible time that an activity can begin. Before any activity may begin, all activities preceding it must be completed. The EST is determined by a forward pass through the network, allowing the expected time (ET) for each activity to determine the start time for each succeeding activity.

Expected time (ET) The average duration time, where:

$$ET = \frac{4(M) + 1(B) + 1(W)}{6}$$

Event Indicates the start or completion of one or more activities, and is shown graphically as a circle, square, or other symbol. Theoretically, an event is an instant of time.

Float (slack) Latest start time (LST) minus earliest start time (EST) for a given activity.

Latest start time (LST) Latest possible time an activity can begin and still be completed by the time the next activity must begin. The LST is calculated by beginning at the end of the network and working backwards through the network.

Most likely estimate (M) Length of time probably needed.

Network A graphical representation of a project plan, which shows activities and the relationships between them.

Project Two or more activities with a given aim. A project must have a beginning, budget, schedule of completion, and an end.

Worst (most pessimistic) estimate (W) Time required if all goes wrong.

Advantages of AND

The advantages of using AND are quite numerous and include:

- It forces management to plan a project before it begins.
- It requires an analytical approach to planning.
- It separates the planning and scheduling functions.
- It permits the planner to concentrate on the relationship of items of work, without considering their occurrence in time.
- It allows the planner to develop a more detailed plan, since one is concerned with how the work will be performed, not when.
- It results in a more realistic schedule.
- It clearly shows the dependency relationships between work tasks.
- It facilitates control of a project.
- It simplifies maintenance of the plan and schedule.
- It informs management of the current status of the project.
- It focuses management attention on critical items of work.
- It gives management the ability to assess consequences of anticipated changes to the plan.
- It makes it easy to relate other functions of project control to the basic planning and scheduling functions.
- It meets contractual requirements of government and private industry.

Construction of AND

To actually construct the AND the following steps are necessary:

1. Basic rules of network logic.
   a. Before any activity may begin, all activities preceding it must be completed.
   b. Arrows imply logical precedence only. Neither the length of the arrow nor its compass direction on the drawing have any significance.
2. Procedure for developing the network.
   a. Select an appropriate team.
   b. Identify and list all activities.
   c. Determine the relationships among activities:
      What activities must immediately precede this activity?

What activities cannot be started until after completion of this activity?

What activities can be done concurrently with this activity?

d. Draw the network.

e. Schedule activities.

Contact people who normally do the activity for realistic time estimates.

Calculate the ET, EST, LST, and slack time for each activity.

Calculate the critical path.

f. Determine major milestones and target dates in the project effort.

Note: A variation of this diagram is the Gantt chart, which defines only the time line and milestones of a project.

REFERENCES

Brassard, M. and Ritter, D. *The memory jogger*. (1988). GOAL/QPC. Methuen, MA.

Brassard, M. and Ritter, D. (1994). *The memory jogger II*. GOAL/QPC. Methuen, MA.

Duncan, A. J. (1986). *Quality control and industrial statistics*. 5th Ed. Irwin. Homewood, IL.

Ford Motor Co. (1987). *Team oriented problem solving*. Power Train Operations. Ford Motor Co. Dearborn, MI.

Gitlow, H., Gitlow, S., Oppenheim A., and Oppenheim, R. (1989). *Tools and methods for the improvement of quality*. Irwin. Homewood, IL.

Grant, E. L. and Leavenworth, R. S. (1980). *Statistical quality control*. 5th ed. McGraw-Hill Book Co. New York.

Gulezian, R. (1991). *Process control statistical principles and tools*. Quality Alert Institute, Inc. New York.

Ishikawa, K. (1982). *Guide to quality control*. Asian Productivity Organization. Kraus International Publications. White Plains, NY.

Kerzner, H. (1995). *Project management: A systems approach to planning, scheduling, and controlling*. 5th ed. Van Nostrand Reinhold. New York.

Mizuno, S. (Ed.). (1988). *Management for quality improvement: The 7 new QC tools*. Productivity Press. Portland, OR.

Montgomery, D. C. (1985). *Introduction to statistical quality control*. John Wiley and Sons. New York.

Stamatis, D. H. (1995). *Total quality service*. St. Lucie Press. Delray Beach, FL.

Tague, N. R. (1995). *The quality tool box*. Quality Press. Milwaukee, WI.

6

Special Tools and Techniques for the Future World of Quality

This chapter discusses some of the special tools and techniques that in the future world of quality, they will be very helpful in both implementation and monitoring the process. Specifically, the topics of needs assessment, ideation, the five whys, force field analysis, affinity chart, storytelling, nominal group technique, survey, and audit are explored.

I. NEEDS ASSESSMENT

The industrial needs (of both the organization and employees) in a given company are not static, but undergo change over time. Therefore, it is important that management periodically review themselves. The forces for change are many and are constantly with us, occurring as a result of changes in employee and/or corporate expectations, changes in technology, changes in knowledge, and a variety of changing social patterns. To adequately assess the perceived needs of a system requires the involvement of representative segments of the corporate culture. By seeking needs information from the representative groups, management can be more responsive to the "wishes" and real "needs" of all the employees.

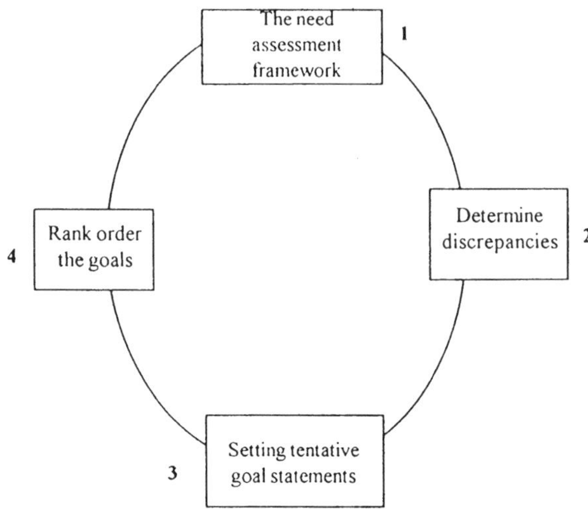

Where:

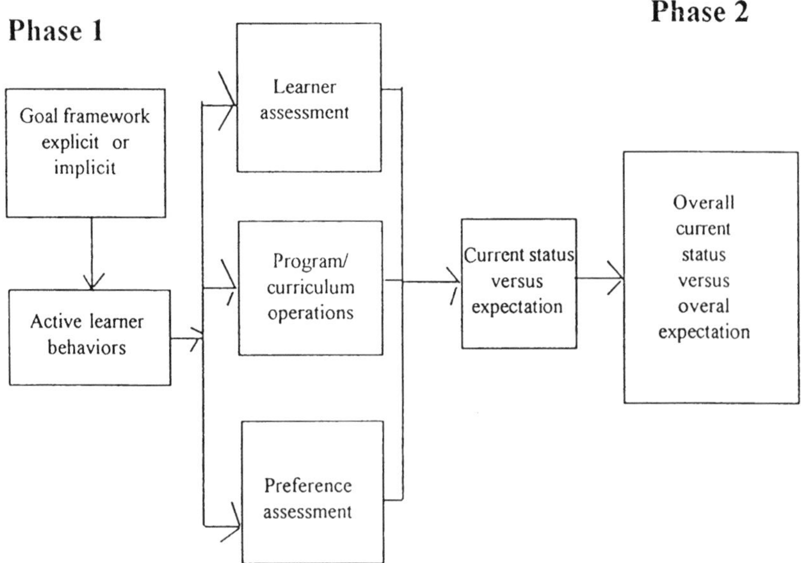

Figure 6.1 The four phases of a typical needs assessment.

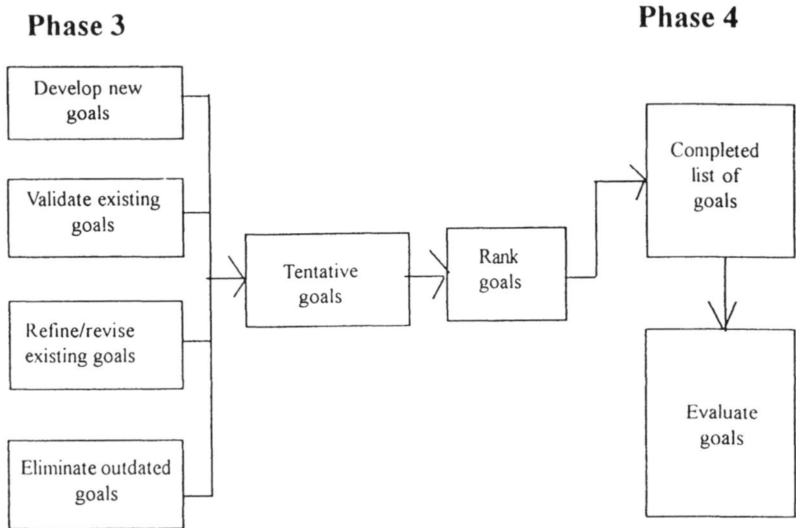

Figure 6.1 (Continued)

The identification of needs involves a discrepancy analysis, sometimes called a gap analysis, that identifies two opposite positions and the difference, if any, between them. An example will clarify the point. A discrepancy analysis might use either of the following two questions: 1) Where are we now and where should we be? or 2) How important is this and how well do you feel this is being done? By comparing the answers to these questions it is possible to ascertain where significant discrepancies exist and where they do not exist.

Obtaining industrial needs from the employees is not an end in itself, but instead represents a very valuable source of information regarding the current status of the corporate system. A needs assessment can provide the corporate system with some of the best available information on the immediate characteristics of the corporate quality system, training programs, and those areas where attention and resources should be directed, both internally and externally to the organization. For a detailed concept and application explanation see Kaufman (1979). Here, however, a much more simpler model for a needs assessment is presented, based on four distinct phases as shown in Figure 6.1.

A. Phase 1. The Needs Assessment Framework

Here the corporate goals and objectives are set. The existence of the goals and objectives are the primary function of this phase.

B. Phase 2. Determine Discrepancies

Essentially this process involves a comparison of "what is" with "what should be" in the organization, process, and in the three areas of the training setting. For the components of the organization and the process, the discrepancy phase identifies the current operational status and the expected operational status. The most important guideline in determining such discrepancies is to insure an unbiased objective accounting and an active participation of all involved. On the other hand, in the areas of training setting, Phase 2 focuses on employee achievement, program and/or training operations, and preference assessment.

C. Phase 3. Setting Tentative Goal Statements

Given the information collected and analyzed in Phase 2, management must analyze and interpret this data within the goal framework. Some of the alternatives are:

1. Validate existing goal statements.
2. Develop new goals as indicated and supported by the discrepancy analysis.
3. Refine or revise existing goal statements.
4. Eliminate outmoded or outdated goals.

D. Phase 4. Rank Order the Goals

The final phase of the process involves determining the priority of goal statements. Essential to the effective conduct of the needs assessment is the sincere commitment of management at all levels. It is this commitment that facilitates the process and brings it to closure. The preference assessment is designed to gather a reasonable amount of judgmental input from the various employee groups, processes, and systems. The instrument used for such a task is usually a survey instrument asking questions relative to the concern of the management team. The questions focus on "what is" and "what should be." The actual development of the questions must utilize a cross-sectional or cross-functional team of employees for a correct and valid content.

Once the needs assessment has taken place and the priorities have been determined, the next step is evaluation. There are many models one can follow in evaluating a program evaluation. Excellent sources are Madaus, Scriven, and Stufflebeam (1983); Bank, Henderson, and Eu (1981); and Scriven (1991). For our purposes Brinkerhoff's (1987) model will be used with some minor modifications. The essence of the model is that it corresponds to the decisions

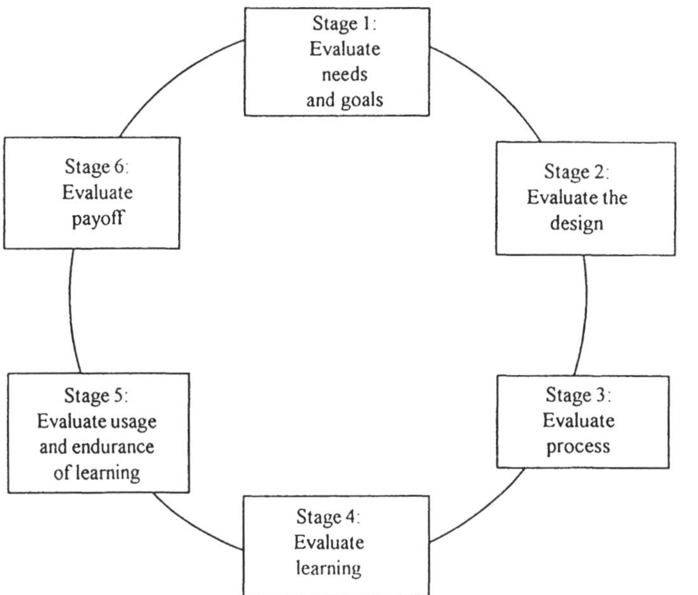

Figure 6.2 The six steps of a typical needs assessment.

necessary for programs to proceed productively and defensibly through the stages, Thereby enabling and facilitating quality efforts (Figure 6.2).

Stage 1 evaluates the value and importance of problems and/or opportunities in the organization.
Stage 2 aims at the production of a defensible program design and might access a given design's practicality.
Stage 3 assesses the significance of planned and unplanned departures from the design.
Stage 4 reveals that sufficient skill, knowledge, and attitude were in fact acquired.
Stage 5 assesses how much and how well the acquired skill, knowledge, and attitude are being translated into intended on-job behavior changes.
Stage 6 shows the value of organizational effects and their relationship to training.

In order to implement this six-stage model a series of relevant questions must be asked. Typical questions and concerns that may help in the process are shown in Table 6.1.

Table 6.1 Questions/Concerns for the Six-Stage Model

Evaluation stage	Some key evaluation questions	Some useful procedures
Goal setting (What is the need?)	How great is the need, problem, or opportunity? Is it amenable to the organization's solutions? Is the difference worth making? Is the work likely to pay off? Are the criteria for success realistic? How would the organization know if the work paid off or not? Is this approach to the problem the best we can come up with?	Organizational surveys and/or audits, performance analyses, records analysis, observation, benchmarking, document reviews.
Program design (What will work?)	Given our culture in the organization, what might work best? Is this particular design better than something else? What is wrong with design A? Is the selected design good for our organization?	Secondary research, expert reviews, panels, checklists, site visits, pilot tests, participant review.
Program implementation (Is it working?)	Has it been installed as it is supposed to be? Is it working on schedule? What problems are cropping up? What really took place? Did the employees like it? What did it cost?	Observation, checklists, feedback from both employees and supervisors, records analysis.

Immediate outcomes (Did they learn it?)	Are the employees using it? How well are the employees adjusting to the new method, approach and so on? What did we learn? How are we going to avoid similar problems?	Knowledge and performance tests, observation, simulations, self reports, work sample analyses.
Intermediate or usage outcomes (Are they keeping and/or using it?)	Are the employees still using it? How well are the employees using it? Is it working as expected? What parts of the program the employees like best, or worst? Why? What parts of the program do they use the most or least? Why? What part of the program is the most or least appropriate for the organization? Why?	Self, peer, and supervisory reports; case studies; surveys; site visits; observation; work sample analyses.
Impacts and worth (Did it make a worthwhile difference?)	What difference does using it make? Has the need been met? Was it worth it? How can we improve next time?	Organizational audits, performance analyses, records analysis, observation, surveys, document reviews, panel reviews and hearings, cost benefit comparison.

II. IDEATION

A long-standing tool to inspire creativity in the world of quality, in fact in all team activities, has been the use of the brainstorming technique. In the last couple years, however, a new twist of the brainstorming methodology has been utilized and it is called *ideation.*

The brainstorming activity focuses on the process. However, that focus sometimes created a problem in the creativity process. Ideation, on the other hand, focuses on the results and therefore eliminates the confusion of the process. Whereas brainstorming breaks down barriers, allows participation from all the involved participants, and is based an a ritualistic approach, ideation goes a step further in the sense that it comes up with ideas without worrying about the process. In fact, because ideation does not worry about the process, the participants are more likely to push the development of ideas into concepts.

Ideation is very flexible and uses less rubrics in deciding on a particular result. It is more efficient than brainstorming and thus may be used in a variety of activities. The actual application of ideation is the same as the brainstorming minus the rules. While brainstorming typically involves a facilitator and the employees, in a typical ideation approach outsiders may be involved. These outsiders may be from cross-functional and multidisciplined areas. Their function is to give their perspective and to break as many barriers as possible.

To have excellent results in any ideation process, the make up of the participants should have at least the following profile characteristics:

Those who envision: These people add direction, inspiration and momentum to the discussion. They focus on the end result and present a vision of what they want to create. They are very good for strategic planning activities, since they are capable of describing the ideal future in 5–10 years.

Those who modify: These people examine the components of problems and bring stability and thoroughness to the process. They prefer to take things one step at a time and build on what they already know.

Those who experiment: These people like to test carefully and receive input to confirm ideas. They like to troubleshoot and answer questions on how to use products within their intended markets and how to find other possible uses.

Those who explore: These people would excel in taking a product and incorporate new ideas for improvement or enhancement of the product. They thrive on the unknown and have a sense of adventure.

A typical creative usage analysis using ideation may follow the steps identified in Figure 6.3.

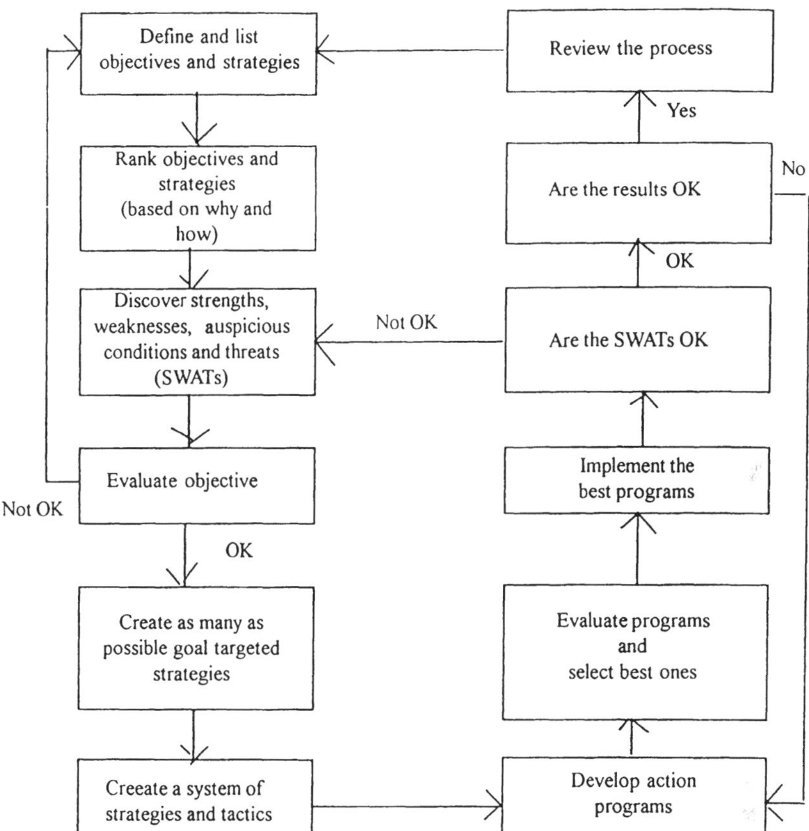

Figure 6.3 A typical creative usage analysis.

III. FORCE FIELD ANALYSIS

Force field analysis (FFA) is a systematic way of identifying and portraying the forces (quite often people) for or against change in an organization. The specific forces will be different depending upon the area where they are applied. Typical steps in conducting a FFA are:

1. Identify the actual driving forces or factors working for change. (Define the current situation.)
2. Prioritize driving forces by voting or use of some consensus achieving technique. (Define/base the desired position on the results of an FFA.)

Desired Change or Improvement
--------------------------------->

<table>
<tr><th>Driving forces</th><th></th><th>Restraining forces</th></tr>
<tr><td>Any conditions, incentives, resources,</td><td>--------> | <--------</td><td>Any conditions, disincentives, limitations, processes</td></tr>
<tr><td>processes, key individuals and groups,</td><td>--------> | <-------</td><td>key individuals and groups, problems, opportunities</td></tr>
<tr><td>problems and opportunities and other</td><td>--------> | <------</td><td>and other factors which act in opposition to the proposed</td></tr>
<tr><td>factors which act in support of the</td><td>-------> | <-----</td><td>action. What is working against this innovation?</td></tr>
<tr><td>proposed change. What is working for</td><td>--------> |</td><td></td></tr>
<tr><td>this innovation?</td><td>--------> |</td><td></td></tr>
</table>

The Five Whys

When identifying underlying causes, it can also be useful to ask five sequential "whys" to get to the heart of a problem. For example: What is the problem? "Deliveries not complete by 4:00 p.m."

Why does this happen? Routing of trucks is not optimized.

why? Goods are loaded in the trucks based on size, rather than location of delivery.

Why? Computer automatically defines the order based on large items first.

Why? Large items are delivered first.

Why? Current workflow prioritization puts large items first on the delivery schedule, regardless of location and proximity.

Affinity chart

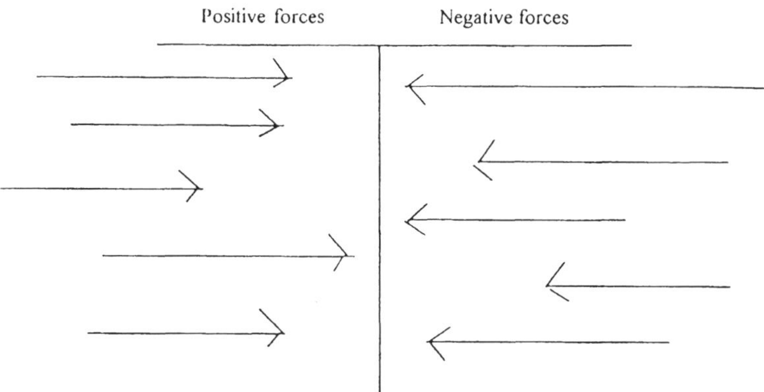

Figure 6.4 A visual view of the a typical force field analysis.

3. Identify current restraining forces or factors working against change. (Define the worst possible situation.)
4. Prioritize restraining forces. (What are the forces for change? What is their relative strength?)
5. Identify potential driving forces or factors that realistically favor change but are not yet currently operating. (What are the forces against change? What is their relative strength?)
6. Identify potential restraining forces or realistically possible forces that would work against change if they were present. (What forces can you influence?)
7. Determine if there are actions that can be taken to:
   - increase the effect of the current driving forces
   - add to the driving forces or cause potential driving forces to happen
   - decrease or eliminate the current restraining forces
   - ensure that the potential restraining forces do not occur or do not impact the issue being investigated
   - eliminate specific action that can be taken relative to each of these forces that you cannot influence
8. Begin to develop some strategy statements about proposed actions. Statements should begin to answer: Who, What, When, Where, How, How much, and What are the expected results?

A typical visual representation of a force field analysis is shown in Figure 6.4.

IV. THE FIVE WHYS

When identifying underlying causes, it can also be useful to ask five sequential "Whys" to get to the heart of a problem. For example: What is the problem? "Deliveries not complete by 4:00 p.m." *Why* does this happen? The routing of trucks is not optimized. *Why*? Goods are loaded in the trucks based on size rather than location of delivery. *Why*? The computer automatically defines the order based on large items first. *Why*? Large items are delivered first. *Why*? Current workflow prioritization puts large items first on the delivery schedule, regardless of location and proximity.

V. AFFINITY CHART

An affinity chart is the organized output from a team brainstorming session. (It differs from brainstorming in that the affinity chart uses cards as headers that can be changed and organized in piles for discussion.) It was created by Kawakita Jiro and is also known as the KJ method. The purpose of an affinity chart is to generate, organize, and consolidate information concerning a product, process, service, and/or complex issue or problem. The chart is used when the answer to the following three questions is a positive "yes."

1. Is the issue complex and hard to understand?
2. Is the problem uncertain, disorganized, or overwhelming?
3. Does the problem require the involvement and support of a group or a team?

The actual construction of the chart takes seven basic steps:

1. Choose a group leader.
2. Choose the problem and if possible state it in a question form.
3. Brainstorm and record all ideas. Each idea should be written on its own index card or note.
4. Arrange the cards into like categories.
5. Name each category with a header card.
6. Draw the affinity chart. Arrange headers with the appropriate generated ideas and circle or box them together.
7. Discuss the categories.

VI. STORYBOARD

Quality improvement storyboards and storybooks use the steps in the FOCUS-PDCA strategy to help teams organize their work and their presentations so others can more readily learn from them. They reduce variation in the process by focusing the learning experience on the content rather than the method of telling. A typical structure for a storyboard is shown in Figure 6.5. For a very detailed discussion see Forsha (1995a,b). In addition, storyboards form a permanent record of a team's actions and achievements, and all the data generated and storyboards can function as the working minutes of the team.

VII. NOMINAL GROUP

The nominal group technique (NGT) is one of many structured group processes that have been designed and developed. It is a special-purpose technique that is useful for situations where individual judgments must be tapped and combined to arrive at decisions that cannot be reached by one person. The NGT is a problem-solving or idea-generating strategy, and is not typically used for routine meetings.

The NGT was developed by Andre L. Delbecq and Andrew H. Van de Ven in 1968. It was derived from social-psychological studies of decision conferences, management–science studies of aggregating group judgments,

Find a process to improve	Clarify current knowledge of the process
Organize a team that knows the process	Understand causes of process variation
Select the process improvement	
Team Roadmap	

Figure 6.5 The structure of the storyboard.

and social-work studies of problems surrounding citizen participation in program planning. Since that time, the NGT has gained extensive recognition and has been widely applied.

The NGT takes its name from the fact that it is a carefully designed, structured, group process that involves carefully selected participants as independent individuals rather than in the usual interactive mode of conventional groups. It is a well-developed and tested method that is fully presented in the work of Delbeck, Van de Ven, and Gusstafson (1975), and Delbeck and Van de Ven (1986).

The NGT has four phases in addition to an introduction and a conclusion. The participants are physically present in groups of 8–12 and the session is controlled by a process consultant (facilitator) and an assistant. During the introduction, the facilitator attempts to familiarize the participants with the process and make them feel at ease with what will transpire during the meeting. The facilitator usually discusses very briefly at least the following items:

1. the purpose of the session and the importance of the process
2. the steps of the NGT
3. how the results will be used and the next steps

The facilitator then reads a carefully worded task statement. This is the task that the participants should respond to during the structured group session. It is usually simple and direct. If the facilitator is asked what is meant by the task statement, he or she usually avoids introducing bias that occurs by giving examples. Instead, the facilitator often asks several participants to give their interpretation of the task statement. Additionally, the facilitator often asks several participants to directly respond to the task statement. The process of forcing the participants to clarify the task statement themselves is called *self-priming* and can be very effective. When the responses appear to coincide with the objective and the remainder of the participants appear to have grasped the task, the facilitator proceeds to the first basic step of the NGT.

The first phase is called silent generation and typically takes about 10–15 minutes. During this phase, the group members are instructed to write their responses to the task statement. Both the facilitator and the assistant also write during this period. Even if a majority of participants appear to stop writing before ten minutes has elapsed, the period is not shortened. If some talking occurs, the facilitator tactfully asks for cooperation in permitting others to think through their ideas.

Like each of the steps in the NGT process, silence is purposefully designed. It is based on the notion that silent generation focuses attention on a specific tasks, frees the participants from distractions, and provides them with an opportunity to think through their ideas rather than simply to react to the comments of others. In this sense, it is a search process that yields contributions of greater quality and variety. Participants are motivated by the tension of seeing those around them working hard at the group task. they are forced to attend for a longer time to the task, rather than rushing immediately to consider the first idea that is suggested to the group. They are freed from all of the inhibiting effects of the usual face-to-face interaction of unstructured groups. Judgment of ideas cannot take place during this early and crucial portion of the group process.

Next comes the round-robin phase. The facilitator interrupts the process, yet emphasizes that there is no need to stop generating. (Any additional ideas should be added to the silent generation lists.) The facilitator calls on participants one-by-one to state one of the responses they have written. Participants may pass at any time and may also join in on any subsequent round. A participant may propose only one item at a time, and either the facilitator or an assistant records each item as it is offered. The only discussion allowed is between the facilitator and the participant who proposed the item. The discussion is limited to seeking a concise rephrasing for ease of recording. As each participant responds, the facilitator repeats verbatim what has been said and the assistant records the concise phrase on a sheet. This phrase goes on until all the ideas generated by the group are listed and displayed.

The round-robin phase permits the leader to establish an atmosphere of acceptance and trust. He or she does not unduly rephrase or evaluate the contributions, and they are equally and prominently displayed before the group. It is essential that the leader be open and that there is no valuative behavior. Each idea and each participant receive equal attention and acceptance. There is little opportunity for the process to be dominated by strong personalities, to be inhibited by possible sanctions or conflicts, or to be suppressed by status differences. The process separates ideas from their authors and permits conflicting and incompatible ideas to be explicitly tolerated. It provides a written record of the group's efforts as a basis for the next step.

The third phase is called clarification. Once all the items have been recorded, the facilitator goes over each one in order to ascertain that all participants understand the item as it has been recorded. Any participant may offer clarification or may suggest combination, modification, deletion, and so forth of items. However, evaluation is avoided. The facilitator moves rapidly from one measure to the next, keeping up the pace of the process. During this step, the underlying logic behind items may be brought out, there may be some expressions of differences of opinion, and the group may conclude that some items can be eliminated or combined because of duplication.

Pace is important to this step and the facilitator's job is to keep the group moving rapidly through the list of items. Although in this phase the group is more like an interacting one, the facilitator seeks to control lengthy discussions, arguments, and "speech making." Again the effort is to separate ideas from their authors, to clarify rather than to evaluate, and to ensure full opportunities for participation.

It is important to point out that the clarification aspect of the NGT is perhaps the primary determinant for the resulting quality of the list of items. If there is a great deal of overlap from item to item and if there is ambiguity on the part of the group members as to exactly what each item means, the next step, which involves voting and ranking, will be invalid. Experience has shown that a certain amount of combination is necessary.

The fourth phase—voting and ranking—provides the participants with an opportunity to select the most important items and to rank those items. The participants are provided with between five and nine blank 3′×5′ cards. Usually, participants are provided with eight cards, but the number can vary depending on how many responses are generated during the round-robin phase. Each participant is asked to select the eight most important items from the list displayed before him or her. Typically, the list will contain 20–30 items. To avoid any confusion in handling their judgments, participants are asked to write the items out, one per card, in an abbreviated fashion, in the center of the blank cards. They are also asked to write the sequential list number of the item in the upper left-hand corner of the card. When all have completed this step, they are asked to spread the eight cards out in front of them and to rank and weight the items. Typically, they are given the following instruction:

> From the eight cards, choose the most important item, write the number 8 with a circle around it in the lower right-hand corner of the card, and set the card aside. Another way of phrasing this to assist some in deciding which in most important is: Which of the eight items would you use to guide future actions relative to this topic if you could only use one?

The ranking process continues:

> From the remaining seven cards, choose the least important item, write the number 1 with a circle around it in the lower right-hand corner of the card, and set the card aside. Another way of phrasing this to assist some in choosing the least important item is: If you could use only six of the seven items in front of you, which one would you drop off?

The process continues in this fashion until all the cards have been ranked. At this point of the process, tabulation of the votes takes place. The facilitator has three alternatives:

1. Invite the participants to take a ten-minute break (possibly for refreshment) while the facilitator and the assistant tabulate and display the results.
2. Invite the participants to watch the tabulation process take place.
3. Invite the participants to fill out a brief questionnaire that has been prepared by the coordinator for the specific purpose of evaluating the reaction of the participants to the process, obtaining suggestions from the participants as to next steps, determining the likelihood of implementation, and so on.

The tabulation process involves sorting the cards by sequential item number from the original list and recording the weight given to each.

The fourth step in the NGT process permits the participants to express their individual evaluations of the items in a way that is free of social pressure. It provides a constructive method for dealing with conflicts, and leads to a clear expression.

VIII. SURVEY STUDIES

Survey research studies large and small populations by selecting and studying samples chosen from the interested populations to discover the relative incidence, distribution, and interrelations of the selected variables. This type of survey is called a sample survey.

Surveys are not new. They have been around since the eighteenth century (Campbell and Katona, 1953), but surveys in the scientific sense are a twentieth-century development. To be sure, surveys are considered to be a branch of social scientific research, which immediately distinguishes survey research from the status survey. The procedures and methods of survey research have been developed mostly by psychologists, sociologists, anthropologists, economists, political scientists, and statisticians, all of which have a tremendous influence in how we perceive the social sciences.

On the other hand, in the quality profession, even though we use survey instruments, the process of surveying is not quite clear. By definition, a survey links populations and samples. Therefore, the experimenter is interested in the accurate assessment of the characteristics of whole populations. For example a typical survey study may be conducted to investigate how many suppliers of type A material qualify for the approved supplier list, given characteristics such as: delivery, price, quality, capability, and so on. Another example where a survey may be used by the quality professional is in the area of identifying and/or measuring the current culture, attitude, and general perception of quality in a given organization.

In using a survey, it must be understood that samples are used—very rarely, if ever, are populations used—by which they will infer the characteristics of the defined universe. To be successful, such undertaking must depend on random samples and unbiased questions. For the mechanics on how to construct a questionnaire see Stamatis (1996a) and Kerlinger (1973).

A. Types of Surveys

Surveys can be classified by the following methods of obtaining information:

1. *Interviews.* They are very expensive, but they do provide for individualized responses to questions as well as the opportunity to learn the reasons for doing or even believing something.
2. *Panel techniques.* A sample of respondents is selected and interviewed, and then reinterviewed and studied at later times. This technique is used primarily when one wants to study changes in behavior and/or attitudes over time.
3. *Telephone surveys.* At least from a quality application, they have little to offer other than speed and low cost.
4. *Mail questionnaires.* These are quite popular in the quality field. It is used to self-evaluate your own quality system, culture of the organization, and so on. Its drawback, unless used in conjunction with other techniques, is the lack of response, inappropriate (e.g., leading, biased, or ambiguous) questions, and the inability to verify the responses given.

B. Methodology for a Valid Survey

It is beyond the scope of this book to address the details of survey methodology (for details see Kerlinger, 1973, and the references therein); however, the most fundamental steps are:

1. *Specify and clarify the problem.* The experimenter should not expect just to ask direct questions. It is imperative that the experimenter should also have specific questions to ask that are aimed at various facets of the problem. In other words, plan the survey.
2. *Determine the sample plan.* The sample must be representative of the population in order to be effective. For a good source of how to do this see Kish (1965).
3. *Determine the schedule and if appropriate and applicable, what other measuring instruments.* This is a very difficult task and it must not be taken lightly. Here the task is to translate the research question into an interview instrument and into any other instruments constructed for the survey. For example, one of the problems in a quality survey may be: How are permissive and restrictive attitudes toward the new quality vision of our organization related to perceptions of both employees and management? Among the questions to be written to assess permissive and restrictive attitudes, one may be: How do you feel about quality?
4. *Data collection.* The experimenter's concern is how the information is going to be gathered. What method is to be used? And to what extent should appropriate precautions planning be designed for eliminating spurious answers and responses.
5. *Analysis.* In this step of the survey, there are three issues that concern the experimenter: 1) coding, the term used to describe the translation of question responses to specific categories for purposes of analysis, 2) tabulation, the recording of the numbers of types of responses in the appropriate categories, and 3) statistical analysis, the determnation of percentages, averages, rational indices, and other appropriate tests of significance. The analysis of the data are studied, collated, assimilated, and interpreted. Never, should the experimenter give the results without the rationale, assumptions, and interpretation of the data and results.
6. *Reporting.* This step reports the results of the survey to all concerned parties.

IX. AUDIT

This section gives only a cursory overview of the audit, rather than an exhaustive discussion, and the reader is strongly encouraged to see Mills (1989),

Arter (1994), Keeney (1995a,b), Russell (1995), Stamatis (1996b), and Parsowith (1995) for more detailed information.

Whereas the survey is a methodology to find and project the results of a sample to a population, an audit is a methodology that uses samples to verify the existence of whatever is defined as a characteristic of importance. That characteristic may be defined in terms of quality, finance, and so on.

An audit from a quality perspective based on ISO 8402 is a systematic and independent examination to determine whether quality activities and related results comply with planned arrangements, and whether these arrangements are implemental effectively and are suitable to achieve objectives (ANSI/ISO/ASQC A8402, 1994). This definition in no uncertain terms implies that an audit is a human evaluation process to determine the degree of adherence to prescribed norms and results in a judgment. The norms, of course, are always predefined in terms of criteria or standards or both.

The definition of the norms is always done by the management of the organization that is to be audited. On the other hand, it is the responsibility of the auditor conducting the audit in any organization to evaluate the compliance of the organization to those norms. For this reason the audit is usually performed as a pass/fail evaluation rather than on a graded scale.

Any quality audit may be internal or external and in one of three forms:

1. *First party audit.* This audit is conducted by an organization on itself and may be done on the entire organization or part of the organization. It is usually called an internal audit.
2. *Second party audit.* This audit is conducted by one organization on another. It is usually an audit on a supplier by a customer, and it is considered an external audit.
3. *Third party audit.* This audit is conducted by an independent organization (the third party) on another organization. It can only be performed at the request or the initiative of the organization seeking an impartial evaluation of the effectiveness of their own programs. This audit is mandatory for those organizations that seek ISO 9000, ISO 14000, and/or QS-9000 certification. It is always an external audit.

The requirements to start any audit fall into three categories:

1. *Start.* To start an audit, it is imperative to know what to do. An auditor must gain some knowledge about the organization and the audit environment before starting the physical audit.
2. *Documentation.* Documentation implies "Write it down." The auditor has the responsibility to be prepared, fair, impartial, inquisitive, honest, and observant.

3. *Evaluation.* An auditor's ultimate responsibility is to evaluate the quality system of a given organization based on the organization's appropriate standards and follow-up (if this is a surveillance audit).

On the other hand, the actual audit process follows a sequential path that has four phases:

Phase 1. Preparation (preaudit). This phase includes selecting the team, planning the audit, and gathering pertinent information.
Phase 2. Performance (on site visit). This phase begins with the opening meeting and finishes with the actual audit.
Phase 3. Reporting (postaudit). This phase includes the exit meeting and the audit report.
Phase 4. Closure. This phase includes the actions resulting from the report and the documentation information.

REFERENCES

ANSI/ISO/ASQC A8402. (1994). *Quality vocabulary.* ASQC. Milwaukee, WI.

Arter, D. R. (1994). *Quality audits for improved performance.* 2nd ed. Quality Press. Milwaukee, WI.

Bank, A., Henderson, M., and Eu, L. (1981). *A practical guide to program planning: A teaching models approach.* Teachers College, Columbia University. New York.

Brinkerhoff, R. O. (1987). *Achieving results from training.* Jossey Bass. San Francisco.

Campbell, A. and Katona, G. (1953). The sample survey: A technique for social-science research. *In*: L. Festiger and D. Katz, *Research methods in the behavioral science.* Holt, Reinhart, and Winston, New York.

Delbecq, A. L. and Van de Ven, A. H. (1986). *Group techniques for program planning.* Green Briar Press. Middleton, WI.

Delbecq, A. L., Van de Ven, A. H., and Gustafson, D. H. (1975). *Group techniques for program planning: A guide for nominal and delphi processes.* Scott Foresman. Glenview, IL.

Forsha, H. I. (1995a). *Show me: The complete guide to storyboarding and problem solving.* Quality Press. Milwaukee, WI.

Forsha, H. I. (1995b). *Show me: Storyboard workbook and template.* Quality Press. Milwaukee, WI.

Kaufman, R., and English, F. W. (1979) *Needs assessment: Concept and application.* Educational Technology Publications, Englewood Cliffs, NJ.

Keeney, K. A. (1995a). *The ISO9000 auditor's companion.* Quality Press. Milwaukee, WI.

Keeney, K. A. (1995b). *The audit kit.* Quality Press. Milwaukee, WI.

Kerlinger, F. N. (1973). *Foundations of behavioral research.* 2nd ed. Holt, Reinhart and Winston, Inc. New York.

Kish, L. (1965). *Survey sampling.* John Wiley and Sons. New York.

Madaus, G. F., Scriven, M., and Stufflebeam, D. L. (Eds.). (1983). *Evaluation models.* Kluwer-Nijhoff Publishing. Boston.

Mills, C. A. (1989). *The quality audit.* Quality Press. Milwaukee, WI.

Parsowith, B. S. (1995). *Fundamentals of quality auditing.* Quality Press. Milwaukee, WI.

Russell, J. P. (1995). *Quality management benchmark assessment.* 2nd ed. Quality Press. Milwaukee, WI.

Scriven, M. (1991). *Evaluation Thesaurus.* 4th ed. Sage Publications. Newbury Park, CA.

Stamatis, D. H. (1996a). *Total quality service.* St. Lucie Press. Delray Beach, FL.

Stamatis, D. H. (1996b). *Documenting and auditing for ISO 9000 and QS-9000.* Irwin Professional. Burr Ridge, IL.

7

Unusual Yet Powerful Tools Used in TQM

This chapter discusses some of the unusual charts used in quality. Our intent is to familiarize the reader with their existence, rather than develop their rationale and their detailed explanation. For more information on any of these charts, the following sources may be consulted: Gulezian (1991), Crowder, S. V. (1987, 1989), Hunter (1986), Lucas and Saccucci (1990), DeBruyn (1968), Duncan (1986), Grant and Leavenworth (1988), Montgomery (1991), and Walker, Philpot and Clement (1991).

I. MOVING AVERAGE AND MOVING RANGE CHARTS

Moving average charts are useful in cases where a single observations as opposed to a sample is observed at each time point. This chart is basically an arithmetic mean calculated on the basis of a series of observations by successively dropping an observation at the beginning and adding one at the end and recomputing the mean. This is so, because measures of variability must be based on at least two observations, a moving measure of variation is needed, whether an individual chart or a moving average chart is used. An

Table 7.1 Moving Average and Moving Range

Observation number	Sample	Moving total	Moving average	Moving range
1	13.6			
2	15.5			
3	13.8	13.6 + 15.5 + 13.8 = 42.9	42.9/3 = 14.3	15.5 − 13.6 = 1.9
4	14.1	15.5 + 13.8 + 14.1 = 43.4	43.4/3 = 14.5	15.5 − 13.8 = 1.7
5	13.9	41.8	13.9	0.3
6	14.0	42.0	14.0	0.2

example of calculating the moving average and moving range is shown in Table 7.1.

The conventional approach for calculating control limits for a moving average chart and the moving range follows along the same lines as for mean and range charts.

II. EXPONENTIALLY WEIGHTED MOVING AVERAGE CHART

An exponentially weighted moving average (EWMA) is a form of average that weights past data but applies a smaller weight to data as they get older. A such, a control chart based on EWMA falls between a basic mean chart and a CUSUM chart (described later), where these either have no memory or a complete memory.

An EWMA is based on a statistic of the form:

$$\text{EWMA} = wX + (1 - w)B$$

where X is an observed value of the variable measured at a given point in time, B is the EWMA calculated at the immediately preceding time point, and w is a constant between zero and one that determines the degree of memory of the EWMA. The closer w is to 1, the more weight is given to the most recent observation, and the closer w is to 0 the less weight is given to a current observation. (When w is close to 1 the points plotted on a control chart appear similar to a Shewhart chart, and when near 0 the plotted points resemble a CUSUM chart.

Control limits for an EWMA can be calculated for given values of w for either individual observations or subgroup averages. In this respect they follow the same formulae as the $\bar{X}$ and R charts, and individual moving range charts.

EWMA charts are very useful for detecting small process shifts and can be enhanced to provide a quick initial response for critical processes. For more detailed discussion and applications the reader is encouraged to see Hunter (1986), Montgomery (1991), and Van Gilder (1995).

III. WEB OR RADAR CHART

In today's complex world, many factors affect customer satisfaction with a product or service. As a result, an organization is forced to evaluate products and services on multiple attributes. These multiple attributes may be graphed on what is called a web or radar chart. This chart is specifically designed to represent multiple data. They are simple, concise, and can be read quickly. What is interesting about this chart, is the fact that the analysis is continuous. The attributes are plotted in a circular fashion, making the overall and individual relationships easily recognized. A pictorial representation of a radar chart is shown in Figure 7.1.

There are seven steps in constructing the chart.

1. Assemble the right team.
2. Select and define the rating categories.
3. Construct the chart.
   a. Draw a large circle on a flip chart with as many diameters as there are categories.
   b. Identify each end of the diameter with each of the categories from Step 2.
   c. Identify the scale for each category. Make sure that "0" is in the center, with an increasing of the scale in an outward direction.
4. Rate all categories either individually or by the team.
5. Connect the ratings for each category.
6. Interpret and use the results.
   a. the overall ratings identify gaps within each category, but not the relative importance of the categories themselves.
   b. Post the results, update when necessary.
7. Use the results. Work on the biggest gap in the most critical category.

IV. EVOLUTIONARY OPERATION

From a quality perspective, most process-control techniques measure one or more output quality characteristics; if these quality characteristics are satisfactory, no modification of the process is made. However, in some situations

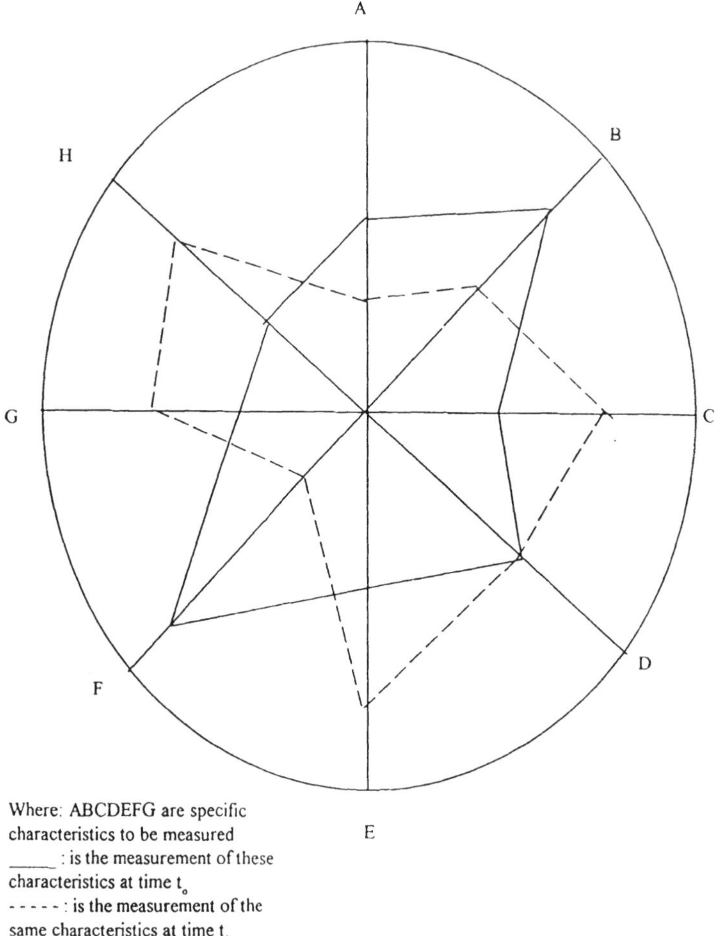

Figure 7.1 A typical radar chart.

where there is a strong relationship between one or more controllable independent process variable and the observed quality characteristic (dependent variable), other process control methods can sometimes be employed.

Evolutionary operation (EVOP) was proposed by Box (1957) as an operating procedure. It is designed as a method of routine plant operation that is carried out by operating personnel with minimum assistance from the quality or manufacturing or engineering staff. EVOP makes use of principles of experimental design, which is usually an “off-line” quality control method.

Essentially, EVOP consists of systematically introducing small changes in the levels of the process operating variables. The procedure requires that each independent process variable be assigned a "high" or "low" level. The changes in the variables are assumed to be small enough so that serious disturbances in product quality will not occur, yet large enough so that potential improvements in process performance will eventually be discovered. This approach, of course, is similar to the full factorial designs (see Chapter 9), except that this approach is based on progressive adjustments called "cycles."

In testing the significance of process variables and interactions, an estimate of experimental error is required. This is calculated from the cycle data. By comparing the response at the center point with the 2^k points in the factorial portion we may check on the change in mean. That is, if the process is really centered at the maximum, then the response at the center should be significantly greater than the responses at the 2^k peripheral points.

In theory, EVOP can be applied to k process variables. In practice, only two or three variables are usually considered at a time. For more details and examples see Montgomery (1991), Box (1957), and Box and Draper (1969).

V. STANDARDIZED CHARTS FOR ATTRIBUTE CONTROL

Frequently, sample sizes on attribute charts in general, and proportion defective (p) and yield (q) charts in particular, will vary to a significant extent. This often results from those situations where attribute analysis is employed for daily, weekly, or monthly analysis in a final test context. In such cases, sample sizes will often vary by more than ±25% from the average sample size (n) used to calculate control limits. (It must be emphasized that opinions vary regarding the percent variability of any n from $\bar{n}$ considerably among quality professionals. However, many conservative statisticians employ a ±10% variability from $\bar{n}$ as a determinant for employing options other than the standard p chart.) In these or similar instances, one of the following options may be employed:

1. Calculate a separate set of control limits for each sample; or
2. Calculate a set of control limits for those individual (I) samples where the sample size (n_i) is within ±25% of n, and separate limits for those samples which fall outside this range; or
3. Calculate three sets of control limits, each corresponding to the minimum, average, and maximum expected/observed sample sizes.

While all of these approaches are statistically sound, all present problems from the standpoint of the day-to-day interpretation of the plotted data. Varying control limits for all or some samples frequently tends to make all but highly trained statisticians uncomfortable. The calculation of limits each time a new sample is drawn becomes tedious, and the proper identification of runs, trends, and cycles becomes difficult when sample sizes (and therefore limits) vary widely.

An additional problem is frequently encountered in the interpretation of attribute charts in those cases where: 1) sample sizes periodically are very large; and 2) at the same time, proportion defective/defect rates are very low. In these cases, the day to day problems noted previously are compounded, in that

Extreme changes in sample sizes will cause points to be plotted off the chart, or so close together (i.e., tightly) that interpretation of point to point drift is difficult.

Rounding error, even when four decimal places are employed, will cause plotted values to appear on or inside control limits when in fact they are outside of the limit.

Meaningful changes in process defective rates or yield levels appear to be insignificant as they occur within the context of the combination of large sample sizes and already small defective rates.

For example, let us assume that two days of final inspection data yield the following results:

n	np	p	q
101,256	47	0.00046	0.99954
83,411	40	0.00047	0.99953

where:

$\bar{p}$ = 0.005
$\bar{q}$ = 0.9995
n = sample size
np = number of defective parts
p = proportion of defective parts
q = $1 - p$ = yield or the proportion of good parts

In most cases, these p or q values would be plotted as the same value; even if they were not, the "feel" for the magnitude of difference between the values would commonly be lost.

One may suggest that this problem could be solved by scaling the control chart with a finer definition; perhaps converting the entire range of the scale from possibly 0.0010 (or 0.9990) to 0 (or 1.000). The problem when this procedure would become evident when a third day's data resulted in:

$n = 4{,}376$	$np = 2$	$p = 0.00046$	$q = 0.99943$

Recalling that the control limits are a function of the sample size, the control limits for the three samples would be calculated as follows:

Sample	n	np	p	UCL_p	LCL_p
1	101,256	47	0.00046	0.00071	0.00029
2	83,411	40	0.00047	0.00073	0.00027
3	4,376	2	0.00046	0.00151	–0.00051 (0)

As shown by this data, the UCL for the p chart (or the LCL for a yield chart) would be off the chart if the scale was adjusted to show differences between plotted values of, for example, 0.00046 and 0.00047. Note that this would result even though the p values for samples 1 and 3 are equal.

Finally, let us return to the issue of rounding error under these conditions. Utilizing standard rounding procedure, the following values would be obtained at varying sample sizes for a q (yield) chart where $\bar{q} = 0.9995$, the observed yield was 0.9994, and where values were to be rounded to four places:

n	Calculated LCL	Rounded LCL	Interpretation of control chart
460,000	0.999401	0.9994	In control
485,000	0.999404	0.9994	In control
510,000	0.999406	0.9994	In control
535,000	0.999408	0.9994	In control
560,000	0.999410	0.9994	In control
585,000	0.999412	0.9994	In control
610,000	0.999414	0.9994	In control
635,000	0.999416	0.9994	In control

Two aspects of these data are of some significance. First, under many circumstances, all of the observed yield values from this example would probably have been plotted on the LCL, and would not have been considered to be an out-of-control point/value. Second, and perhaps more importantly, note that although the sample sizes shown are significantly different, the reflected discrepancies between the observed q value (0.9994) and the calculated LCL values reflect very little difference in terms of a "perceived" context; this is true regardless of the control condition decision. The typical individual responsible for maintaining control charts would simply not see that a change or difference of 0.000004 in yield (where n = 535,000 and 585,000) was very significant.

A useful option to the varying limits chart, which will also obviate many of the problems previously noted, is the standardized-value attribute chart. Recall that, for example, the standard p-chart control limits are calculated as:

$$\bar{p} \pm 3\left(\frac{\bar{p}\bar{q}}{n}\right)^{1/2}$$

where

$$\bar{q}\text{ (yield)} = 1 - \bar{p}$$

It is important to remember that the formula component

$$\left(\frac{\bar{p}\bar{q}}{n}\right)^{1/2}$$

represents the standard error of the sampling process. That is, a value corresponding to the expected sampling error at any sample of size n. Control chart convention places the control limits at ±3 standard error increments above and below the process average ($\bar{p}$), which is the origin of the constant (3).

In a standardized-value chart, we are interested in transforming each observed value (p, for example) into a standardized value. This value represents the number of standard errors that the observed value falls away from the process average.

In order to calculate a standardized value for a p chart, for example, we would employ the formula:

$$P_s = \frac{p - \bar{p}}{[(\bar{p}\bar{q})/n]^{1/2}}$$

where:

P_s = the standardized p value
p = the observed sample proportion defective
$\bar{p}$ = the process average proportion defective
q = $1 - p$ (yield)
n = sample size

The resultant standardized value will represent the deviation of p from $\bar{p}$ in terms of standard error units. Values of p equal to $\bar{p}$ will result in a standardized value of 0. Values of p which would fall exactly on the control limit(s) reflect standardized values of 3 (or –3). Therefore, the standardized-value chart allows for a control chart where the process average will always have a scale value of 0 with control limits of +3 and –3 (assuming that tighter control is not desired).

For example, let us assume that:

$\bar{p}$ = 0.0026
p = 0.0031
n = 59,593

Then

$$P_s = \frac{(0.0031 - 0.0026)}{[(0.0026 \times 0.9974)/59{,}593]^{1/2}} = 2.397$$

The standardized value would be plotted directly on the control chart, and obviously would fall between the process average of 0 and the upper control limit of +3.

The advantages of this form of chart are:

1. The process average value and control limits never have to be recalculated in terms of the control chart. The p value employed in the calculation of the P_S values would, of course, have to be modified as appropriate. However, the chart configuration and scaling would remain constant.
2. The standardized values will reflect meaningful differences between plotted values when sample sizes are relatively large and process defective rates are fairly small. For example, returning to the yield data previously presented, the standardized values would appear as:

n	q	Standardized q	LCL
460,000	0.9994	–3.03	–3.0
485,000	0.9994	–3.12	–3.0
510,000	0.9994	–3.19	–3.0
535,000	0.9994	–3.27	–3.0
560,000	0.9994	–3.35	–3.0
585,000	0.9994	–3.42	–3.0
610,000	0.9994	–3.49	–3.0
635,000	0.9994	–3.56	–3.0

Note that in this case, the changes in the chart at different sample sizes for equal yields would be reflected by plotted point (q_s) differences of 0.03 to 0.56. This compares to a standard chart where the plotted points remain the same, but the control limits vary. For this data, the control limits varied from 0.999401 to 0.999416, or only 0.000015. In most cases, the individual utilizing this chart would have a much easier time interpreting the standardized chart than the typical varying limits chart. This assertion is based upon the fact that:

1. The standardized values provide a more accentuated degree of difference. That is, the "appearance" of change will be more noticeable.

2. Operators generally have been trained to describe out-of-control conditions such as runs and trends in the context of the plotted points, not control limit changes. Many operators and supervisors for that matter, would not have recognized that the eight identical yield values (0.9994) actually represented a trend condition. A quick glance at the standardized values, however, not only show that the points are outside of the control limits, but that the process defective values (were they occurring in a sequential process) were, in fact, trending further and further away from the process average. Standard trends and runs (which occur within the control limits) would be just as obvious when displayed by standardized values, and just as difficult to observe with a varying limits chart.

Finally, the standardized value approach may be used with any of the common attribute charts. For a yield (q) chart, the standardized value formula would appear as:

$$q_s = \frac{q - \bar{q}}{[(\bar{q}\bar{p})/n]^{1/2}}$$

For a u chart, the formula would appear as

$$u_s = \frac{u - \bar{u}}{(\bar{u}/n)^{1/2}}$$

Although this same procedure could be employed with np and c charts, it is unlikely that this need would arise given that these charts require equal sample sizes. This condition would occur relatively infrequently within an associated context of large sample sizes and small discrepancy values. For more information on standardized charts, see Montgomery (1991), Duncan (1986), and Charbonneau and Webster (1978).

VI. CUMULATIVE SUM CHART

An alternative to the basic control charts discussed in Chapter 5, there is an additional chart called the Cumulative Sum or CUSUM chart. This is a chart in which the observations are successively cumulated, and each new sum is plotted at each point in time.

The control limits for CUSUM charts are not parallel as in the simple mean chart. Instead, a so called "V-mask" is used, which is a "V" placed on its side such that it is pointing in the direction of the passage of time. The sides of the V-mask are the control limits which incrementally are moved with successive observations. A typical CUSUM chart is shown in Figure 7.2.

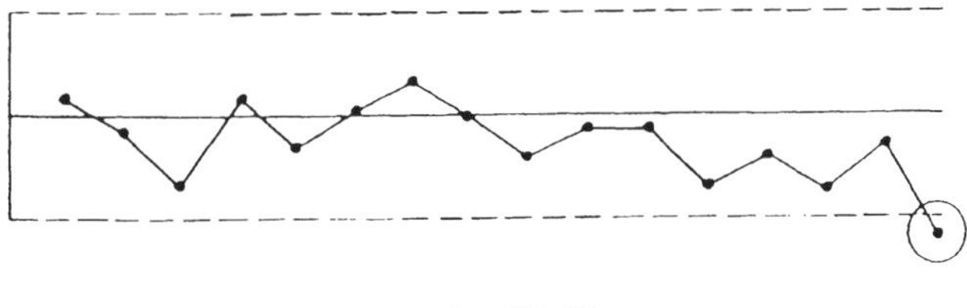

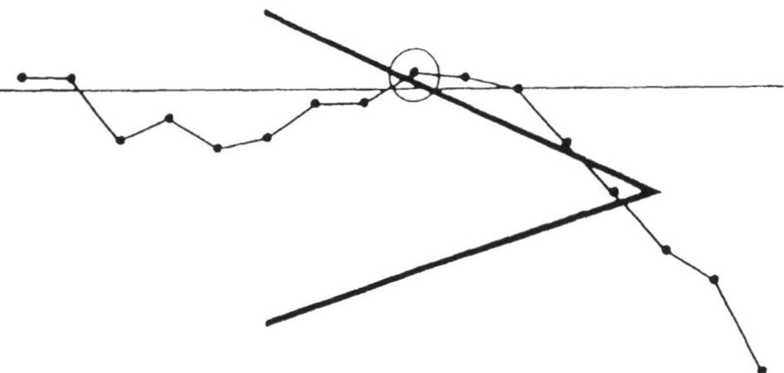

Figure 7.2 CSUM chart.

Inputs into the construction of the V-mask are the lead distance between the point of the mask and the lost plotted point and the angle between the sides of the "V." These inputs account for the probability of a Type II error associated with an important shift in mean along with a chosen probability of a Type I error, which is considered but fixed with a traditional chart.

Unlike the traditional chart which has no memory and the moving average which considers the immediate history, the CUSUM chart remembers everything, so to speak, where all observations contained in the CUSUM are given equal weight. A CUSUM chart will pick up a process shift sooner than the ordinary control chart. Because of this sensitivity, the CUSUM chart is more effective in detecting relatively small shifts in the process mean, on the order of $0.05s_{\bar{x}}$ to about $2s_{\bar{x}}$. Over this region the CUSUM control chart will detect shifts approximately twice as quickly (or with smaller sample size) as the corresponding Shewhart chart. In addition, the process shift is often easy to detect visually by the change in slope of the plotted points. Finally, it is often relatively easy to detect the point at which the shift occurs simply by visual examination of the plotted data, noting where the change in slope occurred.

Details regarding the construction of a CUSUM chart can be found in Ewan (1963), Page (1961), Lucas (1973), Johnson and Leone (1977), Duncan (1988), and Montgomery (1991).

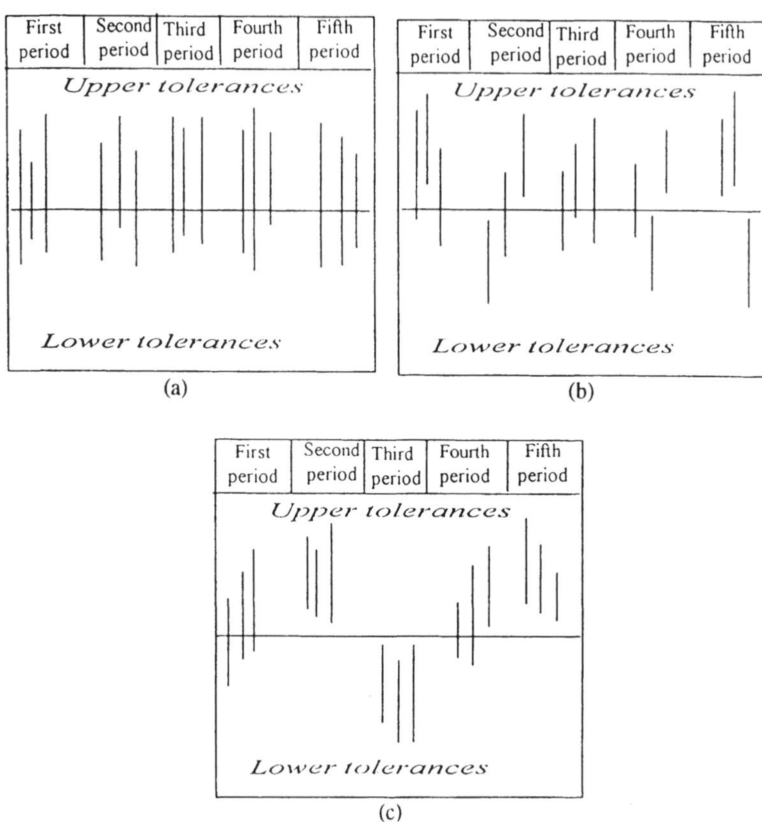

Figure 7.3 Multi-vari chart. (a) Within-piece variation. (b) Piece-to-piece variation. (c) Time-to-time variation.

VII. MULTI-VARI CHART

The basic multi-vari chart uses a vertical line to show the range of variation within a single product piece. When generating clues for problem solving efforts, one looks at the combination of the two key elements of variation and time as shown in Figure 7.3. The figure shows three families of variation in relation to the product tolerance. Specifically, Figure 7.3a displays a case in which the within piece variation is excessive when compared to the tolerance limits. Figure 7.3b shows that piece to piece variation is excessive as indicated by the different positions of the vertical lines. Finally Figure 7.3c depicts a situation in which the main family of variation in the process is time to time.

The multi-vari chart is a very useful tool for detecting different types of variation that are found in all kinds of processes, both in manufacturing and service. Its ability to pinpoint the different types of variation goes far beyond that of the control chart. This ability—to depict different families of variation—helps uncover the cause of the variation, not just identify it as either common cause or special cause. This is especially true in machining dimensional measurements.

The actual mechanics of the multi-vari charts is very similar to those of individual X chart. However, the reader is encouraged to read Bhote (1991) and Nemeth and Zaciewski (1995).

REFERENCES

Bhote, K. R. (1991). *World class quality*. AMACOM. New York.

Box, G. E. P. (1957). Evolutionary operation: A method for increasing industrial productivity. *Applied Statistics*. 6: 235–339.

Box, G. E. P. and Draper, N. R. (1969). *Evolutionary operation*. John Wiley. New York.

Charbonneau, H. and Webster, G. (1978). *Industrial quality control*. Prentice Hall. Englewood Cliffs, NJ.

Crowder, S. V. (1987). A simple method for studying run-length distributions of exponentially weighted moving average charts. *Technometrics*. 29(8): 405.

Crowder, S. V. (1989). Design of exponentially weighted moving average schemes. *Journal of Quality Technology*. 21(4): 155–162.

DeBryn, C. S. V. (1968). *Cumulative sum tests: Theory and practice*. Hafner Publishing Co. New York.

Duncan, A. J. (1986). *Quality control and industrial statistics*. 5th ed. Irwin. Homewood, IL.

Ewan, W. D. (1963). When and how to use Cu-Sum charts. *Technometrics*. 5: 346–351.

Grant, E. L. and Leavenworth, R. S. (1988). *Statistical quality control*. 6th ed. McGraw-Hill. New York.

Gulezian, R. (1991). *Process Control: Statistical principles and tools*. Quality Alert. New York.

Hunter, J. S. (1986). The exponentially weighted moving average. *Journal of Quality Technology*. 18(4): 203–210.

Johnson, N. L. and Leone, F. C. (1977). *Statistics and experimental design in engineering and the physical sciences*. John Wiley and Sons. New York.

Lucas, J. M. (1973). A modified V-Mask control scheme. *Technometrics*. 15: 464–472.

Lucas, J. M. and Saccucci, H. S. (1990). Exponentially weighted moving average control schemes: Properties and enhancements. *Technometrics*. 32(1): 1–29.

Montgomery, D. C. (1991). *Introduction to statistical quality control*. 2nd ed. John Wiley and Sons. New York.

Nemeth, L. and Zaciewski, R. D. (1995). The multi-vari chart: An underutilized quality tool. *Quality Progress*. Oct.: 81–83.

Page, E. S. (1961). Controlling the standard deviation by CUSUMs and warning lines. *Technometrics*. 3: 248–351.

Van Gilder, J. F. (1995). Application of EWMA to automotive onboard diagnostics. *Automotive Excellence*. Fall: 19–23.

Walker, E., Philpot, J. W., and Clement, J. (1991). False signal rates for the Shewhart control chart with supplementary runs tests. *Journal of Quality Technology*. 23(2): 247.

8

Regression Analysis: The Foundation of Multivariate Analysis

This chapter introduces an advanced statistical technique to the quality professional. Specifically, the topic of regression and some of its derivative techniques are discussed. In addition, a short discussion is given on the issues and concerns of all statistical techniques. The intent of the discussion is awareness rather than competency in implementing the techniques themselves. Consequently, the reader is encouraged to pursue on their own the individual topics. For more details and understanding, see Duncan (1986); Judge et al. (1985); Studenmund and Casidy (1987); Wittink (1988); Gibbons (1985); Cohen and Cohen (1983); Box and Draper (1987); Hanson (1988); Makridakis, Wheelwright, and McGee (1983); Kreitzer (1990); Mirer (1988); Thomas (1990); Mullet (1988); and Pedhazur (1982).

I. REGRESSION ANALYSIS

Regression analysis is the grandfather of all multivariate analytical techniques. It is a methodology that provides many individual techniques to identify variation and relationships. For the selection of the individual techniques see Figure 8.1. Specifically, what regression analysis usually does, is to find an

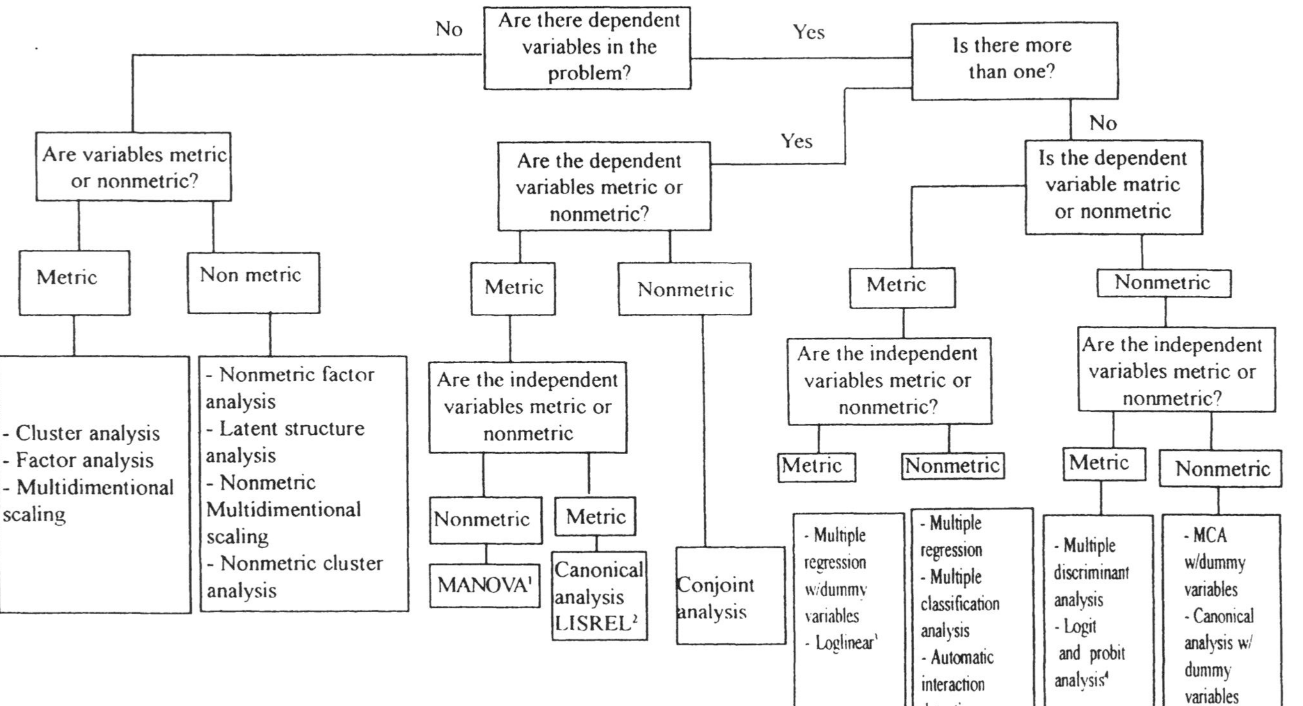

Figure 8.1 Selection of motivariate techniques.

equation which relates a variable of interest (the dependent or criterion variable), to one or more other variables (the independent or predictor variables), of the following form:

$$y = a + b_1x_1 + b_2x_2 + b_3x_3 + \ldots + b_nx_n + e$$

where

y	=	dependent variable
a	=	y intercept
b_1, b_2, b_3, b_n	=	the coefficient of the variables
x_1, x_2, x_3, x_n	=	the independent variables

The major thing to recognize in regression analysis is that the dependent variable is supposed to be a quantity such as how much, how many, how often, how far, and so on. (If you are using a computer package, be warned! The computer will not tell you if you have defined the variable of interest wrong, it is up to you—the experimenter.)

Most regression models will leave you with an equation that shows only the predictor variables which are statistically significant. One misconception that many people have is that the statistically significant variables are also those which are substantive from a quality perspective. This is not necessarily true. It is up to the experimenter to decide which are which.

Regression analysis is one of the most underutilized statistical tools in the field of quality. It provides a tool offering all of the analysis potential of the analysis of variance (ANOVA), but with the added ability of answering important questions ANOVA is ill equipped to address.

It is true, that there are many statistical techniques for determining the relevance of one measure to another. The strongest of these techniques are analysis of variance, goodness of fit, and regression analysis. Each is capable of answering the same basic question of whether or not variation in one measure can be statistically related to variation in one or more other measures. However, only regression analysis can specify just how the two measures are related. That is, only regression can provide quantitative as well as qualitative information about the relationship.

For example, suppose that you need to address the question of whether or not a particular nonconformace is related to factor *X*. Suppose further that you use ANOVA and find a statistically significant relationship does exist. That is all! You have gone as far as ANOVA can take you. Regression analysis offers additional information which neither ANOVA nor goodness of fit can provide, namely, how much of factor *X* will change the specific nonconformity. This information provides an objective basis for the infamous "what if" problems, so central to Lotus-type simulations.

A second advantage of regression analysis is its ready application to graphic imagery. Regression is sometimes referred to as curve or line fitting. The regression output, indeed, yields an equation for a line which can be plotted. The old adage about "a picture is worth a thousand words" holds especially true for regression lines. The image of actual data plotted against predicted values is instantly accessible to even the staunchest statistical cynic. A regression which has been well specified provides its own pictorial justification.

So why would anyone not use regression analysis? There are at least two reasons why regression analysis might not be the first choice of technique for a quality professional and/or researcher. Regression analysis assumes (indeed requires) the error terms to be normally distributed. The normality requirement is sometimes more than a researcher is willing to take on. Nonparametric techniques, which do not make such a restrictive assumption, are better suited to the temperament of these individuals. In general, however, the normality assumption is not outrageous if a sufficiently large sample can be obtained.

The second reason regression might be avoided is paradoxically related to regression's strong suit, quantification of relationships. In some cases one might wish to determine whether or not two measures are significantly related, yet it does not make immediate sense to quantify the relationship between qualitative measures.

To understand regression, there are a few terms that must be understood before undertaking the actual analysis; once understood, they facilitate the understanding of multivariate analysis. The terms are:

Multicollinearity. The degree to which the predictor variables are correlated or redundant. In essence, it is a measure of the extent that two or more variables are telling you the same thing.

R^2. A measure of the proportion of variance in, say, the amount consumed that is accounted for by the variability in the other measures that are in your final equation. You should not ignore it, but it is probably overemphasized. There are a variety of ways to get to the final equation for your data, but the thing to recognize for now is that if you want to build a relationship between a quantitative variable and one or more other variables (either quantitative or qualitative), regression analysis will probably get you started.

II. DISCRIMINANT ANALYSIS

Discriminant analysis is very similar to regression analysis except that here the dependent variable will be a category such as good/bad, or heavy, medium, or light product usage, and so on. The output from a discriminant analysis will be one or more equations which can be used to put an item with a given profile into the appropriate slot.

As with regression, the predictor variables can be mixed bag of both qualitative and quantitative. Again, the available computer packages will not be of any help in telling you when to use regression analysis and when to use discriminant analysis. A discriminant analysis is a perfect tool for an organization to use when there is a question of which supplier to audit or survey as opposed to those that they need not bother.

As with regression, you need to be concerned with statistical versus substantive significance, multicollinearity and R^2 (or its equivalent). Used correctly, it is a powerful tool since so much quality data is categorical in nature.

III. LOGISTIC REGRESSION

Logistic regression does the same things as regression analysis as far as sorting out the significant predictor variables from the chaff, but the dependent variable is usually a 0–1 type, similar to discriminant analysis. However, rather than the usual regression-type equation as output, a logistic regression gives the user an equation with all of the predicted values constrained to be between 0 and 1.

Most users of logistic regression use it to develop such things as probability of success from concept tests (a very good tool for product development). If a given respondent gives positive success intent, they are coded as "1" in the input data set; a negative intent yields a "0" for the input. The logistic regression can also be used instead of a discriminant analysis when there are only two categories of interest.

IV. FACTOR ANALYSIS

There are several different methodologies which wear the guise of factor analysis. Generally, they are all attempting to do the same thing. That is, to find groups, chunks, clumps, or segments of variables that are corrected within the chunk and uncorrelated with those in the other chunks. The chunks are called factors.

Most factor analyses depend on the correlation matrix of all pairs of variables across all of the respondents in the sample. An excellent application of factor analysis is in the areas of customer satisfaction, value engineering, and overall quality definition for an organization; after all, quality is indeed a bundle of things. The question is, however, what makes that bundle, and this is where factor analysis can help. Also, as it is commonly used, factor analysis refers to grouping the variables or items in the current study or the questionnaire together. However, *Q*-factor refers to putting the respondents together, again by similarity of their answers to a given set of questions. Two of the most troubling spots in any factor analysis are:

Eigenvalue. Eigenvalues are addressed here from a practical approach and in no way represents a mathematician's and/or a pure statistician's explanation. All that is necessary for the user to know about eigenvalues in a factor analysis is that they add up to the number of variables that you started with and each one is proportional to the amount of variance explained by a given factor. Analysts use eigenvalues to help decide when a factor analysis is a good one and also how many factors they will use in a given analysis. Usually that comes from a visual inspection of the graph output called "spree."

Rotation. Rotating an initial set of factors can give a result that is much easier to interpret. It is a result of rotation that labels such as "value sensitive," "quality," "basic characteristics," and so on are applied to the factor.

Factor analysis should be done with quantitative variables, although it is possible to conduct an experiment with categorical variables. As with most multivariate procedures, that seems to be the bottom line for factor analysis: Does it make sense? If yes, it is a good one; otherwise, it is probably not, irrespective of what the eigenvalues say.

V. CLUSTER ANALYSIS

Now the points of interest are respondents, instead of variables. As with factor analysis, there are a number of algorithms around to do cluster analysis. Also, clusters are usually not formed on the basis of correlation coefficients. Cluster analysis usually looks at squared differences between respondents on the actual variables you are using to cluster. If two respondents have a large squared difference (relative to other pairs of respondents) they end up in different clusters. If the squared differences are small, they go into the same cluster. Cluster analysis may be used in conjunction with a quality function deployment in defining customer requirements.

VI. GENERAL LINEAR MODEL

The general linear model (GLM) procedure uses the method of least squares to fit general linear models. Among the many options within the GLM family of statistical methods, one may use the following, especially if unbalanced data is present:

Analysis of variance (especially for unbalanced data)
Analysis of covariance
Response-surface models
Weighted regression
Polynomial regression
Partial correlation
Multivariate analysis of variance (MANOVA)
Repeated measures analysis of variance

Even though the GLM approach and application to data analysis is very complicated, with the use of computer software the applicability of GLM in quality applications is within reach. There are many reasons why one may want to use the GLM model; the following nine are the most important as identified by the SAS Institute in explaining the features of their specific software (SAS, 1985):

1. When more than one dependent variable is specified, GLM can automatically group those variables that have the same pattern of missing values within the data set or within a group. This ensures that the analysis for each dependent variable brings into use all possible observations.
2. GLM allows the specification of any degree of interaction (i.e., crossed effects) and nested effects. It also provides for continuous-by-continuous, continuous-by-class and continuous-nesting effects.
3. Through the concept of estimability, GLM can provide tests of hypothesis for the effects of a linear model regardless of the number of missing cells or the extent of confounding data. GLM can produce the general form of all estimable functions.
4. GLM can create an output data set containing predicted and residual values from the analysis and all of the original variables.
5. The MANOVA statement allows you to specify both the hypothesis effects and the error effect to use for a multivariate analysis of variance.
6. The REPEAT statement lets you specify effects in the model that represent repeated measurements on the same experimental unit, and provides both univariate and multivariate tests of hypotheses.
7. The RANDOM statement allows you to specify random effects in the model; expected mean squares are printed for each Type I, Type II, Type III, Type IV, and contrast mean square used in the analysis.
8. You can use the ESTIMATE statement to specify an **L** vector for estimating a linear function of the parameter **Lb**.
9. You can use the CONTRAST statement to specify a contrast vector or matrix for testing the hypothesis that **Lb** = 0.

VII. STEP REGRESSION

Step regression is a methodology that may be used for exploratory analysis because it can give you insight into the relationships between the independent variables and the dependent or response variable. There are five ways that the experimenter may want to use step regression. They are:

1. *Stepwise.* This method is a modification of the forward technique and differs in that variables already in the model do not necessarily stay there. As in the forward selection method, variables are added one by one to the model, and the F statistic for a variable to be added must be significant at the preset level. After a variable is added, however, the stepwise method looks at all the variables already included in the model and deletes any variable that does not produce an F statistic significant at the preset level. Only after this check is made and the necessary deletions accomplished can another variable be added to the model. The stepwise process ends when none of the variables outside the model has an F statistic significant at the preset level, and every variable in the model is significant at the preset level equal to the level or when the variable to be added to the model is one just deleted from it.
2. *Forward.* The forward technique begins with no variables in the model. For each of the independent variables, the forward approach calculates F statistics reflecting the variable's contribution to the model if it is included. These F statistics are compared with the preset significant value that is specified in the model. If no F statistic has a significant level greater than the preset value, the inclusion of new variables stops. If that is not the case, the variable with the largest F statistic is added and the evaluation process is repeated. Thus, variables are added one by one to the model until no remaining variable produces a significant F statistic. Once the variable is in the model, it stays.
3. *Backward.* The backward elimination technique begins by calculating statistics for a model, including all of the independent variables. Then the variables are deleted from the model one by one until all the variables remaining in the model produce F statistics significant at the preset level. At each step, the variable showing the smallest contribution to the model is deleted.
4. *Maximum R^2 (MAXR).* As a general rule this approach of step regression is considered to be superior to the stepwise technique and almost as good as all possible regressions. Unlike the previous three techniques, this method does not settle on a single model, instead, it tries to find the best one-variable model, the best two-variable model, and so on. One of the greatest shortcomings it is the fact that the use of MAXR does not guarantee to find the model with the largest R^2 for each size.
5. *Minimum R^2 improvement (MINR).* This method closely resembles the MAXR but the switch chosen is the one that produces the smallest increase

in R^2. Of note is the consideration that for a given number of variables in the model, MAXR and MINR usually produce the same "best" model, but MINR considers more models of each size.

VIII. CORRELATION ANALYSIS

By far the simplest and most often used analysis in the quality profession, correlation analysis is the simplest of the regression methodologies and its focus is to define quantifiably relationships between variables. The relationship is always identified with the letter r, and is called the correlation coefficient. Whereas the scatter plot shows pictorially the relationship, the correlation analysis defines the relationship with a numerical value. The numerical value varies from $r = -1$ (perfect negative relationship: as one variable increases the other one decreases), to $r = 0$ (no relationship), to $r = +1$ (perfect positive relationship: as one variable increases the other one increases). In using correlation analysis, care must be exercised not to assume that just because there is a relationship between the variables, there is also a causation. That is a major error and it should be avoided. A reality check may help in avoiding this assumption. Instead of the correlation coefficient, the experimenter should use the coefficient of determination, which of course is r^2. The r^2 identifies the true explanation of the relationship. For example if $r = 0.8$, then the $r^2 = 0.64$, which is much less.

IX. CONJOINT ANALYSIS

Marketing professionals have used conjoint analysis for over 15 years to focus on the relationship between price differentiation and customer preferences. Quality professionals can and should use the conjoint analysis especially in the areas of supplier selection, customer satisfaction, as part of the QFD approach and product development. Whereas it is beyond the scope of this book to address the step-by-step mechanics and methodology of conjoint analysis, in this section we are going to give some potential applications in the field of quality. The reader is encouraged to see Sheth (1977) and any advanced marketing research book for a detailed discussion.

Conjoint analysis, a sophisticated form of trade-off analysis, provides useful results which are easy for managers to embrace and understand. It aims for greater realism, grounds attributes in concrete descriptions, and results in greater discrimination between attribute importance. In other words, it creates a more realistic context.

Some of the potential applications are:

1. Design or modifying products and services. One needs to develop suitable mathematical formulations which translate subjective evalu-

ations of new product or service concepts into objectives (e.g., short-term and long-term profits, or rate of return) of the firm. Such formulations then enables the determination of optimum new products subject to several technological, economic, social, or other constraints. Specification of the objectives as well as constraints can be quite complex, particularly for the case of public goods or services. For initial formulations of this problem see Rao and Soutar (1975).

2. Comparing the importance and value of technical product features. Engineers tend to fall in love with innovative ideas. Conjoint analysis, on the other hand places the importance and value of specific features in perspective. A very strong application in this area may be to study the balance between precision performance of a product and durability and/or longevity.

X. CONCERNS WITH STATISTICAL TECHNIQUES

A. Missing Data

Many times we run a specific statistical technique using canned software programs. However, when the analysis is all done we noticed that our results were based on much fewer responses than anticipated. Why? Because, the packages assume a multivariate procedure and drop any response from the analysis which fails to answer even a single item from your set of independent or dependent variables. What do you do?

The question is very important, however, there is no definitive answer. Some approaches of resolving this dilemma are:

1. Use only the responses with complete information. This has the effect of you basing your study on much less respondents than planned for. As a consequence, the validity of the study may be questioned.
2. Use the mean as a substitute for any value that is missing. Most software packages will do this if you tell them to; however, there are some reprocussions if you do this:
   a. If there is a given response that, say, 90% is not available, the mean answer of the remaining 10% is substituted and used as if that is what the other 90% responded. In effect, your response distribution is forced to be more leptokurtic.
   b. A response with few or no data is included in the analysis, unless the experimenter overrides the substitution of means. In either case, the response distribution is forced to be more leptokurtic.

An alternative that seems to be gaining favor is to use only those responses who are corresponding to the dependent variable and then substitute the response's own mean on the ones that there is no response. For more information on missing data see Cohen and Cohen (1983), Cox and Snell (1981), and Milliken and Johnson (1984).

B. Significance Testing

Software packages generally test two types of statistical hypotheses simultaneously. The first type has to do with all of the independent variables as a group. It can be worded in several equivalent ways: Is the percentage of variance of the dependent variable that is explained by all of the independent variables (taken as a bunch) greater than zero or can I do a better job of predicting/explaining the dependent variable using all of these independent variables than not using any of them? Anyway, you will generally see an F-test and its attendant significance, which help you make a decision about whether or not all of the variables, as a group, help out. This, by the way, is a one-sided alternative hypothesis, since you cannot explain a negative portion of variance.

Next, there is usually a table which shows a regression coefficient for each variable in the model, a t-test for each, and a two-sided significance level. (This latter can be converted to a one-sided significance level by dividing by two, which you will need to do if, for example, you have posited a particular direction or sign, a priori, for a given regression coefficient.) Now here is the funny thing. You will sometimes see regression results in which the overall regression (the F statistic) is significant, but none of the individual coefficients are accompanied by a t statistic which is even remotely significant. This is especially common when you are not using stepwise regression and are forcing the entire set of independent variables into the equation, and can result from correlation between the independent variables. If they are highly correlated, then as a set they can have a significant effect on the dependent variable, whereas individually they may not.

Finally, one of the most troubling issues in quality today is the interpretation of significance. That is because both researchers and practitioners do not understand the concept fully. Let us then examine this little more closely. In everyday language the term significance means importance while in statistics—and this is very important—means *probably true*. The implication of these simple definitions is the fact that a particular study may be true without being important. As a result, when a quality engineer says that a particular result is highly significant, it means that the result is very probably true; it does not mean that the result is highly important.

Significance levels show you how probably true a result is. The most common level is 95% with no special significance, other than the result has a 95% chance of being true. This is also misleading, however, due to the fact that most software packages show you 0.05, meaning that the finding has a 5% chance of not being true, which is the same as a 95% chance of being true. The 95% level comes from academic publications, where theory usually has to have at least a 95% chance of being true to be considered worth reporting. However, under no circumstances is the 95% level of significance sacred. It may be set at any level by the experimenter. In fact, in the business world if something has a 90% chance of being true ($p = 0.1$), it certainly cannot be considered proven, but it may be much better to act as if it were true rather than false.

To find the exact significance level subtract the number shown from one. For example, a value shown in the computer output as 0.01 means there is a 99% ($1 - 0.01 = 0.99$) chance of being true. A strong warning about this significance is the fact that *it must be set before the experiment and analysis have taken place.* Do not fall in the trap of seeing the significance on the computer printout and then changing the parameter of the significance. For more information on significance see Lehmann (1986), Fuller (1987), and Rousseeuw and Leroy (1987).

C. Wrong Signs

This happens all the time. For example, you know that overall satisfaction and convenience are positively correlated. Yet, in multiple regression, the sign of the coefficient for convenience is negative. How come?

There are a couple of things which can be going on. First the t statistic for the coefficient may not be statistically significant. We interpret this as an indication that the coefficient is not significantly different from zero and, hence, the sign (and magnitude for that matter) of the coefficient are spurious. Fully half the time and for a truly nonsignificant effect, the sign will be wrong.

The second thing that can be happening is the partialling effect noted above. It could be that the slope is negative given the effect of the other variables in the regression (partial) even though all by itself the variable shows a positive correlation and slope (total). For more information on signs see Hays (1981), Snell (1987), Box and Draper (1987), and Gibbons (1985).

REFERENCES

Box, G. E. P. and Draper, N. R. (1987). *Empirical model-building and response surfaces*. John Wiley and Sons. New York.

Cohen, J. and Cohen, P. (1983). *Applied regression/correlation analysis for the behavior sciences*. 2nd ed. Lawrence Erlbaum Associates, Publishers. Hillsdale, NJ.

Cox, D. R. and Snell, E. J. (1981). *Applied statistics: Principles and examples*. Chapman and Hall. New York.

Duncan, A. J. (1986). *Quality control and industrial statistics*. 5th ed. Irwin. Homewood, IL.

Fuller, W. A. (1987). *Measurement error models*. John Wiley. New York.

Gibbons, J. D. (1985). *Nonparametric statistical inference*. 2nd ed., rev. and expanded. Marcel Dekker. New York.

Hanson, R. (1988). Factor analysis: A useful tool but not a panacea. *Quirk's Marketing Research Review*. May: 20–22, 33.

Hays, W. L. (1981). *Statistics*. 3rd ed. Holt, Rinehart Winston. New York.

Judge, G. G., Griffiths, W. E., Hill, R. C., Lutkepohl, H., and Lee, T. (1985). 2nd ed. John Wiley and Sons. New York.

Kreitzer, J. L. (1990). To progress you must first regress. *Quirk's Marketing Research Review*. Jan.: 8–9, 22–26.

Lehmann, E. L. (1986). *Testing statistical hypotheses*. 2nd ed. John Wiley and Sons. New York.

Makridakis, S., Wheelwright, S. C., and McGee, V. E. (1983). *Forecasting: Methods and applications*. 2nd ed. John Wiley and Sons. New York.

Milliken, G. A. and Johnson, D. E. (1984). *Analysis of messy data*. Van Nostrand Reinhold Co. New York.

Mirer, T. W. (1988). *Econometric statistics and econometrics*. 2nd ed. Macmillian Publishing Co. New York.

Mullet, G. (1988). Multivariate analysis—some vocabulary. *Quirk's Marketing Research Review*. Mar.: 18, 21, 33.

Pedhazur, E. J. (1982). Multiple regression in behavioral research. 2nd ed. Holt, Rinehart and Winston, New YOrk.

Rao, V. R. and Soutat, G. N. (1975). Subjective evaluations for product design decisions. *Decision Sciences*. 6: 120–134.

Rousseeuw, P. J. and Leroy, A. M. (1987). *Robust regression and outlier detection*. John Wiley and Sons. New York.

SAS Institute (1985). The GLM procedure. In: *SAS User's Guide: Statistics*. Version 5 Edition. Cary, NC.

Sheth, J. N., Ed. (1977). *Multivariate methods for market and survey research*. American Marketing Association, Chicago.

Snell, E. J. (1987). *Applied statistics: A handbook of BMDP analyses*. Chapman and Hall. New York.

Studenmund, A. H. and Casidy, H. J. (1987). *Using econometrics: A practical guide*. Little Brown and Company. Boston.

Thomas, T. (1990). New findings with old data using cluster analysis. *Quirk's Marketing Research Review*. Feb.: 9–18, 36.

Wittink, D. R. (1988). *The application of regression analysis*. Allyn and Bacon. Boston.

9

Design of Experiments

The design of experiments (DOE) enables researchers to plan or design an experiment that simultaneously alters a number of variables in an experimental system to see how they affect and interact to affect responses. The process is obviously handled with predefined statistical parameters that the experimenter will use to draw objective, reliable conclusions by controlling experimental variation and estimating its magnitude to guarantee accuracy. The DOE approach also allows experimenters to keep the size of the experiment no larger than necessary to meet objectives. Specifically, this chapter addresses the concepts of DOE from at least three distinct approaches without the heavy statistical and mathematical equations; these approaches are classical design, Taguchi design, and Shainin methodology.

I. DOE VERSUS ONE AT A TIME

DOE is not new. The basic principles were established in the early 1920s and the methodology has continued to evolve and improve in the decades since. In spite of its history, however, DOE is not widely used. Practitioners prefer the one-factor-at-a-time approach in which practitioners alter the setting of one variable, hold the other constant, and then measure the resulting response.

One-factor-at-a-time experimentation has many shortcomings. It leaves too much to chance, does not control for experimental variation, does not uncover interactions among variables, seldom optimizes a multivariable system, and compromises critical experimental principles. Furthermore, contrary to popular belief, it also requires larger experiments than necessary.

One of the main reasons for this preference for the one-factor-at-a-time experimentation is the lack of education. Typically, people who are expected to use DOE, have little to no exposure to statistics in their formal training. At best, they may have exposed themselves to an introductory statistics course. But few, if any, statistics courses have much to say about DOE, especially about how to use the methodology.

The other thing that prevents use is that the up-front planning required is counterproductive, especially to engineers. Instead of building on the known, DOE requires the practitioner to plan to jumble the sequence of testing for the sake of randomization. Then, you sort it out. The idea of messing things up and then sorting them out does not seem sensible. Many engineers and scientists who have little or no training in DOE consider it irrelevant. They are used to the concept of the controlling experiment—which means one factor at the time. Many consider this approach to be a very pure part of the scientific approach. As a result of this kind of thinking many have a perception that DOE requires more samples for experimentation than necessary. This, of course, can be a legitimate cost/benefit issue that the statistician needs to address.

Finally, another major problem is lack of management support. Managers seldom, if ever, embrace what they do not understand. The problem is that managers are too busy fighting fires and meeting budgets to think about supporting planned multivariate experiments. They want to know if you can fix the problem on a short-term basis by turning a few knobs, instead of asking whether you understand how the process works, so you can control variability in the system into the future. It is a fire-fighting mode of operation versus deeper process understanding. A lack of understanding can even breed antagonism, as some top level technical managers may believe that if you have enough process data, a bright numbers person should be able to unlock the process secrets without experimentation.

II. CLASSICAL DESIGN

Where in Chapter 8 we discussed the regression methodology and its derivatives, here we are concerned with experimental design. An experimenter uses this approach to plan experiments that will provide the most information with minimum available resources. This usually involves two basic processes: treatment selection and error control.

Treatment selection is the choice of experimental variables, the treatments, and the settings of these variables (i.e., the levels), to include in the experiment. Changes in the response variable due to changes in the treatment level are called effects of the factor or treatment. The factors may be continuous (quantitative) or discrete (qualitative).

Combinations of the factors make up the treatments in the experiment. In fact, an experiment containing all possible combinations of levels of factors is called a factorial experiment. It is very expensive and time consuming for any one to do experimentation with all combinations (also called a full factorial design). As such most people use portions of the full factorial called fractional factorial. The degree of the fraction used depends on the time and resources available.

One of the classical methodologies in experimental design is the analysis of variance (ANOVA). It is used when statistical inferences are to be made about a set of more than two means. This ANOVA can be used with both balanced and unbalance designs (for unbalanced designs it is more appropriate to use the GLM method; see Chapter 8.)

Under the ANOVA methodology there are several ways of doing the experiment depending on how, where, and why the data is collected and whether or not the design of the experiment is randomized. In this section, we address the most common ones without the mathematical background. The reader is encouraged to see Montgomery (1991), Hicks (1982), Duncan (1986), Freund and Littell (1981), Dayton (1970) and Box, Hunter, and Hunter (1978) for more details.

A. Completely Randomized Design

This design assumes that experimental units of a population are randomly assigned to subpopulations often called treatments. An equivalent description assumes a set of independently drawn samples from each of several populations. The null hypothesis of interest is that each of the subpopulations have the same mean (assuming equal variance).

The general model for the completely randomized design is:

$$y_{ij} = m + t_i + e_{ij}$$

where

m = grand mean of treatment populations
t_i = effect of treatment i
e_{ij} = experimental error

B. Randomized Block Design

This design assumes that a population of experimental units can be divided into a number of relatively homogeneous subpopulations or blocks. The treatments are then randomly assigned to experimental units such that each treatment occurs equally often (usually once) in each block and that each block contains all treatments. Blocks usually represent naturally occurring differences not related to the treatments. In the analysis the variation among blocks can be partitioned out, thereby usually reducing the error mean square.

The classic of blocks is agricultural fields divided into smaller, more homogeneous subfields. Other examples include days of the week, identical measuring or recording devices, and operators of a machine.

The general model for the randomized block design is:

$$y_{ij} = m + b_i + t_i + e_{ij}$$

where

m = grand mean of treatment populations
b_i = effect of block i
t_j = effect of treatment j
e_{ij} = experimental error

C. Latin Square Design

This type of design is a very complicated design in the sense that it allows for two sources of blocks (usually referred to as rows and columns) and treatments are randomly assigned to experimental units such that each treatment occurs once in each row and once in each column. For example, consider an experiment with five treatments, and experimental units are obtained five times a day for five days. This could be considered as a randomized blocks design with days as blocks. However, if it is suspected that there may be differences due to times of day, then times could be considered blocks. Such situations call for a Latin Square design. In this example, the rows are the days and columns are the times of day.

The general model for the Latin square design is:

$$y_{ijk} = m + r_i + g_j + t_k + e_{ijk}$$

where

y_{ijk} = observed response in the ith run and j^{th} position for the kth material
m = grand mean average
r_i = effect of run
g_j = effect of position
t_k = effect of material
e_{ijk} = experimental error

D. Factorial Experiment

While an experiment design is concerned with the assignment of treatments to experimental units, a factorial experiment is concerned with the structure of treatments. However, the factorial structure may be placed into any experimental design.

A factorial experiment consists of all possible combinations of several levels of two or more types of treatments. These types, called factors, may be different ingredients (A, B, C), operating conditions (T, P), biological factors (B, V), or combinations of the above (interactions). Levels may refer to numeric quantities of factors (e.g., pounds of ingredients or degrees of temperature) as well as qualitative categories (e.g., names for the biological factors). Examples of factorial experiments, in this case, are three levels of A B C resulting in $3^3 = 27$ treatment combinations or three varieties using two levels of conditions and two levels of biological conditions resulting in $3 \times 2 \times 2 = 12$ treatment combination. Factorial experiments determine not only the effects of the levels of each factor and their main effects, but also the existence of interactions; that is, how do levels of one factor affect the behavior of the response variable across levels of another factor?

To illustrate the factorial experiment three seed growth methods are applied to lots of seed from each of five varieties of turf grass. Six pots are planted with seed from each method–variety combination. The 90 pots are randomly placed in a uniform growth chamber and the dry matter yields (y) are measured after clipping at the end of four weeks. We view the experiment as if we were interested only in these five varieties and growth methods, regarding variety and method as fixed effects. Thus, the experiment is a 3×5 factorial in a completely randomized design with two factors: method and variety. The following model is assumed for any single measurement of a dependent variable in the experiment:

$$y_{ijk} = m + a_i + b_j + ab_{ij} + e_{ijk}$$

where

y_{ijk}	=	the yield of the kth pot of the ith method and j^{th} variety
m	=	grand mean of treatment populations
a_i	=	main effect of method
b_j	=	main effect of variety
ab_{ij}	=	interaction of method and variety
e_{ijk}	=	experimental error

The method × variety interaction is a measure of whether comparisons of method means depend on the variety being tested. If the interaction is present, it may be necessary to compare method means separately for each variety. If the interaction is not present, we can make an overall comparison of methods averaged over all levels of variety.

E. Split Plot Design

This design is a specialized designed for a factorial experiment. While one factor is more readily applied to large experimental units or main plots, another factor can be applied to smaller units or subplots within the larger unit. The model for a typical split plot design is:

$$y_{ijk} = m + b_i + t_j + d_{ij} + h_k + (th)_{jk} + e_{ijk}$$

where

m = overall mean
b_i = effect of replication i
t_j = effect of main plot treatment j
d_{ij} = experimental error for main plot treatment
h_k = effect of subplot treatment
$(th)_{jk}$ = interaction between factors t and h
e_{ijk} = experimental error associated with subplots

F. Nested Designs

Nested designs are most frequently used to estimate a mean of a large population whose units are classified in a hierarchical or nested manner. Typically, these samples are taken in several stages. The first stage is selection of main units, the second stage is selection of subunits from each main unit, the third stage is selection of sub-subunits from the subunits and so on. Normally, the classification schemes at each stage are considered random effects, but in some cases a classification scheme may be considered fixed, especially one corresponding to the first stage of sampling.

The model for a typical nested design is:

$$y_{ijk} = \mu + a_i + b_{ij} + e_{ijk}$$

where

y_{ijk} = the yield for the kth duplicate of the jth sample from the ith sample
m = overall mean of the sampled population
a_i = effect of being i^{th} level of treatment dimension
b_{ij} = effect of being i^{th} level of nested factor
e_{ijk} = experimental error

G. Repeated-Measures Design

This is an experimental design that is very commonly used in the pharmaceutical industry. It is usually a completely random design with the response measured on each experimental unit at several points in time.

H. Split-Split Experiment

As in any split-plot design the foundations are those of the nested design. However, in the split-split plot design a special combination occurs in that (usually) two main effects are confounded with blocks. As a consequence, we are very much interested to find and define the error terms. For example: If we consider two or more factors then our design may be:

	Design	Analysis
1.	Complete randomized block $y_{ijk} = \mu + R_k + A_i + B_j + AB_{ij} + e_{ijk}$	Factorial ANOVA
2.	Incomplete, confounding: Main effect—split plot $y_{ijk} = \mu + \underbrace{R_i + A_j + RA_{ij}}_{\text{whole plot}}$ $+ \underbrace{B_k + RB_{ik} + AB_{jk} + RAB_{ijk}}_{\text{split-plot}}$	Split-plot ANOVA

III. AFTER THE ANOVA

The F test that results from an analysis of variance tests a null hypothesis that the means in a set are all equal. The conclusion of such a test is not often a satisfactory end to the analysis. The experimenter usually wants to know more about the differences among the means, for example, which means are different from which other means, or if any groups of means have common values. The way to answer these questions is through multiple comparisons of the means. One method of multiple comparisons is to conduct a series of t tests between pairs of means. [This is the method known as least significant difference; see Steele and Torrie (1980). Of course, other mean comparisons do exist, such as the Duncan, Waller-Duncan k-ratio, Sheffee's test, Newman Keuls and many others. For explanation of these tests see Duncan (1986), Hicks (1982), and SAS (1985).]

In addition to the means analysis, the experimenter should be concerned with two more issues: confounding and pooling of variables. Confounding happens when two or more variables interact, but the interaction is hidden. Therefore, the experimenter must be aware of the possibility of confusing the interaction with a factor or vice versa. Sometimes when the experimenter knows that the interaction is unimportant, one may use that knowledge to study further the effect of a factor and/or a different interaction all together. For more on confounding see Box (1950), Duncan (1986), and Box, Hunter, and Hunter (1978).

Pooling, on the other hand, has to do with sums of squares and degrees of freedom. As you pool (group together) variables with low significance, the remaining variables become more significant. There are at least three schools of thought on the issue of pooling.

1. Never use it, but remember that in some cases a precise test will not be possible.
2. Always use it, but remember that the risks of falsely rejecting or incorrectly accepting the null hypothesis will have to be modified since the basis for our tests is much more complicated than if we never pool.
3. Use pooling selectively, when it is beneficial.

A good rule of thumb when to use pooling is when the degrees of freedom for the original residual mean square are less than 6. On the other hand, if the variance ratio for the sum of square that is to be pooled is high, although not too high to be significant, the danger of incorporating nonchance variability in our residual term may be fairly high. For more on pooling see Duncan (1986), Box, Hunter, and Hunter (1978), and Dayton (1970).

IV. TAGUCHI DESIGN

The Taguchi approach for quality has become popular over the past decade in the United States. Kirkland (1988) reported that Japanese companies (e.g., Sony and Toyota) have practiced elements of the Taguchi approach since the early 1960s. These companies have led the successful invasion into the U.S. electronics and automotive markets and their success is primarily due to aggressive use of the Taguchi approach in their overall quality.

The methodology used by Taguchi may be used in any product design or manufacturing operation, regardless of volume or type of production or market served. Port (1987) identified the basic tenant of Taguchi's contribution to quality as a design of a product robust enough to achieve high quality despite fluctuations on the production line. This sounds simple, however, the mathematical formulas associated with robust quality are too intensive to be discussed here. For a detailed analysis of the topic see Roy (1990), Ross (1988), and Taguchi (1986, 1987). The goal of this section is to address the basic philosophy of Taguchi and explain some of his ideas as they relate to the quality issue.

On the question of how Taguchi is able to do what he claims to do (i.e., a robust design), the answer is very simple: he does it by consistency throughout. He feels so strongly about consistency of performance that in fact he uses it as the sole definition of quality. Please note that his definition of quality is quite different from that of many people in the field of quality. By the same token, one must admit that the aspect of quality that engineers may affect can be generally measured in terms of performance of the product or process under study.

When Taguchi speaks of a product or process that consistently performs the intended function (at target), it is considered to be of quality. When the performance is consistent, its deviation from the target is minimum. As a consequence, the lower the deviation of performance from the target, the better the quality. The secret of Taguchi's philosophy and approach is to select an appropriate quality characteristics for performance evaluation. For example, if we want to hire an engineer with an excellent command of the language, then we look for one who has a better track record in English. On the other hand, if we are looking for an athlete, we put more emphasis on the candidate's ability and/or performance in sports activity. In selecting the best combination of product or process parameters that are consistent with his definition of quality, Taguchi looks for a combination that produces the least average deviation from the target. In doing this, he neglects the interaction of the variables and this has become one of the strongest arguments against the method (Box, Hunter, and Hunter 1978).

Taguchi (1987) points out that in all cases, interaction is not considered in his book. This is not because there is no interaction, but rather since there can be interaction, we perform experiments only on the main effects, having canceled interactions. If the interactions are large, then no assignment will work except experiments on a certain specific combination. Good results are obtained using an orthogonal array if interactions have been omitted. This is because minimizing interaction is not a matter of assignment, but should be dealt with by techniques of the specific technology and appropriate analysis technique.

Taguchi uses a statistical quantity that measures deviation from the target and he calls it *mean square deviation* (MSD). For a robust design of product or process, the combination with the least MSD should be selected. For convenience of analysis and to accommodate a wider range of observations, MSD is transformed into a signal-to-noise (S/N) ratio before analysis.

So far we looked at Taguchi from an overall point of view. Now let us address some specific philosophical issues.

1. An important dimension of the quality of a manufactured product is the total loss generated by the product to society. Barker and Clausing (1984) interpret Taguchi to mean that any product's quality can be defined as the monetary loss that a product imparts on society once it is shipped. The

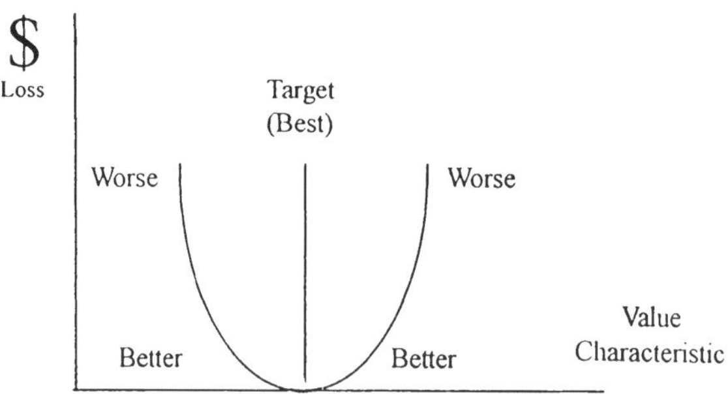

Figure 9.1 The loss function.

idea of quality as a loss to society is unique in that quality is typically defined in positive terms. Taguchi's definition implies that a product's desirability is inversely related to the amount of societal loss. Wood (1988) believes that in addition to the above, loss to society should include raw materials, energy, and labor consumed during the production of the product. However, Taguchi specifically means losses such as failure to meet ideal performance, and harmful side effects caused by the product. Taguchi has shown this idea in a graphical form by plotting a parabola, which he calls the *loss function*. An example of the loss function is shown in Figure 9.1. This concept was used by Kacker (1986) to redefine the aim of quality control as a concept to reduce total societal cost and its function to be discovering and implementing innovative techniques that produce a new savings to society.

2. In a competitive economy, continuous quality improvement and cost reduction are necessary for staying in business. In today's competitive market environment, a business must earn a reasonable profit to survive. Profit is a function of reducing manufacturing costs and increasing revenue. Market share may be increased by providing high quality products at competitive costs. Customers want high quality and low cost. Taguchi believes that the customer is willing to pay a little more for a higher quality product. He hints (but does not say explicitly) that quality must have a reference frame of price to be meaningful. It would not be fair to compare the quality of a Ford to a Rolls Royce. The Rolls is perceived to be a much higher quality car and at substantially higher price. Yet, because of their price differential, Ford sells a significantly higher volume of vehicles. Taguchi insists that companies determined to stay in business must use high quality and low cost in their business strategy. The quest for increasing quality at decreasing cost must be a never ending proposition.

3. A continuous quality improvement program includes incessant reduction in the variation of product performance characteristics about their target values. A product's quality cannot be improved unless the associated quality characteristics are quantifiable. Kackar (1986) points out that continuous quality improvement depends on knowledge of the ideal values of these quality characteristics. However, because most products have numerous quality characteristics the most economical procedure to improve a product's quality is to concentrate on its primary or performance characteristics. Performance characteristics are values such as the life span of a tire or the braking distance of an automobile. These are measurable quantities. Taguchi believes that the smaller the performance variation about the target value, the better the resulting quality. In contrast to this, the classical approach has been to have target values which for all practical purposes are ranges. This has lead to the erroneous idea that the quality within specification intervals is equal. This concept has been known as the "goal post philosophy." Taguchi suggests that the target value be defined as the ideal state of the performance characteristic. All performance characteristics may not be measurable on a continuous scale and subjective evaluation may be necessary. Taguchi recommends the use of categorical scale such as poor, fair, good, or whatever fits the product and/or process under study.

4. The customer's loss due to a product's performance variation is often approximately proportional to the square of the deviation of the performance characteristic from its target value. Taguchi proposes that customer's economic loss due to the performance variation can be estimated using a quadratic approximation. The derivation of the loss function is fully developed and discussed by Kacker (1986), Ross (1988), and Taguchi (1986, 1987). For our discussion here, it is only important to recognize that as a performance characteristic moves away from the target value (in either direction) the customer's (or societal) monetary loss increases quadratically (Figure 9.1).

5. The final quality and cost of a manufactured product is determined to a large extent by the engineering designs of the product and its manufacturing process. Earley (1989) pointed out that Taguchi's method of experimental design is becoming increasingly popular because they are speedy, dependable, and cost effective. Taguchi believes that a product's field performance is affected by environmental variables, human variables in operating the product, product deterioration, and manufacturing imperfections. He believes that countermeasures caused by environmental variables and product deterioration can only be built into the product at the design stage. The implication that Taguchi makes here is that the manufacturing cost and manufacturing imperfections in a product are a function of the design of the manufacturing process.

Table 9.1 Off-Line and On-Line Characteristics

	Off-line	On-line
Applied for	Product and/or process optimization	Manufacturing stage
Using	Orthogonal arrays, signal-to-noise ratios, quality loss function	Loss function
Emphasizing	Efficient, experimentation and/or simulation, reducing variability, low cost, robustness of products	
Determining		Checking interval, adjustment limit, inspection needs
Preventing		System downtime, reduce the variability of a process, reduce the quality loss of a process

Process control built into the process design will significantly reduce manufacturing imperfections. Process controls are expensive, but are justifiable as long as the loss due to manufacturing imperfections is more than the cost of the controls. Optimally, both manufacturing imperfections and the need for process controls should be reduced.

6. The performance variation of a product (or process) can be reduced by exploiting the nonlinear effects of the product (or process) parameters on the performance characteristics. Taguchi's basic premise of product or process robustness is best acquired by implementing quality control in all the steps of the product development cycle. To do this, a two-pronged approach is utilized. First, off-line quality control methods are used. These are technical aids for quality and cost control in product and process design. They may include prototype testing, accelerated life tests, sensitivity tests, and reliability tests. These tests are used to improve product quality and manufacturability through evaluation, and specification of parameters should be given in terms of ideal values and corresponding tolerances. Second, on-line quality control methods are used. These are technical aids for quality and cost control in the actual manufacturing process and/or customer service stage. Taguchi indicates that by stressing off-line quality control in the development process of a product the potential of on-going loss associated with a product is dramatically reduced. Table 9.1 shows some of the characteristics of both off-line and on-line.

To assure off-line quality, Taguchi introduced a three-step approach:

1. *System design.* This is the process of developing a prototype design using engineering knowledge. In this stage, the definition of the product or process parameters is established.
2. *Parameter design.* This is the stage where product and/or process parameters are established for an optimized sensitivity.
3. *Tolerance design.* This is the scientific determination of the acceptable tolerance around the ideal setting.

7. Statistically planned experiments can be used to identify the settings of product (and process) parameters that reduce performance variation. Taguchi established that statistically planned experiments are essential to successful parameter design. Statistically designed experiments have previously been used in industry; however, Taguchi's contribution brings this experimentation to a new height by providing a quick yet accurate way of determining optimization. Taguchi believes that his form of experimentation is the only method for identifying settings of design parameters while paying attention to costs.

V. TAGUCHI'S DEFINITION OF QUALITY CHARACTERISTICS

Selecting the appropriate quality characteristic is one of the first steps in quality engineering. To do the selection properly one must be familiar with the types of quality characteristics.

1. Measurable. Units of measurement are on a continuous scale. Under this measurement, there are three classifications.
   a. Nominal target (i.e., the best). This is a specific target value (e.g., temperature, pressure, speed, dimension, etc.; Figure 9.1)
   b. The smaller the better. Here, zero is the ultimate target (e.g., tool wear, noise level, contamination, etc.; Figure 9.2)

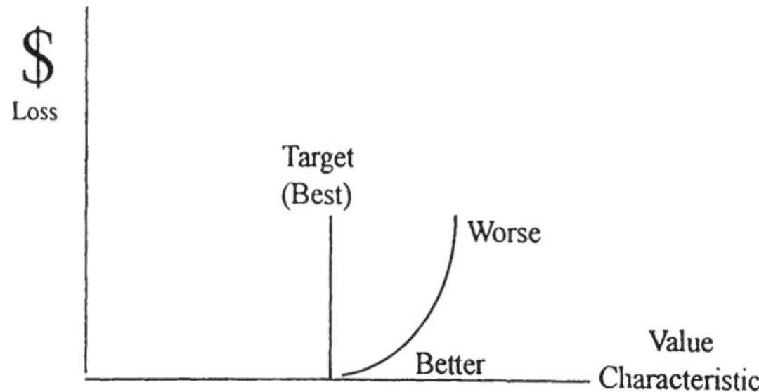

Figure 9.2 Loss function curve: the smaller the better.

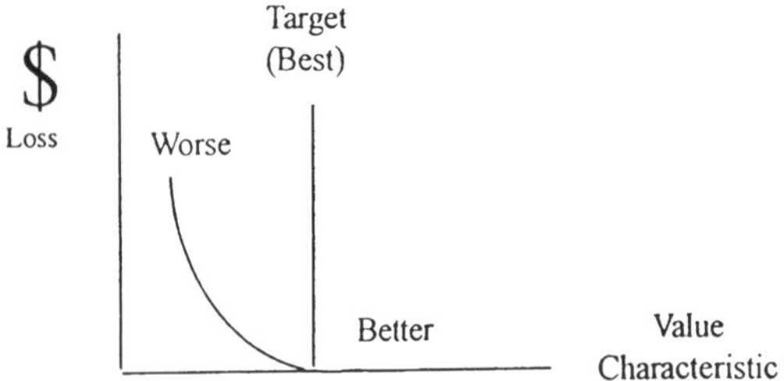

Figure 9.3 Loss function curve: the larger the better.

c. The larger the better. Here, infinity is the ultimate target (e.g., strength, life, efficiency, etc.; Figure 9.3)

2. Attribute. Units of measurement are not on a continuous scale, rather they may be classified on a discretely graded scale (usually subjective). Examples are appearances, porosity, good, bad, etc.
3. Dynamic. They are functional (on line) quality characteristics of a system. Their behavior follows the following diagram:

signal input ——> system ——> output

VI. A COMPARISON OF CLASSICAL AND TAGUCHI DESIGNS

The aim of quality control is to identify and eliminate variation from a product/process. The tools used for this purpose are *statistical process control* (SPC) and *design of experiments* (DOE). One uses DOE to uncover special causes which can then provide a basis for bringing the process under control. There is a clear relationship between SPC and DOE:

- SPC can produce significant reductions in system noise which may lead to more efficient and reliable DOE.
- DOE results may point to the existence of important relationships which may be responsible for out of control conditions.
- DOE may be very useful in improving already controlled processes.
- DOE may be useful in bringing unstable processes into a state of statistical control by revealing key variable effects/relationships.

Within the scope of DOE there are two approaches: the classical and the Taguchi approach. Each one has a goal of the improvement of the product/process, however, the assumptions and the way at which they arrive to this conclusion is a little bit different. Table 9.2 shows some of the differences.

Table 9.2 Design of Experiments: Classical Versus Taguchi Approaches

	Classical	Taguchi
Philosophy	Emphasis on statistical techniques, mathematical models	System, parameter and tolerance design with a goal
Purpose	Improving the efficiency of acquiring information	Improving the efficiency and quality engineering
Objective	To detect causes for meeting specifications	Strive for robustness in a cost-down fashion
Attitude	Cause detection	Removal of the impact of causes
Content	Model identification	Discovery of optimum parameter level combination
Stage of manufacture	Later in production	To dampen noise effects and reduce variation
Way of approach for quality improvement	Tolerance design: (upgrading materials) problem solving, failure analysis, cause detection	Parameter design: (start with low grade materials and optimize design parameter levels) avoid future problems through parameter design optimization
Design	Multiple statistical techniques	Orthogonal arrays, linear graphs, inner/outer array design
Interactions	Try to account for all interactions	Minimize interactions between control factors (prefer main effects)
Methodology	Strictly follow statistics, assumption of equal variability	Development of cost effective methods
Technical differences	Emphasis on F test, interactions treated defensively, assumption of multivariate distribution	Parameter design for noise (S/N), prefer main effects, emphasis on selecting characteristics with good additivity, loss function approach to tolerance design.

VII. TYPICAL TAGUCHI APPROACH TO EXPERIMENTATION

Obviously there are many ways that one may pursue DOE, including the Taguchi approach. However, the following approach is one that we believe is a simple one, yet systematic enough to provide the experimenter with consistent, reliable and replicable results.

1. Define the problem.
   - be realistic
   - temper the scale or size of the problem
   - narrow the problem to find the weakest link
2. Determine the objective.
   - identify the measurable characteristic(s)
   - make it as close to the end use of the product as possible
3. Identify all factors that induce variability.
   - use brainstorming
   - use the cause-and-effect diagram
4. Eliminate all factors that can not be measured.
   - use brainstorming
   - if you cannot eliminate any factors, then find the means of measuring the variable.
5. Separate factors into groups that are either controllable or noncontrollable (noise); focus the experiment on the controllable factors.
6. Identify controllable factors.
   - establish the number of levels (usually 2 or 3)
   - determine the value of each level (be bold here: You are looking for discrimination values)
7. Identify interactions between controllable factors; keep interactions to a minimum and emphasize the factors.
8. Determine the number of parts run per experiment (remember the cost of running experiments).
9. Choose an orthogonal array.
   - consider the cost and time required to do experiments
   - reduce the number of controllable factors, if necessary
10. Identify the noise or noncontrollable factors.
    - limit to the most important factors (no more than 3)
    - establish the number of levels (usually 2 or 3)
    - determine where the levels will be set (be bold)
11. Choose an appropriate linear graph
12. Choose an orthogonal array; if feasibl use a full factorial design.
13. Run the experiment.
14. Verify the experiment and make appropriate and applicable changes.

VIII. SHAININ APPROACH

Dorian Shainin in the last ten to fifteen years has been using a nontraditional DOE approach with some spectacular results. It is only fitting then, that at least in this section we recognize his contribution—even though the analysis is indeed very cursory. The reader is strongly advised to read Bhote (1991) for more detailed information.

Shainin created seven DOE tools (Table 9.3) that can diagnose and greatly reduce variation, leading us beyond zero defects, beyond the milestone of $C_{pk} = 2.0$, to near zero variability (Bhote, 1991). These tools are focusing on reducing variation of a process based on a predetermined priority that Shainin calls the "Red X" (the number one cause), the "Pink X" (the second most important cause) and the "Pale X" (the third most important cause).

What follows then, is a full factorial analysis if there are only four or less variables, or a variables search if there are 5–20 variables. The next step is the validation run of the better versus the current (B vs. C) analysis, and finally the optimization, which he calls "realistic tolerance parallelogram plots." This optimization graphical technique helps the experimenter to evaluate the relationship of the dependent and independent (Red X) variables.

Once the DOE has been performed, validated, and optimized, then Shainin introduces SPC for maintenance. Two of the most famous and at the same time controversial items within the scope of his SPC are the issues of positrol (a positive control) and precontrol. The role of positrol is to ensure that important variables (the what) that are identified and reduced with DOE tools, stay reduced with a "who, how, where, and when" plan and log. On the other hand, the precontrol chart is a chart that divides the specifications in four equal parts, each part designated by a different color. The green color, which is located next to the middle of the specification on both sides, signifies acceptable performance, the yellow, which is located next to the specification,represents a cautioned performance, and the red zone represents points out of the specifications. The color zones are used in addition to identifying the performance of the process, as an incentive/penalty tool. It penalizes poor quality by shutting down the process so often that problem solving with the use of design of experiments becomes imperative. It rewards good quality by sampling less frequently.

In discussing Shainin's approach to DOE, it would be unfair if we did not address the concept of multi-vari charts. For Shainin, the multi-vari chart is a stratified experiment to determine whether the major variation pattern is positional, cyclical, or temporal (time related). If the greatest variation is temporal, the several factors contributing to positional or cyclical variations can be eliminated or given a much lower priority for further investigation. Typical examples of these kind of variation are:

Positional: variation within a single unit, by location, or from machine to machine

Table 9.3 The Seven DOE Tools Used by Shainin

Tool	Objective	Where applicable	When applicable	Sample size
Multi-vari chart	Reduce a large number of unrelated, unmanageable causes to a family of smaller and related causes, such as time-to-time, part-to-part, within part, machine-to-machine, test position-to-test position and so on Detects nonrandom trends	Determines how a product and or process is running, a quick snapshot, without massive, historical data that are of very limited usefulness Replaces process capability studies In some white collar applications	At engineering pilot run, production pilot run, or in production	Minimum 9–15 or until 80% of historic variation is captured
Components search	From hundreds or thousands of components/subassemblies, homes in on the Red X, capturing the magnitude of all important main effects and interaction effects.	Where there are 2 differently performing assemblies (labeled "good" and "bad") with interchangeable components	At prototype, engineering pilot run, production pilot run, production, or field	2

Paired comparisons	Provides clues to the Red X by determining a repetitive difference between pairs of differently performing products	Where there are matched sets of differently performing products (labeled "good" and "bad") that cannot be dis-assembled	Same as components search	5–8 pairs of "good" and "bad" product
Variables search	1. Pinpoints Red X, Pink X and so on. 2. Captures the magnitude of all important main effects and interaction effects 3. Opens up tolerances of all unimportant variables to reduce cost	Where there are 5–20 variables to investigate Excellent problem prevention tool	Excellent in R&D, develop-ment engineering, and in production for product and/or process characterization Also pinpointing Red X after multi-vari or paired comparisons	1–20
Full factorials	Same as variables search	Practical only where there are 2–4 variables	Same as variable search	1–16
B versus C	Validates superiority of a new or better (B) prod-uct or process over a current one with a desired statistical confidence (usually 95%) Evaluates engineering changes Reduces cost	Follows one or more of the above 5 tools When problem is easy to solve, B versus C can bypass above tools In some white collar applications	In prototype, pilot run, or production	Usually 3 Bs and 3 Cs
Realistic tolerance parallelogram (scatter plots)	Determine optimum values (levels) for Red X, Pink X variables and their maximum allowable tolerances	Following above six tools	In pilot of product and or process	30

Source: Used with permission from K. R. Bhote. *World Class Quality*. AMACOM. New York.

Cyclical: variation between consecutive units from the same process or between groups of units, and batch-to-batch or lot-to-lot variation

Temporal: variations from hour to hour, shift to shift, and so on.

The idea of the multi-vari is that by plotting the results, the experimenter can determine if the largest variation is within the unit, unit to unit or time to time. The chart itself is a snapshot in time and should not be confused with a control chart.

IX. A COMPARISON OF THE THREE APPROACHES TO DESIGN OF EXPERIMENTS

A cursory summary of the three DOE approaches is shown in Table 9.4.

X. TYPICAL STAGES IN INDUSTRIAL EXPERIMENTATION

To carry out a successful experiment for quality improvement one must be systematic. The steps involved in any experimentation endeavor are:

1. Define the problem. A clear statement of the problem to be solved.
2. Determine the objective. Identify output characteristics (preferably measurable and with good additivity).
3. Brainstorm and identify factors. It is desirable (but not essential) that inputs be measurable. Group factors into control factors and noise factors. Determine levels and values for factors. Discuss what characteristics should be used as outputs.
4. Design the experiment. Select the appropriate orthogonal arrays for control factors. Assign control factors and interactions to orthogonal array columns. Select an outer array for noise factors and assign factors to columns.
5. Conduct the experiment or simulation and collect data
6. Analyze the data by:

Regular analysis	S/N analysis
Average response table	Average response table
Average response graph	Average response graph
Average interaction graph	S/N ANOVA
ANOVA	

7. Interpret results. Select optimum levels of control factors. For nominal the best use mean response analysis in conjunction with S/N analysis. Predict results for the optimal condition.

Table 9.4 A Comparison of the Three DOE Approaches

Characteristic	Classical	Taguchi	Shainin
Principal techniques	Fractional factorials EVOP	Orthogonal arrays	Multi-vary, components search Paired comparisons, variables search, full factorials, B vs. C Scatter plots and other tools
Effectiveness	Good in the absence of interactions Poor if interactions are present Limited optimization	Good in the absence of interactions Very poor if interactions are present Very limited optimization	Extremely powerful regardless of interactions Retrogression rare Maximum optimization
Cost/time	Moderate (8–50 experiments)	Moderate, if no interactions (8–36 experiments for inner array alone) High, if interactions present (several trials of same experiment: 64 to over 300 experiments for inner and outer arrays combined)	Low (3–30 experiments)
Complexity	Moderate ANOVA required 3–5 days of training	High Inner and outer arrays multiplied S/N and ANOVA required 3–10 days of training	Low Mathematics "embarrassingly" simple 1–2 days of training
Statistical validity	Low Saturated designs, with confounding of main and interaction effects	Poor Highly saturated designs, with extreme confounding of main and interaction effects	High Clear separation of main and all low and high order interactions

Table 9.4 (Continued)

Characteristic	Classical	Taguchi	Shainin
Statistical validity (cintinued)		S/N only effective if ratio of mean to standard deviation is constant	
		The objective to make a design "robust" against noise factors is worthy, but the means to achieve this are poor	
		Nonrandomization, a glaring flaw	
Versatility	Low	Poor	High
	Only two tools available	Only one tool available	20 tools available to tackle a wide range of problems
Scope	Requires hardware	Can be used at design stage if formula governing input and output variables is known	Requires hardware
	Main use in production	Main use in preproduction	Can be used at prototype, preproduction, and production stages
Ease of implementation	Moderate	Poor	High
	Statistical knowledge and computers required	Statistical knowledge and computers required	Minimal statistical knowledge required
	Engineers discouraged by complexity	Engineers "turned off" by complexity and modest results	Engineers, production, direct labor, and suppliers gravitate to its use, encouraged by excellent results

Source: Used with permission from K. R. Bhote. *World Class Quality*. AMACOM. New York.

8. Always, always, always run a confirmation experiment to verify predicted results. If results are not confirmed or otherwise unsatisfactory, additional experiments may be required.

REFERENCES

Barker, T. B. and Clausing, D. P. (March 1984). *Quality engineering by design—The Taguchi method.* Presented at the 40th annual RSQC conference.

Bhote, K. R. (1991). *World class quality: Using design of experiments to make it happen.* American Management Association. New York.

Box, G. E. P. (1950). Problems in the analysis of growth and wear curves. *Biometrica.* 6: 362–389.

Box, G. E. P., Hunter, W. G., and Hunter, J. S. (1978). *Statistics for experimenters.* John Wiley and Sons. New York.

Dayton, C. M. (1970). *The design of educational experiments.* McGraw-Hill. New York.

Duncan, A. J. (1986). *Quality control statistics.* 5th ed. Irwin. Homewood, IL.

Earley, L. (1989). Kanban car design. *Automotive Industries.* Feb.: 8–11.

Freund, R. J. and Littell, R. C. (1981). *SAS for linear models: A guide to the ANOVA and GLM procedures.* SAS Institute. Cary, NC.

Hicks, C. R. (1982). *Fundamental concepts in the design of experiments.* 3rd ed. Holt, Rinehart and Winston. New York.

Kackar, R. N. (1986). Taguchi's quality philosophy; Analysis and commentary. *Quality Progress.* Dec.: 45–48.

Kirkland, C. (1988). Taguchi methods increase quality and cut costs. *Plastics World.* Feb.: 42–47.

Montgomery, D. C. (1991). *Design and analysis of experiments.* 3rd ed. Quality Press. Milwaukee, WI.

Port, O. (1987). How to make it right the first time. *Business Week.* June: 32.

Ross, P. J. (1988). *Taguchi techniques for quality engineering.* McGraw-Hill. New York.

Roy, R. (1990). *A primer on the Taguchi method.* Van Nostrand Reinhold. New York.

SAS Institute (1985). *SAS user's guide: Statistics.* 5th ed. SAS Institute. Cary, NC.

Steel, R. G. B and Torrie J. H. (1980). *Principles and procedures of statistics.* 2nd ed. McGraw-Hill Book Company. New York.

Taguchi, G. (1986). *Introduction to quality engineering.* Asian Productivity Organization. Available in North America through Kraus International Publications. White Plains, NY.

Taguchi, G. (1987). *System of experimental design.* Vol. 1 and 2. Kraus International Publications. White Plains, NY.

Wood, R. C. (1980). The prophets of quality. *The Quality Review.* Winter: 12–17.

10

More Tools for Modern Quality

This chapter presents some more leading edge tools for modern quality. Specifically, it discusses the failure mode and effect analysis, quality function deployment, benchmarking, cost of quality, geometric tolerancing, and cross tab analysis.

I. FAILURE MODE AND EFFECT ANALYSIS

An failure mode and effect analysis (FMEA) is a technique of a very strict methodology to evaluate a system, a design, a process, and/or a service for possible ways in which failures can occur. For each failure, an estimate is made of its effect on the total system and of its seriousness. In addition, a review is made of the action being taken (or planned) to minimize the probability of failure or to minimize the effect of failure (Stamatis, 1995).

This simple but straightforward approach can be very technical (quantitative) or very nontechnical (qualitative), and utilizes three main factors for the identification of the specific failure. The three factors are: 1) occurrence—how often the failure occurs, 2) severity—how serious the failure is, and 3) detection—how easy or difficult is to detect the failure. The complication of the approach is always dependent on the complexity of the problem as defined by the following (Juran and Gryna, 1980).

1. *Safety.* Injury is the most serious of all failure effects. In fact, in some cases, it is of unquestionable priority and, of course, at this point it must be handled either with a hazard analysis and/or failure mode and critical analysis (FMCA).
2. *Effects on downtime.* How are repairs made? Can repairs be made while the machine is off-duty time or while the machine is operating?
3. *Access.* What hardware items must be removed to get at the failed component? This area will be of great importance as environmental laws are changed to reflect world conditions for dis-assembly.
4. *Repair planning* (including repair time, maintenance, repair tools, cost, and recommendation(s) for changes in design specifications). Here, The Shingo approach (Poka Yoke), DOE or design for manufacture (DOM) may be considered.

To carry this methodology to its proper conclusion the following prerequisites are necessary:

1. Not all problems are important. This is very fundamental to the entire concept of FMEA, because unless internalized we are going to "chase fires" in the organization. We must recognize that some problems are more important than others for whatever the reason. The fact is that some problems have indeed higher priority than others. FMEA helps identify this priority.
2. You must know the customer. The definition of "customer" normally is thought of as the "end user." However, a customer may also be defined as a subsequent or downstream operation as well as a service operation. When using the term customer from an FMEA perspective, the definition plays a very major role in addressing problems. For example, as a general rule in the design FMEA one views the "customer" as the end user, while in the process FMEA the "customer" is viewed as the "next" operation in line. This "next" operation may indeed be the end user but it does not have to be. Once you define your customer (internal, intermediate, or external) you may not change it—at least not for the problem at hand—unless you recognize that by changing it you may indeed have changed your problem and/or consequences.
3. You must know the function. It is imperative to know the function, purpose, or objective of what you are trying to accomplish, otherwise you are going to waste time and effort in redefining your problem based on situations. If you have to, take extra time to make

sure you understand the function or purpose of what you are trying to accomplish.

4. You must be prevention oriented. Unless you recognize that continual improvement is in your best interest, the FMEA is going to be a "static" document to satisfy your customer or market requirements. The push for this continual improvement makes the FMEA a "dynamic" document changing as the design and/or process changes with the intent always to make a better design and/or process.

A. The Reason for FMEAs

The propensity of managers and engineers to minimize the risk in a particular design and/or process has forced us to look at reliability engineering to not only minimize but also to define the risk. Obviously the risk is a multifaceted issue, but, from a generic perspective one may define it based on: management emphasis, market pressures, customer requirements, safety, legal and statutory requirements, public liability, development technical risks, warranty and service costs, competition, and so on. These risks can be measured by reliability engineering and/or statistical analyses. However, because of their complexity, the FMEA has extracted the basic principles without the technical mathematics and has provided us with a tool that anybody committed to continual improvement can utilize.

Statistical process control (SPC) is another tool that provides the impetus for the implementation of an FMEA, especially for a process FMEA. SPC provides information about the process in regards to changes. These changes are called common and special causes. From an FMEA perspective we may look at the common causes as failures that are the result of inherent failure mechanisms that can affect the entire population. In this case, this is a cause for examining the design (Denson, 1992). On the other hand, special causes are looked at as failures that result from part defects and/or manufacturing problems and that can affect a relatively small population. In this case, there is cause for examining the process.

Customer requirements is, of course, a very strong influence as to why we may be doing an FMEA. For example: All major automobile companies in their own quality system as well as their supplier certification programs require an FMEA program from their suppliers. The courts—through product liability—may require some substantiation as to what level of reliability your products perform.

International standards such as the ISO 9000 series may define the program of documentation for your design. For example: the Product Liability Directive of EU 1985 stipulates that manufacturers of a product will be held liable, regardless of fault or negligence, if a person is harmed or an object is damaged by a faulty or defective product (this includes exporters into the

European Union market). This liability directive essentially reverses the burden of proof of fault from the injured to the producer. Quality systems incorporating specific tools such as FMEA or fault tree analysis (FTA) or failure mode and critical analysis (FMCA) with safety prevention provisions will be particularly important in protecting a company from unfounded liability claims. Furthermore, proposed safety directives would oblige manufacturers to monitor the safety of their products throughout their foreseeable life (Stamatis, 1992).

Other benefits of the FMEA include:

- improving the quality, reliability and safety of the products
- improving the company's image and competitiveness
- helping increase customer satisfaction
- reducing product development time and costs
- helping select the optimum system design
- helping determine the redundancy of the system
- helping identify diagnostic procedures
- establishing a priority for design improvement actions
- helping identify critical and/or significant characteristics
- helping in the analysis of new manufacturing and/or assembly processes

Even though all these reasons are well worth the effort of conducting an FMEA, the most important reason for writing an FMEA is the need to improve. Unless this need is part of the culture of the organization, the FMEA program is not going to be successful.

B. Language of the FMEA

To understand the FMEA one must understand its language. There are several terms that one must understand. They are:

Function: the task that a component, subsystem, or system must perform. This function is very important in understanding the entire FMEA process. It has to be communicated in a way that is concise, exact, easy to understand, and without jargon. To facilitate this, it is recommended that an active verb be found to describe the function. The active verb by definition defines performance and performance is what a function is. Examples of this may be found in the following words: position, support, retain, lubricate.

Failure: the problem, the inability of a component, subsystem or system to perform to design intent. This inability can be defined as both known and potential. Of great importance here is that when potential failures in terms of *functional defectives* are identified, the FMEA is fulfilling its mission of prevention. Functional defectives are failures that do not meet the customers requirements, but we ship them those products with the failures anyway because the customer will:

- never know the difference
- never find out
- find out, but whatever is delivered can be used
- find out, but whatever is delivered has to be used because there are no alternatives

Examples of failures are: broken, worn, corrosion, noisy.

Causes of failure: what is the "root cause" of the listed failure. Next to the function, cause of the failure is perhaps the most important section of the FMEA. It is indeed here that we point the way toward preventive and/or corrective action. The more focused we are on the "root cause," the more successful we are in eliminating failures. In this section, we must be careful not to be too eager for solutions because we are going to fall victims of symptoms and short-term remedies, rather than complete elimination of the real problem(s).

Examples for design: wall thickness, vibration, shock load, torque specifications

Examples for process: voltage surge, dull tools, improper set up, worn bearings

Effects of failure, the outcome of the failure mode on the system and/or the product. In essence the effects of the failure have to do with the question of: What happens when a failure occurs? We must understand, however, that the effects of the failure must be addressed from two points of view. The first, local, in which the failure is isolated and does not effect anything else. The second, global, in which the failure can and does effect other functions and/or components (i.e., it has a domino effect). Generally speaking, the failure with a global effect is more serious than the one of a local nature. The effect of the failure also defines the severity of a particular failure; for example, a local failure is a parking light bulb failure, whereas a global failure is power brake failure. In the first case, one can identify a nuisance of a failure, where in the second case a catastrophic failure is eminent.

Product validation. Controls that exist to prevent the cause(s) of the failure from occurring and to validate repeatability for certain processes especially with the Federal Drug Administration (FDA).

Current control(s). Controls that exist to prevent the cause(s) of the failure from occurring in the process phase (e.g., any of the SPC tools, capability, operator(s) training).

Design verification. Controls that exist to prevent cause(s) of the failure from occuring in the design phase (e.g., design guidelines, design reviews, specific specifications).

C. Mechanics of an FMEA

In order to effective, a team must generate the FMEA. The reason for this is that the FMEA should be a catalyst to stimulate the interchange of ideas between the functions affected. A single engineer or any other single person can not do it.

The team should be made of 5–9 persons (preferably five). All team members must have some knowledge of group behavior, the task at hand, some knowledge about the problem to be discussed, some kind of either direct or indirect ownership of the problem, and above all, they must all be willing to contribute. Team members also must be cross-functional and multidisciplined. Furthermore, whenever possible and/or needed, it is encouraged to have the customer and/or the supplier actively participate.

D. Design FMEA

A design FMEA is a systematic method to identify and correct any known or potential failure modes before the first production run. A first production run is viewed as the run that you produce a product and/or service for a specific customer with the intent of getting paid. This definition is important because it excludes, initial sample runs (ISR), trial runs, sometimes the prototype run(s), and so on. The threshold of the first production run is important because up to that point to modify and/or change the design is not a major thing. After that point, however, the customer gets involved through the letter of deviation, waiver of change, or some other kind of formal notification.

Once these failures are identified, they are ranked and prioritized. The leader (the person responsible for a design FMEA) should be the design engineer, primarily because he/she is the most knowledgeable about the design and can best "anticipate" failures. To facilitate the meeting, the quality engineer may be designated as the facilitator.

The minimum make up of the team for the design should include the design engineer and the process (manufacturing) engineer. Anyone else that can contribute or whom the design engineer feels appropriate may also participate. A typical design team may include a design engineer, manufacturing engineer, test/development engineer, reliability engineer, material engineer, and field service engineer (i.e., the customer's voice). Of great importance in the make up of the team is that the team must be cross-functional and multidisciplined. However, remember that there is no such thing as "THE" team. Each organization must define their optimum team participation, recognizing that some individuals may indeed hold two or more different positions at the same time.

The focus of the design is to minimize failure effects on the system by identifying the key characteristics of the design. These key characteristics sometimes may be found as part of: 1) customer requirements, 2) engineering specifications, 3) industrial standards, 4) government regulations, and 5) product liability.

The objective of the design FMEA is to maximize the system quality, reliability, cost, and maintainability. It is important here to recognize that in design we have only three possibilities to look at defects, they are:

Components. The individual units of the design.
Subsystem or subassembly. Two or more combined components.
System or assembly. A combination of components and subsystems for a particular function.

Regardless of what level we are in the design, the intent is the same (i.e., no failures on the system). To focus on these objectives, the design team must use consensus for their decision, but more importantly, they must have the commitment of their management.

The timing of the design FMEA is initiated during the early planning stages of the design and is continually updated as the program develops. As a team, you must do the best you can, with what you have, rather than wait until all the information is in. By then, it may be too late.

E. Process FMEA

A process FMEA is a systematic method to identify and correct any known or potential failure modes before the first production run. Once these failures are identified, they are ranked and prioritized.

The leader (the person responsible for the process FMEA) should be the process/manufacturing engineer primarily because this person is the most knowledgeable about the process structure and can best "anticipate" failures. To facilitate the meeting the quality engineer may be designated as the facilitator. The minimum make up of the team for the process includes the process/manufacturing engineer, design engineer, and operator(s). Anyone else that can contribute or the process engineer feels appropriate may also participate. A typical process team is the process/manufacturing engineer, design engineer, quality engineer, reliability engineer, tooling engineer, and operator(s).

The focus of the process FMEA is to minimize production failure effects on the system by identifying the key variables. These key variables are the key characteristics of the design, but now in the process, they have to be measured, controlled, monitored, and so on. This is where SPC comes alive.

The objective of the process FMEA is to maximize the system: quality, reliability, and productivity. Of note here is that the objective is a continuation of the design FMEA—more or less—since the process FMEA assumes the objective to be as designed. Because of it, potential failures which can occur because of a design weakness are not included in a process FMEA. They are only mentioned if those weaknesses effect the process.

The process FMEA does not rely on product design changes to overcome weaknesses in the process, but it does take into consideration a product's design characteristics relative to the planned manufacturing or assembly process to assure that, to the extent possible, the resultant product meets customer needs, wants, and expectations.

Another important issue in a process FMEA, is the fact that it is much more difficult and time consuming than the design FMEA. The reason for this is that in a design FMEA we have three possibilities of analysis while in the process FMEA we have six and each one of them may have even more. The six major possibilities for a process FMEA are:

- manpower
- machine
- method
- material
- measurement
- environment

To show the complexity of each one of these possibilities let us take the "machine" for example. Some of the contributing failures may in fact be one of the following categories (the list is by no means complete).

- tools
- work station
- production line
- the process itself
- gauges
- operator(s)

Again, just like in the design, the process team must use consensus for their decision, but more importantly they must have the commitment of their management. To facilitate this consensus and gain the commitment of their management, the group in both design and process FMEAs must set specific improvement goals. To do that, the leader of the group must use the following specific behaviors:

1. Focus the team on result areas. The leader should stress the importance of goalsetting for focusing team efforts, and discuss possible result areas the team may choose.
2. Review existing trends, problem areas, and goals. The leader should share with the team data that could shed light on the team's performance and indicate possible areas for improvement.
3. Ask (actively solicit) the team to identify possible areas for investigation. The leader should actively ask for complete participation from all team members. The more the participation, the more the ideas. Brainstorming, affinity charts, and force field analysis are some of the tools that can be used by the leader to facilitate an open discussion by all.
4. If possible, ask the team to identify the goal. The leader, after a thorough discussion and perhaps ranking, has the team agree on the selected goal. This will enhance the decision and will insure commitment.
5. Specify the goal. It is the leader's responsibility, once the goal has been set, to identify the appropriate measures, time, and amount of improvement. As a general rule, cost is not considered at this stage. The cost is usually addressed as part of another analysis called value engineering (see Chapter 12).

In the final analysis, regardless of what level the FMEA is being performed the intent is always the same (i.e., no failures on the system).

The timing of the process FMEA is initiated during the early planning stages of the process before machines, tooling, facilities, and so on are purchased. The process FMEA, just like the design FMEA, is continually updated as the process becomes more clearly defined. As for the displaying format, there are many forms available, but no single one is standard for everyone. Typical forms used in both design and process are shown in Figures 10.1 and 10.2. Figures 10.3 and 10.4 show a typical FMCA form. An FMEA form that also includes a control plan is shown in Figure 10.5.

F. Guidelines

The abundant availability and singularity of use in different organizations does not allow for specific discussion of guidelines. Rather a generic discussion will follow. The guidelines are numerical values based on certain statistical distributions that allow us to prioritize the failures. They are usually of two forms: qualitative and quantitative. In the qualitative form, one bases the meaning on theoretical distributions such as normal, log normal (skewed to the left), and discrete distributions for the occurrence, severity, and detection

Potential Failure Mode and Effect Analysis (Design Form)

___ System
___ Subsystem
___ Component
Model ______

Design responsibility ________
Due Date ________
Core Team ________

FMEA Number ______
Page ___ of ___
Prepared By ______
FMEA date (Orig.) _____ Rev: ______

Item/ Function	Potential Failure Mode	Potential Effect(s) of Failure	SEV	CRIT	Potential Cause(s) Mechanism(s) of Failure	OCCUR	Current Design Controls	DETEC	RPN	Recommended Action(s)	Responsibility & Target Completion Date	ACTION RESULTS Action Taken	SEV	OCC	DET	RPN

Figure 10.1 Design FMEA form.

Potential Failure Mode and Effect Analysis (Process Form)

___ System
___ Process
___ Item
Model ______

Process responsibility ________
Due Date ________
Core Team ________

FMEA Number ______
Page ___ of ___
Prepared By ______
FMEA date (Orig.) _____ Rev: ______

Purpose/ Function & or require-ments	Potential Failure Mode	Potential Effect(s) of Failure	SEV	CRIT	Potential Cause(s) Mechanism(s) of Failure	OCCUR	Current process Controls	DETEC	RPN	Recommended Action(s)	Responsibility & Target Completion Date	ACTION RESULTS Action Taken	SEV	OCC	DET	RPN

Figure 10.2 Process FMEA form.

Part Number (1) ____ Assembly Number (2) ____ Responsible engineer (3) ____															Production release date (4)____ Page (5) ___ of ___ Date (6) ____
Line number (7)	Crossreference number (8)	Circut location (9)	Enter the part/ component number/name (10)	Function(s) & specification(s) (11)	Potential failure mode(s) (12)	System effect (0=Unsafe condition) (13)	Unsafe (14)	Cause(s) of failure (15)	Internal or external counter measures (controls) (16)	Severity (17)	Base failure rate λB (18)	Failure mode ratio (19)	Effectiveness (20)	Risk priority number (RPN) (21)	

1. Part number. Enter the part number under consideration
2. Assembly number. Enter the number on the part or drawing or part list
3. Responsible engineer. Enter the name of the responsible engineer
4. Production release date. Enter the date the product is to be released for production
5. Page. Enter the FMCA page number
6. Date. Enter the date the page was worked on. Or, enter the revision date, if it is a revised FMCA
7. Line number. Identify the part for which the FMCA is to be conducted.
8. Crossreference number. Enter the number if there is a crossrefernce with other parts or assemblies.
9. Circuit location. Describe the location of the part on the circuit.
10. Enter the part/component number/name. Enter the appropriate name.
11. Function(s) and Specification(s). Describe the function(s) the part is to perform and the specification(s) required. Make the description as clear and concise as possible. Be sure you include all functions. Include pertinent information about the product specification, such as operating current range, operating voltage range, operating environment and everything else that is applicable and appropriate.
12. Potential failure mode(s). A failure mode is a design flaw or change in the product which prevents it from functioning property. The typical failure modes are a short circuit, open circuit, leak, loosening. The failure mode is expressed in physical terms of what the customer will experience.
13. System effect. The system effect is what a system or a module might experience as a result of the failure mode. List all conceivable effects, including unsafe conditions or violations of government regulations. A typical system effect is a system shut down or a failure of a section of the product.
14. Unsafe. Enter 0 for unsafe end product condition.
15. Cause of failure. The ROOT CAUSE - not the symptom - is the real cause. Examples: Insufficient/inaccurate voltage, firmware errors, missing instructions on drawings.
16. Internal or external counter measures (controls). Identify the controls and or measures established to prevent or detect the cause of the failure mode. Examples: Perform a derating analysis, perform transient testing, perform specific testing, identify specific inspection and manufacturing specifications.
17. Severity. An estimate of how severe the sub-system and or the end product will behave as a result of a given failure mode. Severity levels are being scaled from 1 to 10. Number 10 is to be used for a definite unsafe condition, and number 0 is to be used for a negligible severity (nuisance). Usually this rating, at this stage, is very subjective rating.
18. Base failure rate (λB). A subjective estimate of failure rate (probability of failure in a billion hours). This is also called Inherent failure rate.
19. Failure ratio. A subjective likelihood in comparison to the other failure modes. The sum of all failure rates for a part/component should be equal to 10 percent.
20. Effectiveness. A subjective estimate of how effectively the prevention or detection measure eliminates potential failure modes. A typical ranking is the following:

1	= The prevention or detection measure is foolproof.
2-3	= Probability of failure occurrence is low.
4-6	= Probability of occurrence is moderate.
7-9	= Probability of occurrence is high.
10	= Very high probability. The prevention/detection measure is ineffective.

21. Risk Priority Number (RPN). The product of severity, base failure rate, failure mode ratio and effectiveness.

Figure 10.3 Design FMCA form.

Operation name (1) ___________ Sub-Assembly number (4) ____________ Production release date (7)________
Work station (2) ___________ Supplier (5) ____________________ Page (8) ___ of ___
Responsible engineer (3) __________ Original date (6) __________________ Revised date (9) _______

Line number (10)	Crossreference number (11)	Circut location (12)	Enter the part/ component number/name (13)	Operational steps (14)	Potential failure mode(s) (15)	Cause(s) of failure (16)	Internal or external counter measures (controls) (17)	Severity (18)	PPM (19)	Effectiveness (20)	Risk priority number (RPN) (21)

1. Operation Name. Enter the name of the operation.
2. Work station. Enter the name or number of the work station.
3. Responsible Engineer. Enter the name of the responsible Engineer.
4. Sub Assembly Number. Enter the sub-assembly name or number.
5. Supplier. Indicate where the process is performed.
6. Original Date. Enter the date that the FMCA is due and or completed.
7. Production Release Date. Enter the date the is to be released, for production.
8. Page. Enter the FMCA page number.
9. Revise Date. Enter the date of the revision date.
10. Line Number. Identify the part for which the FMCA is to be conducted.
11. Crossreference Number. Enter the number if there is a crossreference with other parts or assemblies.
12. Circuit Location. Describe the location of the part on the circuit.
13. Enter the part/component number/name. Enter the appropriate name.
14. List all steps of operations in the process. A good tool to use for this is the Process flow diagram.
15. Potential Failure Mode(s). A process-related failure mode is a deviation from specification caused by a change in the variables influencing the process. Examples: Damaged board, misaligned, discolored, missing, bent etc.
16. Cause of Failure. The ROOT CAUSE - not the symptom - is the real cause of the failure. Examples: Transient, human error, machine out of tolerance, ESD equipment failure.
17. Internal or external counter measures (controls). Identify the controls and or measures established to prevent or detect the cause of the failure mode. Examples: Verify tooling to its specification, effective incoming inspection, testing, etc.
18. Severity. A subjective estimate of how severe the sub-system and /or the end product will behave as a result of a given failure mode. Severity levels are being scaled from 1 to 10. Number 10 is to be used for a definite unsafe condition, Number 1 is to be used far a negligible severity (nuisance).
19. PPM. Is the percent failure per 1 million parts.
20. Effectiveness. A subjective estimate of how effectively the prevention or detection measure eliminates potential failure modes. A typical ranking is the following:

    1 =The prevention or detection measure is foolproof.
    2-3 =Probability of failure occurrence is low.
    4-6 =Probability of occurrence is moderate.
    7-9 =Probability of occurrence is high.
    10 =Very high probability. The prevention/detection measure is ineffective.

21. Risk Priority Number (RPN). The product of severity, PPM and effectiveness.

Figure 10.4 Process FMCA form.

Company: ____	Part Name: ____	Authorized Control Plan By: ____
Division/Plant: ____	Part Number: ____	Authorized FMEA By: ____
Department: ____	Control Plan Orig. Date: ____	Control Plan Leader: ____
Process: ____	FMEA Orig. Date: ____	FMEA Leader: ____
Operation: ____	Control Plan Rev. Date: ____	FMEA members: ____
Mechine: ____	FMEA Rev. Date: ____	
Station: ____	Page ____ of ____ Pages	

Char #	Characteristic Description (product & pocess)	Spec	TYPE	IMP	FAILURE MODE	Effects of Failure	SEV	Causes of failure	OCC	CURRENT CONTROLS	DET	RPN	RECOM. ACTIONS	ACTIONS TAKEN	SEV	OCC	DET	RPN	RESPONSIBLE PERSON	Ctrl. Fact	CLASS	Control Method	TOOL	Gage, desc., master, detail	Other *

* Typical other items may include:

GR & R & Date	Cp/Cpk (target) & Date	Reaction plans	Inspection requirements	Sampling requirements

Figure 10.5 A typical FMEA and control plan form.

respectively. In the quantitative form, one uses actual statistical and/or reliability data from the processes. This actual data may be historical and/or current. In both cases the numerical values are from 1 to 10 and they denote set probabilities from low to high.

The first criterion in the guidelines is the occurrence. Under occurrence we look at the frequency of the failure as a result of a specific cause. The ranking that we give to the occurrence has a meaning rather than value. The higher the number the more frequent the failure.

The second criterion in the guidelines is the severity. In severity we assess the seriousness of the effect of the potential failure mode to the customer. Severity applies to the effect and to the effect only. The ranking that is typically given is from 1 to 10, with 1 representing a nuisance and 10 representing a very major noncompliance to government regulations or safety item.

The third criterion in the guidelines is the detection. In detection we assess the probability of a failure getting to the customer. The numbers 1 to 10 again represent meaning rather than value. The higher the number, the more likely your current controls or design verification failed to contain the failure within your organization. Therefore, the likelihood of your customer receiving a failure is increased.

A general complaint about these guidelines is frequently heard in relationship to consensus. For example, "How can a group of people agree on everything?" The answer is that they cannot, but that does not mean that you do not have to use consensus to decide.

If the decision falls in an adjacent category, the decision should be averaged out. If on the other hand, it falls in more than one adjacent category, you should stick to the consensus process. A problem at this point may be that someone in the team may not understand the problem, or some of the assumptions may have been overlooked, or perhaps the focus of the team has drifted.

If no consensus can be reached with a reasonable discussion with all the team members, then traditional organizational development (team dynamic and problem-solving techniques) must be used to resolve the conflict. Under no circumstances should a mere agreement or majority be pushed through in order that an early completion may take place. Remember all FMEAs are time consuming and the participants as well as their management must be patient for the proper results.

G. Risk Priority Number

A risk priority number (RPN) is the product of the occurrence, severity, and detection. The value should be used to rank order the concerns of the design and/or process. In itself, the RPN has no other value or meaning.

A threshold of pursuing failures is a RPN equal or greater than 50 based on a 95% confidence (although there is nothing magical about 50). We can identify this threshold by any statistical confidence. For example, say 99% of all failures must be addressed for a very critical design, what is the threshold? There are 1,000 points available (10 × 10 × 10) from occurrence, severity, and detection; 99% of 1,000 is 990. So anything equal to or over 10 as an RPN must be addressed. On the other hand, if we want a confidence of only 90% then, 90% of 1,000 is 900. So, anything equal or greater than 100 as a RPN must be addressed. Please note, however, that this threshold is dependent on the organization and can change with not only the organization but the product and/or the customer.

If there are more than two failures with the same RPN, then we use severity, then detection, and finally occurrence as the order of priority. Severity is used first because it deals with the effects of the failure. Detection is used over the occurrence because it is customer dependant which is more important than just the frequencies of the failure. (The reader must also understand that nothing prevents him or her from acting on every failure identified. However, if that is the desired action, then the FMEA is superfluous. After all, we use the FMEA to identify and prioritize the importance of each failure and act accordingly.)

H. Recommended Action

When the failure modes have been rank ordered by RPN, corrective action should be first directed at the highest ranked concerns and critical items. The intent of any recommended action is to reduce the occurrence, detection, and/or severity rankings. Severity will change only with changes in design; otherwise, more often than not, the reductions are expressed in either occurrence and/or detection. If no actions are recommended for a specific cause, then this should be indicated. On the other hand, if causes are not mutually exclusive, a DOE recommendation is in order.

In all cases where the effect of an identified potential failure mode could be a hazard to manufacturing/assembly personnel, corrective actions should be taken to prevent the failure mode by eliminating or controlling the cause(s). Otherwise appropriate operator protection should be specified.

The need for ranking specific, positive corrective actions with quantifiable benefits, recommending actions to other activities and following-up all recommendations can not be overemphasized. A thoroughly thought out and well-developed FMEA will be of limited value without positive and effective corrective actions. It is the responsibility of all affected activities to implement effective follow-up programs to address all recommendations (Ford Motor Co., 1988, 1989; Chrysler, Ford, and General Motors, 1995).

I. The Process of Doing an FMEA

To do an FMEA effectively one must follow a systematic approach. The recommended approach is an eight-step method that facilitates the system, the design, the process and the service FMEAs.

1. Brainstorm. Try to identify in what direction you want to go. Is it system, design, process, or service? What kind of problems are you having with a particular situation? Is the customer involved or are you pursuing continual improvement on your own? If the customer has identified specific failures, then your job is much easier, because you already have the direction identified. On the other hand, if you are trying on your own the brainstorm and/or the cause-and-effect diagram may prove to be the best tools to identify your direction of attack.

2a. Block diagram (used only for system and design). Make sure everyone on the team is on the same wavelength. Does everyone understand the same problem? The block diagram will focus the discussion. If nothing else, it will create a baseline of understanding for the problem at hand.

2b. Process flow chart (used only for process and service). Make sure everyone in the team is on the same wavelength. Does everyone understand the same problem? The process flow chart will focus the discussion. If nothing else, it will create a baseline of understanding for the problem at hand.

3. Prioritize. Once you understand the problem and its relationship to other problems, the team must begin the plan for action. For example, in one problem there may be subproblems or areas that need to be addressed first. More often than not, a thorough study of the block diagram or the process flow chart will indicate the appropriate action. This is the stage where we really begin to focus on the problem.
4. Data collection. Begin to collect data of the failures and start filling out the appropriate form.
5. Analysis. Focus on the data and perform the appropriate analysis. Everything is fair, provided it is appropriate and applicable. Here, QFD, DOE, another FMEA, SPC, and anything else may be used to provide the appropriate results.
6. Results. Based on the analysis, the results are derived. The results must be data driven; nothing else will do.
7. Confirm/evaluate/measure. Once the results have been recorded, it is the time to confirm, evaluate, and measure the success or failure

of the recommended action taken. This evolution takes the form of three basic questions:

- Are we better off than before?
- Are we worse off than before?
- Are we the same as before?

8. Do it all over again. Regardless of how you answer Step 6, you must pursue improvement all over again because of the continual improvement philosophy. Your long-range goal is to eliminate completely all failures and your short-term goal is to minimize your failures if not eliminate them. Of course, the perseverance for those goals have to be taken into consideration in relationship to the needs of the organization, cost, customer, and competition.

For a very detailed coverage of FMEA see Stamatis (1995) and Chrysler, Ford, and General Motors (1995).

II. QUALITY FUNCTION DEPLOYMENT

Quality function deployment (QFD) is difficult to define for many reasons. It is sometimes narrowly defined and sometimes broadly defined. Although it was introduced as a concept 20 years ago, it is not yet fully systemized. It is used in different ways in different companies. Some companies, like Bell Labs, are calling it matrix product planning, but there are some very successful projects with no matrices. Let me suggest a working definition. QFD is a philosophy and strategy which enables all members of the organization to participate in the design process. In other words, QFD is a methodology to bring together the various departments within a corporation—executive, marketing, engineering, production, and so on—in a planned manner and cause them to focus on the "voice of the customer."

As a philosophy, QFD teaches that all design of product and process should be rooted in the needs of the customer and that the design should be accomplished efficiently and accurately before production of first production piece or offering of service on a regular basis. As an organizational strategy, QFD involves all segments of the organization in the design of product and process, thus reducing the chances of design errors and assuring the optimal design possible.

As a collection of tools, QFD may include quality tables, matrices, value engineering, fractional factorial experimentation, FMEAs, SPC data, finite element analysis, simulation studies, planning tools, reviewed dendrograms (a chart presenting thought patterns in a design improvement effort), bottleneck engineering (a group of schematic presentations of design improvements

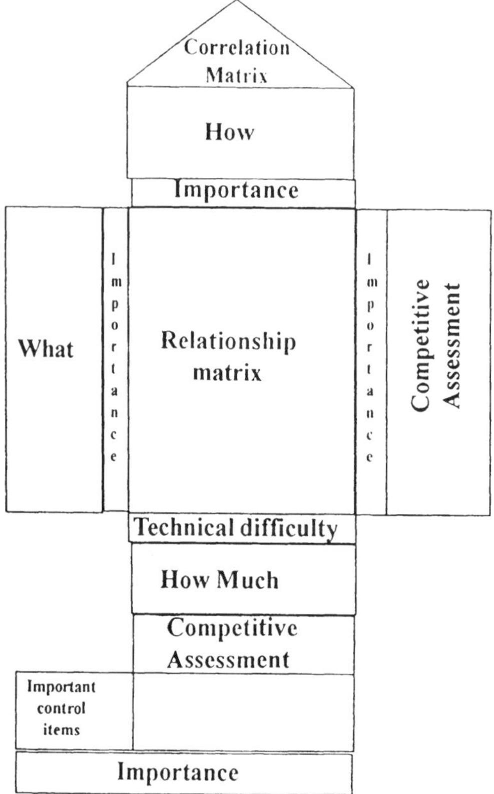

Figure 10.6 House of Quality matrix.

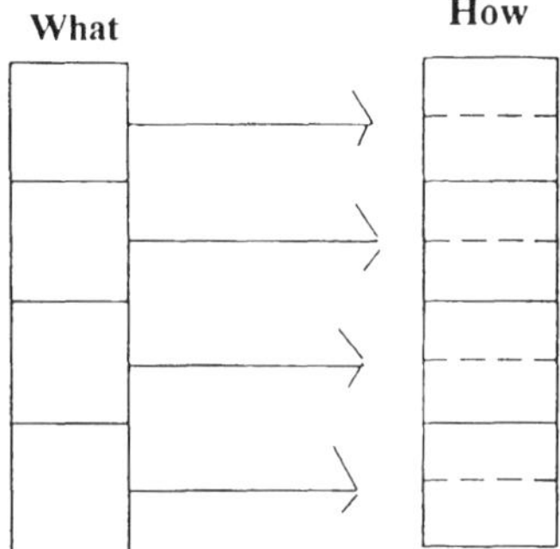

Figure 10.7 The relationship between the "what" and "how".

such as the reviewed dendrogram), and other tools that enhance and document design efforts.

QFD can help us shore up our weaknesses while building on our strengths. It encourages a comprehensive, holistic approach to product development that are generally found lacking in American industry (Eureka and Ryan, 1988).

QFD's primary goal is the overcoming of three major problems (Eureka and Ryan, 1988), namely:

1. disregarding the voice of the customer
2. losing information
3. preventing different individuals and groups from working to different requirements

The implementation of QFD can provide the organization with both strategic benefits as well as operational benefits. Strategic benefits include, but are not limited to: larger market share, improved quality, shortened cycle time, fewer engineering changes, lowered manpower, reduced cost, increased reliability, and reduced process variation. On the other hand, operational benefits include, but are not limited to: combination with other quality technologies, increased process efficiencies, encouragement of a new way of thinking and/or changing paradigms, improved participation, systematic approach to analysis, improved group and team interactions, and product organizations (Sullivan, 1987), enhanced communication, identification of conflicting requirements, and most importantly, it preserves information for the future.

Whereas it is beyond the scope of this book to define and explain the detail implementation process for a QFD, the reader is encouraged to see Cohen (1995), Day (1993), Akao (1990), Bossert (1990), Sullivan (1986), Eureka (1987), and American Supplier Institute (1987, 1989) for a detailed implementation methodology and actual case studies from U.S. companies. However, because QFD is so powerful tool and its use is being expanded in all kinds of industries, the process is summarized here.

The QFD approach translates the customer's requirements, part characteristics, manufacturing operations, and production requirements. Each step is interrelated and tracked through the House of Quality matrix (Figure 10.6).

QFD matrices and charts deploy the customer's requirements from the product planning stage all the way to the shop floor. The QFD methodology involves translating a list of loosely stated customer objectives into "what." These "whats" are the items we wish to accomplish (i.e., the basic customer requirements. Each "what" is then broken down into "how" or the method required to achieve each "what." The "what" and "how" can then be listed. Although the "how" list represents greater detail than the "what" list (Figure 10.7), they are often not directly actionable and require yet further definition.

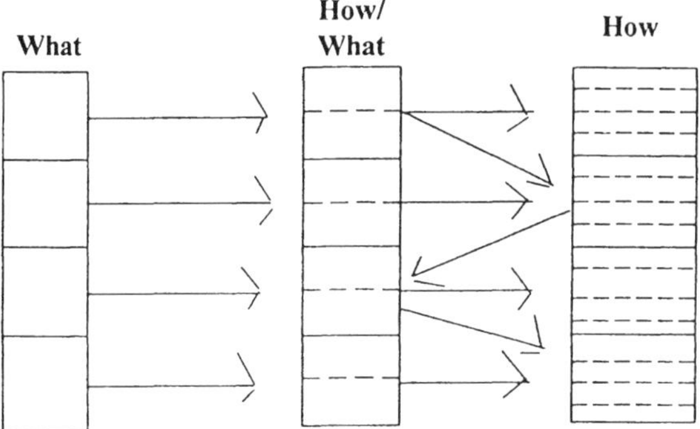

Figure 10.8 The confusing relationship between the "what" and "how."

Unfortunately, this process is complicated by the fact that through each level of refinement some of the "how" affect more than one "what." Attempting to clearly trace the relationships of the "what" and "how" becomes confusing at this point, as shown in Figure 10.8. By restructuring the "what" and "how" lists perpendicular to each other the relationship can be observed clearly and then rated (Figure 10.9). At this point, it becomes necessary to establish "how much" for each "how" (Figure 10.10).

The flow of information is from "what" to the "how" through the relationship matrix, then to the "how much." The "how" and "how much" are then translated into "what" for the next chart, ensuring that the objective values are not lost. This process continuous until each objective is refined to an actionable level (Figure 10.11).

Figure 10.12 shows the four basic charts and their development as the customer's requirements are developed into design requirements then to part characteristics and so on. This process continues until all manufacturing operations and production requirements are defined.

One can notice that by using the QFD methodology, one can indeed prevent "things from falling through the cracks" by forcing interrelationships between each discipline. In fact, both detail and functions now have clear line of responsibility and accountability.

As we already have mentioned, one of the QFD benefits is its ability to preserve knowledge. As engineers and management change assignments the QFD charts remain with the product and/or process, informing the incoming group how the product was developed and produced. This also provides a marshaling point for similar designs, reducing turn around time to develop new products.

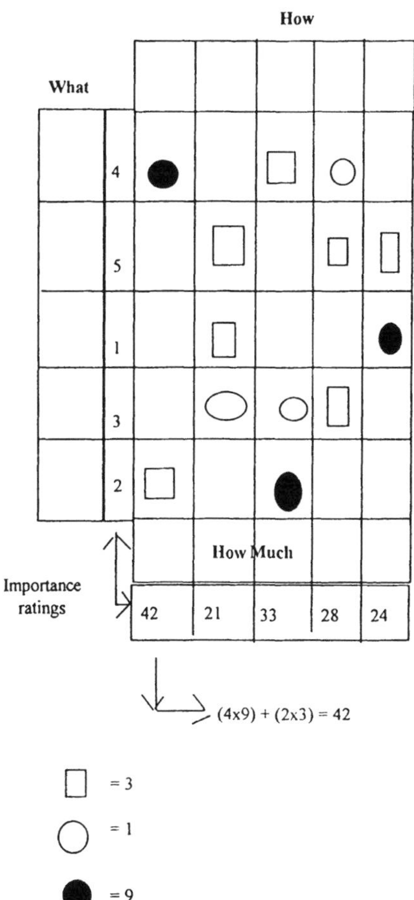

Figure 10.9 A typical rating matrix. Note: make sure that the ratings differentiate to the point of discrimination between each other. You are interested in great differentiation rather than a simple priority.

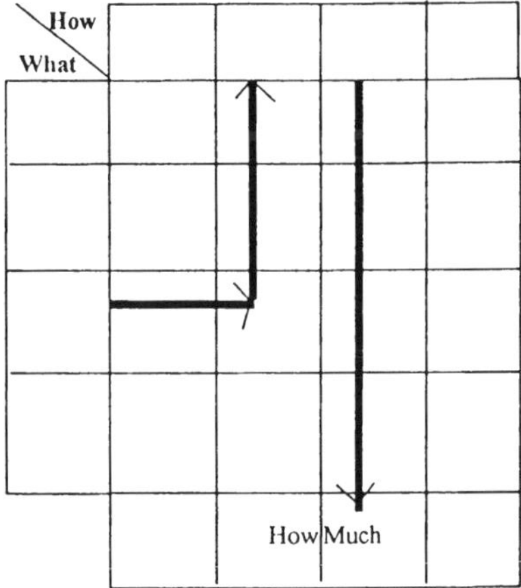

Figure 10.10 The development of "how much."

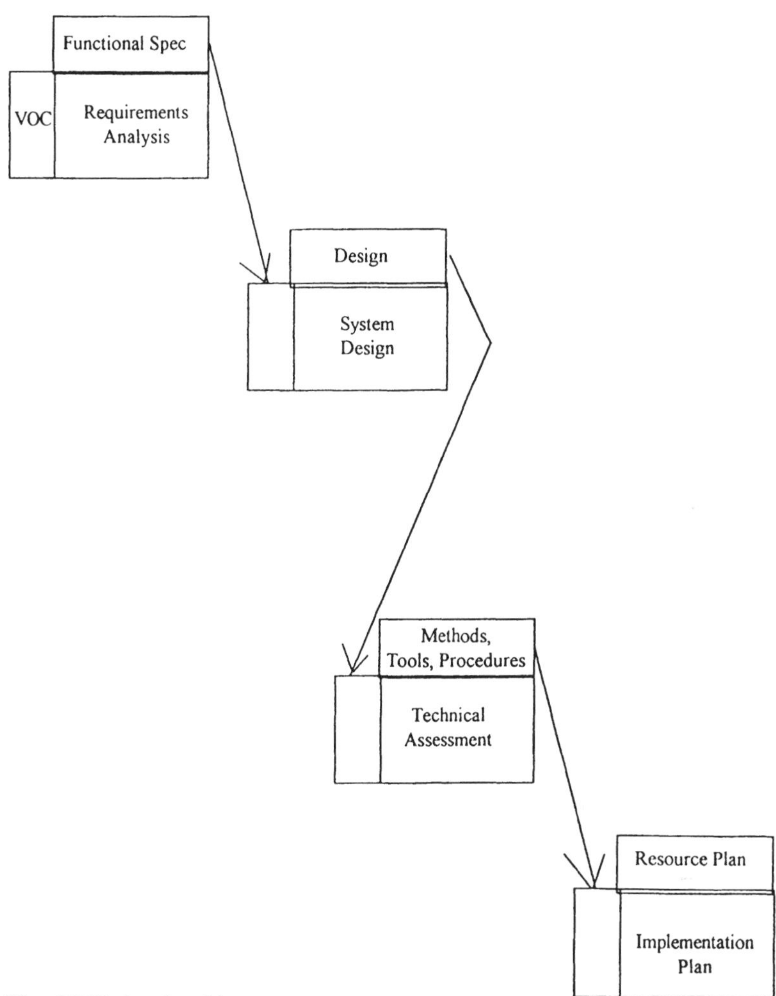

Where VOC is the voice of the customer

Figure 10.11 The flow of information from "what" to "how" to "how much."

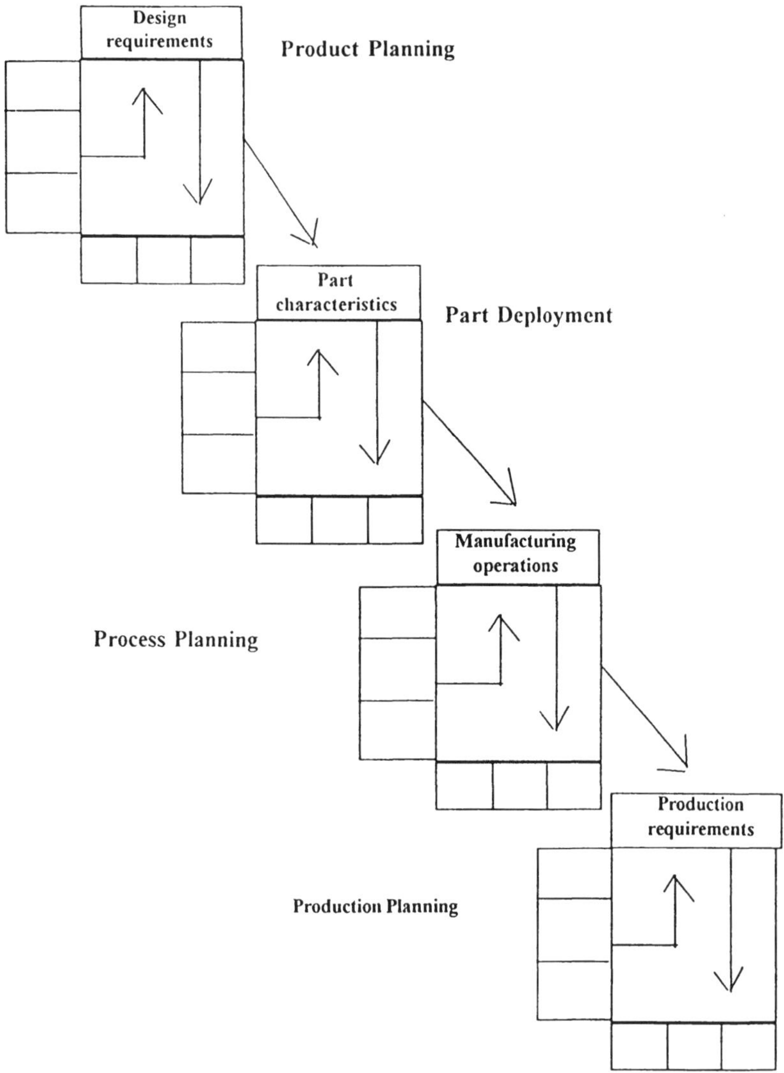

Figure 10.12 The basic charts of QFD and their deployment.

Clausing (1988) summed up the concept of QFD as:

> ... the voice of the customer to the factory floor. It brings all the corporate wisdom, including all the specialized knowledge of the many and diverse corporate specialists, to bear on the product. And it achieves multifunctional consensus, which results in a product that's best for the customer and the corporation. QFD is a major operational practice that helps achieve products that are higher in quality, lower in cost, and developed much faster, and that contain features that are better matched to customer requirements. By using QFD to achieve products that strongly appeal to potential customers, corporations will experience increased market share and growth and improved quality of work life.

III. BENCHMARKING

An additional tool in the tool box of improvement is benchmarking. Benchmarking is a way of comparing the practices of an organization with that of another, with the intent of learning about the process(es) and improving the results of the organization. A very simple self-explanatory model for benchmarking is shown in Figure 10.13. Its basic assumption is a win-win situation for all the parties participating. It is beyond the scope of the book to address all the issues and methodological characteristics of benchmarking; however, we give an overview of the fundamentals. For more information see Watson (1992), Spendolini (1992), Camp (1989, 1995), and Stamatis (1996).

Benchmarking is an approach that when used appropriately one may examine both internal and external excellent performance for both understanding and improvement of a process and/or the entire organization. Formally, benchmarking has been defined a number of ways (Adam and Vande Water, 1995) including:

- as a process for identifying and learning from the best practices in the world
- as a search for and application of significantly better practices that lead to superior competitive performance
- as a process of comparing the business of one organization against another to gain information about "best practices" that when creatively adapted, can lead to superior performance

The performance analysis demanded by the benchmarking process may take a variety of approaches; however, the most common, effective and simple to understand is the hierarchical approach shown in Figure 10.14. In this figure the lowest level indicates an internal comparison of best practices. This

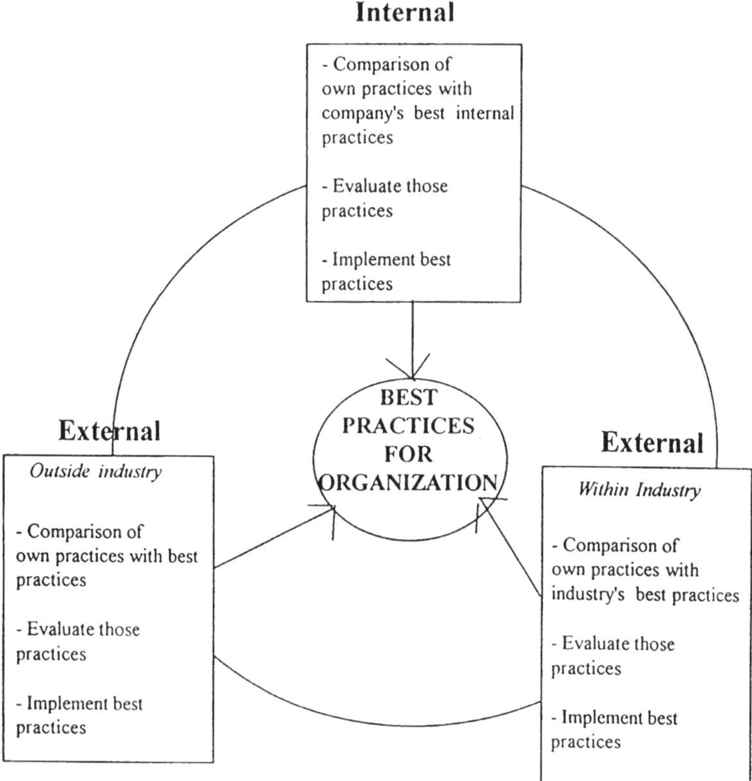

Figure 10.13 Benchmarking model.

activity typically involves comparing functions of departments or business units within the same organization. On the other hand, the top level is reached when organizations throughout the world are considered as potential benchmarking partners. The effectiveness and the level of difficulty of the analysis increases as one progresses up the pyramid.

The creation of a profitable and sustainable competitive advantage should be the focus of any business process. Future growth and market share allocations are dependent on management's ability to identify positive and negative aspects of the competition's product or service, as well as the customer's likes and dislikes. Benchmarking is the vehicle to understand and facilitate this competitive advantage. As a result of benchmarking the organization may continue on the path of continual improvement (the Kaizen approach), or pursue a more dynamic approach to change (the reengineering approach), or

Figure 10.14 A typical performance analysis of benchmarking.

pursue a combination of both. No matter what the selection is, it is important that the organization sooner or later focus on an operating strategy. The basic questions of operating strategy then become a function of one of the three strategies:

1. A business can pursue a cost strategy by being the low cost producer with a standard level of quality.
2. The company can embrace a value strategy by offering more value to the customer than the competition is able or willing to provide while maintaining proximity on cost and quality.
3. A niche strategy that is a hybrid of one of the first two paths.

A typical process for conducting a benchmarking is the following:

1. Identify those processes needing improvement.

2a. Identify a department within your company that conducts the process better than your department.

2b. Identify a firm that is the world leader in performing the process.

3a. Contact the internal manager and arrange for a physical visit of the process in question and/or an interview with appropriate and applicable personnel.

3b. Contact the manager of the company you are benchmarking and arrange for a physical visit of the process in question and/or an interview with appropriate and applicable personnel.

4. Collect the data.
5. Analyze the data.
6. Take action.
7. Evaluate action.

This seven-step process should provide for a customer-focused benchmarking activity that produces deliverables for customers. Underlying the structure is the fundamental requirement of identifying the customer of the process for the subject being benchmarked and identifying this customer's needs. A customer is a person or activity that may benefit from information, a product, or a service. Customers may be internal to the organization, or they may be external to the organization. Regardless of the internal/external relationship, the needs, wants, and expectations of the customer must be targeted for achievement.

Finally, it is because of this customer-driven focus that organizations should target their benchmarking activities based on priorities within:

- the organization's core competencies
- significant problem areas
- areas of specific customer satisfaction and dissatisfaction
- the organization's core processes

IV. COST OF QUALITY

There is at least one area in any organization that quality and improvement may be measured at the same time. That area is the cost of quality (COQ). Within the cost of quality, however, there are two categories that define the structure and reporting of that cost: avoidance costs and total failure costs. These two categories comprise the two operating curves used for achieving the break-even point of the quality costs break-even chart. Avoidance costs and total failure costs are broken down further in two categories in each area.

A. Avoidance Costs

Appraisal Costs

These costs are those expenditures that an organization makes to examine the levels of quality at which products are being produced. If the product quality level are satisfactory, production is allowed to continue, or if the quality levels are unsatisfactory, production is suspended until effective corrective action has been implemented to return levels to satisfactory product quality levels (Beveridge, 1985). Appraisal costs expected to discover and correct problems after the problems surface. Examples are: all inspection and associated expenses; inspector's wages and fringe benefits; laboratory test technicians wages, salaries, and fringe benefits; test samples for destructive testing; laboratory materials consumed in the process testing; laboratory tooling expenses; the portion of direct labor wages devoted to performing statistical process control (SPC), inspections and tests within the process; and process capability studies.

Prevention Costs

These costs are those expenditures a company makes to keep from producing defective or unacceptable product. Prevention costs are intended to be applied before the fact of producing defective product. Examples are: all training; quality assurance wages, salaries, and fringe benefits; cost of design changes incurred prior to releases for production; the portion of product design engineering devoted to quality assurance; and cost of processing changes incurred prior to releases for production.

B. Total Failure Costs

Internal Failure Costs

These are those costs incurred by an organization while it still has ownership of the product. Examples are: scrap, waste in process, rework charges, and repair charges.

External Failure Costs

These are are those costs incurred by an organization after it has transferred ownership of the product to its customer, the customer has received and originally accepted the product. Examples are: warranty costs, returned goods, design error, and marketing error.

V. STARTING A QUALITY COST REPORTING SYSTEM

Kuzela (1984), Bogardy (1987), and Miller (1988) suggested that a good place to start is to review the company's "Chart of Accounts," to determine which of those accounts should be selected and to which quality cost category those selected accounts should be assigned. The reader is encouraged to see Hagen (1986), Grimm (1987), Campanella (1990), and Rust, Zahorik, and Keiningham (1994) for a discussion on the details for constructing a quality cost reporting system. Table 10.1 gives a cursory view of the most standard form based on a typical chart of accounts. By identifying each of the accounts the reader of the report can see the influence of these costs to the total cost of the organization and, as a result, an appropriate management decision will take place to reduce the out of line cost. What is important about the form, is the fact that it has two time periods for comparison purposes.

Table 10.1 Quality Cost Report

	Time Period A	Time Period B	Difference
Appraisal costs			
Account 1			
Account 2			
Account 3			
Account n			
Prevention costs			
Account 1			
Account 2			
Account 3			
Account n			
Internal failure costs			
Account 1			
Account 2			
Account 3			
Account n			
External failure costs			
Account 1			
Account 2			
Account 3			
Account n			
Total failure costs			

Once the established the quality cost report format, the next step is to evaluate where the firms stands regarding the total failure cost experience as related to the company's expenses to obtain the necessary quality level. This is done by performing a break-even analysis. Figure 10.15 demonstrates this relationship. We can also make several observations from Figure 10.15 (Chase, 1986; Smith, 1980; Tersine, 1972; Juran, 1988), including:

- As the rate of avoidance cost expenditures increase, the resultant total failure costs rate decreases.
- Total quality fixed costs remains relatively stable over time.
- The break-even point between avoidance costs and total failure costs occur at a dollar level and a point in time where one dollar expended on avoidance costs equal one dollar experienced in total failure costs.
- Avoidance costs plus total failure costs produces total quality variable costs.

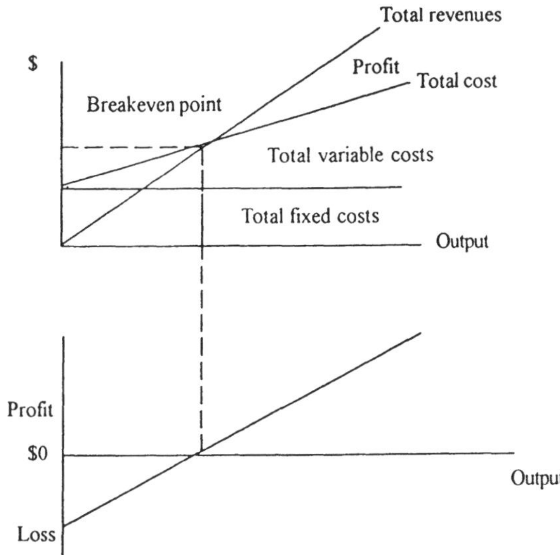

Figure 10.15 Break-even analysis.

- Total quality fixed costs plus total quality variable costs equals total quality costs at any point on the time period continuum.
- The first derivative of the total quality cost curve is the optimal quality cost for the model structure if no changes are made in the total quality fixed costs, rate increase in avoidance costs, or the associated reduction rate of the total failure costs.
- The rate of reduction of total failure costs is a variable dependent on the independent variable, avoidance costs.

So, how can quality costs be used in the modern world of quality? Relating these observations to change in engineering-oriented accounts located in the quality cost report structure should lead the observer to further insights toward the impact on costs that faulty engineering decisions can make, or even more important, the impact of engineering decisions that do not include an understanding of quality costs on the quality levels of products ready for consumption of these impacts.

To have an effective cost of quality some general principles and procedures must be internalized.

- Prior to initiating of a cost of quality system, a strong commitment from top local management is required to ensure its creditability.

- Cost data must be able to be readily extracted (or estimated) from existing data bases. Costs not easily identifiable or unjustifiable should not be included in COQ.
- The system must be developed jointly with the finance community.
- Cost data need not be estimated to the third and fourth decimal places.
- COQ should not be used as strictly "a report card." Its prime value is as a tool to identify opportunities to achieve continuous improvement in quality and ultimately costs.
- All COQ cost basis development work should be done in a team approach with input from all activities responsible for generating quality costs. The team must reach a consensus as to which costs and cost allocations are to be included prior to proceeding with the COQ process.
- Common baselines (e.g., total sales, total operating costs, or per unit costs) are recommended.
- All costs identified should fall into one of the four major categories (prevention, appraisal, internal failure costs, and external failure costs). For typical costs and their categorization see Appendix F.

Other approaches to accurately capture and evaluate the cost of quality may be found in the implementation of activity based costing (*The Economist*, 1990; Cooper, 1990, 1991; Gagne and Discenza, 1992; Sharman, 1990), bath tub curve analysis (Pauly, 1986), failure mode and effect analysis (Josephs, 1987; Stamatis, 1995).

VI. GEOMETRIC DIMENSIONING AND TOLERANCE

Geometric dimensioning and tolerancing (GD&T) is an engineering product definition standard that geometrically describes design intent. It also provides the documentation base for the design of quality and production systems. Used for communication between product engineers and manufacturing engineers, it promotes a uniform interpretation of a component's production requirements (Wearing and Karl, 1995; Krulikowski, 1994; Foster, 1982).

GD&T establishes a nationally recognized means of identifying geometric characteristics that had been overlooked, omitted, or explained by elaborate notes in the traditional method. It provides a standard symbolic engineering language that fully defines geometric requirements. When properly used it eliminates the ambivalence and confusion of drawing specifications

GD&T must be used by companies that want government contracts as well as in the automotive industry. Also, it may be used by manufacturers of other commercial products. Its use is often necessary to remain competitive, as it lowers product cost because:

- Engineering can accurately define what is required.

- Everyone can do their jobs right the first time, resulting in less reported discrepancies, less material review board (MRB) action, fewer drawing changes, better yields, and fewer production delays.
- Inspection setups are not subject to interpretation.
- Datum systems speed up manufacture and inspection and provide an effective basis for SPC.
- The accumulated benefits of GD&T often enable lower and more accurate price estimates for fabrication.

To implement a GD&T program in a given organization, there are at least four steps:

1. Get management support. Unless management is well acquainted with the benefits of GD&T, there is little chance of implementation. The awareness process for the management personnel, may be accomplished with a team that includes representatives of each affected area, or it might also be presented by a qualified GD&T expert.
2. Appoint a company advocate or guru. This person would be responsible to plan and implement GD&T, and act as a clearing house for questions and problems after GD&T is in use. Ideally this person should be a top-level individual that is familiar with the product, is expert in the use of ANSI Y14.5 (ANSI, 1983), has a strong background in manufacturing and inspection processes, and has the ability to advise and instruct others.
3. Document company practices. ANSI Y14.5, the document that formally defines GD&T, is complex and is not a practical reference for drafters, designers, or users of product drawings. To fully understand how the provisions and recommendations of this document apply to product designs requires a background in its derivation and intent. Most companies define their practices in drafting manuals or issue one of the many interpretative books that are available.
4. Conduct GD&T training. Provide training to those that prepare drawings and those that use them.

For more information on GD&T see Whitmire (1991), Chadderdon (1992), Garcia and St. Charles (1988), Foster (1982), Karl, Morisette, and Taam (1994), Wearing and Karl (1995), and Krulikowski (1994).

VII. THEORY OF CONSTRAINTS

The theory of constraints (TOC) is a holistic management philosophy that focuses on the organization's most critical issues—or constraints—and the implications of those constraints on the rest of the system. A constraint is anything that limits an organization's ability to achieve a higher level of performance. TOC overlaps with the TQM philosophy in many areas, such as:

1. It stresses ongoing continuous improvement.
2. It stresses company-wide education and cultural transformation.
3. It seeks employee involvement and empowerment.
4. It establishes clearly defined quality measurements.
5. It breaks down department barriers, stressing system rather than local goals.
6. It calls for a shrinking supplier base and reduction in inventories.

However, it departs from the general concepts of TQM in several critical ways. Perhaps, the most important departure is in the area of accounting. Goldratt (1994) points out that while an organization may have many purposes (e.g., providing jobs, consuming raw materials, increasing sales, increasing share of the market, developing technology, or producing high quality products), these do not guarantee the long-term survival of the firm. They are means to achieve the goal, not the goal itself. If the firm makes money, only then it will prosper. When a firm has money, then it can place more emphasis on other objectives.

To adequately measure a firm's performance, two sets of measurements must be used, one from the operations point of view and one from the financial point of view.

A. Operational Measurements

1. *Throughput*: all the money the system generates through sales; otherwise known as "goods sold"
2. *Inventory*: all the money the system has invested in purchasing things in which it intends to sell
3. *Operating expense*: all the money the system spends to turn inventory into throughput

B. Financial Measurements

1. *Net profit*: an absolute measurement in dollars
2. *Return on investment*: a relative measure based on investment
3. *Cash flow*: a survival measurement

Goldratt (1994) points out that these measurements must be very specific in nature and in case they have to be interpreted, an explanation must follow. For Goldratt, throughput is specifically defined as goods sold; an inventory of finished goods is not throughput. Actual sales must occur so as to prevent the system from continuing to produce under the illusion that the goods might be sold. Not building product into inventory for which there is no demand is a key fundamental in the synchronous manufacturing philosophy. Manufacturing product without demand inflates inventory, ties up machine and labor capacity (operating expenses), and starves a firm's cash flow. The most fundamental concept with regards to these definitions is the fact that the goal of a firm is to increase throughput while simultaneously reducing inventory and reducing operating expenses. A final measurement in this system is the definition of productivity, which is all the actions that bring a company closer to its goals.

TOC further identifies the relationship between dependent events and statistical fluctuations or variations. Traditional management's reluctance to recognize the concept of variation and not to account for it will ultimately affect throughput for a given entity, as well as its effect on cost. Therefore, TOC suggests that rather than balance capacities, the flow of product through the system should be balanced. Activities whose available capacities are less than the demand placed upon them to produce are said to be "bottlenecks."

VIII. CROSSTABULATION ANALYSIS (DISAGGREGATE ANALYSIS)

In the quality profession, one routinely calculates the mean, median, and standard error for any set of numbers. In fact, we may say that these kinds of statistics are so common that more often than not they become the "default statistics" for our analysis. To be sure, this analysis is very useful, but it has a major flaw in it: it tempts us to rely far too much on aggregate analysis, which can be misleading.

The temptation is due partly to the ready availability of the means, which, of course, are aggregate measures. It is further abetted by the common practice of designing standard banners for computer tabulation, which discourages use of crosstabulation that goes beyond these standard setups.

How can aggregate analysis mislead? Simply by leading us to draw unwarranted conclusions from superficial information. (Misleading information may result in erroneous supplier evaluation, auditing quality systems, or even a customer satisfaction survey.) For example, Table 10.2 shows some hypothetical data from a customer survey: six respondents rating two products on a 1–10 scale. Clearly, there is no difference between A and B. In fact if we run a simple correlation we get a value of $r = 0$. That is strange! We expected a positive correlation. This expectation is the kind of superficial error that aggregate analysis can lead us into. The correlation coefficient is a result of disaggregate analysis, based on the interrelation between A and B, separately for each respondent.

Table 10.2 Aggregate Analysis

Rating	A	B
Rating = 10	1	1
Rating = 9	1	1
Rating = 8	2	2
Rating = 7	1	1
Rating = 6	1	1
Mean:	8.0	8.0
Standard error	0.527	0.527

On the other hand, if we run the actual crosstabulation, we find something quite interesting (Table 10.3), that tells us a good deal more than the previous analysis. This crosstabulation analysis identifies four people that gave B a higher rating than A and only two rated A higher than B. Where A was rated higher, the advantage over B tended to be greater than the other way around.

On can clearly see that our original conclusion that "clearly, there is no difference ... ," can lead us further to the thought that "A and B are equal." Only the crosstabulation, the disaggregate analysis, shows us that indeed there is a difference.

Aggregate analysis is based on marginal totals for groups. But even though quality strategy designs the approaches and selects the media to address concerns (e.g., audits, customer satisfaction, supplier evaluation), these concerns should be addressed by disaggregate analysis.

To make the point clearer, let us look at the intriguing parallel between the lazy analysts' reliance on grouped data and the classical gambler's fallacy. In both cases, unwarranted conclusions are drawn from aggregate information. The gambler's fallacy uses aggregates of past events to predict the future,

Table 10.3 The Correlation Coefficient

	A = 6	7	8	9	10
B = 10			1		
9			1		
8		1			1
7	1				
6				1	

confusing a priori and a posteriori data to predict or estimate what goes on inside the black box for which he only has the outside dimensions. Cross-tabulation is the simplest and most effective way to look inside the box and glimpse the actual patterns. For more detailed information see any statistic book and/or any statistical package such as the MINITAB, SAS, and many others.

REFERENCES

Adam, P. and Vande Water, R. (1995). Benchmarking and the bottom line: Translating business reengineering into bottom-line results. *Industrial Engineering*. Feb.: 24–26.

Akao, Y. (Ed.). 1990). *Quality function deployment: Integrating customer requirements into product design*. Productivity Press. Cambridge, MA.

American National Standard Institute (1983). *Dimensioning and tolerancing ANSI Y14.5M-1982*. The American Society of Mechanical Engineers. New York.

American Supplier Institute (ASI). (1987). *Quality function deployment: A collection of presentations and QFD case studies*. ASI. Dearborn, MI.

American Supplier Institute. (1989). *A symposium on quality function deployment*. ASI. Dearborn, MI.

Beveridge, D. (1985). Cost of quality. *Sales and Marketing Management*. Dec. 9: 12.

Bogardy, A. E. (1987). When should management think about cost. *California Management Review*. Mar.: 67.

Bossert, J. L. (1990). *Quality function deployment: A practitioner's approach*. Quality Press. Milwaukee, WI.

Camp, R. C. (1995). *Business process benchmarking: Finding and implementing best practices*. Quality Press. Milwaukee, WI.

Camp, R. C. (1989). *Benchmarking: The search for industry best practices that lead to superior performance*. Quality Press. Milwaukee, WI.

Campanella, J. (Ed.) (1990). *Principles of quality costs: Principles, implementation, and use*. 2nd ed. American Society for Quality Control. Quality Costs Committee. Milwaukee, WI.

Chadderdon, R. A. (1992). GD&T training programs. *Quality*. Sept.: 55–56.

Chase, J. P. (1986). Yanks borrow Japanese keys to quality cost. *Electronics*. May: 95.

Chrysler, Ford, and General Motors. (1995). *Potential failure mode and effect analysis*. Chrysler, Ford, and General Motors. Distributed by Automotive Industry Action Group (AIAG). Southfield, MI.

Clausing, D. (1988). Quality function deployment. In: N. E. Ryan. (Ed.) *Taguchi methods and QFD: Hows and whys for management*. American Supplier Institute. Dearborn, MI.

Cohen, L. (1995). *Quality function deployment: How to make QFD work for you*. Quality Press. Milwaukee, WI.

Cooper, R. (1990). ABC: A need, not an option. *Accountancy*. Sept.: 86–88.

Cooper, R. (1991). ABC: The right approach for you. *Accountancy*. Jan.: 70–72.

Day, R. G. (1993). *Quality function deployment: Linking a company with its customers.* Quality Press. Milwaukee, WI.

Denson, D. (1992). The use of failure mode distributions in reliability analysis. *RAC Newsletter.* Spring: 1–3.

The Economist. (1990) Costing the factory of the future. Mar.: 61–62.

Eureka, W. E. and Ryan N. E. (1988). *The customer-driven company: Managerial perspective on QFD.* American Supplier Institute. Dearborn, MI.

Eureka, W. E. (1987). Introduction to quality function deployment. Paper presented in a training class offered by American Supplier Institute. Dearborn, MI.

Ford Motor Co. (1988). *Potential failure mode and effect analysis.* Ford Motor Co. Dearborn, MI.

Ford Motor Co. (1989). *Potential failure mode and effect analysis.* Ford Motor Co. Dearborn, MI.

Foster, L. (1982). *Modern geometric dimensioning and tolerancing.* NTMA. Ft. Washington, PA.

Gagne, M. L. and Discenza, R. (1992). Accurate product costing in a JIT environment. *International Journal of Purchasing and Materials.* Oct.: 28–31.

Garcia, C. J. and St. Charles, D. P. (1988). Automating GD&T. *Quality.* June: 56–58.

Goldratt, E. M. (1994). *It's not luck.* North River Press. Great Barrington, MA.

Grimm, A. F. (Ed.) (1987). *Quality costs: Ideas and applications.* Vol. 1. 2nd ed. American Society for Quality Control. Quality Costs Committee. Milwaukee, WI.

Hagen, J. (Ed.). (1986). *Principles of quality cost.* American Society for Quality Control. Quality Cost Technical Committee. Milwaukee, WI.

Josephs, F. (1987). *Production management: Concepts and analysis for operation and control.* The Ronald Press. New York.

Juran, J. M. (1988). *Quality control handbook.* 4th ed. McGraw-Hill. New York.

Juran, J. M. and Gryna, F. M. (1980). *Quality planning and analysis.* McGraw-Hill Book Co. New York.

Karl, D. P., Morisette, J., and Taam, W. (1994). Some applications of a multivariate capability index in geometric dimensioning and tolerancing. *Quality Engineering.* 6(4): 649–665.

Krulikowski, A. (1994). *Geometric dimensioning and tolerancing: A self study workbook.* Quality Press. Milwaukee, WI.

Kuzela, L. (1984). Here's how to put quality control to work for you. *Industry Week.* Nov. 12: 21.

Miller, W. H. (1988). Changed focus: Technology fix to cost of quality control. *Industry Week.* Aug. 1: 16.

Pauly, D. (1986). The cost of doing things better. *Society of Manufacturing Engineers.* Sept. 8: 196.

Rust, R. T., Zahorik, A. J., and Keiningham, T. L. (1994). *Return on quality: Measuring the financial impact of your company's quest for quality.* Quality Press. Milwaukee, WI.

Sharman, P. (1990). A practical look at activity based costing. *CMA Magazine.* Feb.: 8–12.

Smith, W. (1980). Let's take a closer look at quality cost. *Industrial Research*. June: 154.

Spendolini, M. J. (1992). *The benchmarking book*. Quality Press. Milwaukee, WI.

Stamatis, D. H. (1992). ISO 9000 standards: Are they real? *Technology*. Aug.: 13–17.

Stamatis, D. H. (1995). *Failure mode and effect analysis. FMEA from theory to execution*. Quality Press. Milwaukee, WI.

Stamatis, D. H. (1996). *Total quality in service*. St. Lucie Press. Delray Beach, FL.

Sullivan, L. P. (1987). *The company-wide benefits of QFD*. Video tape. American Supplier Institute. Dearborn, MI.

Sullivan, L. P. (1986). Quality function deployment. *Quality Progress*. June: 39–50.

Tersine, T. (1972). *Engineering economics*. Viking Press. New York.

Watson, G. H. (1992). The benchmarking workbook: Adapting best practices for performance improvement. Quality Press. Milwaukee, WI.

Wearing, C. and Karl, D. P. (1995). The importance of following GD&T specifications. *Quality Progress*. Feb.: 95–98.

Whitmire, G. (1991). Why use GD&T. *Quality*. March: 41–42.

11

The Full Chain of Quality: Supplier–Organization–Customer

I. OVERVIEW

Figure 11.1 shows the relationship of Supplier–Organization–Customer. The figure shows that quality is a function of the total interaction between the three. This interaction is based on continuous actions that result in products and/or services that satisfy customer perceptions and conform to customer requirements using the principles of total quality. That is:

- customer satisfaction
- constant improvement through feedback
- do it right the first time
- teamwork
- open communication

In essence then, this collaboration of actions make up a system of processes and resources that works together to meet a specific objective.

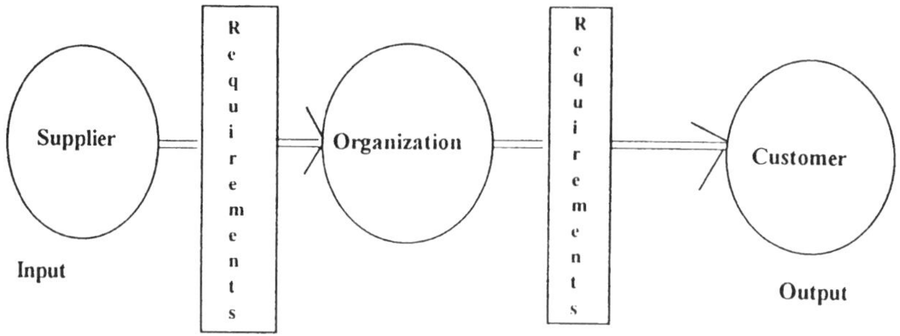

For optimum results:

1. define the scope of the organization
2. Identify the output(s) and customer(s)
3. Determine the requirements for the output(s)
4. Identify the input(s) and supplier(s)
5. Determine the requirements for input(s)

Figure 11.1 The supplier-organization-customer relationship.

II. SUPPLIER

It should go without saying that suppliers play a critical role in any quality process: If what is coming downstream to you is laden with waste, in the form of poor quality or erratic delivery schedules, then your quality efforts, regardless how aggressive, can only suffer. This is true for all the suppliers regardless of their proximity or previous dealings. The criteria must be the same. The reason for this is that one of the most crucial elements of the quality process is predictability—predictability within your own organization and predictability from those who supply you with either products and/or services. And if you cannot get the quality criteria you want with the predictability you need, then terminate your suppliers and call in others.

As you discover the array of suppliers who meet your quality requirements (because they are running their businesses in concert with your needs), you will inevitably find yourself dealing with fewer suppliers. After all, that is part of the goal: to deal with a smaller supplier base who will have a larger share of the business and who constantly demonstrate quality and continuous improvement. This improvement will be demonstrated through a variety of activities and measurables, including: logistic efficiency, customer responsiveness and fulfillment agility (Greene, 1996). However, the result of all improvement will be to provide value both to the customer and supplier.

Creating value based on customer needs has become an important source of competitive advantage, creating a shift from "push" to "pull" strategies for bringing products to the customer. To be sure, approaches, tools, and methodologies available to the quality professional for creating value are constantly changing and evolving at the same time. However, one of the most consistent approaches to quality is the supplier chain. Coordination of activities and management of relationships in the supply chain can be a significant source of competitive advantage and bring additional value to the customer.

Price is one of the main ingredients of supplier relations, but it is not the only one. What is important is the availability of unique benefits that offset the given price in such a way that the competitor cannot match it. The technique or approach to facilitate this has many names, including partnering, supplier development, and value chain analysis. No matter what the name, the intent of the supplier development is to help an organization focus on strategically relevant supplier activities, in order to understand the behavior of costs, variation, and perhaps more importantly the existing and potential sources of differentiation (Ford Motor Co., 1996).

Value can be enhanced and competitive advantage accrued through a low cost, a high differentiation, or a hybrid strategy. Supply chain activities are important sources of added value regardless of strategy. Therefore, recognizing the increasing importance of supply chain management and the growing need for managers with training in this area has been exploding all across industry lines throughout the corporate world. Major corporations such as Motorola, Ford, IBM, and many others actively encourage their supplier base to actively participate in quality improvement programs for mutual benefit.

The curriculum reflects the melding of industry and academic perspective and industry trends. The curriculum takes a cross-functional approach in contrast to more traditional business training, which focuses on business functions as isolated activities. Typical effective management across the supply chain involves the introduction of advanced strategic and operational knowledge. In fact, the core curriculum blends a supply chain management course(s) with engineering, quality management, process control, logistics, inventory, operations research, statistics, logistics, information systems courses, project management, and many more.

The supplier development program may focus on either primary activities (e.g., transportation, scheduling, and production) or support activities (e.g., human resource management and technology development and procurement). However, it must be emphasized that the specific activities in which a company engages are a function of the firm's infrastructure and the industry in which the firm operates. Competitive advantage is gained by performing any one or more of these activities better than or at a lower cost than competitors.

To gain and sustain competitive advantage, a firm must not only understand its own value chain, but that of upstream and downstream suppliers and customers. Therefore, optimizing the total system performance and cost should be the primary objective. Hence, the concept of supply chain management must be understood. Indeed, because of the tremendous impact that the supply chain plays on the organization, it is increasingly becoming more advantageous to manage and/or integrate select activities across firms in the supply chain and eliminate services that do not add value. These cooperative relationships—between upstream and downstream suppliers and costumers—can create very positive win-win situations.

III. SUPPLIER CERTIFICATION

The purpose of the supplier performance classification system from an organization's perspective is to:

- analytically determine the current best-in-class suppliers
- focus the business with best-in-class suppliers and build a strategic relationship with them concentrating on quality, technology, delivery, cycle time, and total cost
- continually improve in all areas of their business to optimize customer satisfaction

On the other hand, the supplier will also benefit from such a relationship. The rewards for excellent performance and willingness to work for a better, more productive future go far beyond the initial phase of increased business, to the much more significant arena of the long-term strategic relationship. The elements of a long-term relationship not only include long-range purchase agreements but extend to more responsive engineering and quality support that quickly resolves problems and improves manufacturability, a more open and informal relationship that stresses helping and information sharing, and a more secure future which will enable the supplier to make more accurate production and financial forecasts. All of these elements will lead a supplier to a higher level of preparedness and responsiveness to customer's needs and desires.

With a focused effort, sound plans, effective communication, and cooperation, the supplier's quality will continually improve beyond zero defects on the road to zero deviation. Ultimately this win-win relationship between the supplier and the organization will result in reduced total cost and significant financial benefits. In summary, supplier benefits will include:

- increased business
- mutually developed specifications
- improved overall quality
- increased profit

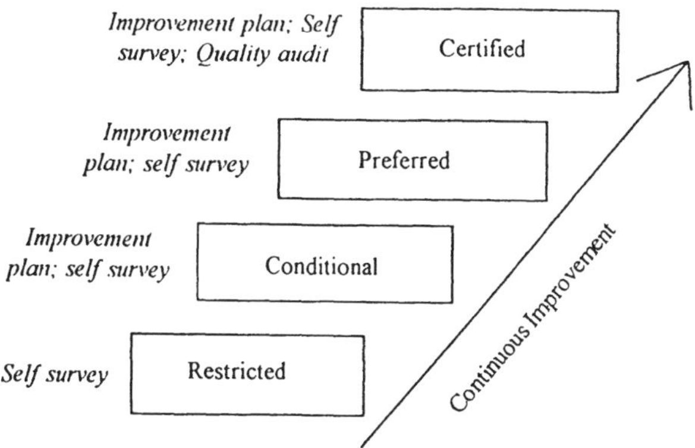

Figure 11.2 A typical supplier performance classification.

- reduced customer inspection, surveys, and audits
- shared market demand forecast
- long-range purchase agreements
- long-term strategic supplier relationships
- engineering, manufacturing, and quality training assistance

Based on a supplier's overall performance in delivering a total quality product, the organization classifies the supplier into one of four categories: 1) certified, 2) preferred, 3) conditional (approved), or 4) restricted. Suppliers will receive increased benefits as they climb from one classification level to the next. A typical supplier performance classification is shown in Figure 11.2.

To become a certified supplier a company must demonstrate a commitment to continuous improvement, have strong leadership from top management, and focus on teamwork. Standardization of procedures, assessment of process capability, reduction of variation, and control of measuring and testing systems are essentials for certified suppliers. Certified suppliers will, by definition, operate with minimum waste and therefore be more cost competitive. It must emphasized that certification may also be obtained through third-party auditing (e.g., ISO 9000 and QS-9000).

IV. ORGANIZATION

Any organization that wants to stay in business for the long term must develop a quality system, develop their supplier base, and most important, must listen and respond to customer needs, wants, and expectations.

A quality system may be based on the ISO 9000, Total Quality Management, or it may, at the very minimum, address the following issues:

1. *Administration.* In this category, it is important to define the organization's quality policies, procedures, and inspection practices. It is also common that under this category the audit and surveillance administration for products and suppliers be developed and appropriate training should be instituted throughout the organizational interfaces.

2. *Quality program management.*

   - organizational structure and functions
   - inspection function locations
   - quality planning
   - inspection/test instruction
   - records and reports
   - corrective action process
   - cost related to quality

3. *Facilities and standards.* This category includes the identification, definition, and implementation practices, and/or responsibilities of:

   - specifications, drawings, changes
   - document control
   - obsolescence control
   - retention policy
   - measuring and test equipment (production, quality), including: the calibration system, calibration control and recall, calibration/service specification and instruction, and certification of measuring equipment and testing equipment.

4. *Control of purchases.*

   - quality evaluation and selection request of active and candidate supplier
   - quality survey of active and candidate supplier
   - procurement document quality requirement review
   - source inspection
   - supplier quality rating

5. *Manufacturing control.* This category should be the most demanding in any organization, since it controls the product being manufactured, and includes:

   - receiving inspection
   - raw material inspection

- in-process inspection
- final inspection
- statistical process control
- protecting product quality
- material review board

A. The Rules for Record Retention

Quite often, the customer and/or the Government (as in the case of nuclear and/or safety items) will dictate the retention period for your records. However, in the absence of guidelines from either the customer and/or the Government, how long must you keep your quality records?

Quality records (paper documents, magnetic media, optical disks, microfilm, e-mail, videotapes, and other media) are an organization's quality memory. Records and the information contained in them can ensure that an organization is in compliance with the appropriate standard(s), quality system and so on. Furthermore, quite often they protect the organization in case of lawsuits. As a result, the answer to the question of "How long must records be retained?" is a question that must be answered accurately and conclusively within all organizations.

Perhaps the legal risks for retaining records or the lack of retaining them in today's litigation intensive society can result in disaster. An organization can now get into just as much legal trouble by retaining data and records too long, as it can by deleting or throwing away records too quickly. On December 1, 1993, Rule 26 of the Federal Rules of Civil Procedure was changed to require voluntary disclosure of documents (including electronic records) that are relevant to federal court cases. That new rule also significantly reduced the time allowed to produce pertinent records. Organizations must be able to demonstrate reasonable and prudent good faith regarding their records management practices. (This has a direct relevancy for quality records, such as: inspection records, reliability data, testing results, and so on.)

So, what is an organization to do? Any organization in today's business environment must have at least a policy for a records retention schedule and demonstrate that in fact this schedule is implemented and is working effectively throughout the organization. The following steps are essential when developing a retention schedule:

1. Top management commitment. The commitment must be communicated throughout the organization in terms of policy as well as action terms. The intent of this step is to demonstrate that senior management supports the development and implementation of a sound record retention program.

2. Conduct a preliminary file purge. The purge is to clean the unneeded duplicates out of files and databases as well as to remove nonrecord materials. No original records should be eliminated at this point. Completing a preliminary file purge is a positive step that will immediately help virtually all users in the organization and make all remaining steps in the retention schedule development process easier and faster to accomplish.
3. Complete a records inventory. Information gathered in the inventory process typically includes identifying records types. Examples are: Descriptions, media, physical locations, dates or date ranges, volume, official or duplicate record, custodians of the records, individuals who did the inventorying, date inventory was completed, and any other pertinent data specific to the organization and/or task.
4. Prepare a master list of record types. Separate the records by type and group them by function. Define the need of the records and their useful life on a preliminary basis.
5. Negotiate the retention period internally. Review all preliminary retention schedules with appropriate personnel to ensure that user needs will be met. Remember that the retention process is always driven by valid user needs.
6. Obtain all essential approvals. To be effective, the final record retention schedule requires approval from management.
7. Publish and distribute the retention schedule. All users of records must be familiar with the retention schedule.
8. Establish controls to ensure that retention schedules are applied and followed. Common—but not unique—practical controls to monitor and ensure retention schedule implementation and compliance include establishing an annual records retention audit conducted by an organization's auditing and/or records management. It is also beneficial to establish procedures that require appropriate approval for changes, modifications and removal of obsolesence records.
9. Review of record retention schedule. A regular review and update of all retention schedules must take place to ensure that current records are kept. To facilitate this review, a records-management software for computer-aided retention management may be used.

For any organization in today's business environment, the most important decision a business must make about its records is not whether or not to destroy them, but when to destroy them. Every organization should develop and implement a records-retention schedule and routinely implement that schedule into the normal course of business.

B. Advanced Quality Planning

The process of advanced quality planning (AQP) involves four basic steps:

1. *Feasibility analysis.* Feasibility analysis uses design and process failure mode effects analysis (FMEA) as its primary tools. The FMEA is a disciplined method of asking the questions "What can go wrong?" and "How can we prevent it from happening in our process?"
2. *Control plan development.* The FMEAs from the feasibility analysis are used as resources for the development of a control plan for the proposed new product. The control plan describes the process, identifies critical process parameters, and describes how those parameters will be controlled. However, all controls should have some means to determine whether or not they are working properly. This might include a gage or meter that can be checked for calibration. Control characteristics on raw materials must also be specified and preferably controlled by the supplier who demonstrates statistical control and capability. In lieu of this, raw material specifications must be determined and procedures developed and implemented to check in-coming raw materials. Characteristics to be SPC charted must be specified and reaction plans to out-of-control conditions described.
3. *Process potential studies.* The process potential study is the experimental part of a new product introduction. It consists of implementing the control plan that has been developed through the above process and evaluating the results through statistical capability studies. If the desired capability is not achieved, changes in the control plan are agreed upon and the study repeated. This process may include a design of experiments to determine optimum process parameters.
4. *Verification of control plan.* The final part of AQP process is to carry out a complete review of the process and the control plan. The review should examine the control plan to ensure that the process is properly understood by all, so that those directly involved in the process know what they are supposed to do, and how to read if something goes wrong. The control plan should also include control characteristics, all inspection characteristics, and specifications. The review should also examine the control plans to ensure that all gages, meters and test equipment used in the process of inspection are properly designated and calibrated on a regular basis. It must be verified that the gages, meters, and test equipment actually do measure what they are supposed to measure and do so with sufficient accuracy. Finally, the control plan needs to be ver-

ified through statistical process capability studies that show that the capability desired has been achieved.

The end result of these steps should be the development of processes that assure the manufacture of products with the greatest possible efficiency and productivity. Advanced quality planning methods can also be a useful tool for the analysis of present in-line products for the purpose of improving their productivity and consistency and achieving the least variation.

If all critical characteristics have been identified and controlled, and the process is shown to be statistically capable, it will not be able to produce material out of specification. Therefore, the goal of AQP is to make the whole production process much more efficient, since time is minimized—if not completely eliminated—for testing, making adjustments and retesting prior to finishing the batch.

In order to implement AQP, training for all participants will need to be done so that all understand the basic statistical principles and philosophy involved. AQP is the first step in a production and management philosophy that emphasizes prevention rather than detection.

The philosophy emphasizes the importance of having suppliers who also have their processes in control and capable. Thus, purchasing needs to become familiar with their supplier's quality systems and their capability to produce materials with little or no variation. Sales needs to develop an intimate knowledge of customer requirements and clearly communicate those needs. Research and development (R&D) needs to realize their role in reducing variation in the products we produce. Since R&D often is the first to use new raw materials, they need also to be aware of and/or request information of the type of variation that can occur from the raw material supplier. They also need to be sufficiently knowledgeable of the process they prescribe to know how the process variables can affect the product. Finally, manufacturing needs to be aware of and able to control the significant variables that affect the product. All of this requires that management devote the resources required to achieve these ends. Therefore, quality is not limited to the manufacturing arm of our company. It is necessary for all of us to become involved and to know how we all fit into the cooperative effort to produce quality products and/or services. It requires the entire organization's commitment to plan quality into our products and/or services. (For more detailed AQP information see Chapter 18.)

C. A Typical Survey Instrument Structure for Certifying a Supplier

As we already have mentioned the idea for a supplier self-survey is to evaluate a supplier's Total Quality System without a representative of the organization actually visiting the supplier's facilities. The questions in the survey are developed by the organization and the complexity as well as the length of the instrument used depends on the industry and the specific needs of the organization. A typical structure of a survey should cover the following items with varying level of detail.

1. *Leadership.* This section examines the total quality commitment and performance of the group leadership. It evaluates how effectively the group management has utilized the total quality process to define and meet customer requirements. It also assesses the degree to which management has instilled the principles of the total quality process in all employees, other company departments, customers, suppliers, and community organizations.
2. *Data gathering and analysis.* This category examines the validity, use and management of data and information that underlie the company's total quality management system. Also examined is the adequacy of the data and information to support a responsive prevention approach to total quality based upon documented facts.
3. *Strategic quality planning.* The strategic quality planning category examines the group's plans for integrating the total quality process into the business. It reviews your long- and short-term plans for implementing and sustaining this total quality process, plus how you intend to measure your progress.
4. *Human resource utilization.* This category examines the effectiveness of the group's efforts to develop the work force and thereby encourage personal and corporate achievement in an atmosphere of safety, health, and well-being.
5. *Quality assurance of products and services.* This category is predominately process and systems oriented. It examines how you design, develop, monitor, control, and improve processes to meet customer needs.
6. *Quality results.* The quality results category examines the results of the systems in place to assure product and process conformance to requirements and the continuous improvement of both. It also focuses on comparing your results with competitors.
7. *Customer satisfaction.* This category examines your knowledge of the customer, overall customer service systems, awareness, responsiveness, and your ability to meet requirements and expectations. Also examined are your methods for measuring customer satisfaction.

When the survey is completed by the supplier and sent back to the seeking organization an evaluation takes place based on predetermined points. Typical scoring is the following:

N/A	Not applicable	Used only when the item does not effect the function of the quality system.
0.0	No system exists	The item is not included in the supplier's quality system.
1.0	Significant deficiency	The item is included in the supplier's quality system but needs significant improvement.
2.0	Improvement needed	The item is implemented but needs some improvement.
3.0	Satisfactory	The item meets all requirements and its effectiveness is verifiable.
4.0	Outstanding	The item exceeds all requirements and its effectiveness is thoroughly demonstrated.

It is very important to recognize that this type of survey is used only when the certification is based on first- and/or second-party auditing. If there is a third-party registration the survey is not usually conducted. Rather, the audit is performed based on the quality system that the organizations and/or supplier is seeking, such as ISO 9000, QS-9000, ISO 14001, and so on.

V. CUSTOMER

No mantra in business is more sacred than "you must listen to your customer." Yet a company can land in a deep trouble by listening singlemindedly to its customers, to the exclusion of others who may be speaking solid, good sense. Examples are everywhere, but one of the most recent and famous is the case of Apple Computers (Levin, 1996). For years, everyone—outside consultants, strategies, analysts—told the company executives to license their technology to the makers of clones, just as IBM did with its PC.

The idea was to drive down machine prices and make Apple's computing style the standard for the masses. Apple had developed some truly innovative concepts: a graphical user interface, plug-and-play capability, easy networking, and so on. In the early days, these concepts made Apple machines easier to use than IBM machines. By the logic of cloning, the more people who become wedded to Apple's standards, the more would eventually buy its software.

Time and again Apple declined to license. The company feared cheaper imitations would drive machine prices down and ruin their profit margins (which is what happened to IBM). Apple drew support for its noncloning strategy from its highly loyal customers.

Whether they realized it or not, the devoted customers must have helped reinforce Apple's hope that customers would continue to pay premium prices for Apple technology and spurn less expensive IBM clones. If this is the message that Apple executives were hearing and believing—and I can only speculate that it was—it is not hard to understand why Apple delayed granting licenses to clonemakers.

Knowing the customer is no substitute for maintaining a keen strategic sense of the industry. No one knew their customers better than oldline department stores; yet somehow department stores were caught flatfooted by catalog companies, factory outlets, home shopping networks, and discounters. Newspaper subscribers may love their morning papers, but the industry struggles to stem declining readership.

Such issues are not simple. With hindsight it is not hard to see that Apple blundered badly by delaying the decision to licence. For most of its history the company was posting solid profits, holding a solid share of the market and above all it had a very satisfied customer base. It is easy to understand how licensing clones might have seemed like a radical course. But that is why directors pay chief executives the big salaries and that is why they fire them.

The moral of the story? Yes, do listen to the customer. However, know your business first and apply the knowledge to satisfy the current customer as well as the future customer.

VI. CUSTOMER SATISFACTION

The growing popularity of customer satisfaction has given rise to the development of different programs for measuring it. Although these programs vary widely in scope and methodology, most are designed to produce a single satisfaction metric based on random sampling of customers at a single point in time. This common approach ignores the fact that in many cases (e.g., equipment and other durable products) customers use products over a relatively long life cycle made up of a sequence of distinct phases. These products may require the investment of significant resources by the customers and the longer the life of the product the greater the potential for customer satisfaction to change (usually decrease) over time. These aspects tend to increase a customer's recollections of his or her experiences with the product throughout its life.

In an organization, customer satisfaction should be a system that involves everyone. Customer satisfaction is a systematic process for collecting customer data (e.g., surveys, complaints, audits, sales calls and so on), analyzing this data to make it into actionable information, driving the results throughout an organization, and implementing satisfaction improvement plans. Therefore, the main components of a customer satisfaction system are the measurement programs coupled with other information provided by the customer (e.g., customer complaints, service information, and so on) and cus-

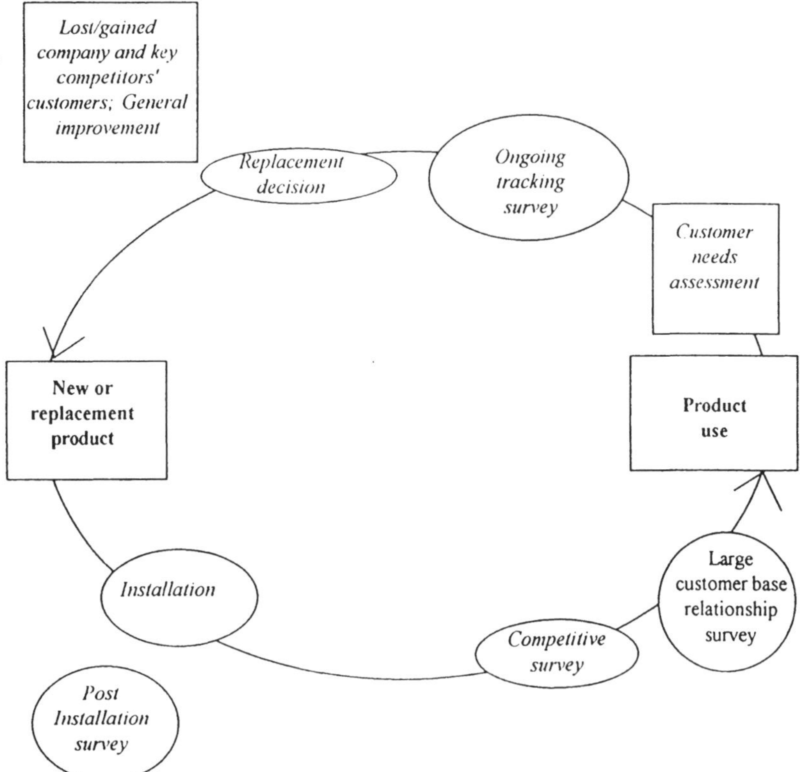

Figure 11.3 The life-cycle phases and the measurement programs to monitor customer satisfaction.

tomer satisfaction improvement plans. The measurement programs should be designed to assess satisfaction levels at different periods during the life cycle of a customer and should clearly point out what aspects of the business need to be improved. The satisfaction measures and analysis obtained are then used to create close-looped customer satisfaction improvement plans which should be devised and implemented by the different functions in an organization. Once improvement plans are implemented, follow-up surveys should be done to track progress. For the construction of a survey instrument see Stamatis (1996) and Kerlinger (1973).

A product goes through different phases during its life: installation, use of the product, and a replacement decision. At the end of the product's life, the customer decides whether or not to replace the product with the same or another manufacturer's new product. One can then think of the customer as having a "life cycle" which ends and may be renewed when the product's life ends. The events preceding the replacement phase can strongly influence customer satisfaction and a customer's willingness to repurchase a company's products. Figure 11.3 shows the life cycle phases and the measurement programs that will allow a company to monitor customer satisfaction and make necessary improvement as the product goes through the different phases.

REFERENCES

Ford, C. (1996). Partnership offers process manufacturing solution. *Managing Automation*. Jan.: 24–25.

Greene, A. (1996). Looking beyond ERP for a supply chain advantage. *Managing Automation*. Jan.: 6.

Kerlinger, F. N. (1973). *Foundations of behavioral research*. 2nd ed. Holt, Rinehart and Winston. New York.

Levin, D. (1996). Apple took advice from wrong group. *Detroit Free Press*. Feb. 8: 1E.

Stamatis, D. H. (1996). *Total quality service*. St. Lucie Press. Delray Beach, FL.

12

Value Analysis

Creating value based on customer needs has become an important source of competitive advantage, and has caused a shift from "push" to "pull" strategies for bringing products to the customer. This chapter addresses the basic concepts and rationale of value analysis. It identifies the four phases of the technique and will provide a common list of techniques utilized in this methodology. Originally developed to identify and eliminate unnecessary costs without reducing product functionality, reliability, durability, or appearance, its use today and in the future will assist team members who strive for improvement. The value analysis is a cross-functional, consensus-based, team- and problem-oriented process whose success depends upon the support and encouragement they receive from their management. For more information the reader is encouraged to see Fowler (1990), Freeze (1994), and Patterson (1995).

I. INTRODUCTION TO VALUE CONTROL

A. The Environment

Among the major problems faced by industry today, two are the cost profit squeeze and ineffective communications. Rising wages and materials costs are squeezing profit against a price ceiling and our communications systems do not seem to be able to help us effect a solution to the problem. We cannot buy labor or material for less than the market cost nor can we sell for more than the consumer is willing to pay. What then is the solution?

It is necessary to apply every known effective technique to learn how to thoroughly analyze the elements of a product or service so that we can identify and isolate the unknown, unnecessary costs. In short, it is necessary to make a direct attack on the high cost of business.

Value control is an effective management tool to seek out and eliminate this hidden cost wherever it may be. It can aid in improving both profit and communications problems, and it can have an effect on operations that will be limited only by your understanding of the techniques and management's willingness to apply them.

Many people are highly skilled at cost analysis and problem solving and think that value control is something we do all of the time. There are many who think that value control is part of every engineers job. Some also think that it is something we have done for twenty or thirty years but we did not call it value control. However, value control is not only different but is a more powerful technique than any used in the past.

Value control is not new in that it has been around for about 25 years, but it has only been within the past five to ten years that it is being widely accepted. It is a broad management tool that considers all of the factors involved in a decision. It goes to the heart of the problem and determines the function to be performed and applies creative problem solving and business operations such as time-and-motion studies, work simplifications, feasibility reviews, systems analyses, and so on. But, in addition, it follows a systematic organized approach that applies unique techniques that identify value control as a special approach to profit improvement.

B. Why Is Value Control Necessary?

It is still possible for one person to know all that is required to operate a small company or design a simple product. However, our increasingly complex society and increasingly complex technology has tended to make most of our managers and technical people specialists in a limited area of activity. It has tended to compartmentalize our operations and, to a large degree, our thinking. The more complex the organization the more the operation becomes fragmented into autonomous units that deal in a small part of the operation and have an effect on only a small part of the profit.

In 1927, one of the greatest technological milestones in the history of man's development was the result of the knowledge of two men. Charles A. Lindbergh sat on the Coronado Beach at San Diego with Donald Hall, Chief Engineer of Ryan Airlines, and established the basic criteria for the Spirit of St. Louis. Two people knew all that was needed to develop an advanced product that even today clearly shows their creative thinking.

Lindbergh established the requirements, Hall provided the technical knowledge and the thirteen Ryan supervisors and employees provided the understanding, know-how, and enthusiasm to develop a product that was designed, built, tested, and won ever-lasting glory for Mr. Lindbergh—all within 13 weeks.

The product was designed to perform a specific function for a specific cost target. There was no communication problem, there was not a cost problem since $15,000 was all they had for the entire Project, and there was no timing problem—any delay was unthinkable.

Consider the design of an advanced aircraft today. The cost in men and materials is almost beyond comprehension. Hundreds of thousands of people in dozens of industries in several states work in vast industrial complexes for years before the product takes to the air. Similarly, the automobile has created a similar situation to the degree that it is a basic national industry and affects people in every corner of the country and in many cases abroad. Is it any wonder that value control is developing on the management scene. It is the only technique that is specifically designed to consider all of the factors involved in decision making—product performance, project schedules, and total cost.

Value control is a program of involvement. It makes use of experience from engineering, manufacturing, purchasing, marketing, finance, and any and every area that contributes to the development of a product. it can be used to keep cost out of a product and it can be used to get cost out of a product. It can do this because cost is everywhere and is contributed to by everyone in the organization.

Cost is the result of marketing concepts, management philosophies, standards and specifications, outdated practices or equipment, lack of time, incomplete or unobtainable or inaccurate information, and dozens of other reasons. Every company has at least some of these problems and it often requires completely new ideas to change them.

To prevent and eliminate unnecessary cost we must know how to identify cost. We must be able to identify a problem and be willing to improve the situation. This means change—change in habits, change in ideas, change in philosophies. We know we must change to keep up with the world. value control enables us to take a good look at all of the factors that must go into making a successful change that will be for our benefit.

Value control requires special skills, It is not cost analysis, design reviews, or something we do as part of our job. It is different because the basic philosophy is different. It is not concerned with trying to reduce the cost of an item or service; it is concerned with function and methods to provide the function at the lowest overall cost.

Value control concerns people and their habits and attitudes. It is to a large degree a state of mind. It accepts changes as a way of life and makes every effort to determine how this change can be made to provide the most benefit. It is a function-oriented system that makes use of creative problem solving and team action. The team is designed to provide an experienced, balanced, and broad look at a subject without being constrained by past experience. It requires trained people who understand the system and its application.

II. HISTORY OF VALUE CONTROL

Value control was originated at the General Electric Company. In 1947 Mr. Harry Erlicher, Vice President of Purchases, noted that during the war years it was frequently necessary to make substitutions for critical materials that quite adequately satisfied the required function and often resulted in an improved product. He reasoned that if it was possible to do this in wartime it might be possible to develop a system that could be applied as a standard procedure to normal operations to increase the company's efficiency and profit.

Mr. L. D. Miles was assigned to study the possibility and the result was a systematic approach to problem solving based on function which he called value analysis. The program was so successful that shortly thereafter the Navy started to use the system to help them to get more hardware in the face of a rapidly shrinking budget. The Navy called their program value engineering.

Value analysis, value engineering, value management, value assurance, and value control are all the same in that they make use of the same set of techniques developed by Mr. Miles in 1947. In many cases, however, the title tends to describe how the system is being applied. Value analysis is generally considered to apply to removing cost from a product. Value engineering and value assurance are applied during the program development phase to keep cost out of a product. Value management and value control are overall programs that recognize that value techniques can be applied at any stage of a program. They strive to apply value techniques to control value in all areas of operations.

The Navy program developed so successfully that it was picked up by the Department of Defense and is now considered to be the key element in Government's cost reduction program. In addition, value techniques are now being used in industry and government throughout the free world. They are being applied to aircraft, engines, automobiles, washing machines, dryers, TV sets, and all sorts of consumer and industrial products as well as construction projects and management planning. In addition, several states are applying value techniques to increase efficiency of operations.

III. THE VALUE CONCEPT

The value process is a function-oriented system. It makes use of team action and creative problem solving to achieve results and is specifically designed to simultaneously consider all of the factors involved in decision making: performance, schedule, and cost.

In order to obtain the experienced, balanced, broad look at all facets of the project, a carefully selected team is organized to satisfy the specific requirements needed. Selection of the team must consider personality as well as technical competence of the candidate. The team must not only have the technical know-how required, but the members must be compatible and know how to work together. The Value Manager acts as a coach to guide the team through the system to obtain maximum benefit from their activities.

The Society of American Value Engineers define the term "value engineering" as: "Value Engineering is the systematic application of recognized techniques which identify the functions of a product or service, establish a monetary value for that function and provide the necessary function at the lowest overall cost." The program described by this definition is not a cost reduction program, but rather it is a profit improvement program in that it recognizes all of the factors contributing to product cost. It recognizes that the lowest cost product may induce high warranty problems that may adversely affect profit. In order to increase profit it may, therefore, be necessary to increase product cots in some cases. Overall cost is the prime concern.

A value program is implemented by applying all of the known techniques of problem solving and cost reduction plus a large body of special skills. The primary objective is to identify and remove unnecessary cost.

Unnecessary cost is cost which can be removed without affecting product function. It has been estimated that the average consumer product may include over 30% unnecessary cost. This unintentional cost is the result of habits, attitudes and all other human factors, and it is contributed to by everyone in an organization.

Freeze (1994) reports that the most effective usage of value analysis is when it is used in the design stage of the product development process, when over 70% of the product life-cycle cost gets "locked-in." The reason for that high amount is the fact that during the design process the value function is defined as:

$$\text{Value} = \frac{\text{function}}{\text{cost}}$$

Therefore the objective of the value analysis is to improve value by either:

1. reducing cost while increasing function
2. reducing cost while maintaining function or performance
3. increasing function while maintaining cost
4. increasing function while increasing cost by a proportionally smaller amount
5. decreasing function (only in an overspecification situation) while decreasing cost by proportionally greater amount

Maximum value is probably never obtained. Therefore, the relative value of any product depends upon the effectiveness with which every usable concept, material, process and approach to the problem has been identified, analyzed and implemented. After all, the purpose of value analysis is to bring better value combinations into focus with less expenditure of resources.

IV. PLANNED APPROACH

Value control achieves results by following a well-organized planned approach. It identifies unnecessary cost and applies creative problem solving techniques to remove it. The three basic steps brought to bear are:

1. identify the function (what does it do?)
2. evaluate the function (what is it worth?)
3. develop alternatives (what else will do the job?)

A. Function

Function is the very foundation of value control. The concern is not with the part or act itself but with what it does: What is its function? It may be said that a function is the objective of the action being performed by the hardware or system It is the result to be accomplished, and it can be defined in some unit term of weight, quantity, pounds, time, money, space, etc. *Function is the property that makes something work or sell.* We pay for a function, not hardware. Hardware has no value, only function has value. For example, a drill is purchased for the hole it can produce, not for the hardware. We pay to retrieve information, not to file papers.

Defining functions is not always easy. It takes practice and experience to properly define a function. It must be defined in the broadest possible manner so that the greatest number of potential alternatives can be developed to satisfy the function. A function must also be defined in two words, a verb and a noun. If the function has not been defined in two words, the problem has probably not been clearly defined.

Function definition is a forcing technique which tends to break down barriers to visualization by concentrating on what must be accomplished rather than the present way a task is being done. Concentrating on function opens the way to new innovative approaches through creativity. Some examples of simple functions are: produce torque, convert energy, conduct current, create design, evaluate information, determine needs, restrict flow, enclose space, and so on.

There are two types of functions: basic and secondary. The basic function describes the most important action performed. The secondary function supports the basic function and almost always adds cost.

B. Value

After the functions have been defined and identified as basic or secondary, we must evaluate them to determine if they are worth their cost. This step is usually done by comparison with something that is know to be a good value. This means the term value must be understood.

Aristotle described seven classes of value: economic, political, social, aesthetic, ethical, religious, and judicial. In value control, we are primarily concerned with economic value. Webster defines value in terms of worth as:

> *Value*: (1) A fair return or equivalent in goods, services, or money for something exchanged; (2) the monetary worth of something; marketable price; monetary value.

Webster in turn defines worth in terms of value:

> *Worth*: The value of something measured by its qualities or by the esteem in which it is held.

If we set a value, we determine its worth. If we determine the worth of something, we set a value on it. The two terms can be used interchangeably and value is defined for our purpose as: *Value is the lowest overall cost to reliably provide a function.* By overall we mean all costs that affect the function such as design and development expenses as well as manufacturing, warranty, service, and other costs.

Value (V_E) is broken down into three kinds, each with a specific meaning:

Use value (V_U): a measure of the properties that make something satisfy a use or service.

Esteem value (V_e): a measure of the properties that make something desirable.

Exchange value (V_X): a measure of the properties of an item that make it possible for us to exchange it for something else.

These measures may be in dollars, time, or any other measurable quantity. However, on occasion it is necessary to rank a series of functions by their relative value, one to another.

Value is not constant; it changes to satisfy circumstances at a given time. As circumstances change, values may change. This is usually true of monetary, moral, and social values. Value can therefore, be expressed as the relationship of three kinds of value:

$$V_E = V_U + V_e + V_x$$

Do not confuse cost with value! Cost is a fact; it is a measure of the time, labor, material that go into producing a product. We can increase cost by adding material or labor to a product, but it will not necessarily increase its value. Value is an opinion based on the desirability or necessity of the required qualities or functions at a specific time.

The relationship of cost to value provides an index of performance (P).

$$P = \frac{V_E}{C} = \frac{V_U + V_e + V_x}{C}$$

Maximum performance of our resources can be obtained when cost is low and value is high or when P is greater than one. However, P is usually less than one and an indication that we are not getting good value for our expenditure of funds. It is a direct indication of unnecessary cost.

C. Develop Alternatives

Function has been defined as the property that makes something work or sell and value as the lowest cost to reliably provide a function. The performance index (P) has identified the problem. Now, what else will do the job? Develop alternative ways to perform the function.

In order to develop alternatives, we make maximum use of imagination and creativity. This is where team action makes a major contribution. The basic tool is brainstorming. In brainstorming, we follow a rigid procedure in which alternatives are developed and tabulated with no attempt to evaluate them. Evaluation comes later. At this stage the important thing is to develop the revolutionary solution to the problem.

Free use of imagination means being free from the constraints of past habits and attitudes. A seemingly wild idea may trigger the best solution to the problem in someone else. Without a free exchange of ideas, the best solution may never be developed. A skilled leader can produce outstanding results by brainstorming and by providing simple though stimulations at the proper time.

V. EVALUATION, PLANNING, REPORTING, AND IMPLEMENTATION

The creative phase does not usually result in concrete ideas that can be directly developed into outstanding products. The creative phase is an attempt to develop the maximum number of possible alternatives to satisfy a function. These ideas or concepts must be screened, evaluated, combined, and developed to finally produce a practical recommendation. It requires flexibility, tenacity, and visualization, and frequently the application of special methods designed to aid in the selection process. The process is carried out during the evaluation and planning phases of the job plan and is covered in detail in Section VI of this chapter.

The recommendation must be accepted as part of a design or plan to be successful. In short, they must be sold. They must show the benefits to be gained, how these benefits will be obtained, and finally show proof that the ideas will work. This takes time, persistence, and enthusiasm; the details of a recommended procedure are covered in later sections of the text.

A. The Job Plan

These are the basic features that make value control an effective tool. All are applied in a step approach to a value study. The approach is called the job plan which demands specific answers to the following questions:

What is it?
What does it do?
What does it cost?
What is it worth?
Where is the problem?
What can we do?
What else will do the job?
How much does that cost?

The plan is broken down into six steps:

1. Information phase
2. Creative phases

3. Evaluation phase
4. Planning phase
5. Reporting phase
6. Follow-up phase

Each step is designed to lead to a systematic solution to the problem after consideration of all of the factors involved.

B. Application

Although indoctrination workshops are usually conducted with existing hardware or systems, the greatest opportunity for saving is in the prevention of unnecessary cost. The function techniques apply, but modifications are required that depend on the user's understanding and ingenuity in applying them to conceptual ideas. Systems, procedures, manufacturing methods, and tool design are some of the areas other than hardware where functional techniques have been used successfully. Figure 12.1 shows how the overall savings varies with time in the application of function analysis techniques from concept to hardware, system, plan, or any other type of project implementation. Figure 12.1 also indicates that the costs to change increase as a project develops and the net savings decrease. Once a product or service is in production or use, the cost of tools, hardware, forms, and time necessary to achieve the product stated at any given time cannot be recovered.

In order to be effective, value control needs trained people working as a team. A team needs a coach who, in this case, is the value manager. The team provides the technical expertise necessary and the value manager provides the know-how to apply this knowledge for effective results. It also requires management to provide the necessary support and a creative environment. In short, success is up to everyone in an organization. People must

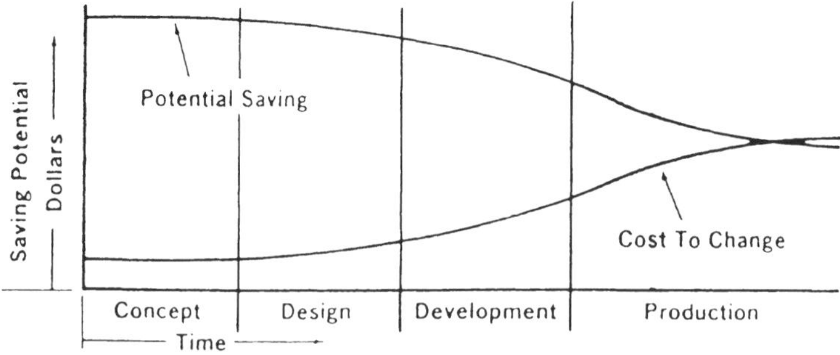

Figure 12.1 Relationship of savings potential to time.

be trained, they must understand the system, they must understand the application, they must be aware of cost and how to handle it, and management must support their activity by their active participation.

Success means change: change in methods, change in procedures, change in attitudes. With this approach, value control will become an effective profit maker.

VI. VALUE CONTROL: THE JOB PLAN

A. Information Phase

What is it? Collect all data, drawings, blueprints, costs, parts, flow sheets, and process sheets. Talk with people, ask questions, listen, and develop ideas. Become familiar with the project. Discuss, probe, analyze.

What does it do? Define functions. Determine basic function(s). Construct a FAST diagram.

What does it cost? Do a function/cost analysis.

What is it worth? Establish a value for each function. Determine overall value for the product or source.

Where is the problem? Analyze the diagram. Locate poor value functions. Pinpoint the areas for creativity.

What can we do? Set goal for achievement.

B. Creative Phase

What else will do the job? Brainstorm the poor value target functions, use imagination, create alternatives, develop unique solutions, and combine or eliminate functions. Look for revolutionary ideas. Do not overlook discoveries obtained by serendipity.

C. Evaluation Phase

Select the best ideas. Screen all creative ideas. Evaluate carefully for useful solutions. Combine the best ideas and categorize into basic groups. Screen for best ideas.

How much does it cost? Generate relative costs. Analyze potential. Anticipate roadblocks.

D. Planning Phase

Develop the best ideas. Develop practical solutions. Obtain accurate costs. Review engineering and manufacturing requirements. Check quality and reliability. Talk with people. Resolve anticipated roadblocks. Develop alternative solution. Plan your program to sell. Show the benefits.

E. Reporting Phase

Present ideas to management. Show before and after costs, advantages, and disadvantages, nonrecurring costs of development and implementation, scrap, warranty, and other forecasts and net benefits. Plan your recommendation to sell. Make recommendation for action.

F. Implementation

Assure proper implementation. Be certain that the change has been made in accordance with the original intent. Audit actual costs.

VII. VALUE CONTROL TECHNIQUES

1. Define functions.
2. **Identify/overcome roadblocks**
3. Use specialty prod/processes
4. Bring new information
5. Construct FAST diagram
6. Cost/evaluate FAST diagram
7. Use accurate costs
8. Establish goals
9. All info from best source
10. **Use good human relations**
11. Get all the facts
12. Blast-create-refine
13. Get costs on key tolerance
14. Put costs on main idea
15. **Use your own judgment**
16. **Spend company dollars as if your own**
17. Use company's services
18. **Use specifics not generalities**
19. Use standards
20. Use imagination
21. Challenge requirements
22. Use vendor services

(Bold numbers indicate techniques that apply at every phase of the job plan as well as in most other activities.)

Job plan	Applicable techniques
Information phase	1,2,4–11,13–19,21,22
What is it?	
What does it do?	
What does it cost?	
What is it worth?	
Where is the problem?	
What can we do?	
Creative phase	10,12,15,20,21
What else will do the job?	
Brainstorm	
Create alternatives	
Evaluation phase	2,3,7,8,10,11,13–22
Review suggestions	
Refine results	
Evaluate carefully	
Planning phase	2,3,5,7,8,10,11,13–22
Develop best ideas	
Develop alternative solutions	
Plan program to sell	
Reporting phase	2,10,15,18
Present ideas to management	
Make recommendation for action	
Implementation phase	2,10,15–20,22
Assure proper implementation	

REFERENCES

Fowler, T. C. (1990). *Value analysis in design*. Van Nostrand Reinhold. New York.

Freeze, D. (1994). *Value analysis/value engineering*. Freeze and Associates, Ltd. New York.

Patterson, J. (1995). Adding value by managing supply chain activities. *Marketing News*. July 17: 6.

13

Problem Solving

Problem solving is a fundamental process that remains the same regardless of the problem; however, the complexity of the process changes with the nature of the problem. Problem solving is basically the scientific approach simplified so that it can be applied to any problem; everyone can and should use it to resolve problems and/or discover opportunities.

I. OVERVIEW

Problem solving is creating change to bring actual conditions closer to conditions that are desired. A problem is a discrepancy between current conditions and desired conditions, and the goal of problem solving is to reduce or eliminate the discrepancy. A typical problem-solving process is shown in graphical format in Figures 13.1 through 13.5. Figure 13.1 presesnts an approach for the initial steps of concern analysis. Figure 13.2 presents the actual problem solving process in the most basic format of what, where, when, and magnitude. Figures 13.3 and 13.4 present a format for identifying and defining decision-making alternatives and risks, respectively. Finally, Figure 13.5 presents a methodological approach to planning for problem solving. No

Process Steps	(1) Major or complex concern	(2) Sub concern breakdown	(3) Potential Impact	Timing	Trend	(4) Determine Process
(1). List concerns (2). Break down into precise, manageable sub-concerns	A.					
(3). Prioritize: - Impact: $ lost, organizational effects - Timing: Whose deadlines. Can we do anything? - Trend: Is it stable, predictable, consistent?	B.					
(4). Determine process: - Past: Cause problem solving - Now: Choice decision making - Future: Implement planning	C.					

Figure 13.1 Process of problem solving—concern analysis.

(1). Problem Statement:					
Description and or questions	**What is a problem**	**What is not a problem**	**(3). Differencers about the IS (only for new and revelant data)**	**(4). Changes about the differences**	**Date**
What: is the nonconforming object? is the nonconformance? *Where:* is the nonconformance object observed? on the object is the observed nonconformance? *When:* in the life cycle of the object does the nonconformance occur? are the nonconforming objects observed? If so, How and when? what is the pattern? *Magnitude:* How many objects are nonconforming? how many objects are nonconforming? what is the trend, if any?					
(5). Potential causes from changes and differences:		**(6). Examine potential causes against each is and is not for most likely cause (Make sure you list as many inconsistencies and assumptions):**			
(7). Document Cause either through a Cause and effect diagram or:	Who: ______ How: ______ Where: ______ When: ______				

Figure 13.2 Process of problem solving—problem solving.

(1). Decision statement:											
(2). Objectives			(5). Alternatives								
			A			B			C		
(3). Required from objective sheets or planning summary sheets			Data:	(6). Yes/No		Data:	(6). Yes/No		Data:	(6). Yes/No	
Process Steps	*Desired*	(4) Value	*Data:*	(7). Score	(8). Value score	*Data:*	Score	Value score	*Data:*	Score	Value score
(4). Value the desired objective from 10-1 scale. (5). Estimate alternatives (6). Test alternatives against required objectives (yes/no) (7) Score alternatives against each desired objective (8). Multiply value x score = Value Score (9). Add all value scores											
				(9). Total value score			(9). Total value score			(9). Total value score	

Figure 13.3 Process of problem solving—decision making alternatives.

(1). Potential alternative	Alternative A			Alternative B				
Process Steps	**(2). Risks**	**(3). Likelihood**	**(4). Impact**	Risks	Likelihood	Impact	**(5). Assess impact**	**(6). Prioritize and select best alternative**
(1). List highest value scored alternatives (2). List risks if we go with a particular alternative (3). Assess likelihood of each risk occuring based on high, medium or low level (4). Assess impact of each risk if occurs based on high, medium or low level (5). Identify items with both likelihood and impact as high (6). Select best alternative for benefit versus risks								

Figure 13.4 Process of problem solving—decision making risks.

(1). Plan statement											
Process steps	(2). Components and or steps	(3). Set in chronological order	(4). Precise potential problems of vital components	(5). Precise potential problems of vital components	(6). Likelihood	Impact	(7). Probable cause(s)	(8). Likelihood H-M-L	(10). Plan for contingent action against the precise potential problem's effects	(9). Plan for Preventive action against cause(s)	(11). Feed back points assign individually, Date and system
(2). List components or steps of plan											
(3). Number components in chronological order											
(4). Identify vital components (root cause(s) that will most likely cause us to fail											
5). Identify precise potential probability of vital components											
(6). Assess each precise potential probability for likelihood and impact											
(7). Probable cause(s) of precise potential problems											
(8). Assess likelihood of each probable cause											
(9). Plan for preventive action											
(10). Plan for contingent action											
(11). Feed back points (who, when and how of progress reporting)											

Figure 13.5 Process of problem solving—planning.

matter how one defines and/or approaches the problem solving process, the fact remains that there are two basic aspects to it.

1. Differentiate between decision making and problem analysis. Decision making consists of determining goals and choosing courses of action to reach those goals. Problem analysis consists of identifying factors that impede goal achievement and determining the forces that bear on those factors. We must understand that this process is basically of the form:
   a. Symptoms are perceived.
   b. Causes are analyzed.
   c. Solutions are designed.

   Furthermore in the identification process we must recognize that problems fall into three types and as such they need different handling: 1) those which we have control over, 2) those which we have no control over but which we can influence through someone who does have control over them, and 3) those which we have no control over and which we cannot influence through any one who does have control.
2. The essential elements of planning depend on accurate problem analysis and shrewd decision making. These elements are determining if a particular problem is significant, setting realistic goals, describing the major forces that affect the problem, and showing how a specific set of interventions can ameliorate the problem. Because of the very important role that the problem-solving concept has in the quality area, we develop it quite extensively. Specifically, we address the issue of team building with its roles and responsibilities and then we discuss specific tools that are used for problem identification and solution. [The reader may wonder why we treat the discussion of teams first and then the tools. It is imperative that we all understand that problem solving is a team effort and the selection of the specific tool can make a difference (Nickols, 1996)].

   Within any group the members assume different roles, several of which have been determined to be necessary for a smoothly functioning effective group (team). These roles may be assumed by separate members or shared by various members at different points, and, in many cases, one or more of the individual members may fulfill more than one role. Because every group has both task and maintenance functions, some of the essential roles are task related in that they help a group accomplish things, and some are maintenance related in that they facilitate the participation of the members.

A. Task Roles

Initiator: proposes tasks, goals, or actions; defines group problems; suggests procedures

Information seeker: asks for factual clarification; requests facts pertinent to the discussion

Opinion seeker: asks members for clarification of their values, (beliefs, feelings, attitudes, assumptions, perceptions) pertinent to the topic under discussion

Information provider: offers facts; gives expression of feelings; gives opinions

Clarifier: interprets ideas or suggestions; defines terms; clarifies issues before the group; makes sure dissenting members understand and resolve differences before the group moves on

Summarizer: pulls together related ideas; restates suggestions; offers decisions or conclusions for the group to consider

Reality tester: makes critical analysis of ideas; tests ideas against data to see how the ideas would work; considers the consequences of decisions and actions

Focuser: defines the position of the group with respect to its goals; points to departures from agreed-on directions or goals; raises questions about the directions pursued in group discussions

B. Maintenance Roles

Harmonizer: attempts to reconcile disagreements; reduces tension; gets people to explore differences

Gatekeeper: helps to keep communication channels open; facilitates the participation of others; suggests procedures that permit sharing remarks

Consensus tester: asks to see whether the group is nearing a decision; sends up trial balloons to test possible solutions. Working and reasoning together, the team members arrive at a joint agreement (i.e., consensus) which all members can accept and feel committed to. It is a course of action which each member can live with and can willingly agree to a decision of a sufficiently high level of quality and high level of acceptance by those who will implement it. A consensus tester must make sure that the decision is:

1. Participative: All members join in the discussion. It is not rule by a vote of the majority, it is not a decision imposed by a dominant member, and it is not made by yielding to an expert member of the team.

2. Decisive: At a designated point, the team must make a decision. No member may leave the group; he or she may disagree with the team decision but willingly agrees to implement it.
3. Synergistic: The action of two or more individuals to achieve an effect of which each individually is incapable. The results achieved by the members of the team working together usually exceed the results that could be expected by the sum of the individual efforts. (The whole is greater than the sum of its parts.) If the team results are not better, go back and redefine the goals

Encourager: is friendly, warm and responsive to others, as indicated by facial expressions or remarks which indicate the acceptance of others' contributions

Mediator: conciliates differences in points of view, offers alternative solutions

Tension reliever: drains off negative feelings through jesting, use of humor to alleviate tense situations within the group

II. FOUNDATIONS OF PROBLEM SOLVING

All problem solving techniques are based on the following:

1. The most powerful action in problem solving is becoming more aware of the problem. This is the so-called paradox of change: starting off by trying to change conditions seems to exacerbate the problem, while accepting the full measure of the problem begins to solve it.
2. Problems may have many causes, not just one. Every situation can be described as a field of forces (i.e., various psychological, social, political, economic, and cultural factors) held in dynamic balance. To produce change, we first must see clearly what forces are at work and how they are balanced. The we must search pragmatically for the most effective place(s) to intervene in the field of forces in order to change the balance, rather than search for the one cause that seems most logical.
3. Valid decisions depend on adequate information. Adequate means accurate, clear, and complete. To draw an analogy, the recipe (process) depends on the ingredients (information).
4. Working with others can improve the process. A group of informed people working on a problem can compress into a few hours the mental work that might take months for only one person. In addition, the members of a group tend to risk more novel approaches than do individuals working by themselves.

5. Getting good results from a valid decision requires that those who must carry out the decision understand and be committed to it. A decision might be technically sound, but politically unreliable, or those who are responsible for implementing it may not be committed to it or capable of doing so.
6. The change agent must develop a supportive environment:
   a. The people who experience the problem should share authority for making any decisions for change.
   b. Those who participate in the problem-solving process should have a trusting relationship and should communicate openly about the problem.

Unless the root cause of the problem at hand is identified and quantified the process of problem solving will not be successful. An excellent source for more information on root-cause analysis is Wilson, Dell, and Anderson (1993).

III. CRITICAL VERSUS CREATIVE

Problem solving is, in many ways, simply a process of information management. Social scientists have discovered that one of the most significant barriers to effective problem solving is that people fail to make use of information that they already have. Like the computer, the brain consists of two components: a storage unit and a processing unit. Although the storage unit can hold a great amount of information, the capacity of the processing unit is quite limited. The average person can effectively manage no more than about seven independent variables of information at one time. When a problem begins to exceed this level of complexity, people overlook elements, perceive the wrong elements, assume wrong elements, or fail to make the right combinations of elements.

Another significant barrier to effective problem solving is that people often fail to use their creative faculties in searching for answers. The processing unit of the brain consists of both critical and imaginative functions. The critical function analyzes, compares, evaluates, and selects relevant information. The imaginative function generates, visualizes, abstracts, and foresees combinations of information. Both synthesize information, but the imaginative function creates ideas whereas the critical function delivers judgments. Both functions must operate mutually for successful problem solving.

For most individuals, judgment grows with age while creativity dwindles. The formation of adult habits and a critical self-image often limit creative abilities, while education and a career often promote the development of only judicial faculties. This overemphasis on judgment creates a tendency to see only the negative side of situations. In creative problem solving, one must turn off the critical function and use the imaginative function. Otherwise, premature judgement may limit creative possibilities or even eliminate any ideas that are generated.

IV. SYSTEMATIC CREATIVITY

Creativity is mental activity characterized by both subjective and objective thinking. It is a process of alternating back and forth between what we sense and what we know. A key to productive creativity is to control the alternation of subjective thinking so that the critical and imaginative functions complement each other.

V. THE URGENT VERSUS THE IMPORTANT

All of us have heard the expression: "Management is in the mode of fighting fires." Firefighters are always in a reactive mode, responding to problems and engaged in short-term activities. In many cases the problems solved by management could be successfully addressed by others in the organization—perhaps those with more direct and/or indirect knowledge. Certainly, it is easy for a manager to solve a short-term problem or dispute, but this raises some basic questions of the entire problem-solving process:

- Is that the best use of management's time?
- Can and/or should others in the organization be empowered to handle that and similar problems?
- What are the long-term effects of tactical decision making at a lower level in the organization?

These questions point to perhaps the most significant issue confronting management, namely, a careful and studied consideration of basic duties and responsibilities. It is very easy to respond to the urgent, the day-to-day crises. We do understand that doing so involves a successful and public exercise of authority and instant gratification. It is a good feeling when a problem is solved, but in solving a day-to-day problem in a reactive mode, an even more serious problem may be created (Helps, 1995): company strategic issues are not being addressed. The manager spending a typical day, every day, reacting to the urgent does not manage, rather he or she reacts to situations. One may even say that urgency forces the manager to ignore important issues regarding the organization, customers, suppliers, and certainly himself or herself. Some of the important issues that may be overlooked because of the crush of daily business are:

- Development of a differentiated strategy. What unique benefits, product, or service are provided by your company that justifies its existence?
- Development of a written business plan to implement the differentiated strategy. It is not enough to have a strategy. It is necessary to breath life into a strategy.
- Development of a program to systematically upgrade the competencies of employees. Employees need the skills to more effectively and productively deal with both technological and nontechnological change. Remember. In today's environment, every employee is in direct position to win and lose customers.
- Formal systematic reviews of the relationship with important customers and suppliers. Any relationship needs periodic renewal, even if there is little overt evidence of a problem. Customers and suppliers are the lifeblood of any organization and need to be nurtured.

How does one break the vicious cycle of being too preoccupied with "the urgent" to find the time to address the "important?" The first step is recognition of the problem and realization of the true long-term costs of short-term firefighting. The cycle is especially vicious when a manager is so effective at firefighting that others in the organization bring him or her an ever increasing number of problems to be solved. It's "upward delegation," a phenomenon seldom discussed or even mentioned in the business literature. (In some cases, managers are so well known in firefighting applications that they make a career hopping from company to company to solve day-to-day problems; they are known as the turn-around managers.)

The second step is the will to break the cycle. The third step is development of organizational skills and understanding to enable day-to-day problem solving to occur at all levels of the organization. The final step is: do it.

The process of problem solving is a proactive approach and helps in both identification and differentiation of the vital few from the trivial many (i.e., the Pareto principle); however, within the vital category we must differentiate the urgent versus the important.

VI. STEPS IN PROBLEM SOLVING

The problem-solving process is based on the scientific approach. It is basically a cyclical feedback system and, in fact, one may even consider it as a continuous system with no real beginning or end. Furthermore, the completion of each step affects the definitions of the previous steps. The steps are:

1. Assess the situation.
2. Identify the problem.

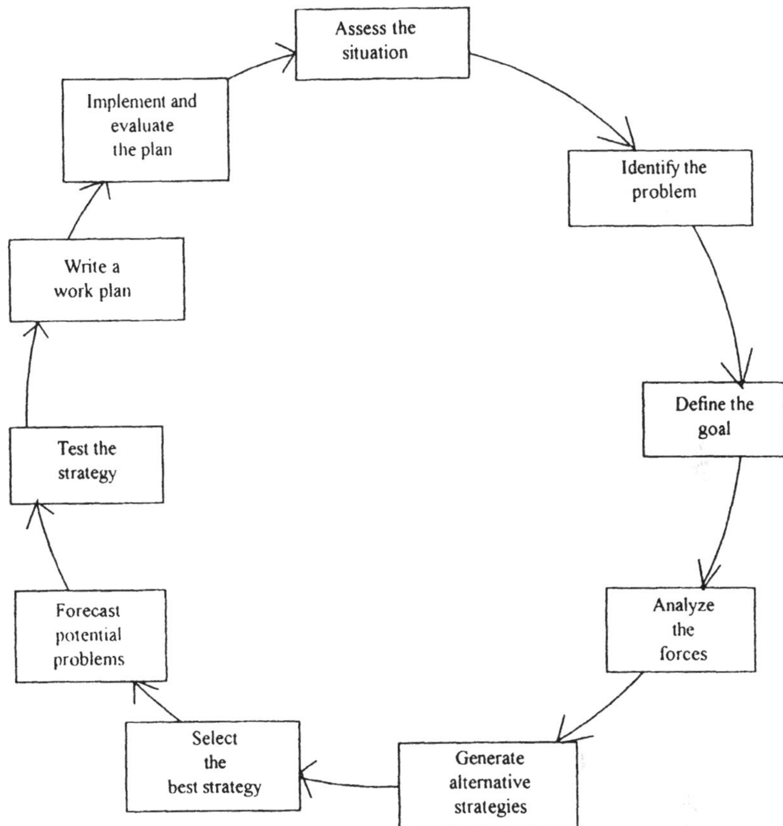

Figure 13.6 Steps in problem solving.

3. Define the goal.
4. Analyze the forces.
5. Generate alternative strategies.
6. Select the best strategy.
7. Forecast potential problems.
8. Test the strategy.
9. Write a work plan.
10. Implement and evaluate the plan.

These steps are shown in Figure 13.6. (It is of paramount importance for the reader to recognize that these 10 steps are not the only correct approach to problem solving. In fact, there are many approaches of equal value and effectiveness. What is important is the fact that a systematic approach exists that is consistent and repeatable.) Let us examine each of the steps little more closer.

A. Assess the Situation

Before launching a strategy for change, it is important to assess the problem (challenge) of the situation in terms of whether action really is needed and whether it actually will have some impact on the problem. To assess the situation, the following questions may start the process.

- What is occurring that requires change?
- What will result if nothing changes?
- Can any significant changes actually be effected?
- Can the relevant information be obtained?
- Does the situation deserve the effort (right now) compared with other priorities and interests?
- Are the persons involved in the situation committed to making a change?
- At which step in the process should the effort begin?
  - To remove or reduce a deficient condition, begin with Step 2 and go through all the steps.
  - To develop an improved strategy (rather than change a deficiency), begin with Step 5, although Steps 3 and 4 can provide some useful tools.
  - If there is a plan of action that has not been tested or put into practice, begin with Step 7.
  - If a tested strategy simply has not been put into practice, begin with Step 10.

B. Identify the Problem

One of the most crucial and difficult steps in the process is identifying the actual problem. Problems usually are obscure, disguised, or locked inside some emotional distress, attitude conflict, or misleading outgrowth of another situation. Another major difficulty is in determining the standard by which the deficient condition is measured. Unless we are clear about our standards, we cannot be clear about our problems. This step may take a long time and may include several revisions, but the effort is well spent. A problem that is well stated is half solved.

To help you identify the problem, the answers to the following questions may be of help:

- What is specifically desired that is not happening?

- What are the standards or values that apply to the situation?
- What is happening (described in objective and observable terms?)
    - Who is involved?
    - Where does it occur?
    - When does it occur?
    - What is the extent of the problem (i.e., how many or how much)?

At this point the problem should be summarized in one comprehensive and concise statement.

C. Define the Goal

The definition of the goal is a statement of what is to be done about the problem. It should be expressed in measurable terms so that the results to be achieved are in the form of observable and/or behavioral outcomes. Abstract or subjective statements of outcome are impossible to assess. To define the goal (the desired condition), the following must be defined and/or answered:

- *Results* (outcome): What is expected? Is it measurable? Is it of noticeable difference?
    - What is specifically is to result?
    - Who will be involved?
    - When will the results be achieved?
    - Where will the result occur?
- *Criteria*: The measures for acceptability.
    - What are the quantitative standards that indicate the minimum level of goal achievement?
    - What are the qualitative standards that indicate the minimum level of goal achievement?
- *Conditions*: The parameters of the effort. Under what circumstances the outcome will be the desired one? What are the limits of the goal?
    - What limitations or restrictions in terms of time and/or money are to be imposed?
    - What resources in terms of people and/or equipment are required?

At this point the goal should be summarized in one comprehensive and concise statement, and then checked to make sure that it describes an outcome rather than a strategy. It is important not to confuse the ends with the means at this point.

D. Analyze the Forces

This step is to collect, organize, and analyze all the relevant information regarding the current situation as a foundation for creating a creative and realistic plan for change. At this point, it is very helpful to involve people who are familiar with the situation. This step has two aspects: past circumstances that have influenced the information of the problem, and present factors (forces) that affect the achievement of the goal.

Past Circumstances

To have a clear picture of the circumstances from which the issue evolved, answer the following questions:

- What are the past decisions, occurrences, and factors that created the present problem situation?
- What was the context in which these circumstances occurred?
- Were any payoffs or benefits derived from these past decisions or actions? If so, what were they and who benefited from them?

Present Factors

Force-field analysis (FFA), developed by Lewin (1951), is a tool for organizing and analyzing information as a basis for a change effort. Any situation can be considered as a dynamic balance of forces working in opposite directions. Forces moving toward change (i.e., helping forces) are opposed by an equal number of forces moving in the opposite direction (i.e., hindering forces). No change will take place in the situation unless an imbalance of these forces is created. The procedure involves identifying the problem, determining the goal that the team or individual wishes to achieve, listing the helping forces and the hindering forces, and assessing each force in terms of its strength and vulnerability to change.

An important aspect of this procedure is brainstorming, a team process designed to produce a large number of ideas in a short period of time. While someone writes down what is said, the members of the team spontaneously and quickly express their ideas, more or less by free association. No comments or criticisms are permitted; anything and everything offered is noticed. Each participant is encouraged to say whatever he or she wishes, no matter how unusual or unrealistic it may appear.

Next, summarize the problem in the middle of a sheet of paper. Summarize the goal at the right side of the sheet. Then use brainstorming to create a list of present helping forces and enter these on the left half of the sheet. Make a similar list of present hindering forces on the right half of the sheet. Be as specific as possible.

The next step is to eliminate repetition and clarify items. Then, for each of the two lists, rate each item in terms of its strength, with the strongest being rated 10 and the rest rated on a scale of 1 to 10, compared with the strongest. Next, for each of the two lists, rate each item with a strength of 5 or above in terms of how vulnerable it is to change efforts. Start by rating the easiest to change as 10; then weigh the other items on a scale of 1 to 10, in contrast with the easiest. This will provide a picture of how easily each force can be controlled or influenced. Finally, identify and list the items for additional information as needed and proceed to obtain that information.

Once the analysis is completed, alternative strategies for creating change can be developed. In general, this change can occur through any of the following alternatives:

- Change the strength of any force.
- Change the direction of any force.
- Withdraw hindering forces.
- Add new helping forces.

In this step, it is best to begin by working with hindering forces. Increasing helping forces often increases resistance (it is a law of physics that every action has an equal and opposite reaction.) Strategies for change that are directed toward reducing hindering forces generally are more effective.

E. Generate Alternative Strategies

The first step is to review and revise the goal if the intervening steps have helped to clarify it. The next steps are quite unlike the systematic problem analysis. They require quite a different orientation—an openness to the absurd, spontaneous, and poetic resources of the preconscious. In these steps, creativity and invention are employed and logic and proportion are suspended. A typical approach to this step is the following:

Fantasizing

Attention is focused on the specifications of the goal and the people involved. For about two minutes, everyone fantasizes (visualizes) freely about a solution to the problem. The fantasies are then shared and compared to see what patterns are present as well as what elements are different from others.

Brainstorming

First, considering those hindering forces that are strong and vulnerable, brainstorm actions to remove or minimize them. The process then is repeated with the helping forces that are strong and vulnerable. When ideas no longer flow freely, repetition is eliminated and statements are clarified.

Synthesizing

The list of items is synthesized by identifying logical combinations. All items that have an organic or logical connection (e.g., credit checks or interest charges) are identified with the same letter. Each combination is then defined by a brief description of the strategy to be used, and, where appropriate, one or more are linked to provide more comprehensive possibilities.

F. Select the Best Strategy

This step uses a matrix to compare alternative strategies with decision-making criteria. This enables decision makers to be as precise as possible about the relative value of any one strategy or combination of strategies. There are two alternative procedures.

Fixed-Criteria Procedure

A "quick and dirty" distinction can be made by selecting the criteria of most benefit and least cost. Cost/benefit criteria are listed vertically and alternative strategies are listed horizontally. Using the rating system employed in the force-field analysis, assign a value of 10 to the alternative with the highest benefit and lowest cost and then rate the others on a scale of 1 to 10. Select the alternatives with the best combinations of benefit and cost.

Goal-Criteria Procedure

Review the goal and identify each element that can be described as a criterion or measure. Segregate into two categories (IN/OUT and WANTS) using criteria by which to measure alternatives. IN/OUT criteria represent minimum conditions that an alternative must satisfy to be considered further (i.e., alternatives meet these criteria or are tossed out). Alternatives that do meet minimum conditions are then evaluated by WANTS criteria (i.e., which is preferable?). The key measure is the comparison among the criteria. If the relative importance of the criteria listed under WANTS differs, assign each a weight of 10 and weigh each against that criterion using a scale of 1 to 10.

Screen out alternatives using the IN/OUT criteria; then compare the alternatives against each WANT criterion in turn and assign a rating. Use the rating mechanism described previously. Multiply the weight for each criterion by the rating for each alternative. Compare the resulting scores. If the alternative with the highest score has face validity, a tentative decision can be made. If it does not, the next highest alternative should be reviewed.

G. Forecast Potential Problems

The next step is to test the feasibility of the selected course of action. Again, it will help if people who are involved, affected by, or have technical knowledge about the situation participate. Looking at the preferred course of action, carry out the following steps:

- Brainstorm a list of things that could go wrong and list every idea.

- Rate each potential problem in terms of probability. Using 10 as a rating for certainty, assign each item a score from 1 to 10.
- For each item that received a rating of 5 or more (seems probable), relate it in terms of threat. Using 10 as a rating for catastrophe, assign each item a score from 1 to 10.
- For items with ratings of 4 or more on both scales, seek preventive actions; if you cannot prevent the problem, seek a contingency action to keep the problem from having a serious impact.
- For an alternative with preventive or contingency actions, if no crippling potential problems seem likely, the preferred course of action should be sound.
- If problems still seem likely, return to Step 6 (select the best strategy). Select another alternative and repeat the steps for forecasting and analyzing potential problems.

H. Test the Strategy

Before beginning to carry out any strategy for change, it is important to test the strategy. Testing may reveal more potential problems and also may clarify the extent to which the ability and commitment exist to carry out the strategy. The result in most instances is a refinement of the strategy that will increase its effectiveness.

The test to be used will be dictated by the nature of the strategy. If the strategy is interpersonal, role playing may suffice. If the strategy is technical (change in policy, procedures or methods), a brief trial period with a small number of people may suffice (pilot testing). The test should give some indication of the plan's feasibility.

I. Write a Work Plan

The next step is to develop a work plan that delineates the activities necessary to carry out the strategy. This should account not only for the activities that directly relate to implementing the strategy but also for any contingency to prevent potential problems. To complete this step, complete the following:

- List all the tasks required to carry out the selected course of action.
- Order the tasks in chronological sequence.
- Write a plan that (at minimum) accounts for the following:
    - Tasks—what needs to be done (in very specific terms)?
    - Primary responsibilities—who is going to carry out the tasks (specific identification with perhaps alternatives)?
    - Deadlines—when are the tasks going to be accomplished (milestone and/or a Gantt chart)?

J. Implement and Evaluate the Plan

The process at this point may seem overwhelming; however, if the chosen strategy seems to be right, the following may help:

- Act as if you can carry it out. Simply go ahead and do it.
- Forgive and remember. If errors are made while the strategy is being carried out, forgive the lapse, remember the goal, and carry on.
- Evaluate and revise. Be aware of the consequences of any action, and if the plan is not progressing, either revise it or return to Step 1.

REFERENCES

Helps, H. (1995). The urgent vs. the important. *P & G Graphlines*. Aug.: 10–11.

Kepner, C. H. and Tregoe, B. B. (1965). *The rational manager*. McGraw-Hill. New York.

Lewin, K. (1951). *Field theory in social science*. D. Cartwright (Ed.). Harper and Row. New York.

Nickols, F. W. (1996). Yes, It makes a difference! *Quality Progress*. Jan.: 83–87.

Smith, G. F. (1994). Quality problem solving: Scope and prospects. *Quality Management Journal*. Fall: 25.

Wilson, P. F., Dell, L. D. and Anderson, G. F. (1993). *Root cause analysis: A tool for total quality management*. Quality Press. Milwaukee, WI.

14

Teams

In the modern world, teams are the way to handle the quality transformation that most organizations are seeking. By implementing and cultivating a culture conducive to team environment, it is more likely that a given organization will increase productivity, build employee satisfaction and lower costs throughout the organization. The reader is strongly encouraged to see Moran, Musselwhite, and Zenger (1996), Weiss (1996), Parker (1996), Roming (1996), Opper and Fresco-Weiss (1992), and Sholtes (1989) for more detailed information on teams.

I. OVERVIEW

When two or more people with a common interest join forces to accomplish a common goal, a group is formed. Since groups are a vital part of our everyday world, our group participation capability strongly influences our success or failure in life. Related to group dynamics, group participation skills demand constant attention if we want a high level of group involvement and productivity. We can enhance our group participation capability considerably by developing an awareness of few strategies, understanding their optimum

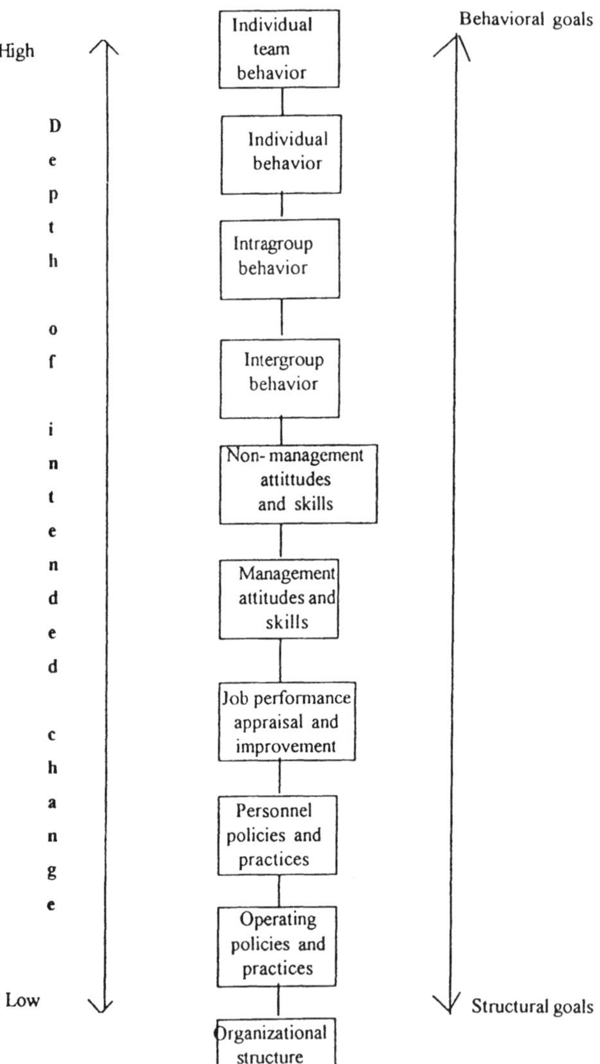

Figure 14.1 The development process of teams.

use, and applying them in appropriate situations. This kind of a process development will result in a team environment (Figure 14.1). The first step in beginning to study the dynamics of the team process and team development is the awareness of work issues (tasks) and personal issues (process) people must address when they join forces as a team.

With the understanding of these two different but interrelated activities, success of any team can be greatly enhanced. These interdependent dynamics are represented as follows:

Work issues (tasks)	Personal issues (process)
Those things which have to do with getting the job done:	Those things dealing with how people:
Planning	Think
Organizing	Feel
Implementing	Respond to one another
Following up	

To understand the team process and its development, we must become aware of the stages of team development. We can turn our attention to the evolutionary steps from a groups inception to the development of a group's final product. We can summarize this development into three stages as follows:

A. Orientation

Here we explore the basics of our existence by asking questions of the type: Why are we here? What is to be accomplished? How will we proceed? In addition, we address work issues for the group action in terms of personal issues and strategiesof:

1. What are our goals?
2. What are our priorities?
3. How much will I involve others in the team?
4. How much will I involve myself in the team?
5. How will I get the participants relaxed, acquainted etc
6. How will I solicit active participation?

B. Power and Influence

Here we find that the team begins to organize itself and members wrestle for control. The work issues here involve: 1) Who does what-the individual roles and responsibilities, and 2) How the roles and responsibilities will get done. The personal issues on the other hand involve: 1) How much control do I want to exert over the team? and 2) How much control will I permit others to have over me? The strategies in dealing with this stage are addressed with the following:

1. Dealing with differences
   - listening to others
   - tolerating others opinions
   - negotiating
   - keeping an open attitude
2. Managing emotions
   - expressing what you feel
   - expressing what you think
   - being sensitive to others and their feelings

C. Team Production and Feedback

Here the team begins to work as a unit as team members give and receive support. The work issues here involve: 1) follow through with execution of responsibilities, and 2) making sure things are being done right. The personal issues involve: 1) How much help, reinforcement, constructive feedback members are willing to give and receive, and 2) What strategies will enable all team members to work efficiently while experiencing good feelings towards the group? The latter includes:

- feedback (reward work)
- building a team spirit
- creating symbols and slogans representing group and member efforts
- dealing with group or member needs and concerns

II. TEAM ROLES AND FUNCTIONS

There are several roles and functions in any given team environment. For a detailed discussion of all the roles and functions see Chapter 15. In this section we will summarize only the two most important ones. The first, is the facilitator who is responsible for at least (but remember that the facilitator is not considered a member of the team):

1. setting up the meeting room
2. formalizing and publishing the agenda
3. verifying that the logistics are adequate
4. helping the team select a recorder
5. keeping the team working on the agenda
6. encouraging all the team members to actively participate
7. preventing any one person and/or clique from dominating the discussion

8. pointing out any ground rule violations
9. monitoring time utilization
10. helping the team focus on its task
11. encouraging team members to perform the team process functions

Second, the participants must be willing to contribute their energy, thoughts, and concerns as well as actively participate towards the team goal. That is true whether the task at hand is in a meeting format or a problem solving situation or even a task within the work environment (i.e., a process that they themselves control).

In teams, persons meet together to deliberate on a topic of common concern. During the course of an active project, there are many occasions when participants need to discuss their progress or meet with representatives of other teams. If that is the case, a facilitator is mandatory.

III. THE CONSENSUS DECISION PROCESS

Research in team dynamics has revealed that the manner in which teams utilize their member resources is a critical determinant of how they perform. Consensus can be difficult to reach. Therefore, not every decision will meet with everyone's complete approval. Unanimity, however, is not a goal although it may be achieved unintentionally. It is not necessary, for example, that every person be as satisfied as they might be if they had complete control over what the team decides.

What should be stressed is the individual's ability to accept a given decision on the basis of logic, whatever their level of satisfaction and their willingness to entertain such a judgement as feasible. When the point is reached at which all team members feel this way as a minimal criterion, you may assume that you have reached a consensus and the judgement may be entered as a team decision. This means, in effect, that a single person can block the group if he/she thinks it necessary. At the same time, it is assumed that this option will be employed in the best sense of reciprocity. Here are some guidelines to use in achieving consensus (not necessarily in this order):

1. Listen carefully to what each person is saying. This will set an example to others. It will also help you ask a team member to clarify something so that the discussion does not bog down in misunderstandings. Try to restate (or ask someone else to restate) what a team member has said so it is clear to everyone.
2. As a facilitator, do not play the role of an expert, make decisions for the team or always be the center of discussion. Encourage team members to direct their comments to each other, not to you.

3. Try to limit the number of times and amount of time any single person speaks. Especially, be cognizant of the effects of domination from the "resource people" and your "experts." Too much of their input may intimidate others into silence. Limit their input to specific factual items and point of perspective, so that the team may refocus and base its decision on accurate data.
4. While some persons may need a little encouragement to speak, this does not necessarily mean you should single them out and ask them to say something just because they have not participated much. Instead, by preventing others from dominating and maintaining a relaxed atmosphere, you can provide the right circumstances for nonparticipants to open up. Be encouraging and supportive be giving positive feedback to what they say. Supportive means the ability to acknowledge the specific merits of other's ideas and to build upon those ideas such as "Yes, that makes sense because ... ," "That is right, and if we did that we would ...." The supporting skills require a particular mindset and two different supporting actions. The mindset is an assumption that other people's ideas have merit even though the ideas may not initially appear to be useful. The first supportive action is the most crucial—listening. The most supportive thing we can do with others is to pay attention to them. Then, having listened to their ideas, we can point out at least one specific merit of the idea. So, to be supportive we need to:
   a. Assume that the other person has useful ideas, information, points of view, etc.
   b. Listen carefully.
   c. Mention the specific element which you find useful.
5. Encourage informality. People should not rise to speak; however, it may be necessary to use some device to make sure everyone who wishes to speak can do so. This item becomes very high priority where the topic for resolution is favored for conflict and argumentation. Treat differences of opinion as indicative of an incomplete sharing or relevant information on someone's part and press for additional sharing, either about task or relationship, so that the points of disagreement can be clarified. Avoid conflict-reducing techniques such as majority vote, force analysis averaging, bargaining, coin flipping, and the like. View differences of opinion as both natural and helpful rather than as hindrances to decision making. Generally the more ideas expressed, the greater the likelihood of conflict will be, but the richer the array of resources will be as well.
6. Do not allow side conversations to go on unchecked. They distract the rest of the team. One way to handle this is to simply stop talking, if you happen to have the floor. Wait to resume talking until the

side conversations have stopped. Another technique that a facilitator can use is to say "Let us try to focus on one conversation at a time."

7. If a discussion is running smoothly, try not to interrupt it simply because a preset time has been reached. Get the group's consensus on what it wants to do.
8. Avoid arguing. Present your position as lucidly and logically as possible, but consider seriously the reactions of the team.
9. Avoid "win-lose" stalemates in the discussion. Discard the notion that someone must win and someone must lose in the discussion; when impasses occur, look for the next most acceptable alternative for both parties.
10. Avoid changing your mind only in order to avoid conflict and to reach agreement and harmony. Withstand pressures to yield which have no objective or logically sound foundation.
11. View quick agreement as suspect. Explore the reasons underlying apparent agreements. Make sure that people have arrived at similar solutions for either the same basic reasons or for complimentary reasons before incorporating such solutions in the team decisions.
12. Avoid subtle forms of influence and decisions modification; for example, when a dissenting member finally agrees, do not feel that he must be "rewarded" by having his own way on some later point.
13. Be willing to entertain the possibility that your team can excel at its decision task; avoid doomsaying and negative thinking. In a team-oriented environment, trouble may be slowly creeping in, not because they fail to support one another, but because they over support.

The skill of differing is dependent on the use of two prior skills, those of listening and support. In other words, if you have carefully listened to another and mentioned what you feel are the specific merits of that person's ideas, that person is going to be far more willing to accept the areas with which you disagree. Having listened and supported, the next step in differing is to phrase your remarks in a way which suggest that it is a concern you have and which does not suggest that the other person or his ideas are wrong.

For example, having listened and supported, use phrasing such as "What I am having difficulty with is ... ," "What concerns me about this is ... ," and "The problem I am having is ... ." Often differences occur when people are talking about two different things. If you have stuck to a rational process it is likely that when you do differ it will be over something which is specific. For example, a team using the rational process for selecting a course of action may have numerous differences, but they would be related to very specific questions. So, effective differing (i.e., discussion which leads to creative cooperation as opposed to defensive conflict) requires:

1. active listening
2. supporting
3. stating differences as your concern

In addition to differing, the team may fall in the trap of "group-think." Group-think is a total (by all team members) conformity in reaching decisions. To avoid it, some hints are suggested:

1. Avoid reaching agreement too quickly; beware of reaching instant, unanimous agreement.
2. The team leader insists on critical thinking; all members recognize dissent as a way of getting all viewpoints.
3. Bring in an outside opinion.
4. List all alternatives.
5. Question too much team unity and team cohesiveness.
6. Avoid pressure on members to change positions too soon.
7. Assign a decision role to each team member to estimate and evaluate the consequences of the proposed decisions.

IV. GUIDELINES TO ASSIST ARRIVING AT A CONSENSUS

1. Prepare your own position as well as possible prior to meeting with the team.
2. Recognize an obligation to listen to the opinions and feelings of all other team members and be ready to modify your own position on the basis of logic and understanding.
3. Avoid conflict reducing techniques such as voting, compromising strong objections, or giving in to keep the peace.
4. Realize that differences of opinion are helpful in exploring differences

V. TEAM PROCESS FUNCTIONS

A team can perform effectively when its members use the skills to achieve its tasks and maintain the relationships among its members. In team dynamics, the interaction of the members can enhance or detract from the team's effectiveness in meeting its goals. Specifically, these dynamics may be summarized in three functions: 1) satisfying the problem solving needs of the team, 2) team building and vitalizing functions that help satisfy the interpersonal needs in a team, and 3) antifunctions (i.e., satisfaction of personal needs at expense of team).

A. Problem-Solving Needs

This is done through the following:

- Initiating: getting discussion started, suggesting new ideas or ways of looking at problem, suggesting new procedures
- Information seeking: seeking clarification of ideas presented or for facts and authority to support ideas presented
- Information giving: offering examples, facts, or authority relative to problem or ideas presented
- Clarifying: prodding for meaning and understanding of problem or ideas presented, restating problem or ideas in different words
- Coordinating: pointing out the relationship the between problem and several ideas that have been presented
- Summarizing: bringing the team up-to-date by reviewing content of discussion to that point, expressing the sense of group's feelings
- Testing for consensus: checking for readiness of group to reach consensus

B. Team Building and Vitalizing Functions

This is done through the following:

- Encouraging: being friendly and responsive to others, praising others and their ideas, accepting contributions of others
- Mediating: conciliating differences in points of view, willing to compromise
- Gatekeeping: encouraging participation, making it possible for all members to contribute
- Standard setting: suggesting standards for selecting subjects, determining procedures and evaluation, suggesting rules and ethical values
- Following: cooperating with group, being a good listener, accepting ideas of others
- Relieving tension: helping drain off negative feelings by jesting, using humor to decrease stress

C. Antifunctions

This is done through:

- Blocking: going off on a tangent, arguing too much on a point already settled by rest of team, rejecting ideas without proper consideration

- Behaving aggressively: criticizing and blaming others, attacking motives of others, competing with others to influence the team the most
- Seeking recognition: calling attention to self by excessive talking, extreme ideas, boasting, boisterousness
- Special pleading: introducing and supporting ideas related to one's own "pet concerns" beyond reason
- Withdrawing: being indifferent or passive, using excessive formality, doodling, whispering to others
- Dominating: asserting authority to manipulate team or certain members, giving directions authoritatively, interrupting others
- Playboy: displaying lack of involvement in form of cynicism, nonchallenge, and horseplay
- Help seeker: trying to get sympathy response from others, personal depreciation of self beyond reason

D. Facilitator Interventions

As part of evaluating the process the facilitator may ask:

1. What is going on in the team right now?
2. How do you feel about what is happening in the team?
3. Are you satisfied with what the team is doing?
4. What do you want to do about the situation?
5. Does anyone see this as a reason to change personal behavior?
6. What do I see happening right in the group?

Or the facilitator may observe and evaluate passively in the following manner:

1. Team members adapted their own goals to team goals. Little/Much
2. Leadership rotated in group. Some/Often
3. Members communicated their feelings. Reservedly Openly
4. Members understand each other/s meaning. Vaguely/Clearly
5. Members seem to stick together. Little/Much
6. Decisions were made by consensus. Little/Often
7. Members told each other how effective or ineffective they were. Reservedly/Open
8. Members used all resources of each other. Some/Most

REFERENCES

Moran, L., Musselwhite, E., and Zenger, J. H. (1996). *Keeping teams on track.* Irwin Professional. Burr Ridge, IL.

Opper, S. and Fresko-Weiss, H. (1992). *Technology for teams.* Van Nostrand Reinhold, New York.

Parker, G. M. (Ed.) (1996). *1996 handbook of best practices for teams*. Irwin Professional. Burr Ridge, IL.

Roming, D. A. (1996). *Breakthrough teamwork*. Irwin Professional. Burr Ridge, IL.

Sholtes, P. R. (1989). *The team handbook*. Joiner Associates. Madison, WI.

Weiss, D. S. (1996). *Beyond the walls of conflict*. Irwin Professional. Burr Ridge, IL.

15

Meetings

One of the most frequent activity that most practitioners of quality will face, is the meeting. In this chapter we introduce the subject and address—in a limited fashion—the requirements and responsibilities for those who attend them. For more information see Bradford (1976), McDonald (1996), Mosvic and Nelson (1987), Parker (1994, 1995, 1996), Pfeiffer (1991), and Roming (1996).

I. DEFINITION

In meetings, people meet together to deliberate on a topic of common concern. During the course of an action project, there are many occasions when participants need to discuss their progress or meet with representatives of other teams to:

- Identify and agree upon: "What is the problem?"
- Explore different approaches and/or possible solutions to a problem.
- Explore all sides of a question.
- Develop plans of action.
- Involve all participants in reaching a consensus.

- Pool abilities, knowledge and experience.
- Develop innovative ideas through interchange among people, and provide everyone with the satisfaction of participating in the process.

In essence then, a meeting is an active participation of two or more individuals (usually about ten) that come together to identify, clarify, modify, discuss, resolve, and communicate a problem or an opportunity, within or outside the organization.

II. REQUIREMENTS

For the meeting to be effective and productive, some of the following elements must be adhered to.

A. Agenda

An agenda is the map of the discussion. Minimum requirements include: the topic(s) for discussion and time limits (if applicable and appropriate). Other optional requirements may include: old business, new business, correction of the old business, and so on.

B. Active Participants

It is imperative that there is an open discussion in order for the team to function properly. Therefore, all participants must follow some basic rules:

- Say what needs to be said.
- Listen to others: Respect their opinion(s).
- Do not take cheap (personal) shots.
- Discuss a topic enough to gain clarity.
- Stay focused on the task.
- Keep confidentiality.
- Strive for consensus.
- Do not be influenced by an individual's title; everyone's opinions are of equal value.

C. Designated Facilitator

The fundamental role of any facilitator may be summarized in the following four items:

- Moderate the ground rules.
- Help the team focus on the task under discussion.
- Facilitate the team process.
- Remain neutral, do not be a participant.

However, specific responsibilities may be of the following type:

1. Before the meeting:

- Set up the meeting room.
- Post the agenda, facilitator role, and ground rules, and set time limits.
- Verify that the logistics are adequate: rooms, furniture, lighting, equipment, easels, easel pads, felt-tip pens, name cards, sign-in sheets, handouts, and so on.
- Meet the chairperson, resource person(s), observer, recorder, and the leader (if any).

2. During the meeting:

- Start on time.
- Introduce both yourself and the participants. Make sure that you ask and record their expectations for the meeting.
- Announce and clarify the agenda, facilitator role, and ground rules.
- Discuss the purpose and expected outcomes of the meeting.
- Help the group select a recorder. (The facilitator may be the recorder.)
- Keep the team working on the agenda.
- Encourage all participants to take an active part.
- Prevent any one person or clique from dominating the discussion.
- Do not participate in the discussion yourself; you are not a member of the group.
- Point out any ground rule violations.
- Monitor time utilization.
- Help the group members to perform the team process functions.

3. Toward the end of the meeting (with the team members)

- Review the task achievement.
- Consider action plans.
- Evaluate the team process.
- Consider the need for another meeting. If so, help the group set the time and place.
- Close the meeting.

4. After the meeting:

- Secure the group memory, flip chart pages.
- Put the meeting room in order.
- Reflect on your facilitator role results.
- Request feedback on ways to improve the future facilitation of meetings.

D. Designated Time Keeper

The time keeper is responsible for the time. He or she makes sure that the time limits for discussing an item are kept and he or she is also responsible for starting and finishing on time.

E. Designated Recorder

The recorder's commitment is to capture the main points of discussion, and specifically:

- to capture basic ideas on large paper in full view of group
- not to edit or paraphrase
- to record enough of the speaker's idea so they can be understood later
- to remain neutral
- if the recorder gets behind, to stop the discussion
- to listen for key words
- to try to capture basic idea
- not to try to write down every word
- to write fast and large
- not to be afraid to misspell
- to abbreviate words
- to vary color
- to number sheets

F. Resources

For a meeting to take place there are at least three items of concern.

People

A facilitator is sensitive to the needs of team members and is able to involve all participants in the meeting process. The facilitator need not have experience and/or expertise in the subject of the meeting. A coordinator should arrange logistics of the meeting. No more than twenty people should take part in the discussions themselves. Larger meeting are possible, but are rarely productive if discussions are required to deal with issues. Many teams find that 8–12 twelve participants is best for active interchange and full involvement.

Time

The coordinator may spent a few hours sending invitations and preparing background materials, if needed, to be distributed either at the meeting or in advance.

Facilities

The size of the room is less important than how it is arranged. People should never be seated in rows, but rather they should be facing each other in a circle, around a table, in a U-shaped arrangement, or in a square. A blackboard or flip chart is often helpful. The seats and room temperature should be comfortable and there should be adequate ventilation.

Materials

For the meeting to take place there are some basic items that are needed, including:

- a meeting room with tables and chairs
- flip charts
- markers (multicolored)
- masking tape
- audio visual equipment (35mm projector, screen, tape player, overhead projector, extension cords, and/or video)
- activity reports
- agenda
- writing paper

G. Ground Rules

These are the rules that the team has defined as their own operational guidelines for handling the meeting. Typical ground rules are similar to the following:

- Say what needs to be said.
- Listen to others, respect their opinion.
- Prohibit cheap and personal shots.
- Discuss a topic enough to gain clarity.
- Stay focused on the task.
- Leave job titles outside the meeting.
- Keep information confidential.

III. ACTION CHECKLIST

Every facilitator is responsible for the flow and effectiveness of their meeting. To make sure these goals are accomplished the facilitator may use the following:

A. Preliminaries

1. Ask participants what their concerns and backgrounds are to help provide a focus with which to begin the discussion and to identify those with special expertise.
2. Find appropriate resource materials to make available to the team. These can include article reprints, case studies, films, videos, and other relevant information.
3. Obtain name tags if the team is coming together for the first time.
4. Plan the opening of the session. The opening should define the purpose of the meeting, outline the proposed agenda, state the ground rules for working together as a group, and give the roles of the facilitator and chair (if there is one.)

IV. DOING IT—FACILITATING THE MEETING

1. Help everyone feel comfortable and get to know each other. Perhaps ask each person to introduce him or herself and explain their interest in the subject, their background and hopes/expectations for the session. Do not let this process define who has top status; however, stress that everyone has something important to contribute.
2. Briefly review the purpose and background of the meeting, review and amend the agenda as necessary, and explain the ground rules and the roles of the facilitator and chair.
3. Appoint a recorder to take notes (if the team has already had an organizing meeting in which it has clarified how it will make decisions as a group, how leadership will be established, and so on—tasks which are appropriate to groups that will be operating for some time—the recording function should have already been dealt with by the team.) The purpose of the recorder, if the job is not done by the facilitator on the flip charts, is to take notes on the deliberations and report on progress from time to time. For example, when the team reconvenes after a break, the recorder can summarize where the meeting/discussion stands.
4. Listen carefully to what each person is saying. This will set an example to others. It will also help you ask a team member to

clarify something so that the discussion does not bog down in misunderstandings. Try to restate (or ask someone else to restate) what a team member has said so it is clear to everyone.

5. As the facilitator, do not play the role of an "expert," make decisions for the team or always be the center of the discussion. Encourage team members to direct their comments to each other, not to you.
6. Give every individual maximum opportunity to participate through the following:
   a. Try to limit the number of times and amount of time any single person speaks.
   b. Do not let resource people and experts dominate the discussion or intimidate others into silence. They should be asked to provide facts or perspectives when needed so that the team bases its decision on an accurate picture of the situation.
   c. While some persons may need a little encouragement to speak, this does not necessarily meana that you should single them out and ask them to say something just because they have not said much. Instead, by preventing others from dominating and maintaining a relaxed atmosphere, you can provide the right circumstances for nonparticipants to open up.
7. Encourage informality. People should not rise to speak; however, it may be necessary to use some device to make sure everyone who wishes to speak can do so. To avoid having everyone wave his or her hands as soon as a speaker finishes, consider taking hands after only every third or fourth speaker and name the next three or four speakers in order. Or, one speaker can take hands and choose who will follow him or her. The team itself should agree on such a procedure in advance.
8. When conflicts and disagreements arise, help clarify the points of disagreement. Encourage participants to deal with disagreements openly and calmly so that the areas of disagreement can perhaps be narrowed.
9. Do not allow side conversations to go on unchecked. They distract the rest of the team. One way to handle this is to simply stop talking, if you happen to have the floor. Wait to resume talking until the side conversations have stopped. Another technique that a facilitator can use is to say: "Let's try to focus on one conversation at a time."
10. If a discussion is running smoothly, try not to interrupt it simply because a preset time, such as a lunch break, has been reached. Get the group's consensus on what it wants to do.

Table 15.1 Meeting Evaluation Form

Statement for evaluation	Usually	Sometimes	Seldom
Members really listened to each other			
Discussion was focussed on the agenda			
Participation of all members was encouraged			
Dissenting opinions were heard and resolved			
Testing for agreement was used before decisions were made			
Members were protected from personal attack			
Facts and feelings were considered.			
Multiple alternative problem solutions were generated			
Participants expressed themselves openly, honestly, and directly			

A. Evaluation and Follow-Up

1. At the end of the session, try to summarize any conclusions, points of agreement or areas where disagreement still exists. Hit the main points; do not bother being too detailed.
2. If desired, call on an observer who did not take an active part in the discussion to comment on the content and process of the session.
3. Ask participants to write their evaluation, perhaps on a prepared form. The evaluation respond may or may not have the participant's name. There are two schools of thought on this subject.

   a. Without name. It is believed that the participant will be more forthcoming and objective with his or her comments, without endangering his or her position.
   b. With name. It is believed that the participant may withhold important information and present a bias view, because of fear of being singled out. A typical evaluation of a meeting may be of the form in Table 15.1.

V. LISTENING SKILLS

We made several references to the importance of listening in the team environment for problem solving. In this section we will address the issue of listening in a meeting environment. Let us start with a definition. When we speak of active listening we mean hearing plus processing plus responding.

Hearing is to be fully accessible to the other person. Evidence of hearing, processing and responding is seen in:

- eye contact
- tone of voice
- facial expression
- posture
- gestures
- body language

Processing is achieved by:

- listening for feelings as well as for facts
- identifying the central theme
- accepting "hidden" messages
- suspending judgement for a while
- waiting before responding
- being aware when feelings are blocking communications
- reflecting on what is being said

Responding includes:

- rephrasing other persons' words to show understanding
- expressing warmth and support
- showing trust
- using feedback to check meaning

Feedback responses are:

- clarifying
- restating
- responding neutrally
- reflecting feelings
- summarizing
- asking for feedback to check understanding
- using self-disclosure in moderation

Active listening, then, is:

- hearing not only the words but also the feelings
- feeling what other person is feeling (role reversal)
- suspending judgement so as to gain understanding
- fostering an understanding atmosphere

Some techniques for active listening include:

Restatement. This is simply a replay of the statement that the other person has just made. It is a repetition word for word of the statement of the person being questioned.

Paraphrase. This technique is closely related to restatement, but the significant difference here is that understanding is demonstrated by the questioner's putting into their own words what they are hearing from the person whom they are questioning.

Reflection. This is a mirroring technique in that the questioner plays back to the person being questioned, the feelings that he believes are being experienced by the other person. He/she responds not only to the words the other person is expressing, but also the music that he/she hears. Reflection of feeling can be an extremely powerful technique for generating the perception on the part of the person that he/she is being understood.

Summarization. It helps from time to time for the listener to integrate the sets of data from the person being questioned in the form of brief summaries. This fosters a climate of understanding. Some of the active listening skills are:

1. Attentive silence. This is accomplished by good eye contact, head nods, facial expression showing alertness, and interest.
2. Attentive words. This is accomplished by inserting positive words or phrases into the conversation, as the other person speaks. For example, "Yeah," "I see," "Uh huh," "I hear you," "I understand," and so on.
3. Door openers. This is accomplished by asking questions for more details. For example: "Tell me more," "What are you thinking?," "How do you feel about that?," "Then what happened?," "Go on," and so on.
4. Summary of thoughts and feelings (of the other person). Verbal feedback of the others expressed thoughts and expressed or demonstrated feelings enables you to prove your understanding and empathy. Understanding is established when you paraphrasing of the other's words is confirmed to be accurate. Empathy is your acknowledging the other's feelings and being on target. The way that the summarizing is accomplished is by asking a "check out" question at the conclusion of your summary or summarizing in a question form that invites the other to respond and helps to keep the communication channel open. For example: "You seem wary of telling him how you feel. Is that right?" or "Your main point is that this will save

us several thousand dollars. Did I understand you correctly?" In the next section you will find some tips to improve your listening skill in a team environment.

VI. TIPS TO IMPROVE YOUR LISTENING SKILLS

We have a myth in our society that listening and hearing are the same. You may be surprised to know, they are not! Listening is a higher cognitive process than hearing. While hearing is a natural, passive process, listening is a skill, an active process that requires attention, concentration, and an empathetic attitude and training. The following are some tips to help you stay in the listening mode rather than the hearing mode.

1. Search for something you can use. Find areas of common interest. If you adopt a positive attitude toward what the speaker is saying, you will usually find something that will broaden your knowledge. Sorting out elements of personal value is one area of effective listening. Ask yourself, "What is being said that I can use?" "What actions should I take?"
2. Take the initiative. Look at the speaker and concentrate on what has been said. Ignore his delivery and personality and reach for the idea he/she is conveying. Show interest by the use of a noncommittal acknowledgment such as: "Oh I see," "How about that?," etc.
3. Work at listening. Efficient listening takes energy. Practice makes it easier when you become aware of internal distractions.
4. Focus your attention on ideas. Listen for the speaker's central ideas and pick out the ideas as they are presented. Sort the facts from principles, the ideas from examples, and the evidence from opinion.
5. Resist external distractions. Where possible, resist distractions. Concentrate on concentrating, you make it possible to be aware of noises without being distracted by them.
6. Hold your rebuttal. Watch out for emotion-laden words. Begin to recognize certain words that affect you to the point where you stop listening and start forming a rebuttal. One way to deal with this is to quietly analyze the reasons why those words stir you. Another method is to jot down major rebuttal points as questions. Do this briefly, not at length. Both methods can help you open your mind so that you can return to listening with an open mind.
7. Keep an open mind. Ask questions to clarify for understanding. Quick and violent disagreement with the speaker's main points to arguments will more often than not cause a psychological deaf spot.

Give the speaker more rather than less attention. Clarify meaning by restating, in your own words, what you thought was said.

8. Concentrate on the process of ideas. Concentrate on what the talker says. Summarize in your head. Decide how well he is supporting his points and how you would have supported them. After each point is covered, mentally review the progress that is being made toward the theme.

VII. IDEA LIST FOR MEETING

Quite often participation in the meeting environment is not what was expected for many reasons (i.e., domineering leader, talkative supervisor, team dis-functionality, weak facilitator, and so on). To prevent some of these, the following list of innovative ideas collected from various meetings over the years. Each one of the ideas has been used and found useful by some. Of course, that does not necessarily mean that it will suit every meeting and every situation. Mark the items that you would like to try out for a trial period in your meeting. Do not be afraid to modify or change these ideas to fit your situation.

1. Asking the participants at beginning of meeting if they have items they wish to add to the meeting agenda.
2. In some meetings, organizing the priority order of the items at beginning of the meeting.
3. When this is done, making a time table allocating time to be given to each item.
4. Putting the agenda and time allocation in print
5. Organizing the agenda so that quick "easy" items are at the beginning and finished off first.
6. Rotating chairmanship of meetings and/or taking turns in chairmanship in long meetings.
7. Writing the main points of what people say in print.
8. Writing team decisions on separate print.
9. Sending written copies of meeting decisions to all members.
10. Occasionally breaking into subgroup during meetings to discuss different aspects of a problem.
11. Occasionally going round the room to check opinions.
12. Using brainstorming when suitable.
13. Occasionally making decisions to try something out for a certain period of time (conditional decisions.)
14. Using subteams to prepare work on a complicated problem for the entire team.

15. Using tasks groups to deal with team projects.
16. Using subcommittees.
17. When an important item is very controversial, trying to reach consensus and not vote.
18. Checking before discussing an item if the people meaningfully affected by it are present.
19. Writing next to each decision who is responsible for implementation; this emphasizes ownership.
20. Fixing a date to check on the implementation of difficult decisions.
21. Using a process observer (a participant may be used) who follows the way the team operates and at the end of the meeting reports his or her observations.
22. Sometimes devoting time at the end of a meeting to discuss the process of the meeting.
23. Allocating a regular monthly (or bimonthly) meeting for discussing basic problems for which the team never seems to find time.

REFERENCES

Bradford, L. P. (1976). *Making meetings work: A guide for leaders and group members*. University Associates. San Diego.

McDonald, T. (1996). Ready, aim, talk. *Successful meetings*. May: 30.

Mosvic. R. K. and R. B. Nelson. (1987). *We've got to start meeting like this! A guide to successful business meeting management*. Scott, Foresman. Glenview, IL. and University Associates. San Diego.

Parker, G. M. (1994). *Crossfunctional teams*. Jossey Bass. San Francisco.

Parker, G. M. (1995). *Team players and teamwork*. Jossey Bass. San Francisco.

Parker, G. M. (Ed.). (1996). *Handbook of best practices for teams*. Irwin Professionals. Burr Ridge, IL.

Pfeiffer, J. W. (Ed.). (1991). *Theories and models in applied behavioral science: Management Leadership*. Vols. 2 and 3. Pfeiffer. San Diego.

Roming, D. A. (1996). *Breakthrough teamwork*. Irwin Professionals. Burr Ridge, IL.

16

Project Management

Project management is the application of knowledge, skills, tools and techniques in order to meet or exceed stakeholder (customer) requirements from a project. Meeting or exceeding stakeholder requirements means balancing competing demands among:

- scope, time, cost, quality, and other project objectives
- stakeholders (customers) with differing requirements
- identified requirements and unidentified requirements (expectations)

Knowledge about project management can be organized in many ways. In fact, the official *Guide to the Project Management Body of Knowledge* (PMBOK) identifies 12 subsections of project management (Duncan, 1994):

1. project management
2. the project context
3. the process of project management
4. key integrative processes
5. project scope management

6. project time management
7. project cost management
8. project quality management
9. project human resource management
10. project communications management
11. project risk management
12. project procurement management

This chapter addresses how total quality management may be implemented efficiently with an undertaking of a project management approach; it also discusses some of the basic concepts of project management and how the methodology of project management may be used.

I. WHAT IS A PROJECT?

Projects are tasks performed by people, constrained by limited resources, describable as processes and subprocesses, and they are planned, executed, and controlled within definite time limits. Above all, they have beginning and an end. Projects differ primarily from operations in that operations are ongoing and repetitive while projects are temporary and unique. A project can thus be defined in terms of its distinctive characteristics: it is a temporary endeavor undertaken to create a unique product or service. Temporary means that every project has a definite ending point. Unique means the product or service is different in some distinguishing way from all similar products or services.

Projects are undertaken at all levels of the organization. They may involve a single person or many thousands. They may require less than 100 hours to complete or over 10,000,000 hours. Projects may involve a single unit of one organization or may cross organizational boundaries as in joint ventures and partnering. Examples of projects include:

- developing a new product or service
- effecting a change in structure, staffing, or style of an organization
- designing a new product
- developing a new or modified product or service
- implementing a new business procedure or process

Temporary means that every project has a definite ending point. The ending point is when the project's objectives have been achieved, or when it becomes clear that the project objectives will not or cannot be met and the project is terminated. Temporary does not necessarily mean short in duration. It means that the project is not an ongoing task; therefore it is finite. This point is very important since many undertakings are temporary in the sense that they will end at some point. For example, assembly work at an automotive plant will eventually be discontinued and the plant itself decommissioned. Projects are fundamentally different because the project ceases work when its objectives have been attained while nonproject undertaking adopt a new set of objectives and continue to work.

The temporary nature of the project may apply to other aspects of the endeavor as well:

- The opportunity or market window is usually temporary since most projects have a limited time frame in which to produce their product or service.
- The project team seldom outlives the project since most projects are performed by a team created for the sole purpose of performing the project, and the team is disbanded and members reassigned when the project is complete.

On the other hand, a project or service is considered unique if it involves doing something which has not been done before and which is therefore unique. The presence of repetitive elements does not change the fundamental uniqueness of the overall effort.

Because the product of each project is unique, the characteristics that distinguish the product or service must be progressively elaborated. Progressively means "proceeding in steps; continuing steadily by increments" while elaborated means "worked out with care and detail; developed thoroughly" (American Heritage Dictionary, 1992). These distinguishing characteristics will be broadly defined early in the project and will be made more explicit and detailed as the project team develops a better and more complete understanding of the product.

Progressive elaboration of product characteristics must not be confused with proper scope definition, particularly if any portion of the project will be performed under contract.

In contrast to a project, there is also a program. A program is a group of projects managed in a coordinated way to obtain benefits not available from managing them individually (Turner, 1992). Most programs also include elements of ongoing operations as well as a series of repetitive or cyclical undertakings. (It must be noted, however, that in some applications program management and project management are treated as one and the same; in others, one is a subset of the other. It is precisely this diversity of meaning that makes it imperative that any discussion of program management versus project management requires a clear, consistent and agreed upon definition of each term.)

II. THE PROCESS OF PROJECT MANAGEMENT

The process of project management is an integrative one. The interactions may be straightforward and well understood, or they may be subtle and uncertain. These interactions often require trade-offs among project objectives. Therefore, successful project management requires actively managing these interactions, so that the appropriate and applicable objectives may be attained within: budget, schedule, and constraints.

A process from a project management perspective is the traditional dictionary definition: "a series of actions bringing about a result" (American Heritage Dictionary, 1992). In the case of a project, there are five basic management processes:

- *Initiating*: recognizing that a project should be begun and committing to do so.
- *Planning*: identifying objectives and devising a workable scheme to accomplish them.
- *Executing*: coordinating people and other resources to carry out the plan.
- *Controlling*: ensuring that the objectives are met by measuring progress and taking corrective action when necessary.
- *Closing*: formalizing acceptance of the project and bringing it to an orderly end.

Operational management—the management of ongoing operations—also involves planning, executing, and controlling; however, the temporary nature of projects requires the addition of initiating and closing. These processes occur at all levels of the enterprise, in many different forms, and under many different names. Even though there are many variations, it is imperative to understand that the operational management is an ongoing activity with neither a clear beginning nor an expected end.

Finally, it must be understood that these processes (initiating, planning, executing, controlling, and closing) are not discrete, one-time events. They are overlapping activities which occur at varying levels of intensity throughout each phase of the project. In addition, the processes are linked by the results they produce since the result or outcome of one becomes an input to another. Among the central processes, the links are iterated; planning provides executing with a documented project plan early on, and then provides documented updates to the plan as the project progresses. It is imperative that the basic process interactions occur within each phase such that closing one phase provides an input to initiating the next. For example: closing a design phase requires customer acceptance of the design document. Simultaneously, the design document defines the product description for the ensuing implementation phase. For more information on this concept see Duncan (1994), Kerzner (1995), and Frame, 1994).

III. KEY INTEGRATIVE PROCESSES

In project management the key integrative processes are:

Project plan development: taking the results of other planning processes and putting them into a consistent, coherent document.
Project plan execution: carrying out the project plan by performing or having performed the activities included therein.
Overall change control: coordinating changes across entire project.

Although the processes seem to be discrete and different from with each other, that is not the case in practice. In fact, they do overlap and interact in ways that is beyond the scope of this book. A typical summary of key integrative process is shown in Table 16.1.

Table 16.1 Key Integrative Processes

Project plan development	Project plan execution	Overall change control
1. Inputs	1. Inputs	1. Inputs
Outputs of other processes	Project plan	Project plan
Historical information	Supporting detail	Progress report
Organizational policies	Organizational policies	Change request
Constraints and assumptions		
2. Tools and techniques	2. Tools and techniques	2. Tools and techniques
Project planning methodology	Technical skills and knowledge	Change control system
Stakeholder skills and knowledge	Work authorization system	Progress measurement
Project management information systems	Status review meetings	Additional planning
	Project management information system	Computer software
	Organizational procedures	Reserves
3. Outputs	3. Outputs	3. Outputs
Project plan	Work results	Project plan updates
Supporting detail	Change requests	Corrective action
		Lessons learned

IV. PROJECT MANAGEMENT AND QUALITY

Project management is a problem solving methodology. On the other hand, total quality management is a "process project" that requires total acceptance for improvement. For that improvement to occur, TQM must be implemented in the entire organization. As such, TQM fits the profile of project management. Every component of it is designed to facilitate the solving of complex problems. It uses teams of specialists. It makes use of a powerful scheduling method. It tightly tracks costs. It provides a mechanism for management of total quality. It depends on the integration of several skills and disciplines. It encourages monitoring of processes and depends on feedback for evaluation. It requires leaders with clear vision and doable objectives. It requires knowledge of appropriate and applicable tools. And it plans for success.

In addition, project management makes and at the same time facilitates change(s). By definition, projects have a start, middle (work accomplished), and a finish. The finish comes when the objectives for the project are satisfied. Project objectives always address changes which will be made in some current situation. If an organization does not want to make a change, then project management is not an appropriate management method for them. This does not imply that changes should not be made there, only that there is no motivation for change. In such an organization, the introduction of project management would have little support and may be even resistance. For a discussion on change and when change actually takes place see Stamatis (1996a).

Since the implementation of TQM is a project (with a beginning, work changes, and an end), project management is indeed a method that can be used in the implementation process. (It is very important to differentiate the concept of TQM—a philosophy—and the implementation of TQM—a project. Here we are talking about the physical implementation of TQM.) For specific examples of implementation strategies of project management in the implementation of TQM in ISO 9000, QS-9000, service industries, and health care, the reader is encouraged to see Stamatis (1994, 1995a,b, 1996a,b).

V. A GENERIC SEVEN-STEP APPROACH TO PROJECT MANAGEMENT

Much has been written about how to use project management in a variety of industries and specific situations. Many articles and books have proclaimed specific approaches for the best results in a given situation. Rather than dwell on a particular approach, here is a summary discussion of a generic seven-step approach of using project management in a quality orientation for any organization. The seven steps are based on the four-phase cycle of any project.

A. Phase 1. Define the Project

Describe the Project

Describing a project is not as simple as it might seem. In fact, this step may be the most difficult and time consuming. To be successful the project description should include: simple specifications, goals, projected time frame, and responsible individuals as well as constraints and assumptions.

Capturing the essence of highly complex projects in a few words is an exercise in focus and delineation; however, we must be vigilant of avoiding becoming too simple and in the process we fail to convey the scope of the project, while a detailed, complex description may cloud the big picture. The key is clarity without an excess of volume or jargon.

Appoint the Planning Team

After describing the project begin to identify the right players. Too many people on a team can stifle the decision-making process and reduce the number of accomplishments. Cross-functional teams are among the most difficult to appoint. Except in the pure project organization, where the team is solely dedicated to completing the project, roles and priorities can cause conflict. In cross-functional teams the project leader must seek support from the functional managers and identify team goals.

Define the Work

Once the planning team is in place, team members must define the work. Since each member hails from a different department, there will be many different concepts of the project's work content. There are many ways to divide the work for convenient use in planning. Two common ways are the process flow diagram and the work breakdown structure (WBS). The method should be chosen to reflect the most useful division and summarization for the situation. After all, the objective of this step is to define the tasks to be done, not the order of doing it.

B. Phase 2. Plan the Project

Estimate Tasks

Before creating a project schedule, each task must be evaluated and assigned an estimate of duration. There are essentially two ways of looking at this process. The first way is to establish the duration of this task by estimating the time it takes to complete the task with given resources. The second way is to estimate the type and amount of resources needed and the effort in terms of resource hours is necessary to complete the task.

Calculate the Schedule and Budgets

The next step is to construct a network logic diagram or a performance evaluation review technique (PERT) and a budget. The focus of the logic diagram and/or the PERT is to develop appropriate scheduling datelines and more importantly to define the critical path. The focus of the budget is to estimate the costs of the project based on all activities.

The identification of the critical path will identify bottleneck areas as well as opportunities for improvement. Tasks not on the critical path may have a float which can be calculated and may be used to facilitate the efficiency and utilization of resources without affecting the project final date.

C. Phase 3. Implement the Plan

The start of the project can really make an impact on project team members' attendance, performance, and evaluation. Kick-off meetings should convey the following ideas:

- This is a new project.
- Project management is going to be used to manage the project.
- A plan exists, open to all, that is going to be followed.
- The focus is on the starts of activities.
- A realistic status is needed to allow timely decisions.
- The focus will always be on forecasting and preventing problems.

D. Phase 4. Track Progress and Complete the Project

The essence of this step is to bring the project to closure. That means that the project must be officially closed and all deliverables must be handed over to the stakeholders (customers). In addition, a review of the lessons learned must take place and a thank you for the project team is the appropriate etiquette. Key questions of this step are:

- Where are we?
- Where should we be?
- What do we have to do to get there?
- Did it work?
- Where are we now?
- Can the process employees take over?
- Can the process employees maintain the new system?
- What have we learned from the successes in this project?
- What have we learned from the failures in this project?
- What we would have done differently? Why? Why not?

VI. A GENERIC APPLICATION OF PROJECT MANAGEMENT IN IMPLEMENTING TQM

Project management brings together and optimizes (the focus is always on allocation of resources) rather than maximizes resources (concentrating on one thing at the expense of something else; maximization leads to sub-optimization), including skills, cooperative efforts of teams, facilities, tools, information, money, techniques, systems, and equipment.

There are at least two reasons why project management should be used in the TQM implementation process. First, project management focuses on a project with a finite life span, whereas other organizational units expect perpetuity. Second, projects need resources on both part-time and full-time basis, while permanent structures require resource utilization on a full-time basis. The sharing of resources may lead to conflict and requires skillful negotiation to see that projects get the necessary resources to meet objectives throughout the project life.

Since we already have defined the process of TQM implementation as a project, then project management will ensure successes of the implementation process by following the generic four phases of a project's life. A typical approach is shown in Figure 16.1. On the other hand, Tables 16.2 and Table 16.3 show the characteristics of the TQM implementation model and process using project management.

VII. THE VALUE OF PROJECT MANAGEMENT IN THE IMPLEMENTATION PROCESS

Project management is a tool that helps an organization to maximize its effort in implementing a project. Since the process of implementing TQM—or any other quality initiative—is a project, the value of project management can be appreciated in at least two areas. First, planning the process, and second setting reliable, realistic and obtainable goals.

A. Planning the Process

There are four steps that define the planning process from a project management perspective.

1. Identify and prioritize the customer base by contribution to current and future organizational profits.
2. Identify and weight criteria regarding key customers; use to select organizations and assess what changes to criteria or weighting likely to occur in the future.

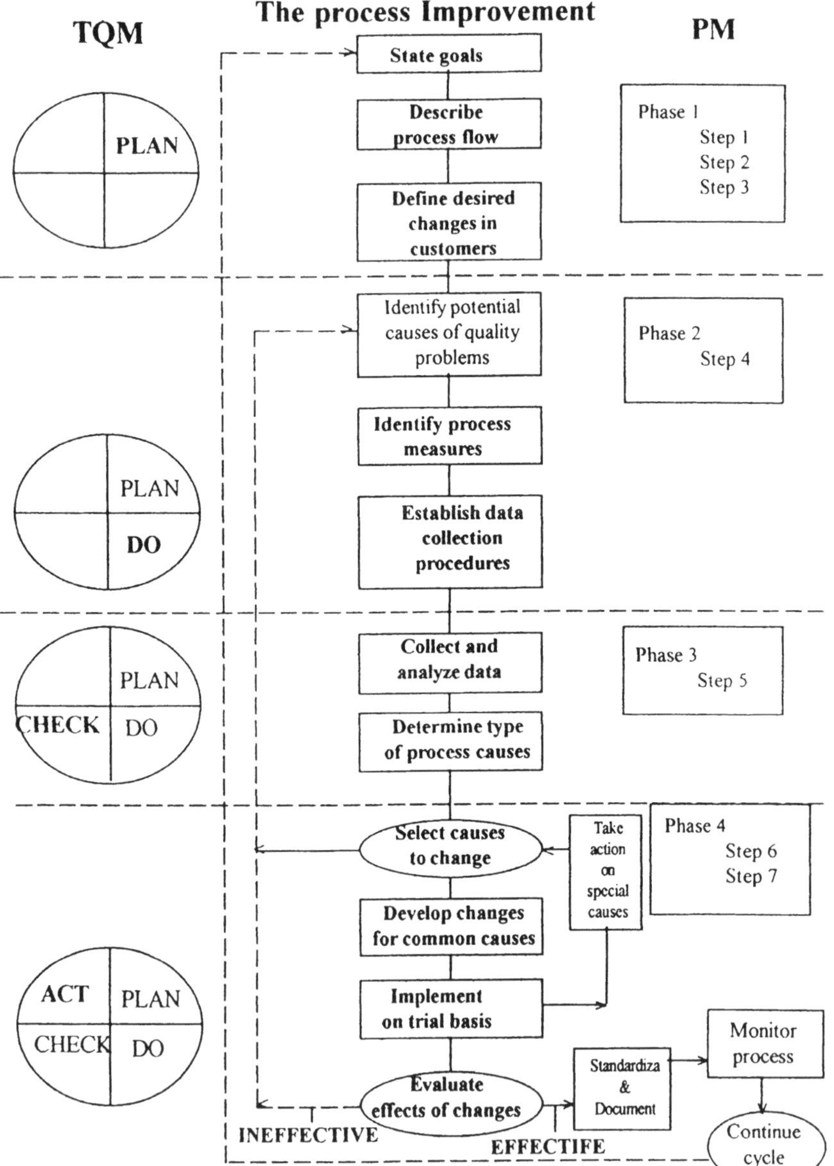

Figure 16.1 The relationship between TQM, process improvement, and project management.

Table 16.2 The Characteristics of TQM Implementation Model Using Project Management

Phase 1 Management commitment	Phase 2 Structure setup	Phase 3 Implementation	Phase 4 Working with employees
	Capture company objectives	Make a goal of TQM total improvement	Provide applicable and appropriate training
Establish a TQM implementation team of one person from each functional area	Define: mission, values, goals, and strategy	Examine internal structure and compare it to the goals of TQM	Prepare the organization for both internal and external audits
Train those selected in TQM requirements	Focus on continual improvement	Determine departmental objectives	Provide and or develop appropriate and applicable methodology for corrective action
	Develop policies and procedures	Review structure of the organization	Continue focus on improvement
	Reconfirm quality management commitment	Review job descriptions	
		Review current processes	
		Review control mechanisms	
		Review training requirements	
		Review all communication methods	
		Review all approval processes	
		Review supplier relationship(s)	
		Review risk considerations and how they are addressed	
		Review all outputs	

Table 16.3 The Process of TQM Implementation Process Using Project Management

Phase 1	Phase 2	Phase 3	Phase 4
Management commitment	Structure setup	Implementation of plan	Working with employees and suppliers
Initiate project	Understand process	Give TQM training	Monitor progress
Management planning and goal setting	Team flowcharting for process understanding and analysis	Executive training	Worker/operator control in process
Departmental commitment	Cause and effect analysis	Departmental training	Define quality system as it relates to current policies and practices (quality manual, procedures, instructions, and so on.)
Quality team selected and actively participates	Critical in-process parameters identified	Identify of shortcomings in the system of quality (specific areas)	Conduct internal audits
Training philosophy and tools of quality	Standard operating procedures review, equipment repair, preventive maintenance, and calibration	Define boundaries of responsibility	Define key characteristics and monitor process variables
Process definition and selection	Process input and measurement evaluation	Define limitations of resources	Apply statistical process control in all key processes
Critical processes and characteristics identified	Static process data collection	Review system for completeness	Initiate and follow-up corrective action
	Process evaluation		

3. Assess the organization's competitive advantages and disadvantages in each area important to decision makers.
4. Establish a long-term strategic objectives by identifying where the biggest gap exists between what is important to key customers and the organization's own strengths and weaknesses relative to competition.

To optimize the output of these four steps, ask the following questions.

- Is there a true management commitment for the project?
- Does the project address needs of the organization's top priority customer groups?
- Does the project address the important needs of the customer(s)?
- Is the organization far ahead of the competition in this area already?
- Does the project truly offer the organization a good chance of making an improvement large enough to change customer behavior?
- Will the project require an investment so large that there is no potential gain?
- How does the project rank on the above criteria in relation to other possible projects?
- Once project is selected, is the team continuously assessing whether or not the project is the best one to move the department and organization toward their goals?

B. Goal Setting

From a project perspective, there are three basic steps in goal setting.

1. Translate corporate strategy into concrete organizational goals that are attainable within a reasonable time.
2. Involve department managers in an internal audit and benchmarking exercise to identify problem areas.
3. With department managers, set specific improvement goals for each department and each team.

VIII. WHY PROJECT MANAGEMENT SUCCEEDS

The single most important characteristic of project management is the consistent ability to get things done. It is a results- or goal-oriented approach, where other considerations are secondary, so the single-minded concentration of resources greatly enhances prospects for success. This also implies that the results—success or failure—are quite visible.

Integrative and executive functions of the project manager provide another inherent advantage in the project management approach that improves the likelihood for success because of the single point of responsibility for those functions. Specific advantages of the single point integrative characteristic include:

- placing accountability on one person for the overall results of the project
- assurance that decisions are made on the basis of the overall good of the project, rather than for the good of a particular department
- coordination of all functional contributors to the project
- proper utilization of integrated planning and control methods and information they produce

The advantages of integrated planning and control of projects include:

- assurance that the activities of each functional area are being planned and carried out to meet the overall needs of the project
- assurance that the effects of favoring one project over another are known
- early identification of problems that may jeopardize successful project completion, to enable effective corrective action to prevent or resolve the problem

Project management is a specialized management form. It is an effective management tool that is used because something is gained by departing from the normal functional way of doing things: people, organizations, and methods. Conflict, confusion, and additional costs are often associated with significant changes of this nature. Poorly conceived or poorly executed project management can be worse than no project management at all. Project management should be used well or not at all. Executives should not permit a haphazard, misunderstood use of project management principles.

Although simple in its concepts, project management can be complex in its application. Project management is not a cure-all intended for all projects. Before project management can succeed, the application must be correct. Executives should not use project management unless it appears to be the best solution. The use of project management techniques seems most appropriate when:

1. A well-defined goal exists.
2. The goal is significant to the organization.
3. The undertaking is out of the ordinary.
4. Plans are subject to change and require a degree of flexibility.
5. The achievement of the goal requires the integration of two or more functional elements and/or independent organizations.

Even though project management may not be feasible, good principles have contributed to the success of thousands of small- and medium-sized projects. Many managers of such projects have never heard of project management, but have used the principles. A wider application of these principles will also help achieve success in smaller projects.

Executives play a key role in the successful application of project management. A commitment from top management to insure that it is done right must be combined with the decision to use this approach. Top management must realize that establishing a project creates special problems for the people on the project, for the rest of the organization, and for the top managers themselves. If executives decide to use this technique, they should expend the time, decision making responsibility, and executive skills necessary to ensure that it is planned and executed properly. Before it can be executed properly, sincere and constructive support must be obtained from all functional managers. Directives or memos are not enough. It takes personal signals from top management to members of the team and functional managers to convey that the project will succeed, and team members will be rewarded by its success. In addition, necessary and desirable changes in personnel policies and procedures must be recognized and established at the onset of the project.

The human aspect of project management is both one of its greatest strengths and most serious drawbacks. In order for project management to succeed, it requires capable staff. Only good people can make a project successful. In the long run, this is true for any organization. Good people alone cannot guarantee project success; a poorly conceived, badly planned, or inadequately resourced project has little hope for success. Great emphasis is placed on the selection of good people. The project leader, more than any other single variable, seems to make the difference between success and failure. Large projects require one person to be assigned the full-time role of project manager. If there are not enough project managers available for full-time assignment to a project, assign several projects to one full-time project manager. This approach has the advantages that the individual is continually acting in the same role, that of a project manager, and is not distracted or encumbered by functional responsibilities.

To conclude, project management is an effective management tool used by business industry and government, but must be used skillfully and carefully. In review, the following major items are necessary for successful results from project management in the field of quality:

- wholehearted executive support and commitment when the decision is made to use this approach
- project management is the best solution or right application for implementing any quality program

- emphasis on selecting the best people for staff, especially the project leader
- apply good principles of project planning and controlling

Effective use of project management will reduce costs and improve efficiency; however, the main reason for its widespread growth is its ability to complete a job on schedule and in accordance with original plans and budget.

REFERENCES

The American Heritage Dictionary of the English Language. (1992). 3rd ed. Houghton Mifflin Company. Boston.

Duncan, W. R. (1994). *A guide to the project management body of knowledge*. Project Management Institute. Upper Darby, PA.

Frame, J. D. (1994). *The new project management*. Jossey-Bass. San Francisco.

Kerzner, H. (1995). *Project management: A systems approach to planning, scheduling and controlling*. 5th ed. Van Nostrand Reinhold. New York.

Stamatis, D. H. (1994). Total quality management and project management. *Project Management Journal*. Sept.: 48–54.

Stamatis, D. H. (1995a). *Understanding ISO 9000 and implementing the basics to quality*. Marcel Dekker. New York.

Stamatis, D. H. (1995b). *Integrating QS-9000 with your automotive quality system*. Quality Press. Milwaukee, WI.

Stamatis, D. H. (1996a). *Total quality service*. St. Lucie Press. Delray Beach, FL.

Stamatis, D. H. (1996b). *Total quality management in health care*. Irwin Professionals. Burr Ridge, IL.

Turner, J. R. (1992). *The handbook of project-based management*. McGraw-Hill. New York.

17

Training and the Learning Organization

One of the essential points in attaining Total Quality Management in any organization is training. It is so fundamental that in this chapter we address some of the issues and concerns that a company must internalize in order for it to establish appropriate and applicable training. This internalization begins with understanding what training and the learning organization are all about, and progresses into the foundations of appropriately designing and evaluating training. Some of the additional topics that we discuss are the adult learner, systems theory and instructional design.

I. TRAINING AND THE LEARNING ORGANIZATION

Train, train, train! It is a very strong message of the Deming and Juran philosophies. It is a fundamental characteristic in the implementation of TQM process. It is also the theme of the 1990s. Corporate America lacks the skills needed to flourish in the high-tech, team empowered, global business world of today (Young, et al., 1995). Henkoff (1993), quoting the Labor Secretary Robert Riech, says that American companies have got to be urged to treat their workers as assets to be developed, rather than as cost to be cut. Some industry leaders such as Motorola, General Electric, Xerox, IBM, Chaparral Steel, Ford, and many others have recognized the link between learning and continuous improvement and have acted on this link. For example, Motorola has created the Motorola University, Ford has developed the Ford Training Center, and so on.

The rationale for training is that any organization improves with learning. Learning about their business, employees, competition, new products/services, and so on. If the organization does not train, how will they improve? Solving problems, introducing a new product, and reengineering a process all depend on the ability to look at things in a different way and then act upon it. Without learning, companies will continue to simply repeat past practices. No real change will occur and improvement programs will fail. Sadly, failed improvement programs far outnumber successes (Garvin, 1993).

So important is this issue of training that Deming (1986) incorporates it as part of the 14 points to management and President Clinton has set a target for corporate America to spend 1.5% of payroll on training; however, spending millions of dollars a year on training in and of itself will not improve a firm's competitive position. Training must be applied, focused, results oriented, and of a just-in-time nature to be an effective change agent.

Fitzen (1993) noted that 1.4% of payroll is spent on training; however, this is a misleading figure as only 10% of the workforce receives any formal employer-provided training. How does this happen? Some companies spend millions of dollars a year on training, some spend very little. In fact, according to the American Society for Training and Development most companies do not offer any training at all. Just 15,000 employers—a mere 0.5% of all companies—account for 90% of the $48 billion spent on training annually (Fitzen, 1993). Those that train often use some antiquated, passive teaching techniques that have failed miserably in the schools of America. Moreover, they lavish most of their training budgets on managers and executives, and only when they personally asked for the specific training (Stamatis, 1997). (We also suspect that corporate America is really trying to educate rather than train; this is a major problem since education and training are not the same.)

Based on Fitzen's reporting, the expenditures on training for 1993 increased 7% over the expenditure in 1992; however, Japanese and European training expenditures are still 3–5 times higher. What is interested about these statistics is the fact that most American companies spent their dollars to train and educate their employees in basic skills and basic writing, reading and math. On the other hand, companies in Japan, Singapore, and Europe spend their training dollars for advanced training needs.

Training needs may be identified throughout the organization regardless of title and/or department. It is not unusual to see executives to be taught principles of effective leadership and strategic planning. Engineers refresh their skills in an effort to keep up with the rapid changes in technology. Production workers participate in seminars on problem-solving, time-management, and quality-improvement techniques. People at all levels are taught basic communication and computer skills. Cultural diversity and sexual harassment issues are reviewed with employees in an effort to improve awareness and sensitivity. Employees are sent to basic skills classes on reading, writing, and math so that they can function better in the work place environment. People at all levels are taught and empowered to perform specific tasks (Arbuckle, 1996). As the jobs and processes change they have to be taught again. Training in and of itself has become a big business.

As President Clinton and other Government officials have recognized, employer-sponsored training has become a critical part of the country's education system. Learning systems in the workplace are the first line of defense against economic and technical changes. The ability of the nation's employers and employees to respond quickly to these changes determines, in large part, the nation's adaptability and competitiveness. The employer's interest in employee education and training is functional. It revolves around the core concern that new information and skills be readily applicable to employee responsibilities in the workplace. Therefore, the employer's ultimate goal in providing workplace learning opportunities is to improve the company's competitive advantage. For the employee workplace learning is supported by a powerful motivator when learning experiences are based on actual job needs. Employees frequently work to increase their proficiency in the expectation that they will trigger immediate reward in terms of achievement, status and earnings (Carnevale, Gainer, and Villet, 1990).

The question is, however, are these goals being met under the current conditions? Opinions are divided on this issue. Schaffer and Thomson (1992) believe that many companies launch massive corporate-wide training programs in the pursuit of activities that sound good, look good, and allow managers to feel good, but add little or nothing to bottom-line performance. Training programs are introduced under the false assumption that if they carry out enough of the right improvement activities, actual performance improvements will inevitably materialize; however, years of effort and money can be wasted on activity-centered programs that only hope to strengthen fundamental corporate competitiveness. Eventually, if results do not materialize, cynicism grows in the organization and potentially useful programs will be discarded

These programs—many of which parade under the banner of "total quality" or "continuous improvement"—assume that results take care of themselves; however, since these training efforts are not focused on specific, measurable operational improvements, within a short period of time they often lose their momentum and die prematurely. Therefore, training like any other successful change program must be focused on demonstrating measurable

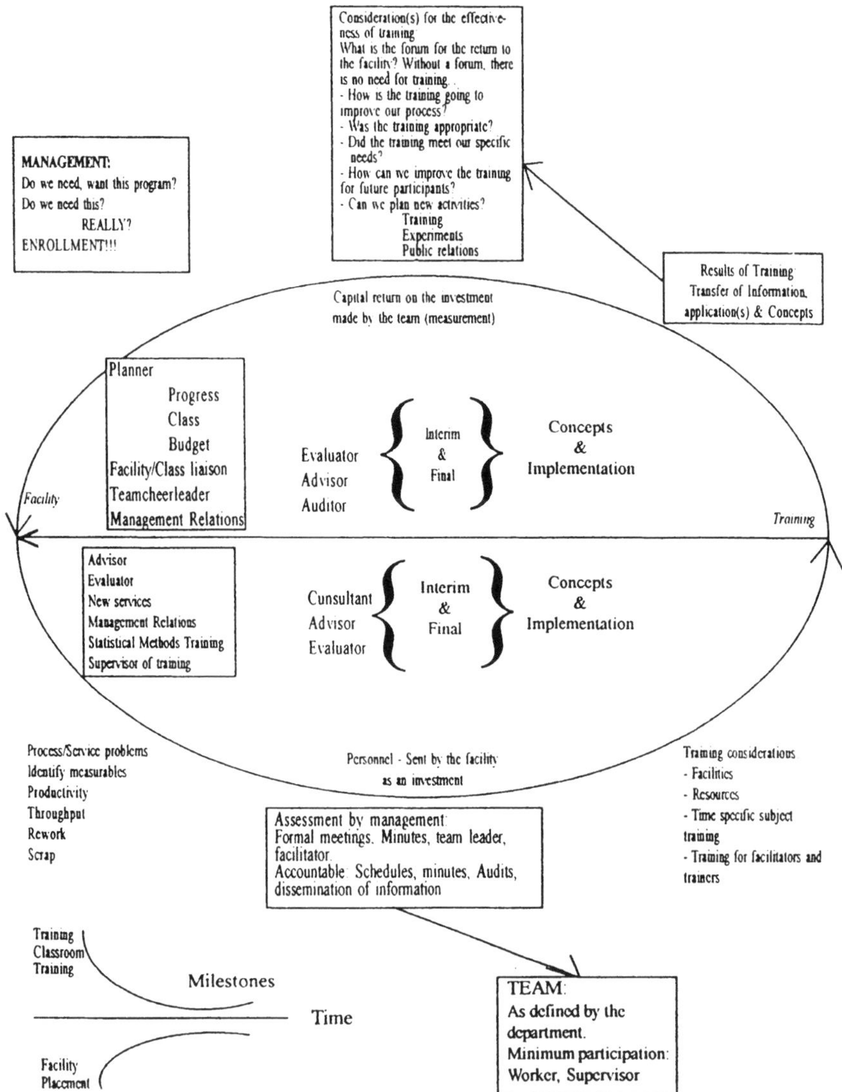

Figure 17.1 A training model that emphasizes results..

results. A typical model for training that emphasizes results is shown in Figure 17.1. This is rarely the case in corporate-mandated, broad-based, shotgun applied programs. The fundamental strength of successful employer-based on training methods is that they are applied. Applied learning works better than learning in traditional classroom formats because it delivers new knowledge in a context meaningful to the learner. New knowledge delivered in the context of work activities motivates learners and can be put into use immediately. Too often, however, employers do not use naturally applied learning methods, but rather transfer the deductive methods characteristic of the school into the workplace. Employers need to use an applied, hands-on approach to developing workplace curriculum. To the extent possible, learning should be embedded in work processes. But as we have seen from many survey findings over the years, most training dollars are spent on management and professionals, not on production and service delivery personnel. Because technical changes have their greatest impact on production and service delivery employees, these employees have the most to learn. Employers must be aware of the economic value these employees represent to the company and teach them the skills required for new and expanding role.

Success in the new global economy depends on the ability to meet the new competitive standards: to run faster in the productivity race while simultaneously delivering to consumers quality, variety, customization, convenience, and timeliness. By these standards, the United States is not the strongest competitor in many sectors of the global economy. Meeting the new competitive standards is no job for an unskilled and untrained workforce. Organizations whose employees learn the fastest how to meet the new competitive standards will have the greatest competitive advantage. But that means their workforce, from top to bottom, need to get back on the learning curve. Evidence already shows that training can leverage competitive advantage in the new economy. Companies that have been successful in delivering quality have invested significantly in training all their employees to deliver quality and to constantly improve it. Xerox, Ford, and Motorola are some of the most well-known examples. These companies use training, strategically and comprehensively, to help build their competitive advantage (Carnevale, 1990).

Already known for product quality, Motorola is ballooning its commitment to employee learning. Motorola's commitment to corporate learning is well documented. Symbolic of Motorola's commitment to workplace learning is Motorola University with its $120 million annual budget (approximately 3.6% of payroll). Headquartered in Schaumburg, Illinois and with 14 branches from Tokyo to Honolulu, it is the benchmark for industry training and a focal point for Motorola's learning program; however, Motorola's leaders fear that by the turn of the century rivals will be matching Motorola's 6σ defect reduction program. Quality will be a given!

In the business wars of the coming decade, they believe that the most important weapons will be responsiveness, adaptability, and creativity. To develop those skills, Motorola is gearing up a new program built around lifelong learning. Motorola plans to dramatically increase training of all employees, from the factory floor to the corner office. The goal is a workforce that is disciplined yet free thinking. The learning campaign will require huge resources. Motorola employees already spend at least 40 hours a year in training and it hopes to quadruple that, perhaps by the year 2000; that equals 160 hours per year per employee on training. Considering that a standard man-year is 2080 hours, that equates to 7.7% of the year or just about one full month (one standard month has 173 hours.) Expenditures at that level may cost Motorola $600 million/year, an amount of money that would go a long way in building a microchip factory (Kelly, 1994). This kind of spending does not happen without strong support and commitment from top management. Motorola's management does not invest such large sums of money in training without believing that they will get a substantial return on their investment.

A mid-1980 study (Kelly, 1994) showed that each training dollar delivered thirty dollars in productivity gains within three years. Motorola claims that problem-solving teams and other work groups have saved approximately four billion dollars. These numbers are indeed very impressive, but it must be recognized that they are soft numbers. After all, what savings might have been made anyway? What is important, though, is that Motorola's leadership believes there is a direct correlation between the dollars it invests in employee development and productivity improvement. Therefore, they are committed to training.

We all must understand that as knowledge becomes old—and in some cases antiquated—we have no choice but to spend on education and training. It is precisely this notion that drives an organization in becoming a learning organization and to focus on life learning for all employees. When that is done, we indeed have become an organization with a competitive advantage. In the case of Motorola and why it is discussed in a somewhat detailed fashion, it is because it binds education to business targets (as opposed to spending the money) so that they can claim that the company has an ongoing training program. For example, Motorola will set a goal to reduce product development cycle time, then create a training plan on how to do just that. At this point, this is not learning for its own sake, but learning with purpose. Learners (employees) are drilled in specific tasks until they get them right; this applies to production skills as well as administration skills.

Motorola's training program stands out and is effective, because is directly linked to the work and then applied to the job at hand. Even the support (soft) skills courses like problem solving are not just taught and then left in the classroom as at many other firms, but are incorporated into all meetings. Motorola does not just train its own employees. It learned early on in its 6σ program that it could not hit quality targets if it got poor quality parts from the supplier base. Training people from outside the organization can have other benefits in the areas of potential new employees and suppliers as well as educating current and potential customers.

For every Motorola success example there are many organizations that look to training as an after thought. Many companies buy new technology without thinking about the repercussions of that technology in their work environment. It is not uncommon to see organizations buying the state-of-the-art technology and at the same time to lose productivity because it causes their employees to feel incompetent, needy, and powerless (Bolman and Deal, 1991). Technology in and of itself does not innovate—people do; they are the key to successful implementation of new technology. If training does not accompany new technology, its success is in jeopardy (Fernberg, 1993).

Another example of training commitment from an organization's management is the story of Selectron of Milpitas, California. Selectron was founded in 1977 with 15 employees for the purpose of assembling printed circuit boards and other high-tech components. The spectacular success of Selectron has been attributed to the philosphy of Selectron's chairman, Winston Chen, that if you want to have high growth with high quality, then training is a big part of the equation. That philosophy has been instrumental in Selectron's employing over 3,500 people and an increase of revenues over 59%/year. In real terms, in 1992 each employee spent an average of 95 hours in training—all done during normal business hours—and in 1993 the figure rose to 110 hours per employee. Selectron uses its own employees and outsiders to teach and train courses ranging from failure mode and effect analysis, concurrent engineering, effective presentations, problem solving, and so on. The need for training for Selectron is a reality because as Chairman Chen says, technology changes so fast that we estimate that 20% of an engineer's knowledge becomes obsolete every year. Training is an obligation and we owe it to our employees (Henkoff, 1993).

Training is indeed an obligation for modern times and the future. Without it, companies will remain hobbled by hierarchical, check-your-brains-at-the-door work methods that make no sense in today's and future economies. It all boils down to if we are interested to build a high-tech, high-performance, flexible organization, then it is imperative that we must train our employees.

Since training is so important, why then do so many organizations not have active and effective programs? One possible explanation is that most of the nearly three million companies that do not have any training are very small and at a low level in Maslow's hierarchy of needs. To such companies making payroll is the major concern and a lot more important than providing training. Many companies view their process as so narrowly defined or low-tech that anyone off the street can walk in and take over without missing a beat. Training for them is not only a priority, but it is not even needed. Others see training as a program to train your competitors' future employees, and still others feel that their training performance is wonderful, when in fact, their training dollars are not as effectively used as they could be. Whatever the case, the realities of the new economy may force many of these companies to reassess their training requirements and/or training methods. One good assessment approach is to ask: What does the specific training add to the bottom line?

Success in the new economy will depend on the ability to meet new competitive standards. The rules of the game have changed. Things are much more complex. In the past, business competed principally on the basis of productivity and price. If you could produce at high volume and with the same or fewer resources than some other company, then your company thrived. On the other hand, in the new economy, competitiveness is based not only on productivity and price, but on a new set of market standards that have become the measure of competitive strength. The are: quality, variety, customization, convenience, and timeliness, and they are all driven by the customer.

A set of profound economic and social changes around the world have put the consumers in the driver's seat. Consumers are richer, smarter, and they demand products and services that exhibit quality, variety customization, and convenience when they want them, and at the lowest price possible. If a given organization is not willing to provide their product and/or service with these characteristics, someone else will.

How does American industry measure up to these new competitive trends? The American productivity rate is still the world standard. In fact, in some measures, it is 50% higher than Japanese productivity (United Kingdom Treasury, 1989). Canada, Germany, France, Norway, and Belgium will catch up and eventually pass the United States by the turn of the century and Japan will surpass American productivity standards by 2003 if present trends continue (Grayson and O'Dell, 1990).

The size of the American workforce will decline soon, especially at the entry level, causing more investment in machine and human capital. On the other hand, as changes are occuring world wide, quality by far leads the list of new competitive strategy. The standardized offerings of mass production have given way to an explosion of choices. Examples are everywhere and range from general manufacturing, automobiles, banking, textiles, and apparel to supermarkets (Womack, Jones, and Roos, 1991; Ingrassia and Patterson, 1989; Noyelle, 1989). Customization has become increasingly important in today's market. Consumers want more choices in the marketplace tailored to their individual needs. As people and machines grow more flexible, the ability to customize increases.

The power of customization was noted by Naser (1987), who observed that in the high-tech markets products that come to market on budget but six months late will earn 33% less profit over five years than products that come out on time but are 50% over budget. It is rare that a single employer or nation is the fastest in all product development races. In the United States, we are generally at the top in terms of getting the big idea, but we have trouble keeping up in the race to get the innovation to market. For example, in the automotive world, the Japanese renew their designs every four years, whereas American companies try to make basic design last up to ten years. Furthermore, American and European companies take roughly five months to move design to market, while Japanese take only 3.5 months (Womack, 1989). With all this said, the United States is not losing every race. In fact, it is the leader in quite a few industries and products. Table 17.1 shows some of the products that America makes best.

The impact of the new economy and quality requirements on the organization and the role of people at work will be dramatic. Capital-to-labor ratios will rise while direct labor requirements will fall as the new world economy takes place. People's responsibilities and skill requirements are becoming less job specific, job assignments are becoming more flexible and overlapping, and employees are spending more time integrating with each other and with customers. Flexible work teams and information networks will become the key to productivity and will also be the key in getting innovations off the drawing board and into consumers' hands quickly.

New technologies, especially those that are flexible and information based, help shape the new economy. The hands-on effort required to extract resources, manufacture products and deliver services is becoming less critical than the information required to respond to markets quickly. Computers integrate producers and consumers into economic networks. As we move into globalization, organizations will encounter numerous roadblocks to adapting flexible technology to worker's skills. New human and machine combinations are certain to challenge the ability of organizations to learn quickly. An unskilled and untrained workforce will not be able to meet the new competitive standards. Work in the new economy calls for a whole new set of skills and a range of knowledge that is both broader and deeper than currently required. Skills and knowledge requirements will depend upon the competitive position each company chooses to pursue; however, in all cases they will be more complex than the skills currently needed to survive.

To ignore the new trends of the globalization and the need for new skills that are necessary for survival would seem foolish; however, most companies are reluctant to admit that their workforce needs any education,

Table 17.1 Some Products that America Makes Best

Product	Company
All-electric plastics injection-molding machine	Cincinnati Milacron
Aluminum foil	Reynolds Metals
Atomic clock	Frequency Electronics, Hewlett-Packard
Ball point pens	A. T. Cross
Balloon and laser angioplastic catheters	C. R. Bard, Eli Lilly, Trimedyne
Bamboo fly-fishing rods	Walt Carpenter
Bed sheets and towels	Burlington Industries, Don River, Dundee Mills, Fieldcrest Cannon, J. P. Stevens, Spring Industries, West Point-Pepperell
Biotech drugs: t-PA	Genentech
Bobcat skid-steer loaders	Melroe
Boots and hunting shoes	L. L. Bean, Timberland
Brain electrical activity mapping system	Nicolet Instrument
Central office switching equipment	AT&T
Charcoal briquettes	Kingsford Products
Combines	Case IH, Deere
Computer operating systems software: MOS-DOS, UNIX, VM, VMS	Microsoft, AT&T, Digital Equipment
Cotton denim	Cone Mills
Cruising sailboats, 37 feet and under	Pacific Seacraft
Crystal	Steuben Glass
Data parallel supercomputers	Thinking Machines
Digital plotters	Hewlett-Packard
Electrodeposition primers	PPG Industries
F-16 jet fighters	General Dynamics
Fast food	McDonald's
Flashlights	Mag Instrument
Flutes	Wm. S. Haynes
FM two-way radios	Motorola
Hay and forage equipment	Ford New Holland
Heating controls	Honeywell
Instant camera films	Polaroid
Jazz music	—
Jeans	Levi Strauss
Jet aircraft: 747 family of planes	Boeing
Jet engines	General Electric
Magnetic resonance imaging scanners	General Electric
Offshore drilling equipment	Cameron Iron Works
Oscilloscopes	Tektronix
Pacemakers	Medtronic
Paper towels	Procter & Gamble, Scott Paper

much less training. In fact, the National Center on Education and the Economy (1990) reports that only 5% of American companies agree that there is skill gap in their workforce. Furthermore, the Massachusetts Institute of Technology Commission on Industrial Productivity repeatedly found managers who said, "We have no training problem here." (Carnevale, 1990). In response to that statement the commission concluded that

> There seems to be a systematic undervaluation in this country of how much difference it can make when people are well educated and when their skills are continuously developed and challenged. This underestimation of human resources becomes a self-fulfilling prophecy, for it translates into a pattern of training for work that turns out badly educated workers with skills that are narrow and hence rapid to obsolescence.

This type of thinking is a remnant of the mass production system, in which work was broken to its smallest task and it was relatively easy to learn. Furthermore, under that system it was easier and cheaper to replace a worker than to cultivate his or her skills. Workers were—and still are to most organizations—a cost to be controlled, not an asset to be developed.

With the new economy we are reaching a point where we need a workforce that knows more so that they can do more. Knowledge (brainpower) is rapidly becoming a critical resource, especially as the workforce shrinks in size. Training can make a demonstrable difference in the ability to meet the new competitive standards. In fact, learning on the job from 1929 through 1989 accounted for more than half the productivity increases in the United States (Denison, 1988; Bishop, 1989). Furthermore, in the same period learning was twice as important as technology in boosting productivity and twice as important as formal education.

Specific examples of appropriate and effective training have been reported in the literature many times and in many ways. The reader is encouraged to see Mercer (1989), Hage (1993), Johnson (1993), Fitzenz (1994), and Perelman (1994) for more information on the value of training in the work environment

II. THE ADULT LEARNER

When we talk about training in the work environment, we talk about adults (i.e., over 18 years old) that need to learn and/or improve specific tasks within a process of the organization. Therefore, to maximize the ability of transfer learning from the classroom/training facility/laboratory to the actual work environment, the trainer must use everything available and within the budget constraints that will maximize the learners' learning. To facilitate this, specific tools and approaches may be used, such as: whole brain thinking (Wonder and Donovan, 1984), left and right brain (Springer and Deutsch, 1981), motivation (Wlodkowski, 1985; Knox, 1986; Stamatis, 1986), and training design considerations (Cross, 1981; Brookfield, 1986; Richey, 1986).

A. Adult Learner Characteristics

When a training program is about to embark, there are four characteristics that are unique to adults and should be taken into consideration (Knowles, 1980; Kidd, 1973; Fox, 1981; Knox, 1977; Cross, 1981; Stamatis, 1986).

1. *The ability to learn.* Adults like youth, vary in their ability to accept new information and adopt new practices. These differences are related to a combination of social, psychological, and physiological factors.
2. *Different approaches to learning.* Adulthood may be characterized by an increasing differentiation rather than increasing similarity among individuals. Learning for adults has many functions. This mix of different individual learning for different purposes makes the approach to learning an important variable in the design of learning experiences for adults.
3. *Different ways adults process information.* Physiological and social factors have a great bearing on the ways adults accumulate, store, and retrieve information. The factors at work in the gathering, storing, and retrieving of information have important implications for the practice of adult education/training.
4. *Different ways adults think.* The way adults acquire content, their cognitive style and their orientation towards achievement have strong effects on their learning behavior and performance.

B. Why Adult Learners Fail to Learn and Participate in the Classroom Environment

Trainers frequently use class participation to stimulate students' interest in the subject and hold their attention. Participation also helps students develop communication skill, eases learning through cognitive and verbal involvement, explores controversial issues through an open forum, and provides a base for evaluating learners.

As important as participation is in the learning process, some adult learners avoid participation in class discussions. The following ideas as to why lack of participation exists may help the trainer to circumvent failure. Even though they are not based on a formal scientific research, these points represent the author's personal experience of over 20 years of training and consulting.

Environmental Conditions

Temperature, lighting, and humidity have a lot to do with the comfort of the learners. If the environmental factors are not appropriately set, lower participation and learning will occur. *Remedy*: make sure that all environmental conditions are appropriate.

Ethnic, Cultural, Linguistic, and Educational Differences

Foreign ethnic participants may be accustomed to different ways of teaching. They may have been taught not to challenge the instructor. Minority learners, on the other hand, may be intimidated because of their values and experiences, which may be different from the majority of the class. Linguistic and educational differences also play an important role in participation in the sense that people with low education and poor linguistic abilities are afraid of being humiliated by their slowness, cumbersome diction, and the like. *Remedy*: make sure that the trainer is sensitive to diversity and act accordingly.

Interpersonal Communication and Group Activities

Especially in the technical classes such as Statistical Process Control (SPC), Design of Experiments (DOE) and many more, the lack of appropriate and applicable prerequisites may intimidate the learners. Also, if the learners feel that there is no way that they may apply the new knowledge, they will withdraw. As a consequence, past experiences in communication or participative activities may contribute to the lack of participation. *Remedy*: make sure that the trainer summarizes prerequisites and encourages sharing of experiences.

Interest in the Subject

Learners who have no interest in the subject or have been forced (either directly or indirectly) to participate will make poor participants. *Remedy*: make sure that the trainer communicates to the learner what is in it for him or her.

Time of Day and Appropriate Sleep

If the training is scheduled for 8:00 AM to 4:00 PM for employees that work the midnight shift expect a very lethargic session, not to mention very little learning. *Remedy*: learn who the participants will be and schedule the training appropriately.

Arrangement in the Classroom

Traditional classroom style of learners sitting in rows of chairs all facing the instructor is a major inhibitor in adult learning. *Remedy*: arrange the room in a circle or a U-shape to facilitate eye contact and participation.

Type of Class Participation

Especially in technical training, learners will tend to participate in activities that are nonthreatening and comfortable. On the other hand, they will avoid activities in which they feel unqualified or threatened by, or that they will never use in their work environment. *Remedy*: establish early on the need for the particular training, the value of the participant's contribution to the goal of the training, and communicate to the learner once again the principle of what is in it for him or her.

Size, Length of Class, and Breaks

Large class sizes in the adult learning environment cause anxiety and inhibits participation. On the other hand, the length of the class meeting plays an important role for both the learning process and participation. From my experience, I have found that after about three hours of training including: whole brain, left-right brain techniques, and combination of lectures, case discussions, small group presentation, question and answer sessions, and video presentations, the adult learner becomes saturated by the subject. The effectiveness of participation decreases as the amount of training increases. *Remedy*: always provide a 7-minute break per hour, with a 3-minute bonus break every second hour. The break brings about a spontaneous recovery which is conducive to effective acquisition of new information and participation.

Age, Experience, Maturity, and Personality

Age, experience, maturity, and personality play a major role in participation. Generally speaking, young and inexperienced learners do not participate as effectively as the older, more experienced learners. Introverted learners are less likely to participate and extroverted ones can hardly wait to say their opinions. *Remedy*: as much as possible, mix the younger and older learners as well as the inexperienced with the experienced learners. Do not assume that your learners fit a certain type of personality unless you have been provided with the results of a personality test.

The Nature and Difficulty of the Training

In the field of quality there are some topics more conducive to participation than others. For example, topics in team building, consensus, corrective action, control charting, and others are more conducive to class participation and discussion. On the other hand, topics on finite element analysis, design of experiments, and many others stimulate less discussion and are more likely to be nonparticipatory. For the adult learner the level of the participation is dependent upon the inherent controversy of the topic and/or the difficulty of the topic. *Remedy*: especially in adult environments, people who are very conscious of being ridiculed or embarrassed do not ask questions or participate; therefore, the trainer should encourage participants to ask questions and more importantly assure them that there is no such thing as a stupid question.

Familiarity with Each Other

Adults who know all or some of the other participants are more likely to participate than those who do not know anyone. *Remedy*: Introductions are very important. Make sure that everyone is introduced, including yourself. A good way of introduction is to form teams of two, then allow five minutes for interviewing each other and one minute of general introduction to the entire group. If there are odd number of participants, you as a trainer may choose to be in a team. If there is a lot of participants (e.g., over 20), use the standard approach to introductions. If there is an even number of participants, you as a trainer may want to introduce yourself.

The Trainer

Both the aversive and the unskilled instructor will contribute to less learning and participation. In fact, both types discourage participation. Typical comments questions for the aversive trainer are: "We have covered this already," "That is in the text material," "Did you do the work sheets?" and "Are you kidding?" On the other hand, the unskilled trainer presents material without humor, has poor listening skills, does not summarize ideas, does not respond to learners, and so on. *Remedy*: make sure as a trainer you have both the ability to teach as well as know the subject matter. Knowing the subject matter only guarantees failure. From the author's experience an excellent trainer is the person who:

1. is enthusiastic about their work
2. sets challenging performance goals for themselves and the learners
3. is committed to what they are doing
4. projects confidence and a positive attitude about the learners' ability to learn
5. always display professionalism
6. treats learners with respect
7. encourages two-way communication and is available for consultation beyond the official training session
8. listens attentively to what learners say
9. is responsive to learners needs
10. gives corrective feedback promptly to learners
11. presents ideas and concepts clearly
12. has excellent communication and presentation skills
13. respects diverse talents
14. creates a climate conducive to learning
15. works collaboratively with management staff
16. is knowledgeable about their work and topic of presentation

17. integrates work examples in the topic of discussion
18. provides an appropriate perspective, including an appreciation of diverse views
19. is well prepared
20. is knowledgeable about adult learning theory
21. provides alternative ways of learning and uses appropriate language
22. stimulates intellectual curiosity
23. encourages independent thinking
24. encourages learners to be analytical listeners
25. always gives consideration to feedback from all learners
26. encourages learners to ask challenging questions
27. treats all questions with dignity and respect

In essence, trainers must have the ability to translate the needs of the learner in the most efficient way. Furthermore, trainers must themselves have a good foundation of the relationship between theory and practice of whatever they are teaching. This notion of theory and practice connection is an old one, as both Plato and Aristotle argued about it (Aristotle, 1975; Irwin, 1995), however, even today we still debate the issue of knowledge as education and knowledge as training. In fact, many organizations confuse this very elementary notion and as a consequence, even though they may think they are getting training, in essence they are getting education.

Another important characteristic of the trainer is the ability to develop a wider perception-based type of knowledge for the learner. This means that the trainer must focus first and foremost on concrete situations to be perceived, experiences to be had, persons to be met, plans to be exerted, and their consequences to be reflected upon; these are the foundation of training. Without such perceptions, no knowledge is formed at all and the expected transfer of that knowledge will certainly not occur. The knowledge that a learner needs (at least in the industrial training environment) is perceptual instead of conceptual, as learning depends primarily on the learners intrinsic motivation (Stamatis, 1986). Therefore, it is the trainer who with his knowledge and experience will create the applicable and appropriate climate for learning to occur. The reader is encouraged to see Kessels and Korthagen (1996), Jonsen and Toulmin (1988), and Nussbaum (1986) for more information on the relationship of theory and practice and applications to the trainer from an educational perspective.

III. SYSTEMS APPROACH

It seems that everyone involved with quality and training keeps talking about systems. But what is a system and how does it relate to quality and training?

The system concept is a set of attitudes and a frame of mind rather than a definite and explicit theory. The application of the concept involves the use of logic, intelligence, and creativity. It is increasingly becoming an important tool for reaching solutions to quality problems. Its major advantage is that it minimizes random trial-and-error methods which are expensive in time, money, and resources. A systems approach forces the recognition of the whole as being distinct from the subsystem elements. It insists that the problem be reviewed in its entirety by taking into account all facts (variables, constants, and so on) and relating them to the whole and to each other (Novosad, 1982).

Systems theory application is still considered an art; however, the mix of human intuition, mathematical precision, computer technology, and the human experience(s) has brought the systems concept to a high plateau of rationality. All systems have common procedural aspects.

- The system concept demands the prior accounting of uncertainty in all strategic decisions.
- The system approach has the feature of being reproducible.
- The systems thinking approach to problem resolution recognizes the situation of avoidance (i.e., when it is appropriate to avoid a particular issue) because of extreme complexity or simplicity.
- Most system thinking approaches are used to generate appropriate and applicable questions.

Perhaps, the most important attribute of the system concept is the fact that it can be used by both a single individual and a team. Its strength is the fact that it can radically challenge the status quo, regardless of its longevity in the work environment. In the quality field and more so in the training environment, problem solving must be accomplished in a systematic manner. We can no longer afford the "first way that works" or "that is the way we have always done it" solutions to our present problem(s). We must revolutionize our thinking process, change our paradigms, and aim for the whole as opposed to suboptimization.

IV. BASIC SYSTEM CHARACTERISTICS

A system may be defined as an arrangement of separate and independent components sharing a relationship for the purpose of attaining some common goal and predetermined objectives. A system can involve people, information, and/or things. At its simplest, a system can be described as a rational and orderly combination or arrangement of such things as objectives, facts, or elements. Two of the most fundamental characteristics of the system are synergy and the systems approach.

A. Synergy

Synergy is the term used to identify the behavior of whole systems unpredicted by the behavior of their parts taken separately. The catalyst for this synergy is coordination. For example, in the formation of teams and self--directed work teams (SDWT) we expect synergy through the coordination of effort on the part of the participants.

B. Systems Approach

A systems approach is a holistic way of thinking that attempts to study the total or synergistic performance of a system before concentrating on the individual parts. This mode of operation recognizes that even if each element or subsystem is optimized individually from a design or functional viewpoint, the total systems performance may be suboptimal owing to the interaction between the parts. Examples of using the system method in quality and more specifically in the training domain are everywhere. Two specific ones, however, are in the implementation process of TQM and the training of management and nonmanagement personnel for specific concepts and tasks.

V. KNOWLEDGE OF A SYSTEM IN TRAINING

Guiding an organization effectively toward continual improvement depends on the organization leaders' developing, basing their leadership on, and communicating to everyone knowledge of the organization as a system of production (i.e., a group of interdependent people, items, processes, and products and services that have a common purpose or aim; Dick and Carey, 1978; Batalden and Stoltz, 1993).

To understand the work of the organization as a system of production and to train for positive results, management leaders and trainers must be able to answer (and help others to answer) three basic questions.

A. Why Do We Make What We Make?

To answer the question as to the aim of the system, requires developing and deepening customer knowledge as well as understanding the social need of the organization.

B. How Do We Make What We Make?

Answering the question as to the means of production requires building knowledge of what is actually created, made, or produced (services/products), how services/products are produced (processes), for whom they are produced (beneficiaries or customers), and with dependence upon whom (suppliers). The reader will notice that in service industries, workers are accustomed to describing what they do everyday or the effects of their actions, not what they make.

C. How Do We Improve What We Make?

This questions requires developing knowledge of what must be done to improve toward a shared vision for the future (plan to improve) and formulating specific plans (design or redesign) based on these improvement priorities. Changes that will improve the overall system are those that increase the system's capacity to deliver services and products that meet the needs and expectations of the customers it seeks to serve.

For the purpose of continual improvement, understanding the interrelationship of the essential elements in the system is critical. Therefore, making wise decisions about what and how to improve requires linking knowledge of the various elements in a system (Ackoff, 1984; Mitroff, 1988; Senge, 1990; Deming, 1986). For example, key processes in an automotive assembly line are those required to produce information, diagnostic services, and the assembly environment. On the other hand, processes such as those relating to transportation, inventory, and quality records contribute to the key processes at various points and may be referred as supportive processes. It is precisely this knowledge of the entire system that will guide the trainers to provide applicable, appropriate, efficient and effective training. Without a systems approach, the results may not produce the desired results.

VI. INSTRUCTIONAL DESIGN

As of 1994, the official academic definition of instructional technology is: "The theory and practice of design, development, utilization, management and evaluation of processes and resources for learning" (Seels and Richey, 1994 p. 1). Instructional design (ID) is the science of understanding and designing appropriate training. The essence of instructional design in the world of industrial training is to use sound adult learning theory so that the learner becomes capable to transfer knowledge and/or skills from the learning environment to the work environment in the most efficient way. It is beyond the scope of this book to elaborate on the details of both theory and practice as to how ID should and/or is used in the training environment; however, here we summarize four of the basic ideas of ID. The reader is encouraged to see Seels and Richey (1994), Briggs (1981), Gagne and Briggs (1979), Reigeluth (1983, 1987), Richey (1986), and Travers (1982) for further understanding of ID principles and the essentials of learning.

A. Function

The training is focused on the tasks, individual steps of performance, and specific performance objectives. The contribution of ID is to find out and apply the most appropriate and applicable technology, and the best learning processes to facilitate both understanding and maximum utilization of these new learned tasks in the organization. This is where the analysis of needs, priorities, and resources is conducted as well as the determination of scope and sequence of curriculum and/or specific delivery of courses.

B. Inductive Learning

The training is focused on discovery based on some kind of investigation. The contribution of ID is to make sure that the trainer is aware of the learners' characteristics and apply appropriate and applicable methodology and techniques to plan, deliver, monitor, manage, and debrief the learning. This is where the determination of structure and sequence is defined as well as the analysis of objectives.

C. Gestalt Approach

The training is focused on a holistic approach, namely, the specific training is related to the whole. The contribution of ID is to find out the most expedient way to relate this particular training to other training and/or to make the connection of relationship of this particular training to the whole organization. This is where the actual preparation takes place, such as the definition of performance objectives, developing the actual material, and selection of media.

D. Define Goals

The training is focused on attainable, realistic and acceptable goals. Furthermore these objectives must be measurable and repeatable. The contribution of ID is to find out the appropriate and applicable measurable (objective) goals and measure their effectiveness. This is where both formative and summative evaluation may take place. Furthermore, the strategy for installation and diffusion of the new learning in the work place will be defined and planned for optimum results.

A typical ID approach to training is shown in Figure 17.2. Obviously, depending on the needs and expectations of the client/customer, all the steps identified in the model may not apply in all situations.

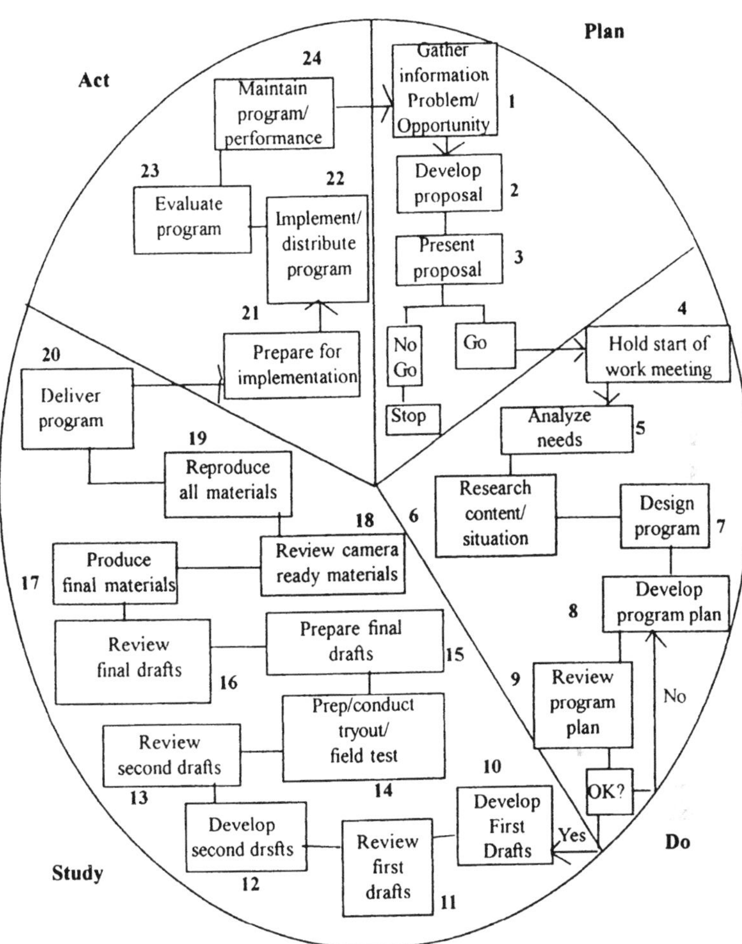

Figure 17.2 Instructional design model. Courtesy of Contemporary Consultants Company, Southgate, MI.

VII. MASTERY LEARNING

Mastery learning is an optimistic theory of human learning that assumes that all learners can achieve mastery of the content that the training has to teach. Another way of putting it is that mastery learning is based on the notion of managing learning rather than managing learners. While research and the testimony of its proponents indicate that mastery learning can work under some conditions, it is not a method to be undertaken without a lot of thought and preparation. We do not know whether mastery learning works with all subjects in all environments. We do know that if it is going to work it requires a dedicated trainer and a committed manager who believe in the concept and who are willing to work at perfecting classroom instruction; furthermore, long-range planning is essential.

Mastery learning is not as simple as it is made to seem by some of its advocates, but it does offer some exciting ideas about organization of instruction that could result in some positive consequences in quality training across all types of organizations.

The mastery learning concept should cause us to examine critically our present notions about how training and learning can best be accomplished, what is the most important content the training development has to teach, and whether everyone would learn the same material to the same level of competence.

Mastery learning has the potential for making a tremendous difference in the training and learning that occurs in our business world and therefore it has the potential for making a difference in the productivity of those businesses as the retraining process becomes effective and more cost conscious.

A. Situation

Imagine a training plan that promises that 95% of the learners in training can learn nearly everything the program has to teach. Imagine an ID that assures that 80% of learners can achieve scores as high as the top 20% now achieve.

The theory suggests that training can provide both equality of education opportunity and also equality of educational outcome. In addition, mastery learning addresses present concerns about basic education as well as current pressures for accountability and minimum competency testing. Advocates further suggest that compensatory learning programs for learners from disadvantaged backgrounds may successfully incorporate the principles of mastery learning as one answer to the present corporate disenchantment with these programs. There is ample evidence that mastery learning works and has the potential for making a difference in training programs.

B. Definition of Mastery Learning

Teaching and learning are natural human phenomena. This is especially true in the parent–child relationship. Some of the principles of mastery learning are central to all teaching and to all learning. A concerned and loving parent teaching a small child to tie his shoe demonstrates many techniques of mastery learning. The parent approaches the task sensitively and systematically, providing guidance, feedback, and correction when and where needed and the parent provides enough time and practice for the child to master the task.

Mastery learning techniques are used in the teaching and learning of psychomotor skills such as typing, driving a car, or flying an airplane, where each skill must be mastered sequentially. For example, an airplane pilot who has mastered all of the skills except one, landing, could not be judged as having achieved final mastery.

There is not one clear definition of mastery learning that encompasses all theories and practices which are called mastery learning by their creators; however, mastery learning may be broadly defined as the attainment of adequate levels of performance on tests that measure specific learning tasks. Mastery learning also describes an instructional model whose underlying assumption is that nearly every learner can learn everything in the training at a specified level of competence given the learner's previous knowledge and attitudes about the subject are accounted for, if the instruction is of good quality and if adequate time on the task is allowed to permit mastery.

The mastery learning model requires concise, testable objectives that clearly describe the criterion for mastery and an accurate preassessment of the learner's knowledge of the task to be undertaken. Methods of instruction usually consist of some large group, some small group, and some one-to-one teaching (including peer tutoring). Various combinations of computer-assisted instruction, programmed instruction, games worksheets, and other activities are components of every mastery learning model. The assessment is followed by prescription for further learning, which provides for progression to new learning tasks for remediation. Enrichment materials are prescribed for learners who finish the tasks ahead of others. A postassessment that measures individual outcome, previously identified in the objectives, is the final stage of the mastery learning process.

C. Historical Background

Awareness of mastery learning, although the use of the term was not used as we know it today, can be found in the writings of Comenius in the seventeenth century and Pestalozzi and Herbart in the eighteenth century. In the early twentieth century, two pioneers in curriculum development, J. Franklin Bobbitt and W. W. Charters, described many of the concepts basic to mastery learning. In 1918 Bobbitt spelled out how to identify major curriculum objectives and how to plan appropriate learning activities as the basis for teaching the activities arranged in hierarchial sequences. Other early pioneers in mastery learning include Carleton Washburne who developed the Winnetka Plan in the 1920s, and Henry C. Morrison who designed a program for the University of Chicago Laboratory School in 1926 (Block, 1971; Block and Anderson, 1975;

Torshem, 1977). These forerunners shared many features of present models of mastery learning:

1. They described in terms of particular educational objectives what each learner was expected to accomplish.
2. They were composed of well-defined learning units.
3. The learning materials were systematically arranged.
4. Mastery of lesser tasks was required before the learners were allowed to move on to more sophisticated tasks.
5. There was a discernable sequence in the materials so that learning was built on previous learning.
6. Ungraded diagnostic progress tests were used as an integral part of the process.
7. Frequent and regular feedback was provided.
8. The instructional task was supplemented with corrective learning material throughout.

The ideas inherent in mastery learning went out of fashion during the 1930s but resurfaced in the 1950s and 1960s as corollaries of programmed instruction and computer-assisted instruction. Some principles of mastery learning found renewed acceptance through John Goodlad and Robert Anderson's work with the nongrade elementary school. Research by Robert Gagne and N. E. Paradise suggested that learning tasks can be sequenced so that mastery of each task is a prerequisite for undertaking other more difficult work. Ralph Tyler's insistence that the program be organized around clearly defined educational objectives provided further direction for teaching for mastery. Keller's "Personalized System of Instruction," which was strongly influenced by behavioristic psychology, also added impetus to a revival of interest in the principles of mastery learning.

The modern notion of mastery learning was born with a model developed by Carroll (1963). Carroll's theory, later transformed into an effective working model by Bloom (1976).

D. Assumptions About Mastery Learning

There are several assumptions about mastery learning that suggest that, even though training has not always done well for every learner, the potential is there for improvement. There is a way for training to assure that all learners learn nearly everything the training has to teach. The following assumptions are encompassed nearly in every mastery learning program.

1. *Nearly all learners can learn equally or nearly equally most standard training tasks.* Quality learning is possible for virtually all learners if the training approaches its task sensitively and systematically, taking into account the present level of each student and providing appropriate help when

and where needed. Very few learning tasks require intellectual capabilities that are beyond those most human beings possess. What one can learn, most others can learn if the quality of instruction is adequate.

2. *Learning, not teaching, should be the primary interest of the training.* Mastery learning assumes that the instruction, not the learner, must be modified and adapted. This may require a more flexible approach to instruction and the way learners are viewed. Under a mastery learning program, training that has traditionally used a selection and classification system based on ability will have to move toward helping all learners develop intellectually. No learner will be assigned to a group or task which is less respectable than any other learners, since mastery of the total training program is assumed to be a possibility and a right for all learners. Nonacademic tasks and low-achieving groups will no longer be acceptable or necessary. Individual differences will be accommodated as all learners achieve mastery under an instructional program where effective trainers approach their learning sensitively and systematically.

3. *The goal of training department is to assure equality of educational outcome as well as equality of educational opportunity.* Few would deny the ideal of equality of educational opportunity. But the notion of equality of educational outcome is contrary to the accepted belief that some learners will excel, others will succeed at an acceptable level, and still others will fail. Mastery learning assumes that if appropriate training has been provided, the difference in individual achievement reaches the vanishing point. All learners become more alike in learning outcomes, namely, that all achieve mastery of the content.

4. *Individual differences do not determine the amount of the training program learned.* Proponents of mastery learning do not deny the existence of intellectual differences. They simply believe that those differences need not set limits on how much of the training program can be learned. Mastery learning advocates believe that differences in achievement result from poor environmental conditions more often than from low intelligence. Furthermore, they believe that learning deficiencies can be remedied without excessive financial expenditure, which is a concern of many corporate managers.

5. *Enough time on the task be allowed to assure mastery.* Time on the task is an important aspect of mastery learning. To achieve mastery the learner must be given enough time to work on the task and enough practice to reinforce the learning. While some learners will require more time for mastery than others, the differences will not be great. With a relatively small increase in instructional time and effort (most researchers suggest 10–20%) nearly all learners can achieve mastery (Harvey and Lowell, 1977). Some programs provide for additional time on task in after-training sessions through peer tutoring, taped instruction by the trainer, work sheets, and programmed instruction.

6. *Most training can be specified in terms of observable and measurable performance.* Designing instructional objectives are specific statements of expected student outcomes. The final component of a mastery model is a test or a series of tests to measure outcome in terms of mastery of skills, concepts, and facts identified in the objectives. If a student fails to meet the objectives at any point in the learning sequence, the learner is provided with remedial or corrective instruction to help achieve mastery. Therefore, identifying essential objectives and measuring how well those objective have been reached is an integral part of the mastery learning process.

7. *Learning is sequential and logical.* Almost every learning task has a base in some prior learning and will have consequences for future learning. Mastery models provide opportunities for logical and sequential learning experiences to facilitate catching up and making sure that views and objectives have been mastered.

8. *Learner motivation is the key to increasing the quality and quantity of training.* Bloom (1976) demonstrated that the extent to which a learner possesses the prerequisites for the learning task accounts for 50% of the variance relevant in achieving on subsequent learning tasks. Another 25% of the variance is attributed to the learner's entry behaviors or the extent to which the student can be motivated to engage in further learning. A learner who has experienced failure in previous learning attempts is not likely to be motivated for further learning, has low self-concept and diminished faith in his or her ability to learn. Mastery learning, when it is handled properly, assures that the learner experiences success at each level of the instructional process. Prior learning is assessed and the student is introduced to the task at a level commensurate with his or her cognitive entry skill. When problems arise, immediate and appropriate help is given as well as time to develop success on the task. Experiencing success at each level is motivation for further learning.

E. Components of an Effective Mastery Learning Model

While the general prerequisites for success with any educational method are knowledgeable and talented trainers, motivated and diligent learners, and appropriate materials, mastery learning has six specific components which are both similar to and different from other methods. Each component is crucial to organizing and teaching for mastery and assuring that all students be given the maximum opportunity to achieve mastery regardless of the subject matter.

Objectives

Objectives are the first component of every mastery learning model. They are clearly stated, specific statements of outcome and goals expected at the completion of each learning task. They must identify specific skills to be gained, key concepts to be understood, and significant facts to be learned, and they must be clearly presented at the onset of the learning process. The criterion for mastery is stated in the objectives. Designing concise, specific, and measurable objectives for each learning task is one of the most difficult undertakings in the entire mastery learning approach. All other components depend on the identification of clearly stated objectives.

Preassessment of the Learner

The preassessment test determines each learners present level of learning so that he or she may be correctly placed in the learning sequence with learning materials that are likely to be most effective. While tests of cognitive entry skills are available and easy to administer, this is not the case with affective entry behaviors. Valid tests for measuring motivation and attitudes are not so readily available and are often more difficult to administer and interpret; however, both the cognitive and affective levels should be measured whenever possible. Preassessment may also include analysis of the learner's previous work, interviews, performance demonstrations, and other information.

Instruction

The instructional component of the mastery learning model is at the heart of the process and is of most immediate interest to trainers. In a well-designed and managed mastery learning model, the goal is to have each learner work at his or her own pace and in continuous progress toward mastery of the learning objectives. Many instructional strategies may be used. The important question is: "Does this strategy help the learner build on his present level of achievement and proceed toward mastery of the objectives?" The trainer's professional judgement usually determines the strategy, but often the learn can offer suggestions about how he or she learns best. Worksheets, individual assignments, and audiovisual materials can all be used successfully in addition to large and small group instruction and peer tutoring. Taped instructions and short lectures help learners progress at their own rate. Whatever method or combination of methods is used, it is extremely important for the trainer to provide regular, frequent, and specific reinforcement to the learner.

Diagnostic Assessment

Diagnostic assessment is used to see how will the instructional component is working. This assessment is conducted during instruction to determine if the learner is progressing as he or she is expected and to assure that no key skills, concepts, or facts are missed or only partly learned. It is used to pace learners and to adapt the instruction as needed.

Prescription

Diagnostic assessment provides information that allows the teacher to prescribe future tasks. The prescription may include specific remediation strategies selected from a broad range of likely alternatives. For example, relocation may be prescribed for a learner if the diagnostic assessment shows inappropriate placement in the learning sequence. The prescription also may provide verification that the learner is making adequate progress or that enrichment material or peer tutoring may be an appropriate strategy. Prescription begins early in the instructional process and continues until nearly all learners achieve mastery.

Postassessment

Postassessment is the final component of the model. It is at this level that the trainer determines what extent the learning objectives have been achieved. Each learner is measured to ascertain whether the crucial skills, concepts, and facts identified in the objective have been mastered. Postassessment results can be used to refine the statement of objectives, to improve the instructional component, and to clarify the criterion for mastery.

F. Implications for Training Practice

Many trainers upon taking a cursory look at mastery learning, see it as just good teaching. Good training in any form demands that:

1. The learner's prior knowledge of and feelings about the subject matter be considered.
2. The trainer has a clear and concise notion of what objectives are to be met and to what level of competency.
3. Frequent reinforcement and monitoring be provided.
4. Assessment determines if the instructional goals have been reached.
5. Remediation and correction be available for those students who need extra help.

Mastery learning does include all of the principles of good training and learning, but it goes significantly beyond the usual definition of good instruction. Mastery learning differs from the traditional view of good teaching in several crucial ways. The most recognizable difference is in the level of specificity and precision required in the design of mastery learning programs. The emphasis of the instruction is on the attainment of clearly defined goals. The level of sophistication necessary to measure the degree to which the goals have been accomplished far exceeds that required in other instruction approaches.

Another important difference is that the planning for mastery learning must be accomplished much earlier in the instructional process than in conventional training. In most cases, all planning and preparation of tests, worksheets, corrective material, and training strategies are completed well before instruction begins. Multiyear goals are a part of the program: Instead of planning from day to day or week to week, as is often the case with conventional training, mastery learning requires that the trainer prepare for at least one complete learning unit and usually for a full program work of more.

Content is laid out in a logical and sequential way. Each student is tested to determine placement in the learning sequence. Help and encouragement are provided along the way. The characteristics of individual children are not viewed as limiting how much or to what level of learning can progress. The emphasis is on adapting the instruction to make it possible for all learners to achieve mastery.

One of the most significant differences between mastery learning and traditional instruction is not in design or plan but in attitude toward human potential which is optimistic and generous. Mastery learning requires that the trainer believes and behaves as if all learners are capable of reaching mastery.

Another difference between mastery learning and other approaches is that mastery learning requires a greater amount of corporate support and management commitment. While it is possible for one trainer to plan some mastery learning units, it is more successful if the approach is used corporate-wide or where several trainers work and plan together and pool resources.

Implementation of mastery learning models will make a decided difference in the lives of trainers. They will be able to see nearly all learners master content that was previously reserved for mastery by only the top students. The training profession will gain more respect from the corporate management as programs are able to fulfill successfully their function of training all learners. New and exciting interactions between trainers will result as information is exchanged and materials are shared. Trainers will undoubtedly work harder, for mastery is not an easy method to implement. For a very detailed training model using the Plan, Do, Study, Act approach, see Figure 17.1.

The model in Figure 17.1 is based on two notions: first, training must be on a just-as-needed basis and second, the training has to be related to the work at hand. The reader will notice that the model is a circular one, so that mastery of skill(s) can be accomplished, since the communication link is indeed open between the training and work environment.

While it is possible to teach all subjects through the mastery learning approach, some subjects lend themselves more readily to the specificity and precision required. This does not mean that subjects other than the basics should be taught by the mastery method. It does mean that teachers will have to work harder to use the mastery model with some content.

Implications for the use of mastery learning in training fits logically into three categories: planning for mastery, training for mastery, and management for mastery.

Planning for Mastery

Since preservice training for most trainers typically has not prepared them for mastery learning, they will have to study, read, attend workshops, and visit schools where mastery models are in operation. In addition to learning about the mastery model, trainers will need to examine their personal attitudes about the method. Mastery learning will require more work, especially in the initial stages, so a trainer who is not committed to the philosophy of mastery learning should not attempt it. The trainer must believe that nearly all learners have the capacity to achieve mastery of all the programs that the training development offer, providing that the instruction is appropriate and enough time is allowed. Only then is the trainer ready to begin using mastery learning.

Once the trainer has an understanding of the process and a commitment to its goals, planning may begin. The first task is to examine carefully the content to be taught in order to refine objectives. The objectives need to be stated in concise, behavioral terms that can be accurately measured. The trainer must then identify the prerequisite component skills that are necessary to achieve the objectives and develop tests to measure mastery of these skills. The criterion for mastery should be established at this stage of planning.

Next, learning units of approximately 2–10 hours must be planned and lesson plans developed for each separate element of the units. Diagnostic tests, corrective and remedial steps for each unit, and enrichment activities for each unit are an integral part of the planning.

Much of the work of mastery learning occurs before the actual training begins. Once the program is under way, and as trainers gain experience and share material and ideas, the work load is reduced.

Training for Mastery

After detailed planning, the next phase is training for mastery. Since most learners will be unfamiliar with the mastery learning concept, a thorough orientation is essential. Since many learners have come to believe that evaluation in education is competition for good grades or position in the class, trainers need to help them understand that the purpose of evaluating education is diagnostic and not competitive. An explanation of how evaluations will be determined is important during orientation, because more training demands that trainers assign some form of acceptable evaluation result.

After a thorough orientation, teaching the elements of the learning units begins. The following sound teaching principles direct the instructional process:

1. Allow the learner adequate time to practice each new skill.
2. Provide frequent, regular, and direct reinforcement.
3. Give learners cues to help them select appropriate responses.
4. See that all learners participate actively in learning tasks.

5. Furnish direct instruction in the learning task.
6. Monitor each learner's work carefully and often.

The trainer needs to administer diagnostic tests at appropriate intervals to determine how each student in progressing. Learners who do not master the work can be assigned remedial and corrective measures until the content is mastered. Learners who master the task can be assigned enrichment materials or work with peer tutoring.

Managing for Mastery

Managing the mastery learning process is an ongoing operation. Since the mastery approach to learning is different from traditional training in so many ways, and since trainers will have a lot of planning and preparation, especially during the early stages, a small start is best. Mastery learning can be initiated by starting with one concept at a time, then moving on to other areas as the trainer and learners feel ready. Since the basic skills areas lend themselves more readily to the mastery approach, it is wise to start in one of these areas before moving to other content.

As mastery learning proceeds, the trainer is responsible for accurate record keeping on the level and progress of each learner. The progress must be reported to the learners at regular and frequent intervals. The trainers must keep the learners motivated and on track toward achievement of goals.

One of the most important management tasks is to provide a classroom environment that is conducive to mastery learning. The environment should emphasize a respect for the training work being done and a need for continued improvement; this is one that is supportive and nurturing, but also one that is businesslike and task oriented.

G. Concerns and Cautions

The evidence from mastery learning indicates that most learners can learn everything the training has to teach and they can learn it at a mastery level with relatively little additional expenditure of instructional effort. Mastery learning seems to fit well with current concerns about training in general. What then is the source of discomfort with it and why has it not been more widely used? Are there problems with this approach about which we need be aware?

The Need for Highly Specified Instructional Goals

For mastery learning to succeed requires specifically stated instructional goals. While most trainers can agree on broad training goals, it is more difficult to agree on specific goals. Furthermore, many trainers have not had the training or experience to tailor their instruction to specific goals required by mastery learning.

Disagreement Over the Concept of Equality

The proponents of mastery learning interpret equality as meaning that learning attain mastery of the same competencies. Opponents argue that equality is the opportunity to develop in different directions according to one's abilities and interests, but not necessarily to achieve the same results. Opponents are concerned that mastery learning with its emphasis on achieving specific instructional goals may be done at the expense of other training areas: for example, achieving competency in shop math or blueprint reading by reallocating time that could be used on management development or utilizing management by objectives (MBO). Until philosophical differences are resolved, mastering learning is not likely to receive wide support from the entire corporate community.

Scarcity of Sophisticated Diagnostic and Assessment Tools

For mastery learning to succeed, more and better instruments for diagnosing learner problems and assessing gains must be readily available to trainers. This is simply not the case at present. Those sophisticated evaluative tools that are available tend to be used by psychologists and researchers who are trained in their use and interpretation. Classroom trainers need diagnostic tools that can be used without the help of specialized personnel.

Lack of Corrective Instruction

A vital component of mastery learning is effective corrective instruction at each step of the way, so that all learners are kept on the path to mastery. If this component is lacking, the entire process fails. At present, we have neither the resources nor well-defined instructional modes to assure that mastery learning will work; however, we are more sophisticated in providing the kind of corrective help needed in basic skill areas than we are in such areas as teaching learners to think creatively or to engage in decision making.

Concern for Trainer Time and Energy

Research suggests and proponents state that a 95% mastery rate can be achieved with as little as 10–20% increase in instructional effort. Research also suggests that even though some learning requires more time for mastery than others, the time difference need not be great; however, many trainers perceive themselves to be working at full capacity now and to increase instructional time and effort only 10–20% seems overwhelming, particularly with the initial effort needed to write specific goals, design appropriate evaluative tools, and plan instructional strategies. Unless a trainer is dedicated to the concept of mastery learning, the enormity of the task is likely to hinder its widespread adoption.

Difficulty of Defining Training for Mastery

Does mastery learning suggest a closed training program? Is a learning ceiling established? After a learner has achieved mastery, then what? What, if anything, is beyond mastery? Does the trainer provide additional content for mastery? If so, are 95% of the learners expected to master this enrichment material, too? What will the training be? Will there be an upper or closed limit to what one can be expected to learn at any level? The sticky problem of definition of training is an inherent concern in working with mastery learning, because before we can intelligently talk about mastering a particular training program, we must come to grips with what is to be mastered. Until some of these questions are resolved, mastery learning is likely to stay outside the mainstream of American training camps.

Problems with Time and Content Variables

In nearly all training, time is a fixed variable while the amount of content mastered is a flexible variable (the amount of content mastered depends, to a large extent, upon what each student is able to learn within a fixed time span.) Mastery learning requires flexible time allotments in order to assure a fixed mastery of content—most of the learners are expected to achieve mastery, although at varying rates. While this idea is uncommonly appealing in terms of what we know about human learning, we have yet to invent any practical means for implementing it in the real world of day-to-day training planning.

VIII. COMPUTER-BASED TRAINING (CBT)

With the proliferation of computers, training and educational courses are available for commercial use. It is possible to have training for computer-aided design (CAD), finite element analysis (FEA), statistical process control (SPC), and many other programs through a computer. Here we focus on some of the ways that computer training may help the quality professional, especially from the mastery learning perspective.

IX. VIRTUAL REALITY

Virtual reality (VR) is a technology that allows the user to interface with a computer in three dimensions as if he or she is inside the computer rather than interfacing with the computer by viewing a two-dimensional screen. The user of virtual reality feels as if he or she is on the inside looking around, rather than the feeling of being on the outside trying to look in. In essence, Virtual reality is a human-to-computer interface that involves real time interaction between the user and a computer-based simulation model. The interactions involve the five human senses: vision, hearing, touch, smell, and taste. The virtual reality interface can provide sensory stimulus in all of the five human senses to try and induce the user into believing that the virtual images presented to him or her are in fact reality.

Virtual reality was invented in 1962 by Mr. Morton Heilig when he was issued patent #3,050,870 for his invention entitled "Sensorama Simulator;" however, it was 10 years ago that Jason Lanier really developed VR as we know it today (Burdea and Coiffet, 1993; *The New Yorker*, 1993). It is created by display and control technology that can surround its user with an artificial environment that mimics real life. The user interacts with objects and settings as he would in the real world. Virtual reality turns you into a full participant in a three-dimensional setting that envelops you completely. It does that by utilizing both computer hardware and software.

The basic hardware used in virtual reality is a computer, a virtual reality helmet, virtual reality gloves, and, in some cases suits, that get the users entire body into the action. The tools need to be modified and, in some cases, designed, depending on the type of virtual reality environment that is being created and just how much detail the user needs.

The goal of VR is to put oneself into a nonexistent or virtual world and let the self interact with that world as much as one would in the real world. It is precisely this ability of interaction that offers a tremendous opportunity to the training arena for any organization and for any situation. It is also the very reason why experts in the computer technology of VR are being called in by engineering and training companies to save time and cost in creating new designs of complex products and services, as well as efficient and effective training programs.

Applications for VR are possible everywhere; however, here we explore three specific applications as they relate to quality and training: VR in education, training and engineering.

A. VR in Education

Education has utilized all media tools from books to video cassette recorders and virtual reality will become another one of those tools. Apple computers have already developed a software teaching aid called "the Virtual Museum" (Whitehouse, 1994). The program allows the user to explore the various rooms of the museum and interact with the displays. The museum has a medical wing, a plant life wing, an astronomy wing, and an environmental wing. After entering the room the user can interact with exhibits in a way that no real museum could possibly duplicate. For example, a visit to the plant room allows the user to view the growth cycle of a particular plant, while a visit to the astronomy room allows the user to see instructional arrows pointing out the name of every star in the night sky. With this type of interaction you can expect virtual reality to become an important media in the education industry. If the old rule that it is not enough to hear, see, and read about something to really understand it, but one should learn by doing is true, then the interactive capabilities of virtual reality will make it a very powerful tool in education (Primentel and Teixeira, 1993).

From a quality perspective VR has a great potential in the classroom for learners who are not yet literate or who have difficulty learning from printed word, and for people whose preferred learning styles are visual, auditory, and kinesthetic. Virtual reality, combined with other technologies, will enable adult learners to interact with experts in virtual worlds that are designed to analyze a learners's individual learning needs and adjust as the learner progresses.

B. VR in Training

Of the areas that VR is being utilized, by far the most promising is that of training. The importance of VR being used in training is that it provides safe, repeatable, and cost-effective training to teach techniques or operations that are dangerous, rare or expensive to perform in real life. It can be expected that the need for very specialized training will create a need for very specialized training tools. Training may be so specialized that a particular task may be the entire focus of the training (Mcllvaine, 1996).

Training will also push the input/output devices incorporated in the training to become more accurate, dynamic and fine tuned to human kinetics. The rationale behind this is that in order to teach difficult to master techniques the input/output devices must have very finely tuned sensors and kinetic feedback. What good would it be to learn how to perform a flight maneuver that would be performed substantially differently in real life? It could be envisioned that dancing, karate, playing a musical instrument, driving a car, or even being taught statistics could be learned via advanced virtual reality systems. Think of the money and the stress saved by a virtual drivers exam, a series of courses on basic statistical process control, or a course on design of experiments. In fact, the benefit of freeing up labor from a stressful job would be compounded with the benefit of being able to test and evaluate a drivers ability under every conceivable condition.

For training that does not require large amounts of physical accuracy, the VR system would become very much like the personal computers of today and a glorified computer-based instruction. The consequences of such a program will result in a system that the virtual tutor will never tire, stick to the curriculum, allow learners to learn at their own speed, and never be prejudiced.

C. VR in Engineering

Engineers have a need to reduce the time it takes to design, develop, and prototype machines. Occasionally, parts are designed and to the shock of the designers it is found that the part can not be fabricated or assembled easily or just does not do the thing it was intended to do. Assemblies may not fit together correctly or aesthetic components may not really work correctly. To check designs engineers make prototypes. This approach however is very expensive and time consuming. Virtual reality can help speed up the development and prototype process by allowing the designer to see the machine part(s) or even the assembly after drawing it.

Among those at the forefront of this development is Britain's National Advanced Robotics Research Centre (NARRC), which has joined with aero-engine manufacturer Roll-Royce to devise a VR system that enables engineers to experience the sensation of moving around and handling the components of a computer-generated simulation of the next generation of engines.

Rolls-Royce produced a complete computer model of a turbofan engine with the aim of eventually avoiding the need to make a mock-up (full scale replica) that would traditionally be used to plan assembly, repair, and overhaul procedures. (Boeing was successful in building the entire 777 plane with VR and saved time, prototyping, and much of the traditional checking before building the plane.)

By turning to NARRC's expertise, Rolls-Royce engineers wearing a virtual reality headset with tiny built-in TV screens are enabled to become immersed in a situation which allows them to walk around a simulated engine, and even to reach into the three-dimensional scene created by the computer in order to remove or replace components almost as they would in the real world. In this way, engineers can, for instance, verify maintenance and repair times which can be quoted to potential customers even before the first engine has been built in reality.

NARRC has researched virtual reality for the past years, mainly in the quest to produce more intelligent robots. Such robots are designed to work in hazardous situations, such as inside nuclear power station reactors or under water. They can carry stereo television cameras capable of rapid movement like the head of a human (e.g., panning at 180°/second) and can send three-dimensional images to a remote operator up to 85 km away.

Now the center has announced the go-ahead for a project to provide a facility for industry to research the benefits of VR and simulation, a scheme that could prove a landmark in industrial design trends. The potential of VR to simulate products at the design stage will rapidly take the technology to the very heart of industrial design and development.

The trigger for adapting VR to industrial applications was the realization that the technique used in VR of selectively switching off data not being viewed held out the prospect of revolutionizing industrial computer-aided design (CAD). The technique allows the operator to move through worlds which were themselves animated. Despite the relatively poor quality of images available in early VR systems, in 1991 the center began producing a variety of immersive and nonimmersive VR models.

Recently, new hardware technology has overcome many of the problems associated with professionally acceptable image quality. Photorealism can now be obtained at much lower cost than a year ago and performance hitherto only available on military machines is featured by some advanced systems aimed at the industrial market. Part of the VR capability is to include several peripheral computers and software for modeling, simulation, special CAD, visualization, and animation.

The potential for visualization which VR offers needs to be supported by computer intelligence and the benefits are many. For example, researchers are developing a laser range-finder system to build automatically a computer model of three-dimensional objects (e.g., pumps, pipes, and valves) directly from sensory data. Based on the work originally carried out at Stanford University, a powerful method of path planning in cluttered two- and three-dimensional worlds has been developed. This technology makes it possible to embed such intelligence into VR worlds, thus aiding, for example, a fitter who is moving parts in highly confined spaces.

This discussion has centered around the engineering aspects of VR; however, the implications of VR—especially those relating to simulation and animation software-are of prime interest in the training area. The benefits of VR in training are astronomical because for the first time we have the opportunity to enhance the training instruction with what appears to be "the" real thing of whatever we are trying to train.

X. EVALUATION OF TRAINING

Money and time are commodities that every organization values and keeps track of. Training encompasses both money and time and therefore any organization, regardless of what it produces, should and must evaluate it. Of course, the question is raised: "How do we evaluate training?"

Many approaches have been proposed (Kirkpatrick, 1959–1960, 1994, 1996; Madaus, Scriven, and Stufflebeam, 1983; Scriven, 1991; Shelton and Alliger, 1993; Bernthal, 1995; Parry, 1996; and Dixon, 1996) and every one of them has advantages and disadvantages. What is important, however, is that every training program offered by the organization must achieve positive results or it should be modified to the point of acceptable results; if not, it should be dropped. The evaluation must be done after the training is completed and not necessarily as an end-course survey, such as:

- To measure effectiveness in a safety training, the organization must measure the effectiveness of that training by a reduction of injuries and lost time.
- To measure the effectiveness of a design of experiments (DOE) training program, the organization must measure the effectiveness by the amount (usage) of DOE experimentation and the improvement of process variation.
- To measure the effectiveness of conflict resolution training, the organization must measure the effectiveness of that training by a reduction of grievances and higher productivity.
- To measure the effectiveness of a statistical process control (SPC) training program, the organization must measure the effectiveness by the usage of SPC in the work environment and the improvement of overall quality as well as a better understanding of the process.

Here we propose a way of evaluation that is both effective and easy to do. The method is a cost-benefit ratio. We believe that unless the training enhances the process and/or the method of doing a particular task in that process, there is no need to do training. Since training is a measure of the transfer of knowledge from the learning environment to the job, the evaluation should measure precisely that transfer. That means that the evaluation will be at some point after the training has been completed. (We have a very low opinion of the end-course evaluation, since it provides irrelevant information about the application of the training. It provides categorical data based on preference criteria which, in most cases, have nothing to do with the transfer of learning to the job.) For more information on achieving results from training see Brinkerhoff (1987). How do we do a cost-benefit analysis? Our explanation will be through an example.

A. Business Problem

A company spent a great deal of money due to defective parts delivered. These charges were considered as avoidable and therefore the decreased profitability was unnecessary.

Cost of business problem: $722,000/year, including:

Cost to settle complaints

Annual number of complaints × average cost to settle complaint

26 complaints/year × $21,000/complaint = $546,000/year

Turnover due to complaints

Annual number of turnovers due to defectives × average cost per turnover

22 turnovers/year × $8,000/turnover = 176,000/year

Total cost: $546,000 + $176,000 = $722,000/year

B. Solution to Business Problem

Train all management staff and employees responsible for process defectives in how to avoid the occurrence of defectives.

Cost of training

Number of staff and employees attending workshop × average hourly rate x number of hours in workshop

465 staff × $16.00/hour × 16 hours = $119,040

Overhead cost

Material + facility + equipment + lost productivity and/or overtime + trainer cost

$1,500 + $200 + $75 + $10,000 + $5,000 = $16,775

Development time

Persons in development × delivery time × hourly rate for development

3 × 48 × $75 = $10,800

Total cost of solution = $119,040 + $16,775 + $10,800 = $146,615

Improvement benefit

Cost of problem before implementing solution – Cost of problem after implementing solution

Cost before implementation = $722,000/year

Cost after implementation:

Settlement cost = annual number of complaints × average cost to settle each complaint

1 complaint/year × $34,000/complaint = $34,000/year

Turnover cost = annual number of turnovers due to complaints × average cost per turnover

1 turnover/year × $14,000/turnover = $14,000/year

Cost of problem after implementation = $34,000/year + $14,000/year = $48,000/year

Improvement benefit = $722,000/year – $48,000/year = $674,000/year

Cost-benefit ratio = improvement benefit/cost of solution = ($674,000/year)/($146,615/year) = 5:1

Obviously this training was very well received and was transferred to the individual work. This type of evaluation is imperative in all training and it is the responsibility of the training department to make sure that not only is it conducted but is also communicated to the department heads and management teams.

XI. THE TRAINER

As already we have mentioned, the work place is being changed by the values and consciousness of both the organization and the work force itself. It is anticipated that between now and the year 2000 the work force will change in many ways because of age distribution, technological needs, and attitudes towards work. Where in the 1980s the dominant age was between 15 and 29, by the year 2000 the largest age group will be 30–44 years old with a rising curve for the 45-year old and over category (Britannica, 1994).

With all these changes anticipated, training will be the forerunner of the transformation. In fact, the changing agent that will be responsible for the transformation will be the trainer. His or her influence will be the most critical attribute of the change and will determine the success of the transition.

In this section we delineate some basic elements the trainer needs to train the employees of the future and to prepare them for that transition, specifically: winning the budget battle and teaching techniques.

The first element that trainers of the future need for a successful program is to win the budget battle. One may consider this attribute as a noncontributor to the transition process, however, it is at this fundamental point that the trainer has to convince the management of the benefits that the specific training will bring to the organization, and that it will outweigh its cost.

With corporate organizations becoming more cost conscious, this stage will indeed be the pivotal point that success will be dependent upon. The significance of this point is so great that we must address some of the steps that the trainer has to take to overcome the objections to the budget dilemma.

1. *The mission of the training must be in congruence with that of the organization.* It is imperative that the training (no matter what kind) should contribute to the organization's overall strategies and goals. The emphasis of the program offered should always be on the corporate mission and how the particular training can contribute to that mission. It is the trainer's responsibility to focus on the mission and then design, develop and then deliver the appropriate and applicable program to fit the particular need. The trainer should not design and develop a program and then try to sell it to the people concerned.

2. *All training must be assessment driven.* Trainers must understand that without proper need assessment their training will not be successful for at least two reasons.

    a. Without assessment, management will not be convinced that the training is necessary.

b. Employees will not participate effectively because they are not personally motivated (i.e., the programs do not fit their expectations).

When the focus of training is generated based on needs assessment, top management will be likely to look at the particular training as a solution to a specific problem rather than as just another frivolous luxury of pure education. One can see that indeed the trainer is a critical element in the total process of training. Furthermore, he or she is the cornerstone that need assessment begins with.

3. *Trainers must plan ahead.* One of the many functions of the trainer is to see opportunities and challenges for the organization in specific situations, before they occur and have the employees prepare for them. To do that effectively a trainer, especially the one of the future, must be prepared to act proactively rather than actively. The needs assessment results, of course, help tremendously in this area. The overall goals of the organization/training department are also important.

4. *Provide a cost/benefit analysis for each program.* One of the most wasted efforts of time is the end-evaluation of a specific program. It does not tell you much except whether or not the chairs were comfortable, the food and beverages were good, and the breaks and lunch were of proper duration—never mind that the majority of the participants never follow the time limits and whether or not the material were easy to understand. If the program is sponsored by a consultancy, then it also provides an addition to their data base (see Section X on training evaluation for a more detailed discussion.)

A successfully implemented training program is proportional to the efforts that a trainer has contributed in knowing, analyzing and marketing the particular training costs for a specific program. It is imperative that the trainer not be insensitive to the costs involved. It is the trainer's responsibility not only to identify these costs but also to sell the particular program as an investment to the organization. It is also the responsibility of the trainer to highlight specific savings and contributions that contribute to a cost-effective program. Furthermore, it is the responsibility of the trainer to include the incurred costs, such as instruction time, materials, travel, facility costs, lost wages for participants, design and development costs, evaluation costs, trainer salary(s), and any other cost that might be a contributing factor.

The focus of the cost/benefit analysis is to establish a benchmark for a more informed decision, not only for monitoring expenditures but also for comparing training alternatives. A simple but effective cost/benefit analysis is:

1. Calculate the cost per trainee.
2. Multiply the number of trainees by cost per trainee.
3. Estimate the benefits/savings expected as a result of the training based on the goals and objectives of the training.

4. Subtract the cost from the benefits to get the dollar savings.

In an environment where organizations are concerned with profitability on a percost center basis, trainers need to know approaches to cost/benefit analysis as well as marketing.

The second element that trainers of the future need for a successful program delivery is an understanding of the learner characteristics (see Section II.A, Adult Learner Characteristics) and teaching/presentation techniques. One may challenge the validity of such an assertion, but one has to realize that the general population is getting older and the new technology innovations demand new approaches to learning as well as skills. In addition, since the learners bring into the training different age characteristics and expectations, it follows that the trainer must be knowledgeable to recognize these differences and to change or modify the method(s) of teaching/presentation for maximum learning and transfer.

Trainers of the future have to make sure that the learners understand what is being taught and can transfer this knowledge outside the training environment. Trainers must remember that while young learners expect to transfer their learning some time in the future, the older learners expect to transfer their learning immediately.

A good rule to follow when teaching/presenting is to follow the nine events of instructions (Gagne and Briggs, 1979). They are helpful, because they offer a systematic approach to learning. The nine events are shown in Table 17.2. We believe that these events are more or less self-explanatory; however, to make these nine events come to life in the training situation we have added one more element to the list: trainer's enthusiasm.

Educational psychology has taught us that people learn among other things not by being told, but by experiencing the consequences of their actions. In other words, it is the learner's response and not the trainer's

Table 17.2 The Nine Events of Instruction

1. Gaining and controlling attention
2. Informing the learner of expected outcomes
3. Stimulating recall of relevant prerequisite capabilities
4. Presenting the stimuli inherent to the learning task
5. Offering guidance for learning (solicit response)
6. Providing feedback
7. Appraising performance
8. Making provisions for transferability
9. Insuring retention

stimulus that determines how successful the learning will be. To facilitate a relevant response from the learners, the following techniques can be used.

1. Always start with questions that can be answered by everyone. It facilitates learning and also serves as a positive feedback for the learners themselves to want to be more participative and attentive.
2. When asking a question never select a respondent immediately. Remember, the intent of the trainer is to encourage/facilitate learning and transfer. By waiting, say, ten seconds, and perhaps by restating the question, you allow everyone in your class to formulate a response. That response, of course, is what you are after and when you select a respondent the opportunity for further development and/or discussion is enhanced.
3. For major (critical) points, provide handouts that contain questions and space for the answer. Allow time for response and then give them feedback before proceeding.
4. Stimulate the group by asking thought provoking questions and have the learners write the answers. When everyone is done call upon individuals and proceed with a feedback and/or discussion of the question.
5. Make your requisition of response fun, by asking the learners from time to time to take a guess on something. This method will allow the learners to focus and/or pay attention to your next point. Besides, it breaks the monotony of seriousness in the class. Another way of implementing fun into the classroom is to ask for a show of hands. This can be used when you as a trainer want to polarize the group and/or to start a discussion on a categorical issue. It is a guaranteed method for group participation.
6. As a trainer you should never lose sight of the learner's expectations and/or needs. As such you can guide the learners through your instruction by providing them with preprinted response sheets that they can complete throughout the lesson. A good instrument for such an exercise is a flowchart. Another method of gauging the success of your eliciting response and the overall learning of the learners is to provide from time to time a self-assessment instrument that is quickly administered and measures knowledge, performance, and/or skills. If this method is used it is imperative that time be allowed for discussion of the self-assessment instrument. It provides benchmarks of progress to both trainer and learners and the instruction can be adjusted accordingly to facilitate both.

The catalyst for eliciting response from an adult learner in a training situation, enthusiasm and initiative on the part of the trainer. Emerson said

"Nothing great was ever achieved without enthusiasm." Certainly, enthusiasm on the part of both trainer and learner is important, but it is the enthusiasm and initiative of the trainer that will make the difference between an outstanding and a so-so program. To cultivate enthusiasm and display initiative, the following tips, may help the trainer:

1. Be confident. If you as a trainer do not show confidence, do not expect your learners to. They trust your expertise and abilities to teach. Do not let them down. Be prepared. (You can never be overprepared.) Expect the learner to learn the skill(s) you are working on because you are prepared. If you do not prepare, it will show.
2. Help the learner to feel confident and to believe that they can do it. Act as though that particular learner is the most important person in the class. Remember, that person is also afraid of what is to come of the training. Take a personal interest in the learning process of the individual learner.
3. Help the learner understand the person for each exercise and try to motivate him or her to do it well. Show the learner why and how the exercise at hand apply to her or his situation. Try to make it applicable to their work environment or appeal to the personal attitude of the "what is in it for me" principle.
4. Give your learner undivided attention during your time with him or her. When they ask you a question act as though he or she is the only person in the room. Take personal interest in their will to learn and reciprocate with understanding and kindness.
5. Do not make a learning period too long. If the learners begin to tire, it is time to stop or, if you are an advanced trainer, you may change the training modality (i.e., from left to right (brain) approach teaching technique or vice versa or even change the method of instruction all together, such as from lecture to video or something else). Psychologists have shown that approximately 17 ±5 minutes of instruction per item is about what the average individual can stand. Anything over that is not recommended. As a trainer you can also design your instruction with left-right brain implications which can extend your training periods; however, unless you know the design principles, do not attempt it—it will backfire.
6. Do not let your learners think of the lessons as punishments. To the contrary, emphasize at all times possible that the lessons are the vehicles that will get them the necessary skills and/or knowledge that they are in the training for. The lessons will help them accomplish the transfer from the training room to their work environment.

7. Help the learners feel secure. Encourage them to ask questions and offer help whenever possible. Make sure that you as a trainer communicate to the learners that when he or she fails to understand something in the instructional process, it is not a personal failure. What he or she needs is more perseverance, dedication, and personal commitment to do better.
8. Frequently point out to your learners what they have learned and the progress they have made. No one likes to be a failure, and adult learners are no exception. They like to know how they are doing in comparison with other learners, standards, and so on. It serves as a feedback to their progress and it may in fact be an impetus for further development especially if they are praised for a good job.
9. Try to enjoy every session with your learners. They will enjoy it more, too. Unless you, as a trainer, make the learning fun, your program will not be successful. Your enthusiasm, initiative, and your overall attitude for the program are very much the mirror of your personal disposition. You will not fool the learners if you try to do otherwise.
10. End every training session on a pleasant note. Make the training session a memorable one and supportive by offering guidelines to homework (evening work), review of the session, and provide specific applications of the skills learned. Always emphasize the accomplishment(s) of the training at the completion of the program.

XII. MATCHING TRAINING MEDIA WITH PROJECT GOALS

Many things have changed over the years. Technological improvements have also had a major impact on how performance solutions are implemented. For example, we have progressed from cumbersome teaching machines to mainframe computer systems to today's powerful microcomputers for delivery of instruction. Traditional concepts of training were often limited to lectures and discussions with audio/visual support materials. The media of choice included lecture, overheads, blackboards, handouts, audiotapes, slides with audio, and sometimes programmed instruction in a printed form. Today, instructional designers have access to a wider variety of implementation strategies and media. They pick from these alternative to effectively accomplish project goals and work within the constraints of the project. Current media alternatives (beyond the traditional) include video, computer-based training (CBT), interactive videodisc, interactive compact disk (CDI), and multimedia CD-ROM, each with its own advantages and disadvantages. Instructional media can teach concepts through the modalities of color, sound, motion visual, or

Table 17.3 Summary of Advantages and Disadvantages of the Most Commonly Used Media

Media	Modality	Advantages	Disadvantages
Print	Visual (color)	Inexpensive except when there is a large number of reproductions and many revisions Consistent presentation	Cannot address auditory or motion concepts Requires a literate audience Static presentation
Audiotape	Auditory	Can address auditory discriminations Medium expense Relatively inexpensive for poor reading audiences Consistent presentation	Cannot address motion and color concepts
Instructor led	Visual (color, motion) Auditory	Can address all modalities Can provide immediate feedback to learner Dynamic presentation	High labor costs Tendencies for inconsistent presentations Sometimes untimely because it requires groups of learners
Video	Visual (color, motion) Auditory	Can address all modalities Can provide immediate feedback to learner Dynamic presentation Consistent presentation	High development and implementation costs Extended development schedule
CBT	Visual (color, motion) Auditory	Can address all modalities Can provide immediate feedback to learner Dynamic presentation Consistent presentation Can accommodate the learner's speed	High development and implementation costs Extended development schedule Requires a literate audience
Multi-media		Can address all modalities Can provide virtual reality Can provide immediate feedback Can accommodate the learner's speed	High development and implementation costs Requires somewhat a computer literate audience

through a combination of these modalities. Some content can be presented through any modality while others require a specific modality. A summary of these feature is shown in Table 17.3 for some of the most commonly used media.

Today's instructional designers often combine the various types of media when teaching knowledge and skills areas to meet the goal of providing high quality, effective, mastery based training.

REFERENCES

Ackoff, R. L. (1984). *Creating the corporate future*. John Wiley and Sons, Inc. New York.

Arbuckle, H. J. (1996). Vital elements to success at roll forming corporation. *Metal Forming*. Jan.: 50– 58.

Aristotle. (1975). *The Nicomachean ethics, Books I-X* (D. Ross, Trans.). Oxford University Press. London. Original work published 1925.

Batalden, P. B. and Stoltz, P. K. (1993). A framework for the continual improvement of health care: building and applying professional and improvement knowledge to test changes in daily work. *Journal on Quality Improvement*. Oct: 424–445.

Bernthal, P. R. (1995). Evaluation that goes the distance. *Training and Development*. Sept.: 41–48.

Bishop, J. (1989). Achievement, test scores and relative wages. Conference paper. American Enterprise Institute for Public Policy Research. Washington.

Bloom, B. S.(1976). *Human characteristics and school learning*, McGraw-Hill. New York.

Block, J. H. (1971). *Mastery learning: theory and practice*. Holt, Rinehart and Winston. New York.

Block, J. H. and Anderson, L.W. (1975). *Mastery Learning in Classroom Instruction*. Macmillan. New York.

Bolman, L and Deal, T. (1991). *Reframing organizations*. Jossey-Bass. San Francisco.

Briggs, L. J. (Ed.). (1981). *Instructional design*. Educational Technology Publications. Englewood Cliffs, NJ.

Brinkerhoff, R. O. (1987). *Achieving results from training*. Jossey-Bass. San Francisco.

Brookfield, S. D. (1986). *Understanding and facilitating adult learning*. Jossey-Bass. San Francisco.

Burdea, G. and Coiffet, P. (1993). *Virtual reality technology*. John Wiley and Sons, Inc. New York.

Carnevale, A., Gainer, L., and Villet, J. (1990). *Training in America*. Jossey-Bass. San Francisco.

Carnevale, A. (1990). Train America's workforce: America and the new economy. *Training and Development Journal*. Nov.: 31–50.

Carroll, J. B. (1963). A model of school learning. *Teachers College Record*. 65: 723–733.

Cross, P. (1981). *Adults as learners*. Jossey-Bass. San Francisco.

Deming, W. Edwards. (1986). *Out of the crisis*. Massachusetts Institute of technology. Cambridge, MA.

Denison, E. (1988). Accounting for United States economic growth 1929–1989. Research paper. The Brookings Institution. Washington.

Dick, W. and Carey, L. (1978). *The systematic design of instruction.* Scott, Foresman and Company. Glenview, IL.

Dixon, N. M. (1996). New routes to evaluation. *Training and development.* May: 82–86.

Encyclopaedia Britannica: 1994 book of the year. (1994). Encyclopaedia Britannica. Chicago.

Fernberg, P. (1993). Learn to compete: training's vital role in business survival. *Managing Office Technology.* Sept.: 14–16.

Fitzenz, J. (1993). Training budget boom. *Training Magazine.* Oct.: 37–45.

Fitzenz, J. (1994). Yes ... you can weigh training's value. *Training Magazine.* July: 55–58.

Fox, R. (1981). Current action principles and concepts from research and theory in adult learning and development. (ERIC Document Reproduction Service NO. ED. 203 008).

Froiland, P. (1993). Who's getting trained. *Training Magazine.* Oct.: 53–65.

Gagne, R. M. and Briggs, L. J. (1979). *Principles of instructional design.* 2nd ed. Holt, Rinehart and Wilson. New York.

Garvin, D. (1993). Building a learning organization. *Harvard Business Review.* July-Aug.: 78–91.

Grayson, C. and O'Dell, C. (1990). *American business: a two-minute warning.* The Free Press. New York.

Hage, D. (1993). Seeking light in the dark. *U.S. News & World Report.* May 31: 60–63.

Harvey, K. and Lowell H. (1977). Bloom's Human Characteristics and School Learning. *Phi Delta Kappa.* 590: 189–193.

Henkoff, R. (1993). Companies that train best. *Fortune.* pp. Mar.: 22: 62–69.

Ingrassia, P. and Patterson, G. (1989). Is buying a car a choice or chore? *Wall Street Journal.* Oct.: 24: C-13.

Irwin, T. (1995). *Plato's ethics.* Oxford University Press. London.

Johnson, R. S. (1993). *TQM: Quality training practices.* Quality Press. Milwaukee, WI.

Jonsen, A. R. and Toulman, S. (1988). *The abuse of casuistry. A history of moral reasoning.* University of California Press. Berkeley, CA.

Kelly, K. (1994). Motorola: Training for the millennium. *Business Week.* Mar. 28: 158–162.

Kessels, J. P. A. M. and Korthagen, F. A. J. (1996). The relationship between theory and practice: back to the classics. *Educational Researcher.* Apr.: 17–22.

Kidd, J. R. (1973). *How adults learn.* Association Press. New York.

Kirkpatrick, D. (1994). *Evaluating training programs: The four levels.* Berrett-Koehler. San Francisco.

Kirkpatrick, D. (1959/1960). Techniques for evaluating training programs. *Journal of the American Society of Training Directors.* Nov.–Dec./Jan.–Feb.: 34–36, 54–57.

Kirkpatrick, D. (1996).Revisiting Kirkpatrick's four level model. *Training and Development.* Jan.: 54–59.

Knowles, M. S. (1980). *The modern practice of adult education: From pedagogy to andragogy.* Follett. Chicago.

Knox, A. B. (1986). *Helping adults learn*. Jossey-Bass. San Francisco.

Knox, A. B. (1977). *Adult development and learning*. Jossey-Bass. San Francisco.

Madaus, G. F., Scriven M., and Stufflebeam, D. L. (1983). *Evaluation models*. Kluwer-Nijhoff Publishing. Boston.

Mcllvaine, B. (1996). What price virtual reality? *Managing Automation*. Jan.: 95.

Mercer, T. (1989). *Turning your human resource department into a profit center*. AMACOM. New York.

Mitroff, I. I. (1988). *Break-away thinking*. J. Wiley and Sons, Inc. New York.

Naser, S. (1987). Do we live as well as we used to? *Fortune*. Sept. 14:, 33–35.

National Center on Education and the Economy. (1990). *America's choice: high school or low wages?* National Center on Education and the Economy. Rochester, NY.

The New Yorker. (1993). Jason Lanier is virtually sure. *The New Yorker*. Dec. 27: 59.

Novosad, J. P. (1982). *Systems, modeling and decision making*. Kendal/Hunt Publishing Co. Dubuque, IA.

Noyelle, T. (1989). *Services and the new economy: Toward a new labor market segmentation*. National Center on Education and Employment—Teachers College. Columbia University. New York.

Nussbaum, M. C. (1986). *The fragility of goodness: Luck and ethics in Greek tragedy and philosophy*. Cambridge University Press. Cambridge, MA.

Parry, S. B. (May 1996). Measuring training's ROI. *Training and Development*. pp. 72–78.

Perelman, L. (1994). Kanban to kanbrain. *Forbes*. June 6: 85–95.

Primentel, K. and Teixeira, K. (1993). *Virtual reality through the new looking glass*. McGraw-Hill. New York.

Reigeluth, C. M. (Ed.). (1987). *Instructional theories in action: Lessons illustrating selected theories and models*. Lawrence Erlbaum Associates, Publishers. Hillsdale, NJ.

Reigeluth, C. M. (Ed.). (1983). *Instructional design theories and models: an overview of the current status*. Lawrence Erlbaum Associates, Publishers. Hilsdale, NJ.

Richey, R. (1986). *The theoretical and conceptual bases of instructional design*. Kogan Page. London.

Schaffer, R. and Thomson, H. (Jan.-Feb.:, 1992). Successful change programs begin with results. *Harvard Business Review*. pp. 80–89.

Scriven, M. (1991). *Evaluation Thesaurus*. 4th ed. Sage Publications. Newbury Park. CA.

Seels, B. and Richey, R. C. (1994). *Instructional technology: The definition and domains of the field*. Association for Educational Communications and Technology. Washington.

Senge, P. M. (1990). *The fifth discipline: the art and practice of the learning organization*. Doubleday. New York.

Shelton, S. and Alliger, G. (1993). Who's afraid of level 4 evaluation? *Training and Development*. June: 43–48.

Springer, S. P. and Deutsch, G. (1981). *Left brain, right brain*. W. H. Freeman and Co. New York.

Stamatis, D. H. (1997). *The nuts and bolts of reengineering*. Paton Press. Red Bluff, CA.

Stamatis, D. H. (1986). *The effects of hierarchical and elaboration sequencing on achievement in an adult technical training program*. Unpublished dissertation. Wayne State University. Detroit, MI.

Stamatis, D. H. (1996). *Total quality service*. St. Lucie Press. Delray Beach, FL.

Torshem, K. P. (1977). *The mastery approach to competency-based education*. Academic Press. New York.

Travers, R. M. W. (1982). *Essentials of learning*. 5th ed. Macmillan Publishing Co. New York.

United Kingdom Treasury. (1989). *Economic progress report: No. 201*. Colibri Press. London.

Whitehouse, K. (May 1994). The museum of the future. *IEEE Computer Graphics and Applications*. pp. 12–15

Wlodkowski, R. J. 1985). *Enhancing adult motivation to learn*. Jossey-Bass. San Francisco.

Womack, J., Jones, D. and Roos, D. (1991). *The machine that changed the world*. Harper Perennial. New York.

Womack, J. (1989). *The U.S. automobile industry in an era of international competition: performance and prospects*. Vol. 1. Working papers of MIT Commission on Industrial Productivity. MIT Press. Cambridge, MA.

Wonder, J. and Donovan, P. (1984). *Whole brain thinking*. W. Morrow and Co. New York.

Young, G., Sansone, D. C., Baily, J., and Blake, K. J. (1995). Personnel: a crisis facing manufacturing in the '90s. *Metal Forming*. May: 38–42

18

Planning for Quality

The theme of quality has been and will continue to be that "unless we practice the systematic doctrine of continuous improvement, there is a high probability that we will not be around to practice any doctrine." This realization has motivated many companies to pursue quality and to develop their own continuous improvement policy. The problem, however, has been that even though their efforts are with good intentions, they are quite often disjointed and ineffectual, ending in failures in the expected quality programs (including TQM and any other continuous improvement programs).

Continuous improvement, by definition, is a system by which we improve processes and products to better meet customer expectations. Organizations that have understood and adopted this philosophy recognize that the system which must be in place to operate that philosophy must incorporate mechanisms that insure continuous improvement. This systematic view of continuous improvement is the driving force behind the advanced quality planning and the quality operating system. Even though both programs were developed by the automotive industry, they are applicable to any organization which is committed to TQM.

I. ADVANCED QUALITY PLANNING

Quality is and will continue to be a key factor in the world of manufacturing and nonmanufacturing alike. It will also continue to be a principal area of competition which will determine not only market share but also the ultimate commercial viability of both products and services. To implement the mantra "quality first" an organization must direct its quality system towards defect prevention and continuous improvement. Defect prevention is a forward-looking strategy that emphasizes development of stable and capable processes. This approach contrasts with defect detection, a past-oriented strategy, that seeks to identify unsatisfactory units produced by unstable and/or noncapable processes.

As important as a prevention policy is, defect prevention alone is not enough. The goal of any organization should be to have their products recognized by their customers as the best available. Obviously, this can only be accomplished through continuous improvement in every aspect of the organization.

In the past, the measures of quality were largely internal—conformance with engineering specifications, shipment rejections, warranty repair rates, and recall campaigns. While these performance indicators still have their place, they have been supplemented by indicators that are focused directly on the customer. Now, things-gone-right, things-gone-wrong, and probe studies are the most significant quality indicators because they measure the satisfaction of the ultimate customer with the products and/or services that the organization may provide. In essence, the true quality movement of today and the future is that customers truly are the focus of everything we do.

Customers are both internal and external to the organization and they both contribute to the overall quality of the products and services that an organization produces. Suppliers play an equally important role in this equation of overall quality. They are responsible for their own quality and they have the responsibility to deliver to their customer what was ordered. It is precisely this responsibility for quality from both the producer and supplier that we address the issue of advanced quality planning (AQP).

Chrysler, Ford, and General Motors (1995) define advanced quality planning as a structured method of defining and establishing the steps necessary to assure that a product satisfies the customer. The goal of product quality planning is to facilitate communication with everyone involved to assure that all required steps are completed on time. The basis for this planning is an embellished Plan-Do-Study-Act model (Figure 18.1). Effective product quality planning depends on a company's top management commitment to the effort required in achieving customer satisfaction. Some of the benefits of advanced quality planning are to (ASQC, 1988):

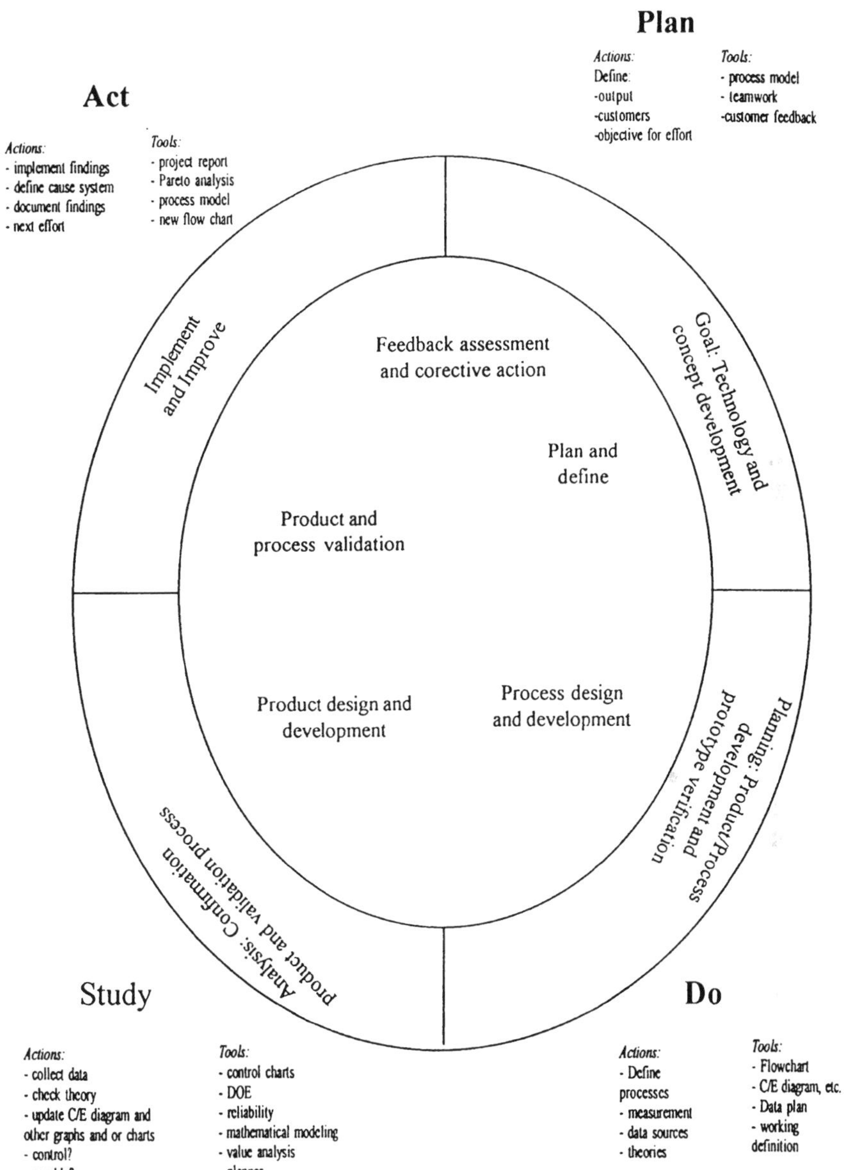

Figure 18.1 A generic AQP model.

- Direct resources to satisfy the customer.
- Promote early identification of required changes.
- Avoid late changes.
- Provide a quality product on time at the lowest cost.
- Reduce downtime.
- Improve productivity.
- Reduce repairs, rework, scrap, and so on.

Some of the tools and analytical techniques about to be described here have been discussed elsewhere and, in some cases, in a much more detail; however, we believe that a logical sequence of their application may make it easier to follow. Obviously, the actual timing and sequence of execution is dependent on customer needs and expectations and/or other practical matters specific to individual organizations.

A. Quality Systems Requirements for All Products

Quality Planning

Quality planning is the cornerstone of defect prevention and includes all of the defect prevention techniques utilized prior to production. Because these techniques are many, to optimize their impact a cross-functional team should be convened for all new and/or changed products. This team should be appropriate, applicable and knowledgeable about the product and/or service as well as knowledgeable about the tools and techniques about to be used. A typical quality planning activity will include the following:

Process Flow Chart. This chart is to be prepared by the producer as a schematic representation of the proposed (or current) process to show the sequence of operations. The flow chart is an essential tool for performing other AQP techniques such as the process failure mode and effect analysis (FMEA) and the control plan.

Failure Mode and Effect Analysis. The failure mode and effect analysis (FMEA) is a powerful tool which assists in the prevention of concerns through a disciplined analysis of possible failure modes. The FMEA process can be applied to the system, design, and manufacturing processes, and is required for all new or changed products and processes. FMEAs are to be considered living documents and should be updated for all system, design, and process changes.

Specific responsibilities for FMEAs are:

- The system and design FMEAs must be initiated by the responsible design activity prior to the finalization of the design (engineering release). This requirement specifically includes outside design sources that develop black box and/or gray box designs.

- The process FMEA must be prepared by the producer's manufacturing engineering activity prior to the commencement of tooling. Ideally, a design FMEA should be available during preparation of the process FMEA; however, even without the design FMEA, a process FMEA may be constructed.

Feasibility. Producers must conduct feasibility evaluations of proposed new product design/products, product/process changes, and major changes in volume. In this context, feasibility is a judgmental assessment of the suitability of a particular design/processes for production, while conforming to all the specified engineering requirements at the specified volume levels. Achievement of satisfactory capability must be considered for all characteristics. Full manufacturing feasibility must be established prior to the commitment of tools or facilities. Performance of feasibility studies concurrently with process FMEAs is particularly beneficial.

Control Plans. Control plans must be prepared by the producer for all new or changed products and for carryover parts released for new parts and/or programs. The optimum method for preparing control plans is with a cross functional team as mentioned above. The purpose of the control plan is to summarize the quality planning for all identified significant process parameters/product characteristics including all critical characteristics designated on drawings and engineering specifications (ESs). No matter what other nomenclature is used, all significant characteristics must be included in the control plan.

During the development of control plans, producers should provide for variables data measurement wherever possible. Variables data generally provide more information as to specific process improvement opportunities and allow more rapid evaluation of process capability. Control plans consist of two sections:

1. *Prelaunch*: the process potential studies that will be carried out prior to the beginning of normal production.
2. *Ongoing*: the process monitoring and statistical methods that will be used to control ongoing production.

Control plans are to be established for all products before the commitment of tools or facilities. A single control plan may apply to group or family of products that are produced by the same process. Sketches may be attached to control plans for illustration purposes. To ensure that control plans are maintained, the latest revision date for related drawings and specifications should be included in each control plan.

Concurrences with control plans are required by both producers and suppliers. Changes to control plans are encouraged as more data about the process becomes available. All changes are to be reviewed with and concurred by the activities that originally concurred with the plan.

Process Potential Studies. These provide a preliminary assessment of the ability of the process to produce products meeting the customer's requirements. Each significant characteristic must be studied. These studies are conducted using as many product measurements as possible. A working guideline would be measurement of one hundred fifty product units from a substantial production trial run (e.g., the production volume from one hour or one shift). Data from the trial run is to be analyzed using a control chart (for stability, sequence trends, and comparison with specifications) and a histogram (for patterns of variation). This guideline assumes the use of variable data. It will not be applicable to all processes and products, so the actual sampling plan and analysis methods should be reviewed with the customer via a control plan.

Process Control Guidelines. The producer must prepare written process control guidelines for all personnel who have direct responsibilities for operation of processes. Theses guidelines will be derived from:

- the control plan
- the customer's engineering drawings and as applicable, engineering specifications and manufacturing standards
- producer expertise and knowledge of the processes

The basic principle for process control guidelines is to provide personnel with the information necessary for ongoing control of the processes. As long as this requirement is met, process control guidelines may consist of process sheets, inspection and laboratory test instructions, test procedures, or other documents normally used by the producer.

Packaging Review. The choice of packaging can have a significant effect on product quality. Producers are responsible for using appropriate packaging considering the various transport methods and routes that may be employed to assure that all products arrive at the point of usage without damage or deterioration.

Planned Preventative Maintenance. Producers should develop and maintain a formal, documented system for the routine maintenance of production processes and equipment. The frequency of such maintenance should be based principally on SPC data, tool wear, manufacturer's recommendations, and any other known reliability considerations.

Prototype Part Quality Initiatives

When the producer is also sourced with the production of prototype products, the customer may elect to review various analyses of quality data that derive from production of the prototypes. These initiatives are intended to learn as much as possible about the new processes and products prior to initiation of ongoing production. Producers to be included in these initiatives will be notified directly.

Incoming Material Control

To control quality, the incoming material must be in control. Therefore, under the AQP, the following are required.

Approved Materials. Materials procured for use in customer's products which require engineering approval must be procured from the customer's approved sources. Any additional sources must be approved by the responsible customer product engineering office prior to use. Suppliers are responsible for complying with requirements in current specifications. Suppliers may obtain specifications directly from the customer's materials engineering. Producers must assure the customer that their products satisfy current requirements for restricted substances and all other pertinent requirements.

Control of Subcontracted Products.

- The producer must transmit all relevant customer drawings and specifications to the subsupplier.
- The producer must ensure that all products and services purchased from subsuppliers conform to customer specifications. The prime method for carrying out this responsibility is to have the subsupplier implement SPC methods. When necessitated by specific quality concerns, subsuppliers should be asked to provide evidence of statistical control and capability with each shipment as appropriate.
- The producer must initiate corrective action with subsuppliers on nonconforming products and maintain records of these concerns with the corrective actions.

All changes in subsupplier sourcing after initial sample approval are considered significant process changes and will require a new initial sample submission.

Process Capability

Process capability is determined by:

- establishing a new control chart
- continuing the control chart begun during the process potential study
- using existing control chart data from essentially the same process

The time period required to establish capability is affected by many factors including the frequency of measurements, production rates, and most importantly, the time required for all process variables to vary within their expected range. A typical customer may expect to see at least thirty production days worth of data to support determination of capability, but the various factors can shorten or lengthen this period. In either case, the documentation should be part of the control plan.

The actual analysis methods should be appropriate and applicable for the process, including: 1) control charts for showing stability, and sequence related trends, and 2) histograms for distribution patterns. When these methods indicate non normal distributions, advanced statistical tools will be needed.

Statistical Process Control (SPC)

A key aspect of defect prevention is using "the voice of the process" (i.e., data from process and product measurements analyzed by appropriate statistical methods) to indicate when actions are necessary (e.g., adjustments, tool changes) and conversely, when the process should be left alone. The approaches to gathering data and analyzing it will vary according to the type of process and such other factors as degree of automation, process capability, and production volume. The launching of SPC and the development of existing programs require the expertise of people trained in the use of statistical methods. While the preferred method is to have such people on the producer's staff, there are situations where outside consultants can be of service.

The process parameters and/or product characteristics to be controlled using SPC are taken directly from the control plan and also incorporate all that has been learned from the AQP process. As more experience is gained through operation of the process, the need for SPC on certain characteristics may decrease. Upstream controls may reduce or eliminate the need to control certain characteristics. Alternatively, customer use of the products may suggest additional characteristics requiring control. In all cases, the control plan should be updated and submitted to the customer for approval.

After control limits have been developed and process capability has been demonstrated, decisions by the producer on process actions and part acceptance must be made according to the customer's requirements. Efforts to increase C_{pk} should continue at all times.

Continuous Improvement

Once processes have demonstrated capability and are operating under SPC, plans should be developed for continuous improvement of the process in the areas of reducing variation and improved productivity. While SPC methods can monitor the progress of improvement actions, other disciplines will be helpful in making process improvements, including:

- design of experiments
- evolutionary operation of processes
- any other method, as long as it is implemented through the Plan-Do-Study-Act cycle

Ongoing Requirements

Once you have controlled the incoming quality of the materials, products, and so on, then the focus is directed towards the ongoing requirements internal to your organization. An active AQP program has several requirements.

Measuring and Testing Equipment. Producers must provide adequate gages and other measuring and testing equipment that are appropriate, applicable, and authorized by the customer in quantities necessary for process control. Whenever possible, such equipment should be designed to provide variable data (as contrasted with attribute data).

Since the use of measuring and testing equipment is itself a process and a source of variation, appropriate statistical studies should be conducted to determine if the measuring and/or testing process is stable and capable. In this context, capability means that the $+-3\sigma \leq 10\%$ of the tolerance for the characteristic being evaluated. Measuring and testing equipment not meeting this requirements must be upgraded or replaced.

Producers must develop and implement a plan to verify the accuracy of this equipment at sufficiently frequent interval (normally once per year) to ensure continued accuracy. This plan should be documented and updated as necessary to reflect relevant engineering changes. Traceability to national and/or international standards on an annual basis should be provided for. When production tools, fixtures, tool masters, or other such devices are used as gages, they must be inspected and documented in the same manner as other gages.

Engineering Specifications Performance Requirements. During production, producers must conduct engineering specifications (ES) tests in accordance with the sample sizes and test frequencies specified in the control plan or in the ES itself. These are minimum requirements and are subject to increase when experience indicates that such action is necessary.

The goal of ES testing is to provide information to the producer that will assist in operating processes so that all customer engineering requirements are met. When either the ES or the customer requirements are inconsistent with establishing a valid process control for a particular process, the producer should propose an alternative approach via a control plan.

When ES tests can be completed during the shift from which the product was taken, the product produced during that shift should be held pending successful test completion. When tests are too lengthy to be of value for product disposition, producers shall maintain positive identification of test samples and the respective shipments. Engineering specification test failure shall be cause for the producer to stop production shipments immediately pending analysis of the process and corrective action. Producers shall immediately notify the customer's purchasing department, and the responsible release activity at the consumer location of the test failure and suspension of shipments.

After an assignable cause of ES test failure has been determined and corrected, producers may resume shipments of product meeting all customer's requirements. Suspect product must not be shipped without sorting or reworking to eliminate the cause of the test failure. Reaction plans in an ES supersede any and all conflicting requirements in this section. Records relating to ES test failures and corrective actions should be available for the customer to review upon request.

When an assignable cause for a test failure cannot be determined, the producer must immediately notify the customer and production will stop pending instructions from the customer.

Product Qualification and Monitoring. Producers are responsible for selecting from a variety of qualification and monitoring methods, the appropriate ones to control all aspects of products produced for a given customer. While SPC will normally be used for all significant characteristics (according to the control plan) there will generally be many other characteristics not covered by SPC. Appropriate controls for thesse other characteristics should be selected from the following:

- *Product audits.* These should be performed on a scheduled basis and, as appropriate, should form an integral part of the SPC system.
- *Periodic layout/laboratory tests.* Periodic layout inspections should use either coordinate measuring machines or surface plate methods. Laboratory tests include physical chemical properties and process variables.
- *Product qualification.* When only attributes data are available, lot acceptance should be based on a predetermined product qualification system or methodology. (Usually this is accomplished with a sampling plan.)

Indication of Product Status. The producer must identify the qualification and test status (i.e., OK, reject, sort, hold for rework, and so on) of product through all stages of the process by means of tags, stamps, routing cards, color codes, or other effective control methods. Positive controls must be developed to ensure that products for a given customer are properly identified and do not become mixed with product for other customers.

Verification of New Setups. When setup changes occur, it is essential to verify that new setups do produce products that meet the customer requirements. Written setup instructions should be available to the people performing these setups. Statistical confirmation of the setup should be obtained by producing products to constitute a subgroup of the size employed for SPC. The products should be measured and/or tested for the significant characteristics and the results entered on the control charts. If theses results are within the control limits, the setup can be approved for production.

Reference Samples. For many commodities, it is useful to retain the last unit of a production run as a reference for the next time that the particular part is to be produced. Similarly, when first piece inspection is used to certify a new setup, the first piece should be retained throughout the production run. When the size of specific commodities makes sample retention impractical, this requirement may be waived by the customer.

Repaired Products. Repairs consist of any actions to the product not part of the basic production process. For certain commodities, unique terminology exists which describes parallel concepts to repairs (e.g., reformulation for chemical processes, reapplication for steel, and so on). Since any action to salvage a product that does not meet customer requirements is both a source of variation and inherently costly, the producer must develop written procedures to correct product nonconformities, which must provide for:

- relevant inspection and/or testing after repair to ensure conformance to customer specifications prior to shipment
- approval via a customer engineering deviation for any repaired product that does not fully meet all customer's specifications
- no repairs visible on the exterior of the product are permissible on those products supplied for service applications; exceptions to this requirement must be approved by the customer

Returned Product Analysis. The producer must analyze parts returned from customer's plants and/or engineering tests. Records of the results of this analysis must be maintained and made available to customer personnel upon request. Producers should initiate corrective actions.

Problem-Solving Methods. When internal (e.g., control charts) or external (e.g., customer concerns) indicators show that a quality-related concern exists, producers must utilize an appropriate problem solving method (e.g., the eight-discipline approach). The approach used requires the producer to provide:

- natural work team members
- interim actions

- identification of the root cause(s) of problems
- permanent actions
- verification of the effectiveness of permanent actions

Heat-Treated Parts. The heat treating process must be controlled according to customer specifications unless heat-treatment characteristics are designated as control items.

Documentation

If quality exists, it must be appropriately documented.

Procedures. Producers must develop, implement, and maintain written procedures for continuous improvement of quality.

Drawings and Change Control. Producers must maintain on file the latest engineering drawings, deviations, and specifications authorized through the customer's purchasing and ensure that the necessary customer engineering documents are available at the time and place of the producer inspection and testing. If the engineering drawings and specifications reference other documents, the producer must obtain and maintain these reference documents. Concurrent with the effective dates of product changes, the producer must ensure that the obsolete information is removed from all points of use. The producer must maintain a record of change effective dates. All changes must have written customer approval prior to being incorporated in production.

Deviation Control. When a customer authorizes an engineering deviation—either through a waiver of change or a letter of deviation—the producer must maintain a record of the expiration date or condition and ensure compliance with the original or superseding specification when the deviation expires.

Records. The producer must prepare and maintain adequate quality systems records, including inspection and laboratory test instructions, gage and test equipment verifications and calibrations, and engineering specifications test methods. The producer must also prepare and maintain quality performance records indicating inspection and test results. These records must be available for review by the customer upon request.

Changes in Manufacturing Procedures. Prior to shipping products manufactured by a changed process, the producer must complete all verifications and tests necessary to ensure that the products continue to meet the customer's requirements.

1. *Production setups and changes.* Where a routine tool setup change has occurred but no change in the manufacturing process has been made, the producer must complete all inspections and tests neces-

sary to ensure that products meet specifications prior to making a production run, but normally need not repeat capability studies, ES tests, or whatever is preauthorized and applicable.

2. *Change approval.* For certain products the customer may impose a "no change without prior approval" restriction affecting the entire product or designated characteristic(s) of the product by appropriate notes on affected drawings or specifications. Prior to incorporating the changes in production when such restrictions apply, the supplier must obtain the appropriate approvals and document them.
3. *Change approval.* On control item products for which the customer does not require preapproval of supplier design, composition, or processing revisions, the producer must obtain appropriate authorization when, in the producer's judgment, the change could adversely affect any significant characteristics of the product.

B. Interface with Customer on Quality Related Matters

Initial Sample and First Production Shipment

Requirements for New Producers. When a producer does not provide any other products for a customer at the specific plant location, the customer will evaluate the supplier's quality system. Based on complete and satisfactorily results of the evaluation the contract will be awarded.

Requirements for Initial Samples. An initial sample is a small quantity of parts (normally six) from a significant production run (typically 300 pieces, but other quantities may be used with the customer's concurrence) which are checked to every dimensional and test requirement on the customer's part drawing and related specification. (Here, it must be emphasized that individual customers may have their own requirements which must be followed.)

Approval of initial samples requires that completion of all statistical studies in the pre-launch control plan and the demonstration of at least a $C_{pk} \geq 1.0$. When this degree of capability can not be demonstrated, steps must be taken to improve the process. If capability still can not be demonstrated, the producer should review the situation with the customer whether the specification can be revised. If the customer can not revise the specification, then the producer must submit a revised control plan for the customer's approval.

Typically, such a control plan require 100% inspection for all nondestructive tests. Approval of the revised control plan then allows approval of the initial sample. Since 100% inspection is both inefficient in totally screening out nonconforming product and is not cost-effective, it should be considered only as an emergency measure and should normally not be established as a permanent part of the process. (The effectiveness of 100% inspection is only 79%.) The revised control plan must include the use of control charts to monitor the processes and to identify opportunities for reducing variation in the process output. Producers must then develop a plan to eliminate the need for 100% inspection For an excellent guide for the transition see Skrabec (1989).

Required gages and test equipment must be available and used in preparing for the initial sample submission. When either the part drawing or a related ES requires customer approval of samples, the producer must obtain such approval in writing prior to notifying the customer of the completion of initial sample inspection and testing.

Requirements for First Production Shipments. A first production shipment approval is based on a review by the customer of the inspection instructions and gaging that apply to the subject product. Normally a sample of 50 units will be subjected to the same inspections and tests that the supplier would use to quality a lot for shipment. Typical situations requiring first production shipment approval are:

- new products
- products from a different manufacturing location utilizing either new or relocated tooling and equipment
- products following a stop shipment order
- products affected by supplier-initiated changes when the customer's drawings or specifications impose a no-change restriction in design, composition, or process without prior customer approval.

Customer Surveys

The customer will survey the producer's facilities to evaluate the quality systems, to review the use of statistical methods, to examine test and inspection records, and to verify the quality of outgoing products. These surveys are conducted to assist producers to evaluate and upgrade the quality system and to promote continuous improvement through reduced variation. (Here, it must be noted that with the proliferation of the third-party auditing the customer survey may be waived.)

C. Requirements for Control Item Products

Control item parts are selected products identified by the customer on drawings and in specifications with an appropriate and applicable designated symbol preceding the part and/or the material number (e.g., an inverted delta for Ford, a shield for Chrysler, and so on). Control item products have critical characteristics which may effect safe product operation and/or compliance with the law. These critical characteristics are also identified on the drawing or in the specification with the appropriate and applicable symbol. When the producer deals with control item products there are certain functions that are required.

Quality Planning

1. Cross-functional teams are essential for all control item parts when so requested by the customer. Producers must take an active role on the team to assure that all aspects of feasibility, processing, and usage are considered during the design phase. Team actions will normally include:

- identification of critical characteristics
- development and/or review the design and process FMEAs
- establishment of design and/or process actions to reduce the potential of certain failure modes
- development and/or review of the preliminary control plan

2. Process failure mode and effects analyses must include all critical characteristics. Efforts must be taken to improve the process to achieve prevention rather than detection. The FMEA must be reviewed with and concurred by the responsible customer representative prior to initial sample submission and/or self-certification. Producers, based on their product and process knowledge, are encouraged to identify additional critical characteristics, both relative to their processes and from characteristics not identified by the customer. When the supplier is responsible for the design (i.e., black and gray boxes), the supplier must also prepare a design FMEA, subject to the above review and approval requirements.

Producer Modification of Control Item Requirements

When data from test results and SPC indicates a high degree of capability ($C_{pk} > 2$), the producer may request a revision to the testing and/or inspection for critical characteristics. This revision is effected by obtaining customer's approval on a control plan showing the proposed revisions. These approvals must be obtained prior to making the process change. The same approach should be used when the producer seeks to substitute upstream process control for finished product testing and/or inspection.

Documentation

Document Identification. The producer's process sheets, process control instructions and similar documents should be marked with the control item symbol to indicate those process steps that affect control item characteristics. When a computerized record system is used, producers may use a symbol other than the designated one.

Product Identification. Producers must insure that the critical characteristic symbol precedes the part number in accordance with customer requirements.

Certification of Control Items. When appropriate and applicable, the customer may require certification for certain control items. These may be also be governmental regulations as for example in the case of "equipment standard parts." For these parts the producer must certify conformance of each part or shipment to applicable customer requirements and governmental regulations.

Lot Traceability. For certain control item parts, producers are required to establish a lot traceability system, that provides for positive identification and record keeping for each lot throughout the major phases of manufacturing, inspection, and testing.

1. *Lot identification.* A producer must assign only one lot code to each lot. When successive subdivisions of a lot are necessary, the principal lot code must be supplemented by subordinate codes to identify each sublot. The producer may ship more than one lot per pallet, but each container on the pallet must contain only the parts from one lot. The packing slip must state the number of containers comprising each lot.
2. *Lot control.* The producer's system must provide that:

- Inspection and/or test and shipping records include the principal and any subordinate lot codes.
- Final inspection reports are crossreferenced to supporting inspection and/or test documents and to the subsupplier's lot code.
- The shipping destination for all parts in each lot can be determined.

3. *System approval.* The producer shall document the lot traceability system and provide a copy to the customer upon request. No changes should be made to the system without written notification to the customer.

Process Changes. Before a producer makes any process change that might affect a critical characteristic, the producer must obtain prior approval from the customer.

Heat-Treated Parts. When characteristics resulting from heat treating are designated with the critical characteristic symbol, the heat-treating process must be controlled per the customer requirements.

D. Typical Analytical Techniques Used in AQP

In a given organization, many tools, methodologies, and techniques are available, applicable and appropriate for usage in an AQP implementation endeavor. Some typical ones are:

- assembly build variation analysis
- benchmarking
- cause-and-effect diagram
- characteristics matrix
- critical path method
- design of experiments
- design for manufacturability and assembly
- design verification plan and report
- dimensional control plan
- dynamic control plan
- mistake proofing (Poka-Yoke)
- process flow charting
- quality function deployment
- system, design, and process failure mode and effect analysis

II. QUALITY OPERATING SYSTEM

One of the pioneer organizations that formally instituted a systematic approach to run a business is Ford Motor Company. Because it came as an extension of quality improvements and was from an automotive company, many misconceptions (including the intent of the program) were introduced and still continue. It is assumed, for example, that this approach is just a quality control item and excludes management. It is also assumed that it is a tool for tracking measurables and not for making improvements and more importantly it is viewed as a Ford requirement. Obviously, these assumptions are not true and organizations that believe them are way off the mark. They just do not understand it.

This systematic approach to run a business has been called the quality operating system, quality business plan, and many other names. For our purposes we call it by the conventional name, which is quality operating system (QOS). To understand this approach, one must understand the definition of QOS.

A QOS is a systematic, disciplined approach that uses standardized tools and practices to manage business and achieve ever-increasing levels of customer satisfaction. A QOS is a way of doing business. The word "quality" in QOS is an adjective and it represents more than the quality department; in fact, it represents the entire organization, including the top management. The emphasis of the QOS is always on the "operating system," since it is a system that makes things happen. It is not a static system by which only the performance indicators are identified. Rather, it is a system in which measurable champions, cross-functional teams, and employee involvement teams work together to improve an organization in many fronts.

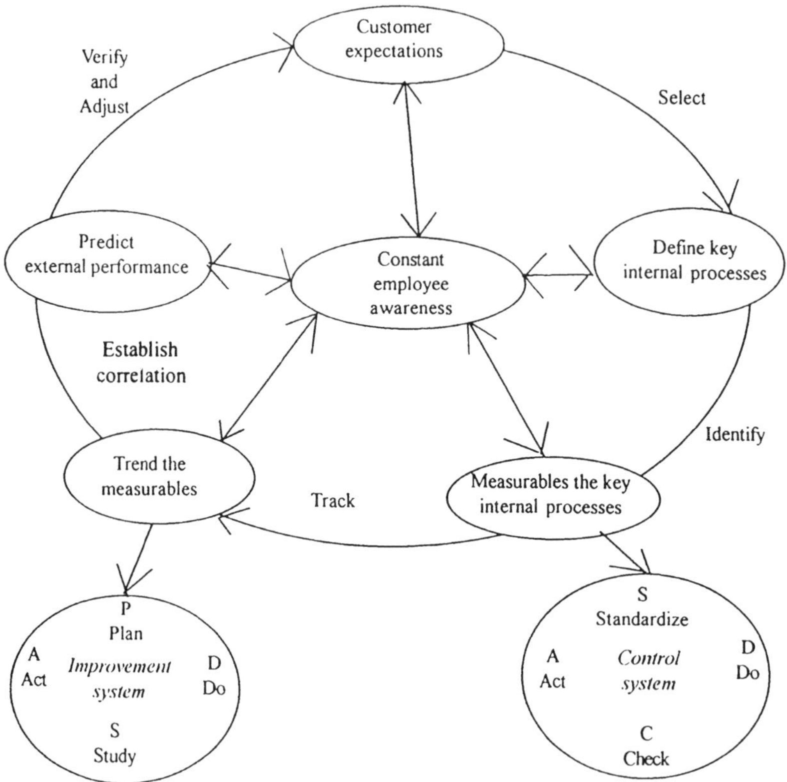

Figure 18.2 A generic quality operating system.

The basic system is shown in Figure 18.2, with two other variations in Figures 18.3 and 18.4. In all the figures, the focus of the QOS is on identifying, then meeting or exceeding customer expectations through employee involvement and awareness. Key processes are identified that support this goal. Measurables are developed for the key processes and their performance is tracked over time. Targets or goals are established for the measurables. Teams are formed and action plans are formulated to achieve these agreed upon goals.

Finally, the QOS establishes relationships (preferably quantifiable) between internal measurables and external performance. The QOS is adjusted on a continual basis in order to improve customer satisfaction. As part of this improvement, it must be understood that in any QOS there are different levels of both expectations and performance; therefore, the feedback from level to level may be quite distinct from each other (Machak and Tolstedt, 1992).

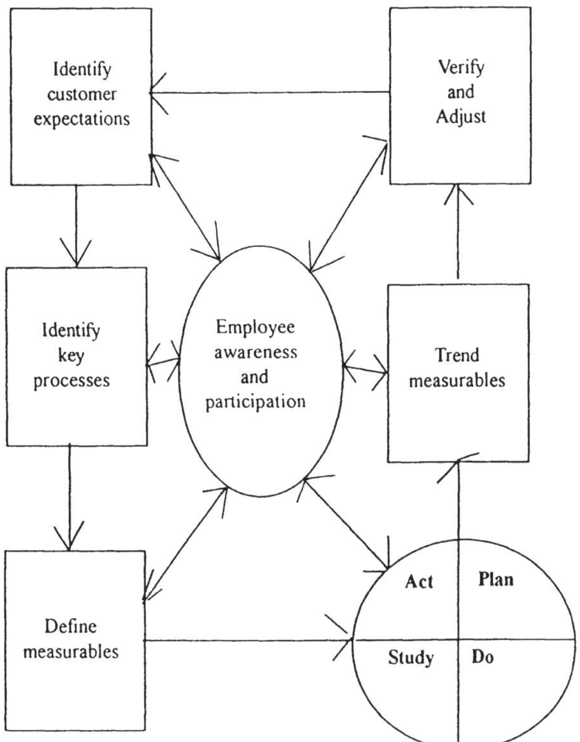

Figure 18.3 A simple model of a quality operating system.

A successful QOS implementation works to unify the concerns of all company systems, and therefore it must be managed by a cross-functional team of top managers, also known as the steering committee. The members act as champions for measurables that they "own." They take actions necessary to ensure that progress is made toward measurable goals and they are empowered to close the gap between where a particular measurable is and where it needs to be.

A. QOS Implementation

A generic approach of implementing a QOS in any organization is based on a seven step approach (Ford Motor Co., 1992a,b, 1993; Kymal and Tolstedt, 1994).

1. Form a Steering Committee
    - Secure management's commitment.
    - Explain the process and goals of QOS.

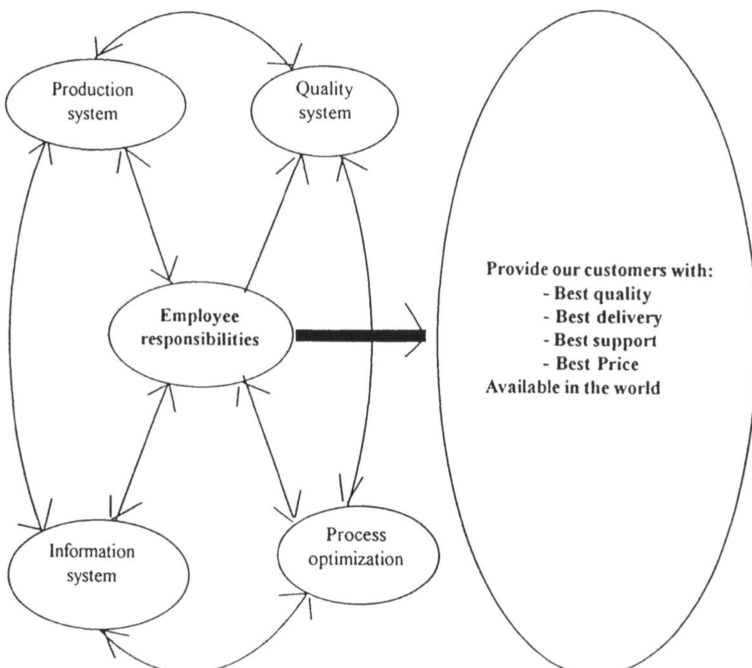

Figure 18.4 An advanced QOS model.

- Explain responsibilities.
- Schedule appropriate meetings.

2. Steering Committee Orientation
   - Make sure everyone understands the mission and business plan of the company.
   - Identify the mission of the committee.
   - Identify all customers.
3. Define Customer Expectations
   - List customer expectations (as many as possible).
   - Group similar expectations.
   - Prioritize the expectations.
   - Benchmark
4. Define Strategic Goals and Objectives
   - Establish goals and objectives.
   - Select internal key processes.
   - Assess training needs.
   - Provide appropriate training to team members.

5. Measurables and Processes
   - List potential key processes.
   - Brainstorm measurables.
   - Identify measurables for key internal processes.
   - Assign responsibility to the measurables.
   - Gather data.
   - Create a plan.
6. Quantify Measurement Selection
   - Apply appropriate tools.
   - Make decisions on consensus.
   - Identify relationships.
   - Track measurables.
7. Action Plan Formation
   - Present improvement opportunities.
   - Identify strengths and weaknesses.
   - Plan for the future.

B. Benefits of a QOS

A QOS is one of a few systems that will identiy results beyond the area of implementation. For example, a good QOS system will realize improvements in customer satisfaction, employee motivation, product quality and productivity, defects per million, premium freight charges, warranty returns, rejects, scrap, and so on. Everyone becomes involved in the process and can see their results. Constant and timely feedback is provided; and perhaps the most important benefit of all is an improvement in the bottom line.

C. Where Can QOS Be Implemented?

If an organization is serious about continual improvement throughout their processes, then there is no limit as to where to apply QOS; however, some diverse examples of potential QOS measurables are:

Claims	Specific product claims
Cost of quality	Prevention costs
	Appraisal costs
	Internal failure costs
	External failure costs
	% of sale
Benchmarking	Number of department using tests
	Number of companies surveyed

Training	Number of of hours per person
	Productivity improvement
	Course effectiveness
Demurrage	Delays in delivery
	Lost time
Cleanliness	Work area
	Bathroom
Process capability	% significant characteristics in control
	% significant characteristics capable
	% control plans completed
Reliability testing	Mean time between failure
	Field failures
Absenteeism	Number of excused hours
	% of work force missing
Engineering specifications	% certified by classification
	In process % test failures

REFERENCES

American Society for Quality Control (ASQC). (1988). *Advanced quality planning.* Quality Press. ASQC Automotive Division. Milwaukee, WI.

Chrysler, Ford, and General Motors (1995). *Advanced product quality planning and control plan: Reference manual.* Chrysler, Ford, and General Motors. Distributed through the Automotive Industry Action Group. Southfield, MI.

Ford Motor Co. (1992a). *QOS: Quality operating system.* Ford-NAAO. Dearborn, MI.

Ford Motor Co. (1992b). *QOS: Quality is the name of the game.* Ford-Vehicle Operations. Dearborn, MI.

Ford Motor Co. (1993). *QOS assessment and rating procedure.* Ford Motor Publications. Farmington Hills, MI.

Kymal, C. and Tolstedt, B. (1994). Quality operating system (QOS): The way to run your business. *OMNEX News.* Winter: 1–2.

Machak, R. and Tolstedt, B. (1992). Quality operating systems: A mechanism for continuous improvement. *OMNEX News.* Spring: 3.

Skrabec, Q. R. (1989). The transition from 100% inspection to process control. *Quality Progress.* Apr.: 35–36.

19

Overview of Reliability

Reliability can be defined simply as the probability that a system or product will perform in a satisfactory manner for a given period of time when used under specified operating conditions. The definition stresses the elements of probability, satisfactory performance, time, and specified operating conditions. These four elements are extremely important, since each plays a significant role in determining system/product reliability.

Probability is usually stated as a quantitative expression representing a fraction or a percent signifying the number of times that an event occurs (successes) divided by the total number of trials. When there are a number of supposedly identical items operating under similar conditions, it can be expected that failures will occur at different points in time; therefore, failures are described in probabilistic terms.

Satisfactory performance indicates that specific criteria must be established which describe what is considered to be satisfactory system operation.

Time represents a measure against which the degree of system performance can be related. One must know the time parameter in order to assess the probability of completing a mission or a given function as scheduled. Of particular interest is being able to predict the probability of an item surviving (without failure) for a designated period of time.

Specified operating conditions include environmental factors such as geographical location where the system is expected to operate, the operational profile, the transportation profile, temperature cycles, humidity, vibration, shock, and so on. Such factors must not only address the conditions for the period when the system or product is operating, but the conditions for the periods when the system (or portion thereof) is in storage mode or being transported from one location to the next

Reliability is an inherent characteristic of design. As such, it is essential that reliability be adequately considered at program inception, and that reliability be addressed throughout the system life cycle. In this chapter, our focus is to give a very cursory overview of the topic so that we sensitize the quality professional to the scope and role of reliability in the pursuit of quality.

I. BASIC ASPECTS OF MECHANICAL DESIGN RELIABILITY

Reliability is a design parameter and, as such, is the responsibility of the designer. Intuitively, one would like to have a high degree of reliability in everything that is designed and built. That means: components, a single unit of a system; subsystems, two or more units of a system; and systems, the whole item under consideration. The reliability attributes of the components which make up the system in its entirety are critical. They are critical from the standpoint that the user's needs for reliability are quite high. It is almost always possible, within the limitations of the present state of the art, to construct mechanical devices of relatively high reliability simply by overdesigning them, provided size, weight, and cost are not problems. Today's technological, sociological, and economic considerations, however, dictate that any device be designed to perform its functions at a minimum cost and a high degree of reliability, and often also at a minimum weight and within a limited space.

These considerations impose opposing demands on the design of a device, forcing the designer to compromise or to make tradeoffs between reliability and the other factors. These other factors, such as performance, cost, weight, and size, can all be expressed in quantitative terms. Therefore, reliability must be quantified to make such trade-offs meaningful.

Utilization of past experience and test data is, in principle, a valid approach for reliability study when data pertinent to the application under consideration are available. Because of the large variety of their use and operating conditions, the establishment of generally valid failure rates for parts and components does not appear to be feasible. Nevertheless, efficient design of devices for reliability requires that the reliability attribute be considered and activity for its achievement be incorporated into the design process. Even if testing of the design is planned, it is desirable to consider the reliability attribute during the preliminary design. This may decrease the amount of testing and the number of modifications needed during the subsequent development process.

For all these reasons, the designer needs methods and techniques for design reliability prediction that, by necessity, must represent an a priori approach and be useful for reliability analyses before test or service data on the design under consideration becomes available.

A system can be considered to be an interrelated assemblage of subsystems, components, and parts. Similarly, a subsystem consists of components and parts, and a component consists of parts. The reliability of an entire system depends on the reliability of its subsystems, components, and individual parts, and on the different modes of interaction and redundancies in functions of these parts. The possible failures of a device, whether drift-type (gradual performance degradation) or sudden (catastrophic), can usually be related to a failure or a malfunction of a part, although the interaction between two or more good parts sometimes can cause system failure.

The solid parts of any device can be considered to consist of various elements of two broad classes: structural and contactual elements. A structural element is defined as a portion of a part whose primary function is to bear loads, either steady, intermittent, cyclic, centrifugal, impactive vibrational, or thermal. A contactual element is a portion of a part whose primary function is to provide a contract surface with another contactual element (usually of another part) for the purpose of providing for static bearing or relative motion of the sliding, rolling, pivoting, impacting, or sealing varieties with or without intermediate lubricant. Any given part may contain several structural and several contactual elements.

The reliability (or rather the unreliability) of a part is usually associated with failure of a structural or contactual element. A given element (structural or contactual) may have several modes of failure. The basic modes of structural element failure include:

1. *Deformation*. Depending on the application, excessive elastic deformation, plastic deformation, or yielding that causes a permanent set, or a given amount of plastic deformation.
2. *Fracture*. Partial or complete separation of material that usually starts at some localized weak point or at a stress concentration (often classified as ductile or sheer-type and brittle or normal-type fractures).
3. *Instability*. Elastic or plastic buckling.

These failures are affected by several factors such as:

- Magnitude, direction, and manner of loading (e.g., steady, intermittent, cyclic, or impactive, or imposed by thermal gradients).
- Environments such as high and low temperatures, hard vacuum, high pressure, and various atmospheres.
- Size and shape of the part and the conditions of its surfaces.
- Material and its properties, including various material treatments and material responses to loads and environments.

Because of all these factors may change with time, from the reliability point of view, all failure modes are time dependent; however, for many applications involving steady loads and certain environments, the changes occurring in materials (and hence in the characteristics of the parts) are relatively small over the time periods of interest and can be neglected. In a such a case, the failure modes listed above can be classified as basic or time dependent (i.e., failure occurs essentially instantaneously whenever certain critical load combinations are exceeded). The time- or cycle-dependent aspects of the structural element failure modes are:

1. Creep, creep rupture, and creep buckling (usually at elevated temperatures).
2. Fatigue.
3. Failures occurring in one of the basic modes after the material has been subjected to creep and/or fatigue.

The modes of contactual element failure include phenomena such as mechanical wear, corrosion, fretting, galling, seizure, surface fatigue, stress corrosion, and others. All these are time and/or use dependent. Excluded from these considerations are failures which may result from causes beyond the control of the designer, such as human errors in manufacture, faulty quality control or assurance, or service use and/or maintenance abuse. While these exclusions are recognized as potent problems in reliability, they are beyond the scope of this chapter; however, they are covered in any reliability textbook.

Each of the failure modes discussed here is associated with certain material properties. The materials produced by current production methods contain various imperfections and defects. The degree of and the nature and amount of imperfections and defects and their distributions affect the material properties. The latter can also be affected by the chemical composition of the material, various metallurgical factors, production processes, various treatments, and other factors. Therefore, if a certain material property (e.g., a static strength property) is determined on a number of nominally identical specimens

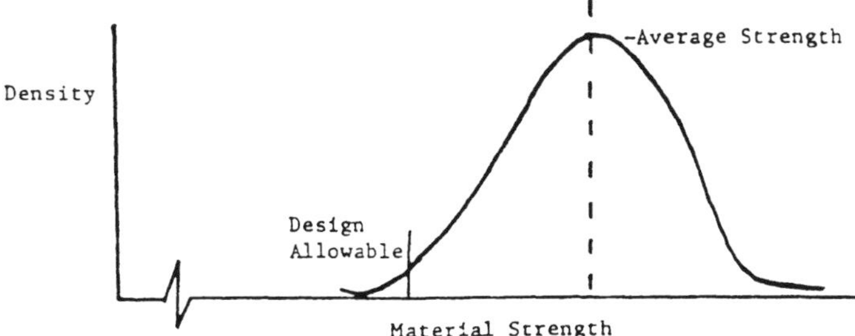

Figure 19.1 Variability in material properties.

of the same material under the same conditions, test results show a certain scatter about an average strength of the material (Figure 19.1).

The scatter or the variability in material property data is important not only for reliability considerations but also for design and for establishing allowable tolerances. Therefore, it is useful to analyze such data statistically and to present them in a statistical form as, for example, by specifying the mean value, the standard deviation, the mathematical distribution that approximates the data, and so on. Some of the design allowable tolerances listed in various materials specifications are based on statistical analysis of a large data sample. For example, the MIL-HDBK-5—Metallic Materials for Flight Vehicles—lists two design allowable tolerances, 99 and 90%. These numbers mean that 99 and 90% of the given material will exceed that value; however, the reality of the situation is that most manufacturers of metal will specify a minimum guaranteed value. Quite often, this value is established from their statistical analysis of their materials property. The minimum guaranteed value is generally the third standard deviation (0.997). This means that 99.7% of the material will exceed that value.

Given this information and the mean or average strength, one can easily develop the distribution of the material property of interest. If it is not possible to get the mean strength, an approximation method can be used. While not precise, it usually presents the worst case condition. The statistical analysis of many metal materials has shown that the mean strength will not be less than about 10% higher than the minimum guaranteed value. This seems to hold for both yield and ultimate. For cast items, the minimum guaranteed value must be degraded by 25%, since the test specimens are not cut from castings, but are pulled from uniform billets.

The scatter for time-dependent or cycle-dependent material properties is usually considerably higher than that for static properties. As an example, Figure 19.2 illustrates the variability in the fatigue properties of a material.

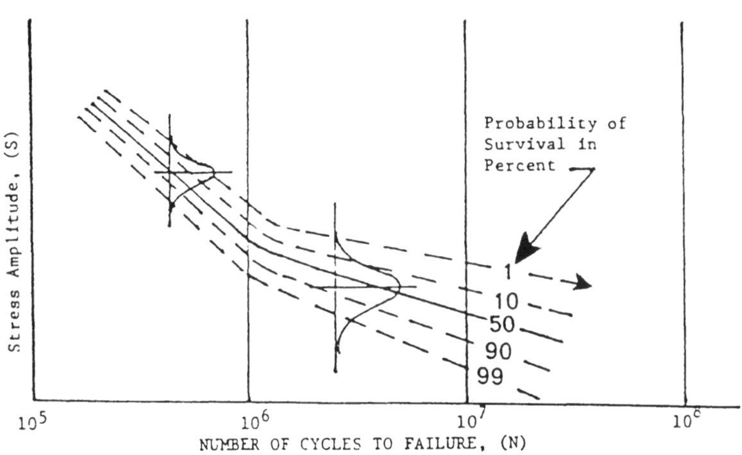

Figure 19.2 Probability-stress-cycle curves for material subjected to fatigue loading.

For a given lifetime, this variability is represented by a distribution in the same manner as for the static material properties. Over a time period or over a number of load cycles, it is customary to represent the variabilities in fatigue properties by a constant probability of survival curve. Figure 19.2 shows that the variability in times to failure at a constant stress amplitude is much higher than the variability in stress at constant life, particularly at the longer lifetimes.

The term "strength" describes the ability of a material or a part to resist failure due to external loading and environment. When a mechanical part is designed against a given failure mode, the relevant material property may, for example, be the ultimate strength, yield stress, modulus of elasticity, creep strength, fatigue strength, or wear or corrosion characteristics of the material.

The relationship between material property and the strength of the part may be quite simple. For a bar in tension, for example, the strength is the cross-section area multiplied by the stress to failure (e.g., the elastic limit, 0.2% offset yield stress, or the ultimate tensile strength, depending upon the failure criterion). Or the relationship might be of a rather complex nature (e.g., plastic buckling under an eccentric load) involving more than one material property, various geometric dimensions of the part, and more complicated loading conditions. The relationships between the strength of the part and the relevant material properties are even more complex for the time- and/or cycle-dependent and for the contactual modes of failure.

Because of the variabilities in material properties and in other factors (e.g., geometric dimensions, production, processes, and the various treatments), the strength of nominally identical parts under the same conditions

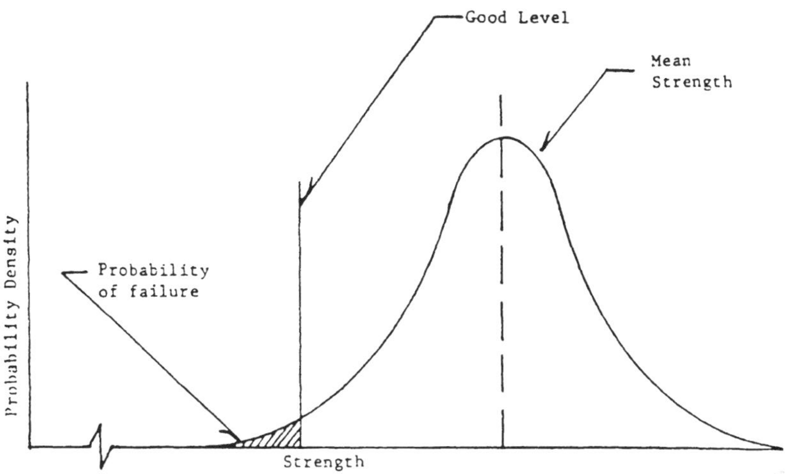

Figure 19.3 Distribution of strength.

usually varies from part to part. This variability can be represented by a distribution curve (Figure 19.3) similar to that of the variability of material properties (Figure 19.1). To estimate the expected strength distribution of a part, all the pertinent variables must be known or assumed, and taken in consideration. Methods for predicting the resulting strength distribution from the distributions that affect it can be found in Carter (1986).

If the probability-density function (Figure 19.3) represents the strength distribution of a part and this part is subjected to a certain given load, then the probability of failure is represented by the area under the probability density function to the left of the load level. The ratio of this area to the area under the entire curve gives the quantitative measure of the probability of failure, and hence the reliability or probability of success can be predicted.

From the reliability as well as from the design point of view, it is desirable to have narrow strength distributions because:

1. A narrow strength distribution results in a higher reliability than does a wide distribution for the same mean strength and for a load level less than the mean strength.
2. A narrower distribution permits the use of parts with lower mean strength (usually resulting in lighter weight)for the same reliability.

In safety devices, where a part is supposed to fail or break at certain load levels or in cases where components are supposed to break or rupture (as in some actuation devices), it is particularly important to have as narrow

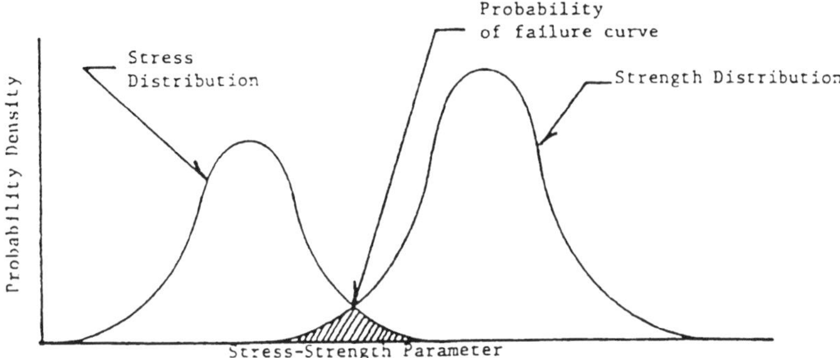

Figure 19.4 Interfering stress-strength probability-density distribution.

a strength distribution as possible. This results in a device with a more predictable (and precise) failure or actuation attribute.

Although the variabilities in part strength can be minimized through careful process control and selection processes, there are always limitations as to the narrowness of strength distributions achievable in mechanical parts. In actual use of a number of nominally identical mechanical parts, each may be subjected to a different load. Usually neither the exact strength of each individual item nor what load level it will have in service is known. If the expected distributions of strength and loads are known or can be estimated, the probability of failure can be predicted by employing the concept of interfering load-strength or stress-strength probability-density distributions, sometimes referred to as interference theory (IT). This concept is widely described in both the reliability and statistics literature. In addition, this concept represents the most promising technique for predicting reliability during the design process. For our purposes, the concept of IT is illustrated in Figure 19.4 with time not considered. The reader will notice that the probability of failure is represented by the area created by the overlap of the two curves. This probability of failure increases—and therefore the reliability decreases—with increasing overlap of the stress and strength distribution curves.

The concept of interfering stress-strength probability-density distributions in its usual form (Figure 19.4) is limited to cases in which it can be reasonably assumed that no significant changes occur in the device, part, or material over the intended design life, and where the failure depends on the instantaneous load only and not on the loading history. Generally, the term "loads" (stresses) used in this concept is not limited to mechanical loads, but it can be used in other contexts (e.g., environment, temperature, electrical current, and so on) that tend to produce a failure. On the other hand, "strength" means all of those factors that resist a failure.

When the concept of interfering load-strength probability-density distributions is used for reliability analysis of a part, either loads or stresses can be plotted, depending on circumstances and sometimes on the analyst's preference. If the loads (forces or moments) are plotted, the load distribution curve represents the probability of occurrence of the various load levels over the expected service life of the part and the strength-distribution curve represents the probabilities of the part being able to sustain certain load levels which, if exceeded, would lead to failure. If stresses are plotted as the parameter, then the strength distribution represents the stresses on the "failure stresses" usually at some critical location on the part, which, if exceeded, would lead to failure, and the stress distribution represents the probability of occurrence of the various stress levels due to the variation in loads. It does not make any difference which of these two alternatives is used as long as all factors affecting these distributions and their interrelationships are taken into consideration. Plotting of stresses may appear to be more general and hence to have wider applicability. On the other hand, plotting of loads permits a better separation of the external loading conditions from the internal properties of the part.

The concept of interfering load-strength probability-density distribution in an expanded form, with the time axis added, is shown in Figure 19.5. Both load and strength distribution may change with time. For simplicity, Figure 19.5 illustrates a case in which the load distribution remains constant and the strength distribution changes with time. This is the case for most common engineering situations.

In principle, such an expanded concept permits predictions of the probabilities of failure with time and of the time-related reliability function, provided that the changing attributes and the variations with time are known. The relationship between traditional design and design reliability approaches is important and must be understood before applying the analytical tools of design reliability.

To avoid failures, the strength of a part must be higher than the applied load. On the other hand, the part should not be over-designed and its strength should not be too high if the design is to be practical. The traditional approach to mechanical design attempts to achieve the proper balance between the external loading and the strength of a part by utilizing factors of safety (and sometimes, margins of safety). These factors of safety can be variously defined and if established from experience, they provide satisfactory designs in many applications. For new designs involving new applications and new materials, however, use of safety factors can be misleading and can result in under-design or over-design, depending upon the circumstances.

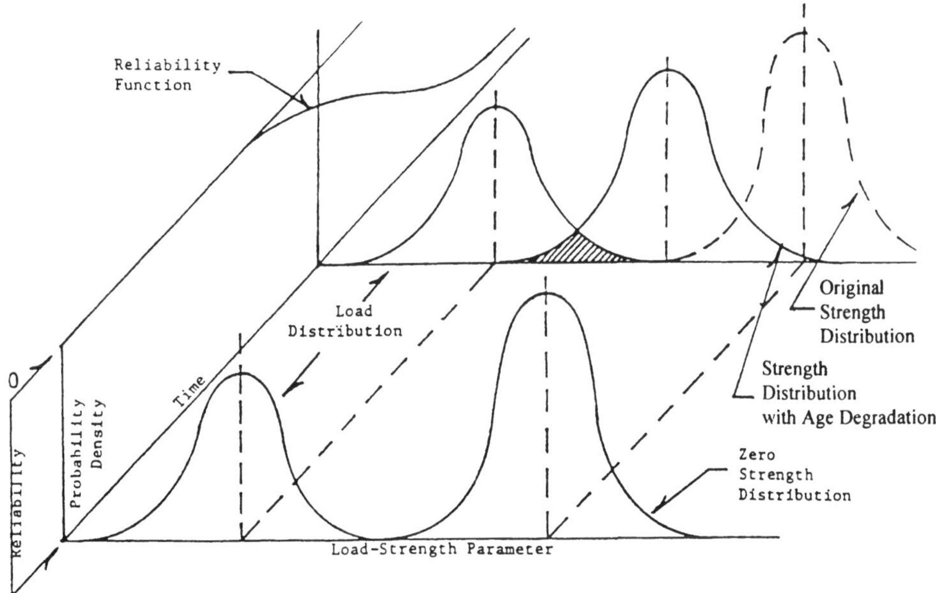

Figure 19.5 Load and strength probability density-distribution over time.

Designing for reliability does not require utilization of safety factors because reliability depends on load and strength distributions and their relative positions on the load-strength parameter axis. From a practical point of view, it is advantageous to relate these two approaches to each other. This has led to some quantitative relationships between factors of safety and reliability as well as to some new definitions of so called reliability safety factors and safety margins.

The illustration between load and strength distributions and factor of safety is shown in Figures 19.6a and 19.6b, whereby the factor of safety FS_m, in this case, is defined as the ratio

$$FS_m = \frac{S_m}{l_m}$$

where S_m is the mean strength and l_m is the mean load. A comparison between Figures 19.6a and 19.6b shows that the same value of factor of safety does not always provide the same degree of reliability.

For given value of mean strength and mean load or for a given factor of safety, narrow distributions give smaller probabilities of failure and therefore higher reliability than wider distributions. This is shown in Figure 19.6b where a small increase in the factor of safety would increase the reliability considerably.

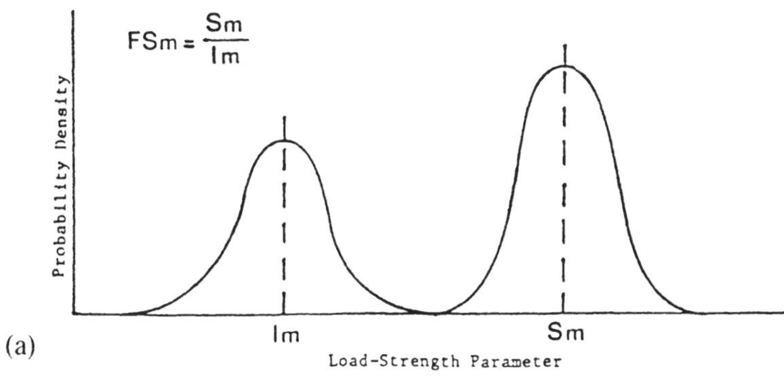

(a)

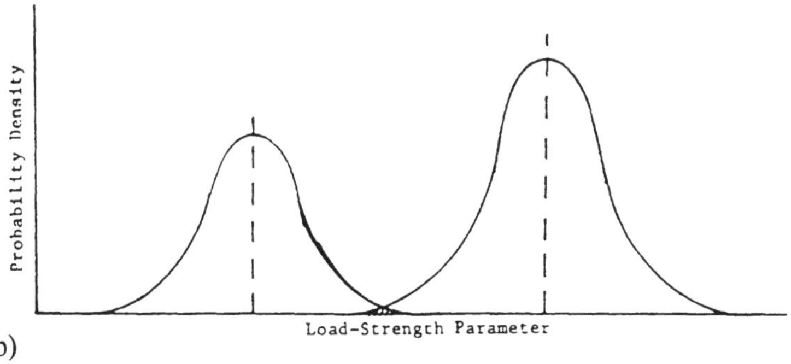

(b)

Figure 19.6 a) Load-strength parameter. b) Relationship between load and strength distribution and a factor of safety based on mean values of load and strength.

The factors of safety that are used frequently in mechanical design, however, are not always based on mean load and mean strength. Because there is scatter in material strength properties, many design computations are based on design allowable tolerances for various materials. These allowable tolerances are usually vaguely defined and represent material property values that are expected to be exceeded by most of the material. Also, the design load is usually taken at some reasonably high value that is not expected to be exceeded in service use. The relationship between load and strength distribution and this concept of factor of safety is represented by

$$FS = \frac{S}{L}$$

where S = strength and L = load (Figure 19.7).

The main difference between the usual design computations and the design reliability approach is the probabilistic nature of the latter. The usual

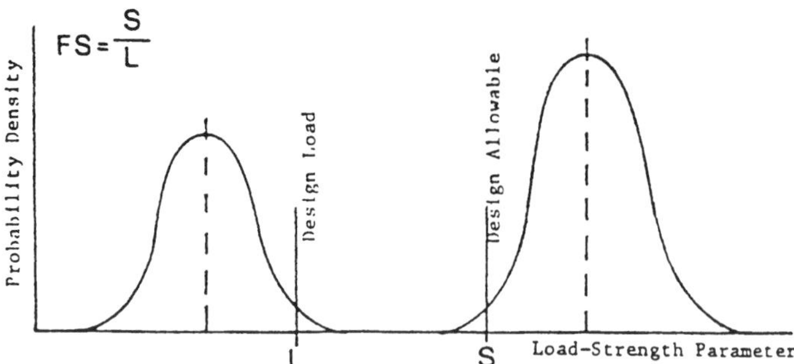

Figure 19.7 Relationship between load and strength distributions and a factor of safety based on design load and design allowable tolerances.

design computations, based on some discrete values for load and strength, are deterministic and all variabilities and uncertainties are included in some general way in the selected factor of safety. In the design reliability approach, the variabilities are taken into consideration for estimating the probability of failure and reliability.

The reliability of a mechanical device depends on the reliability of the individual parts and on the interactions of these parts within the device. Therefore, in a design reliability prediction for a mechanical device, in addition to the increasing number of elements and parts to be considered, the interrelationship and interactions among the elements of a part and among the parts of a device must also be taken into consideration; this complicates such predictions. In practical applications, however, it is sufficient to consider only the predominant modes of failure for a reasonably accurate design reliability prediction.

II. REASONABLE DESIGN RELIABILITY PREDICTIONS

The major limitation to the use of the strength-stress reliability prediction process is the ability to describe the distribution of environment or the load distribution. This limits the accuracy of the resulting reliability prediction, but does not disqualify the methodology as a viable design tool for trade-offs between competing designs and for making the initial reliability prediction during the design process.

The overriding requirement to quantify reliability during the design process to reduce test time and to assess product liability exposure as early as possible has been a forced function toward the development of practical applications of the stress-strength methodology. At present, sound design engineering coupled with reliability considerations are the best way to achieve the desired levels of reliability for basic designs. Although these design reliability techniques are still crude, in that they vary some from classical statistical methods and are handicapped by the lack of some of the needed information, being aware of them and considering them during the design process can help create better and more reliable products.

A. Methodology

We start with the problem of defining the stress or load distribution. While it may be possible through exhaustive analysis and testing to determine the precise distribution of stress or load, it is usually not practical from the standpoint of schedules and available resources. Therefore, the engineer must use all data at his or her disposal and his or her sound engineering judgment to establish the stress-load parameter for the reliability estimate. Before the applied stress-strength reliability method is illustrated, the concept of reliability boundary must be discussed.

B. Reliability Boundary

The reliability boundary (R_b) is the maximum anticipated operating stress-load level and may be presented two ways:

1. *As a discrete point* (e.g., 30 g, +125°F, -25°F, 105 KPSI, or 2900 PSI). When the R_b is presented this way, it is assumed that the device will be operated at that stress-load level 100% of the time. Since this is usually not the case in actual practice, such this method represents a worst-case situation. There are many proponents of this worst-case design concept, but indiscrete and inflexible application sometimes imposes undue constraints upon the designer. This is shown in Figure 19.8.
2. *As a point in the stress-load density function*. For example, the stress load reliability boundary for a pressure vessel could be expressed as a 3σ limit of a normal distribution (e.g., 7.5 KPSI), indicating that a stress load of 7.5 KPSI or more would be experienced only 0.135% of the time. This is shown in Figure 19.9.

In comparison to Figure 19.8, the method shown in Figure 19.9 obviously presents a more accurate picture of what stress-load level to expect. Data to develop discrete distributions is often difficult to obtain in the necessary time frame, therefore, judgmental assumptions are needed.

Now we are ready to discuss the difference this method makes in design philosophy and the resulting reliability estimates.

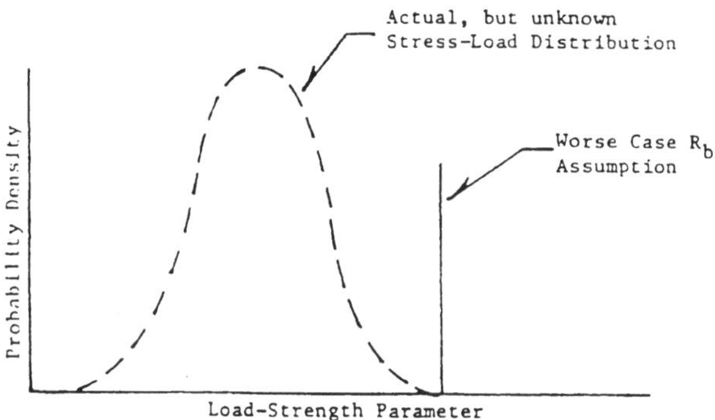

Figure 19.8 Worst-case, discrete point reliability boundary assumptions.

C. Strength Distribution

Strength distribution is an expression of the expected variation in strength of the material as modified by the design application. The SAE Metals Handbook (American Society of Metals, 1985) and MIL-HDBK-5 (Aeronautical System Center, 1995) provide the basis for determining the strength distribution for most common metals. The manufacturers of metals also provide some data. In some cases these references provide statistical values for the mean and negative third standard deviation of strength. In other cases, a minimum

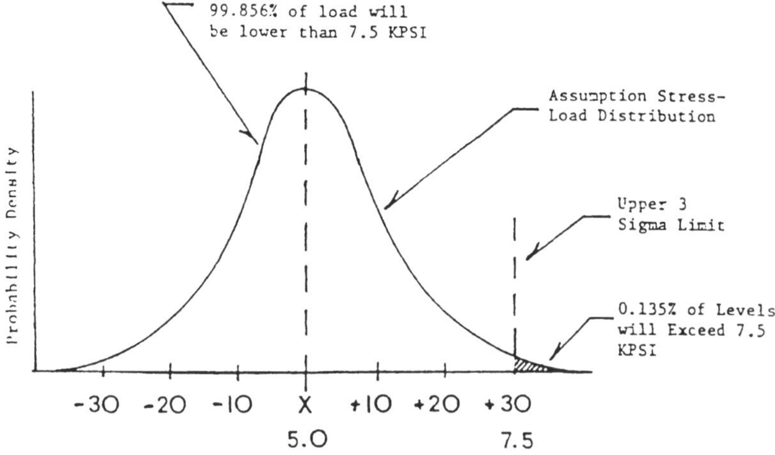

Figure 19.9 Stress load 3σ limit reliability boundary assumption.

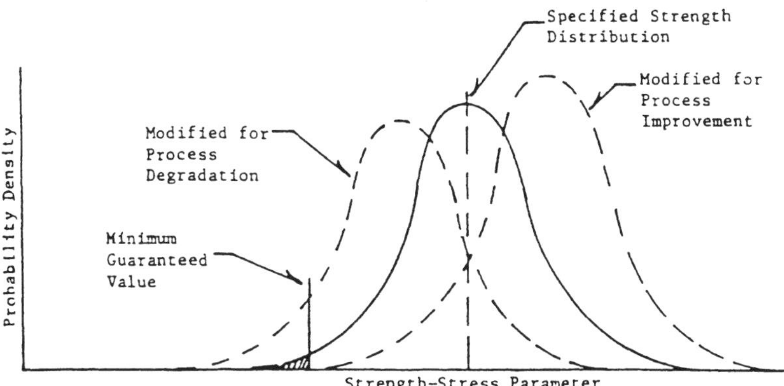

Figure 19.10 Material strength distribution with enhancement and degradation from applied processing.

guaranteed value is specified. The latter is in most cases provided by the manufacturer and is usually the negative third standard deviation. In any case, the values given are for the base material and they must be modified for the particular processes called for in an application (e.g., shot peening increases strength, casting degrades strength, and so on). Many of these factors are provided in the references cited. This is shown in Figure 19.10. When the minimum guaranteed value is used, 0.135% of the material can be expected to have strength less than the value specified.

D. Safety Margin (S_m)

The safety margin of a device is defined as the number of standard deviations of the strength distribution (σ_s), which lie between the reliability boundary, R_b, and the mean strength, $\bar{X}$. This can be expressed mathematically as:

$$S_m = \frac{\bar{X} - R_b}{\sigma_s}$$

This S_m is the same as the X/σ value calculated by applying density function mathematics, where:

$$\frac{X}{\sigma} = \frac{\text{Limit} - \text{Mean}}{\sigma_s}$$

where the limit is R_b.

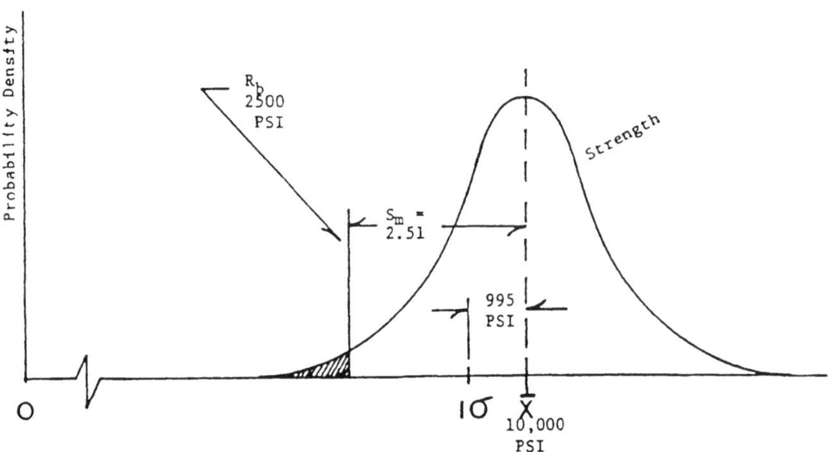

Figure 19.11 Safety margin for pressure vessel burst environment.

As an example, assume we have a reliability boundary, R_b = 7.5 KPSI for a pressure vessel. Through information provided by the material producer and a process degradation factor, the strength distribution for the device is determined to have a mean strength, $\bar{X}$ = 10,000 PSI, and a standard deviation, σ = 995 PSI. The safety margin of the pressure vessel, in reference to the 7.5 KPSI stress-load, (R_b) is given by:

$$S_m = \frac{\bar{X}_s - R_b}{\sigma_s}$$

$$= \frac{7500 - 10000}{995} = 2.51$$

This is shown in Figure 19.11.

Having calculated a safety margin, we can now solve for the percentage of the pressure vessels that will be above or below the reliability boundary. Since we are interested in the success side, we will illustrate the percentage which lie above R_b. To do this we use a statistical method developed for solution to one-limit problems. This method is explained in many statistical texts, and therefore will not be elaborated upon here; however, we will use Table 2.6 for the tail probability.

A safety margin of 2.51 (from Table 2.6) indicates that 0.99396 (= 1.0 – 0.00604) of the pressure vessels will not fail until R_b of 7500 PSI is exceeded. From this we can infer that pressure vessels built to our design will have a reliability of 0.99396 (i.e., 99.396% will not burst at 7500 PSI).

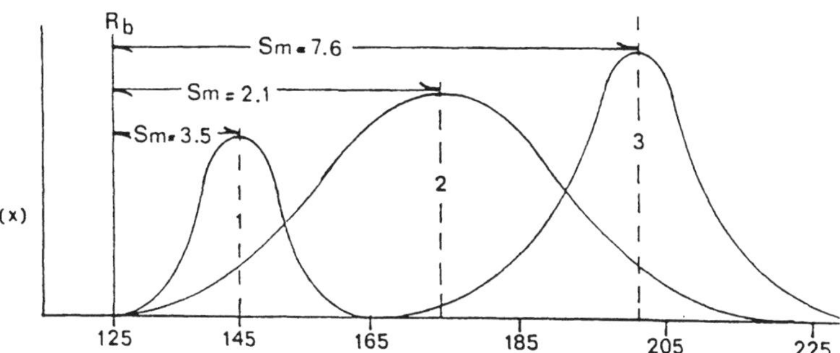

Figure 19.12 Safety margin results when multiple failure modes are of concern.

E. Multiple Failure Mode

For devices that exhibit a single failure mode, the safety margin and reliability are calculated by the technique described above. Before getting into multiple failure modes and their treatment, however, it is important to consider that:

> While the application of this method indicates that a safety margin of 6.32 has statistical meaning, in practice, a population safety margin of 5 or higher indicates that the applicable failure mode will not occur unless the strength distribution deviates greatly from the normal distribution.

This means that once $S_m \geq 5$ is determined, that it should be documented and the failure mode disregarded for reliability purposes. Since most parts and components perform more than one function and/or have more than one critical parameter for each function, and most are made up of many types of materials and parts and require many fabrication processes during manufacture, it follows that they may exhibit a variety of failure modes during operation.

In the conduct of a design reliability analysis, each failure mode of concern must be evaluated individually; a failure distribution must be developed for each failure mode and safety margins calculated for each individual failure distribution (Figure 19.12). In Figure 19.12 each failure mode is described in terms of its own failure distribution and resulting safety margin with reference to the same reliability boundary. If each of these failure modes is independent, then the reliability of the device can be determined using the Z-values from Table 2.6.

$$R = Rf_1 \times Rf_2 \times Rf_3$$

$$= 0.9998 \times 0.9821 \times 1.0 = 0.9819$$

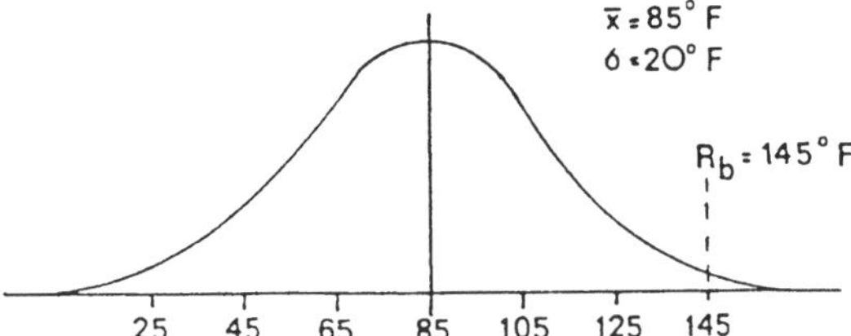

Figure 19.13 Stress distribution for operating temperature.

This shows that this independent evaluation of each failure mode identifies the priorities necessary to improve reliability. For example, a better material or some form of insulation from the environment could increase the safety margin of failure mode 2. If the mode could be eliminated or S_m increased to a number above 5, then the reliability of the device for this environment exposure would be 0.9998.

F. When Stress Distribution Is Known

When safety margins are calculated in reference to a single point or fixed reliability boundary, the resulting reliability estimate is conservative. This is so because it is assumed that the device will always be operated at R_b. To understand the impact of using the distributed parameter for both the stress distribution and the strength distribution, consider the following. Figure 19.13 shows a stress distribution for the operating temperature of a device and the maximum anticipated operating limit (145°F), the +3σ point (normally considered the reliability boundary). On the other hand, Figure 19.14 shows a

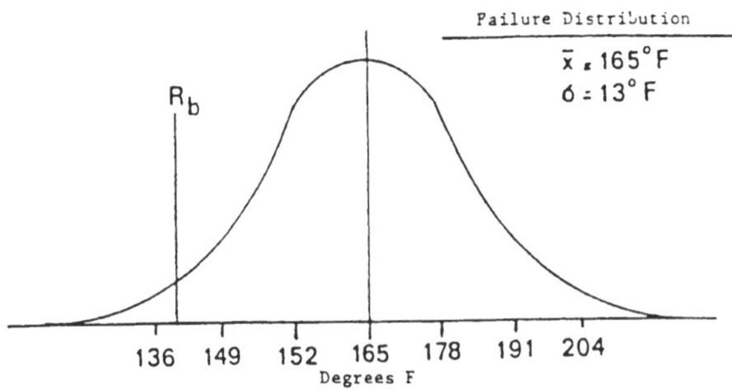

Figure 19.14 Strength distribution for operating temperature.

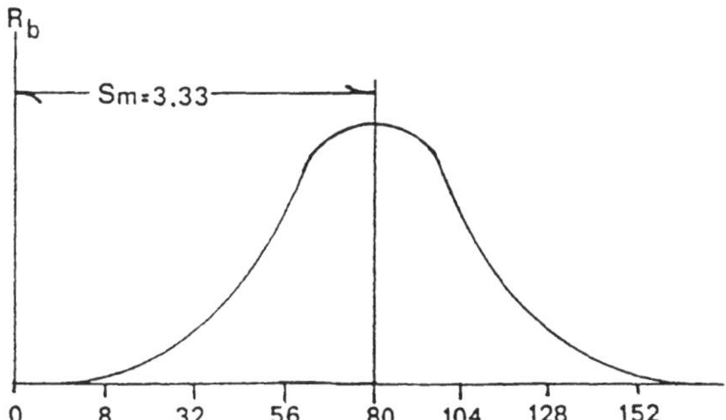

Figure 19.15 Strength and stress difference distribution.

strength distribution developed from material analysis. The graph shows also a safety margin for the device (referenced to the 145°F reliability boundary).

When the discrete reliability boundary (145°F) is used, we find that $S_m = 1.54$, which results in a reliability of 0.938. We know, however, that the 145°F limit is the 3σ limit of the stress distribution and will occur only 0.135% of the time. The question, then, is how does this effect the estimate of reliability for the device in the temperature environment?

If we select random values from the stress and failure distribution and subtract the stress value from the failure distribution and subtract the stress value from the failure value, a positive result will indicate a success; it will do so because the failure limit will be higher than the stress value. A negative answer would indicate a failure since the value for stress would be in the failure domain. With this knowledge, one can calculate a difference distribution and, through the application of the safety margin technique, solve the stress and therefore the success value. When the stress and failure distribution are distributed normally, this difference distribution is also normally distributed and has the following parameter:

$$\bar{X}\,(\text{difference}) = \bar{X}(\text{failure}) - \bar{X}(\text{stress})$$

$$\sigma\,(\text{difference}) = \sigma^2\,(\text{failure}) + \sigma^2\,(\text{stress})^{1/2}$$

From the strength and stress distribution parameter given in the example, we calculate the following.

$$\bar{X}\,(\text{difference}) = 165 - 85 = 80$$

$$\sigma\,(\text{difference}) = (20^2 + 13^2)^{1/2} = 24$$

Therefore, the resulting distribution looks like the one shown in Figure 19.15.

The reliability of the device can be calculated from this difference distribution by setting $R_b = 0$ (Figure 19.15) and calculating S_m. Our example resulted in $S_m = 3.33$ which translates (using Table 2.6) to a reliability of 0.9996. A comparison of this 0.9996 reliability to the 0.9380 predicted when using a discrete number for the reliability boundary illustrates the conservatism and shows the significance of knowing the stress distribution when estimating reliability values.

III. GENERAL RELIABILITY FACTORS

A. The Reliability Function

Whereas in the previous section we discussed mechanical design reliability issues, in this section we take a more general view of what reliability is and discuss some classic issues. Because of the cursory nature of our review, the reader is encouraged to see Blanchard (1986); Kececioglu (1991); Nelson (1983); Shapiro (1990); Moura (1991); Meeker and Hahn (1985) and Kapur and Lamberson (1977); and Lloyd and Lipow (1977).

Reliability as previously defined is simply the probability that a system or product will perform in a satisfactory manner for a given period of time when used under specified operating conditions, and is expressed as:

$$R(t) = 1 - F(t)$$

where $F(t)$ is the probability that the system will fail by time t. What is interesting about $F(t)$ is the fact that it turns out to be the failure distribution function (or the unreliability function). If the random variable t has a density function of $f(t)$, then the expression for reliability is

$$R(t) = 1 - F(t) = \int_t^{\infty} f(t)\,dt$$

Assuming that the time to failure is described by an exponential density function, then

$$F(t) = (1/\theta)e^{-t/\theta}$$

where θ is the mean life, t is the time period of interest, and e is the natural logarithm base (2.7183). The reliability at time t is

$$R(t) = \int_t^{\infty} (1/\theta)e^{-t/\theta}\,dt = e^{-t/\theta}$$

The mean life (θ) is the arithmetic average of the lifetimes of all items considered. The mean life (θ) for the exponential function is equivalent to mean time between failure (MTBF). Thus,

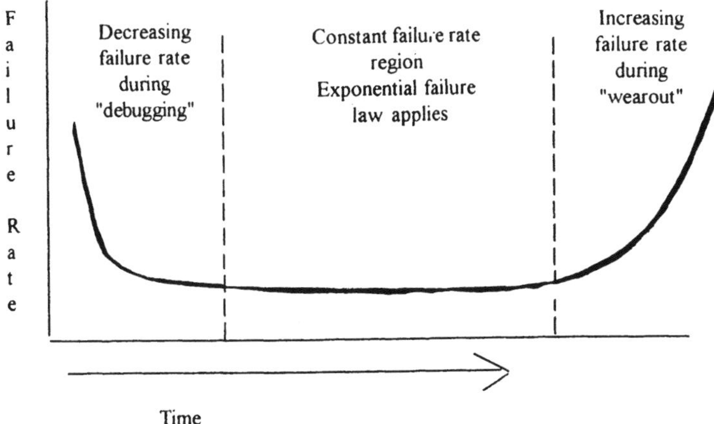

Figure 19.16 Bathtub curve distribution.

$$R(t) = e^{-t/M} = e^{-lt}$$

where l is the instantaneous failure rate and M is the MTBF. If an item has a constant failure rate, the reliability of that item as its mean life is approximately 0.37. In other words, there is a 37% probability that a system will survive its mean life without failure. Mean life and failure rates are related as

$$l = \frac{1}{\theta}$$

Whereas the application of the exponential reliability function is the most commonly assumed in many application, the reader must understand that the failure characteristics of different items are not necessarily the same. There are a number of well-known probability density functions which describe the failure characteristics of different equipments. These include the binomial, exponential, normal, Poisson, gamma, and Weibull distributions.

IV. FAILURE RATE

The rate at which failures occur in a specified time interval is called the failure rate during that interval. The failure rate (l) is represented as

$$l = \frac{\text{number of failures}}{\text{total operating hours}}$$

The failure rate may be expressed in terms of failures per hour, percent failures per 1,000 hours, or failures per million hours. Regardless of how they are reported, the majority of all failures—especially those that follow the negative exponential distribution—follow what is known as the "bath-tub curve distribution" and is shown in Figure 19.16 with some explanation.

V. RELIABILITY COMPONENT RELATIONSHIP

To determine the probability of a system or subsystem or even a component, the experimenter must know their location (whether in series, parallel, or combination networks). These networks are used in reliability block diagrams and in static models employed for reliability prediction and analysis.

A. Series Networks

Series networks are the most common, simplest, and easiest to analyze. They are of the form shown in Figure 19.17. The reliability is simply calculated by multiplying the individual item probability. For example in Figure 19.17 we have three items: A, B, and C, with respective probabilities of 0.9124, 0.7988, and 0.8246. The reliability of this series is:

$$R(t) = R_A \times R_b \times R_C$$
$$= 0.9124 \times 0.7988 \times 0.8246 = 0.6010$$

B. Parallel Networks

A simple parallel network is one where a number of the components are in parallel and where all the components must fail in order to cause a system failure. A typical two-component (A and B) parallel network is shown in Figure 19.18, with their corresponding reliabilities of 0.7891 and 0.9899. The reliability is calculated as:

$$R = R_A + R_B - (R_A \times R_{b)}$$
$$= 0.7891 + 0.9899 - (0.7891 \times 0.9899) = 0.9979$$

The general form for a more complicated (i.e., more than two redundant items) parallel system is:

$$R = 1 - (1 - R)^n$$

Figure 19.17 Series network.

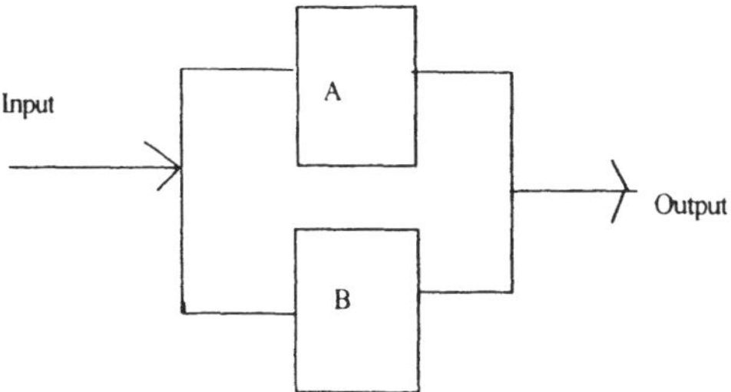

Figure 19.18 Parallel network.

Where R is the reliability of the item and n is the amount of items in the system. Parallel networks are redundant systems used primarily to improve system reliability.

C. Combined Series-Parallel Networks

Various levels of reliability can be achieved through the application of a combination of series and parallel networks. Typical combination networks are shown in Figure 19.19. The reliability of the first network is given by:

$$R = R_A\,(R_b + R_C - (R_b \times R_C)$$

For the second, the reliability is given by:

$$R = [1 - (1 - R_A)(1 - R_{b))]}[1 - (1 - R_C)(1 - R_D)]$$

And for the third network the reliability is given by:

$$R = [1 - (1 - R_A)(1 - R_{b)}(1 - R_C)[R_D][R_E + R_F - (R_E)(R_F)]$$

Combined series-parallel networks (such as in Figure 19.19) require that the analyst first evaluate the redundant (parallel) elements to obtain a unit reliability, and then combine the unit(s) with other elements of the system. Overall system reliability is then determined by finding the product of all "series" reliability. Reliability block diagrams are generated, evolving from the functional block diagram for the system and leading through progressive expansion to individual system components.

VI. MARKOV MODELING AS A RELIABILITY TOOL

A Markov model is a probabilistic mathematical model that can accurately capture the effects of order-dependent component failure and changing failure

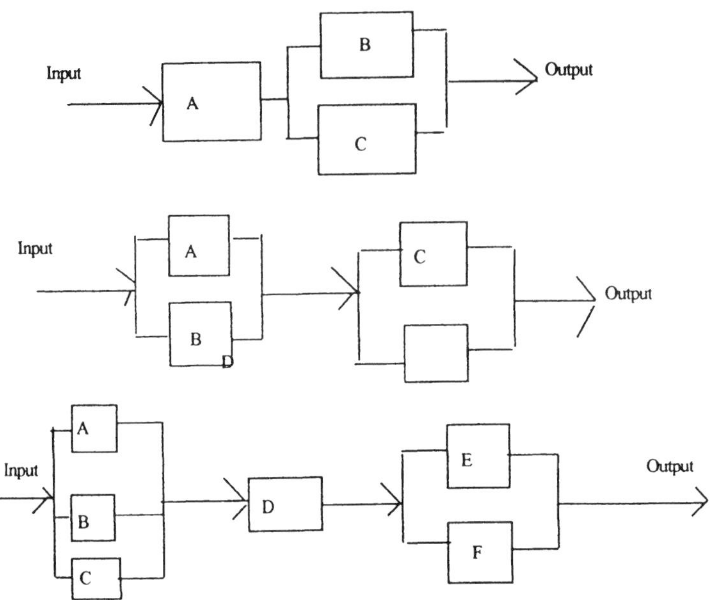

Figure 19.19 Combination network.

rates resulting from stress or other factors. A Markov model is a very flexible tool in modeling dynamic system behaviors. It is not, however, the most appropriate modeling technique for every modeling situation. The first task in obtaining a reliability or availability estimate for a system is selecting which modeling technique is most appropriate to the situation at hand.

There are many reliability modeling tools that can represent system behaviors. The best strategy is to match the modeling tool with the characteristic and required level of detail in the behaviors of the system. Those tools extend from the simplest modeling techniques, such as fault tree and reliability block diagram, to the most complex, such as simulation. In terms of complexity, Markov modeling lies in the middle.

Compared to other modeling techniques, Markov modeling offers greater flexibility in expressing dynamic system behavior. It tells the designer whether or not various reliability goals will be met. If the goals are not met, the reliability bottlenecks of the design are easily identified. This valuable feedback enables one to make appropriate system modifications and to evaluate the impact of those changes by running the revised model through the program. It enables the designers to determine the relative impact of alternative designs on system reliability.

The Markov model can model everything that a fault tree and reliability block diagram can, plus it can model the following (Gebrael, 1996):

- Time-varying failure rate.
- Sequence dependent behavior.
- Repairable systems. (The model can give an estimate of how often the system must be repaired, and an estimate of the system availability.)
- Transient and intermittent faults.
- Standby systems with spares.
- Degraded modes of operation.
- Common-cause failures.

There are two types of Markov models: discrete and continuous. The most useful is the continuous time Markov chain (CTMC), which is divided into three categories:

- Homogeneous: the simplest and most commonly used, and handles constant failure rates.
- Nonhomogeneous: more complex and can handle time-dependent failure rates. The most commonly used—transition rates—are in the form of the Weibull hazard rate function.
- Semi-Markov: the most complex of the three and is capable of representing more complex system behavior than the other two. The state holding times can have distributions that can be completely general (nonexponential) and can also depend on the next state.

Is it practical to use Markov modeling in the reliability area? The answer is a categorical yes. The following are some applications:

1. Reliability Engineering
   - Fault-tolerant system reliability analysis
   - Reliability, availability, maintainability, safety prediction (complex redundancies, time-dependent transitions, sequence-dependent events, common cause failures)
2. Statistics
   - Probability evaluation
   - Renewal and queuing
3. Operations Research
   - Maintainability and availability predictions
   - Waiting line and servicing (queuing) modeling
   - Inventory system analysis
   - Traffic flow modeling
4. Computer and System Engineering
   - Fault-tolerant system design

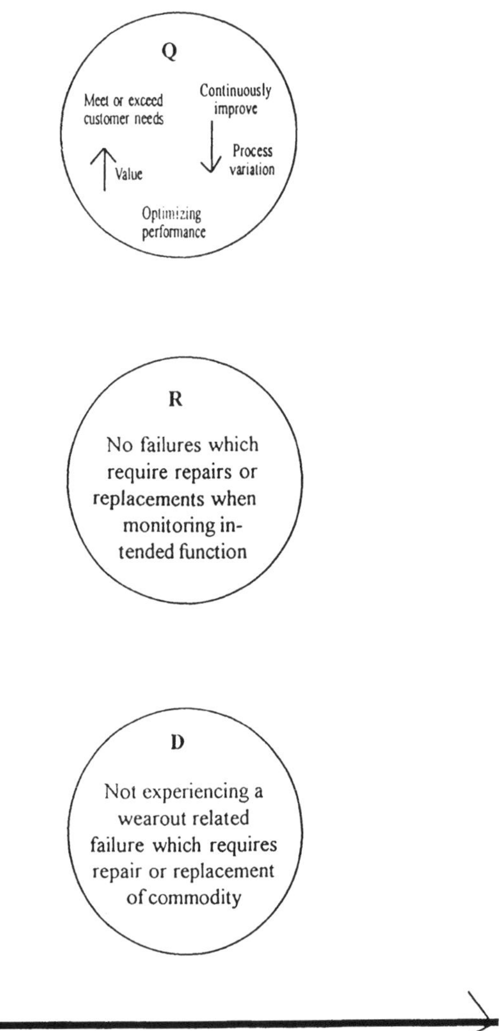

Figure 19.20 A typical overall reliability concept.

  - Computer system performance evaluation
  - State diagram analysis
5. Software Engineering
  - Petri net simulations
  - Markov chain fault modeling

VII. A TYPICAL RELIABILITY PROGRAM PLAN FOR ANY ORGANIZATION

By definition a reliability program must establish and maintain an efficient program to support economical achievement of overall program objectives within the organization. Therefore, to be considered efficient, a reliability program must clearly:

1. Improve operational readiness and mission success of the major end-item
2. Reduce item demand for maintenance manpower and logistic support
3. Provide essential management information
4. Hold down its own impact on the overall program cost and schedule

In essence, then, a reliability program must be:

- Predictable in design
- Measurable on test
- Assurable in production
- Maintainable in the field

An overview of the typical reliability concept is shown in Figure 19.20.

A. Program Requirements

The program must include an appropriate mix of reliability engineering and accounting tasks depending on the life-cycle phase. These tasks must be selected and tailored according to the type of item (system, subsystem, or requirement) and for each applicable phase of the acquisition (concept, validation, research and development, and production.) They must be planned, integrated, and accomplished in conjunction with other design, development, and manufacturing functions. The overall acquisition program must include the resources, schedules, management structure, and controls necessary to ensure that specified reliability program tasks are satisfactorily accomplished.

B. Program Plan Elements

The elements for an effective reliability program are:

- A schedule with estimated start and completion points for each reliability program activity or task.
- A method by which the reliability requirements are disseminated to designers and associated personnel.
- The identification of key personnel for managing the reliability program.
- A description of the management structure, including interrelationship between line, service, staff, and policy organizations.
- A statement of what source of reliability design guidelines or reliability design review checklist will be utilized.
- The procedures or methods (if procedures do not exist) for recording the status of actions to resolve problems.
- A description of the interrelationships between reliability tasks and activities, and a description of how reliability tasks will interface with other system-oriented tasks. The description shall specifically include the procedures to be employed to ensure that applicable reliability data derived from, and traceable to the reliability tasks specified are integrated into the logistic support analysis program (LSAP) and reported on appropriate logistic support analysis records (LSAR).

C. Monitor/Control of Suppliers and Subcontractors

Ensure that both the suppliers and subcontractors have a reliability program that is compatible with the overall program and includes provisions to review and evaluate their reliability efforts.

D. Program Design Reviews

The design review agenda should include all pertinent aspects of the reliability program such, including (when applicable):

- Reliability modeling.
- Reliability allocation.
- Reliability prediction.
- FMEA.
- Reliability content of specification.
- Design guideline criteria.
- Reliability critical items program.
- Other problems affecting reliability.
- Reliability prediction and analysis.
- Test schedule.
- Test profile.
- Test plan, including failure definition.
- Results of applicable reliability/growth testing.

E. Failure Reporting Analysis and Corrective Action System

The organization must establish a closed-loop failure reporting system (a feedback loop), procedures for analysis of failures to determine cause, and documentation for recording corrective action taken.

F. Failure Review Board

The organization must establish a failure review board to review failure trends, significant failures, corrective action status, and to ensure that adequate corrective actions are taken in a timely manner and recorded during the development and production phases of the program. Failure review board members must include appropriate representatives from design, reliability, system safety, maintainability, manufacturing, and parts and quality assurance activities.

G. Reliability Modeling

The purpose of this task is to develop a reliability model for making numerical apportionments and estimates to evaluate system/subsystem equipment reliability.

H. Reliability Allocations (Apportionments)

Once quantitative system requirements have been determined, they are allocated or apportioned to lower levels.

I. Reliability Predictions

This estimates system/subsystem reliability based on component failure rates. This is a hardware oriented estimate of system/subsystem reliability.

J. Failure Modes, Effects, and Criticality Analysis

This is to identify potential design weakness through systematic, documented consideration of the following: all likely ways in which a component or equipment can fail, causes for each mode, and the effects of each failure (which may be different for each mission phase).

K. Failure Mode and Effect Analysis

This is to identify and remove any known and unknown failures from the system, design and process through a systematic methodology based on the frequency, severity, and the detection of the failure.

L. Sneak Circuit Analysis

This is to identify latent paths which cause occurrence of unwanted functions or inhibit desired functions, assuming all components are functioning properly.

M. Electronic Parts/Circuits Tolerance Analysis

This examines the effects of parts/circuits electrical tolerances and parasitic parameters over the tangle of specified operating temperatures.

N. Parts Program

This controls the selection and uses of standard and nonstandard parts (configuration control).

O. Reliability Qualification Test

This determines if equipment representative of production configuration has achieved reliability requirements.

P. Production Reliability Acceptance Test Program

This ensures that the reliability of the hardware is not degraded as the result of changes in tooling, processes, work flow, design, and/or parts quality.

Q. Environmental Stress Screening

This establishes and implements environmental stress screening procedures so that early failure(s) due to weak parts, workmanship defects, and other nonconformance anomalies can be identified and removed from the equipment.

R. Reliability Development/Growth Test Program

This conducts prequalification testing as well and provides a basis for resolving the majority of reliability problems early in the development phase; it also incorporates corrective action to preclude recurrence prior to the start of production.

S. Reliability Critical Items

This identifies and controls those items which require special attention because of complexity, application of advanced state-of-the-art techniques, and the impact of potential failure on safety, readiness, mission success, and demand for maintenance/logistics support.

T. Effects of Functional Testing, Storage, Handling, Packaging, Transportation, and Maintenance

This determines the effects of storage, handling, packaging, transportation, maintenance, and repeated exposure to functional testing on hardware reliability.

REFERENCES

Aeronautical System Center (1995). MIL—HDBK-5, Aeronautical Ssstem Center, Wright Paterson Air Force Base, Ohio.

Ameridcan Society of Metals (1985). *ASM Metal Handbook*. ASM. Material Park, OH.

Blanchard, B. S. (1986). *Logistics engineering and management*. 3rd ed. Prentice Hall. Englewood Cliffs, NJ.

Boyd, M. A. (1995). What Markov modeling can do for you. In: the *7th Annual International Reliability, Maintainability and Supportability Workshop*. NASA Ames Research Center. Moffett Field, CA.

Carter, A. D. S. (1986). *Mechanical Reliability*. 2nd ed. Macmillan Education Ltd., London.

Gebrael, M. G. (1996). Markov modeling as a reliability tool. *Proceedings of the 8th Annual SAE RMS Workshop*. SAE: Reliability, Maintainability, Supportability and Logistics. Dallas.

Kapur, K. C. and Lamberson, L. R. (1977). *Reliability in engineering design*. John Wiley. New York.

Kececioglu, D. (1991). *Reliability engineering handbook*. Vols. I & II. Prentice Hall. Englewood Cliffs, NJ.

Lloyd, D. K. and Lipow, M. (1977). *Reliability: management, methods, and mathematics*. 2nd ed. Defense and Space Systems Group, TRW Systems and Energy. Redondo Beach, CA.

Meeker, W. Q. and Hahn, G. J. (1985). *How to plan an accelerated life test*. American Society for Quality Control. Milwaukee, WI.

Moura, E. C. (1991). *How to determine sample size and estimate failure rate in life testing*. American Society for Quality Control. Milwaukee, WI.

Nelson, W. (1983). *How to analyze reliability data*. American Society for Quality Control. Milwaukee, WI.

Shapiro, S. S. (1990). *How to test normality and other distributional assumptions*. rev. ed. American Society for Quality Control. Milwaukee, WI.

20

Quality Awards

With the proliferation of quality awareness, not only in the United States but all over the world, the need for self-assessment and confirmation of excellence through a third party has accelerated in recent years. In fact, in some cases, that recognition comes directly from the Government itself. Whereas the self-assessment serves the organization internally, the recognition through the third party is known as the "Quality Reward" and it is usually sponsored by a governmental department, such as the Commerce Department.

The self-assessment and quality award are both designed to help organizations continuously improve toward a goal of "world-class/best," but it must be understood that one of the predominant factors of this improvement is the issue of leadership. Nothing changes until management attitudes change.

I. THE MALCOLM BALDRIGE QUALITY AWARD

In the early 1980s, the United States government and business leaders were concerned with inability of U.S. companies to compete with foreign nations. Quality and service needed to improve within our industries for the United States to regain its competitive position. It was during this time that our national political leaders formed quality councils and committees to establish a national quality award program. In August 1986, Congress introduced Bill 5321 to establish a National Quality Improvement Award (DeCarlo and Sterett, 1990; Brown, 1996).

Malcolm Baldrige was the Secretary of Commerce during the Reagan administration. Baldrige was a notable business leader and an avid equestrian. On July 25, 1987, Malcolm Baldrige was killed in an accident while horse riding. In honor of the friend they had lost, the Senate Committee on Commerce, Science, and Transportation renamed Bill 5321 to the Malcolm Baldrige National Quality Improvement Act, and in 1987 the Baldrige Award was voted into public law 100–107.

A. Malcolm Baldrige National Quality Award Organization

The Malcolm Baldrige National Quality Award (MBNQA) is a public/private partnership between private sectors and government. It is supported by private funds, volunteers, and award recipients that participate by sharing information. The MBNQA organization consists of the following divisions: National Institute of Standards and Technology (NIST), American Society for Quality Control (ASQC), Board of Overseers, Board of Examiners, and award recipients. The MBNQA organization main objective is to raise funds to permanently endow the award program.

The National Institute of Standards and Technology (NIST) is an agency of the Department of Commerce that manages the award program. One of NIST's goals is to aid U. S. industry through research and services, and NIST contributes to public health, safety, and the environment. Much of NIST's work relates to technology development and utilization.

The American Society for Quality Control (ASQC) helps in administering the award program under contract to the NIST. The ASQC is dedicated to easing continuous improvement and increased customer satisfaction by identifying, communicating, and promoting the use of quality principles, ideas, and technologies.

The Board of Overseers is the advisory organization on the award for the Department of Commerce. The Board is appointed by the Secretary of Commerce and consists of distinguished leaders from all sectors of the U.S. economy. The Board evaluates the adequacy of the award criteria and processes for making awards.

The Board of Examiners evaluates award applications, prepares feedback reports, and makes award recommendations to the Director of NIST. The Board consists of business and quality experts primarily from the private sector. Members selected by NIST go through a competitive application process. For 1996, the Board consists of about 270 members. Of these, nine serve as judges and 50 serve as senior examiners; the remainder serve as examiners.

Award recipients share information on their successful performance and quality strategies with other U.S. organizations. In the Annual Quest for Excellence Conference, the CEO and other key individuals of companies who have award-winning strategies present their complete strategies in detail.

B. Baldrige Criteria as a System

In short, the Baldrige Criteria System is a universal guideline to help large or small companies improve their organizations. The criteria is consistent with approaches of other quality experts as in Deming, Juran, and Crosby. Companies can follow these quality approaches or develop their own unique approach to quality and customer satisfaction and still be consistent with the Baldrige Criteria.

There are seven categories in the Baldrige Criteria (Table 20.1). These categories are divided into twenty-four examination items. For the examination items there are fifty-two areas to be addressed (NIST, 1995a–c). The Baldrige Criteria System is a series of processes that if followed sequentially can achieve desired results. The word quality has been replaced with the word performance. The sequence of the Baldrige items starts with input process.

Input Process

When a company starts the process of writing an application for the Baldrige Award, they usually start with the input process and gather data. The input process, Item 7.1—Customer and Market Knowledge—helps a company identify who its customers are and what its customers want from a company's product line or service. Heaphy and Gruska (1995) claim that Item 7.1 helps a company decide the customers and market expectations in the short and long term. Customer and Market Knowledge also helps a company develop the proper listening and learning strategies to understand and anticipate customer's needs.

In the Customer and Market Knowledge item, companies need to identify and segment their customers into market, geography, and other categories (Brown, 1995, 1996). A company shows their methods, as in surveys, to identify their customer requirements and priorities, and to continuously evaluate and improve these methods. Also, a company conducts research to identify and find future customers' needs; this includes requirements from both customers and customers of competitors.

Driver

The next sequence of items in the Baldrige System is to get a driver that will lead and take control of the total system. Heaphy and Gruska (1995) point

Table 20.1 1996 MBNQA Criteria—Item Listing

Item	Point value	
1.0 Leadership		90
1.1 Senior executive leadership	45	
1.2 Leadership system and organization	25	
1.3 Public responsibility and corporate citizenship	20	
2.0 Information and analysis		75
2.1 Management of information and data	20	
2.2 Competitive comparisons and benchmarking	15	
2.3 Analysis and use of company-level data	40	
3.0 Strategic planning		55
3.1 Strategy development	35	
3.2 Strategy deployment	20	
4.0 Human resource development and management		140
4.1 Human resource planning and evaluation	20	
4.2 High performance work systems	45	
4.3 Employee education, training and development	50	
4.4 Employee well being and satisfaction	25	
5.0 Process management		140
5.1 Design and introduction of products and services	40	
5.2 Process management: Product and service production and delivery	40	
5.3 Process management: Support services	30	
5.4 Management of supplier performance	30	
6.0 Business results		250
6.1 Product and service quality results	75	
6.2 Company operational and financial results	110	
6.3 Human resources results	35	
6.4 Supplier performance results	30	
7.0 Customer focus and satisfaction		250
7.1 Customer and market knowledge	30	
7.2 Customer relationship management	30	
7.3 Customer satisfaction determination	30	
7.4 Customer satisfaction results	160	
Total points		1000

out that senior executives have the role in creating a successful quality culture that affects the customer, employees, stockholders, suppliers, and the community.

Item 1.0—Leadership—defines the organizations's mission and values, and how leaders give direction on the organization's future. Leadership is so important that Heaphy and Gruska (1995) point out that leaders "must make fundamental changes in the way they do business to achieve performance excellence." Senior executives must be personally involved with customers, employees, and suppliers. They review the customer satisfaction and quality data regularly. Executives show TQM management styles that stress empowerment and continuous improvement.

Item 1.2—Leadership System and Organization—describes how the company's customer focus and performance expectations integrate into the company's leadership system and organization. A company must show evidence that quality and focus on the customer integrate into all parts of the company's structure. Every manager and employee's behavior must show and reinforce the values of the company. Rewards for employee behavior are consistent to the organization's vision and values. Management reviews quality and customer satisfaction performance in their review meetings.

Item 1.3—Public Responsibility and Corporate Citizenship—describes the company's performance policies in areas of public health and environmental protection. As a corporate citizen in a community, a company encourages support and employee involvement with community schools, associations, and charities. Also, a company shows that it is a leader within its community in environmental protection, public safety, and ethics.

Data/Measures

The third sequence of items in the Baldrige System is to define the company's measures of success; this is defined in Item 2.0—Information and Analysis, and Item 7.3—Customer Satisfaction Determination. Item 2.0—Information and Analysis—addresses how success in the organization is measured and how data are used to make business decisions. Item 2.1—Management of Information and Data—addresses the importance that the company measures all the right things. Data gathered from customer surveys are converted into useful information and then connected to key business drivers so that the right people can make the right business decisions.

Item 2.2—Competitive Comparisons and Benchmarks—selects data that is used to improve the general company performance. A company needs to compare its key processes and benchmark against their best competitors. Benchmarks are used to obtain new ideas and to stretch company goals driven toward process improvement.

Item 2.3—Analysis and Use of Company-Level Data—focuses on how the analysis is used to gain understanding of: 1) customer markets, 2) operational performance and company capabilities, and 3) competitive performance. Data collected from all parts of the company are integrated to support reviews, improve and prioritize business decisions, and correlate quality improvements with the company's financial performance.

Item 7.3—Customer Satisfaction Determination—helps define the measures of success sequence in the Baldrige system. Specifically it describes how a company determines customer satisfaction related to company competitors. "Hard" measures of customer buying behavior along with "soft" opinions are gathered through surveys and focus groups. Tailoring the hard and soft survey measurements to markets or buying preferences of customers can give insight into future market behavior. With continual measurement and evaluation of the Customer Satisfaction Index (CSI), improvements within a company can related to the company's financial information.

Goals/Strategies

The next sequence of items in the Baldrige system identifies the Goals and Strategy. Item 3.0—Strategic Planning—examines the processes for annual operating plans and longer term business plans. Item 3.1—Strategy Development—shows that the scope of a company's ideas is based upon customer data, complete business environments, and employee and supplier information. The plans of key goals and strategies, based on the company's future customers, are communicated to all employees. Company operating plans are written quickly and efficiently. Performances against these plans are reviewed regularly.

Item 3.2—Strategy Deployment—shows how the key business drivers are translated into action plans by the company and then deployed. Operational measurements, customer requirements, and competitor performance are worked into annual and longer term goals with continual benchmarks of world-class companies. A systematic approach of cascading the goals down to a company's individual units is described in this section, along with how to deploy strategies to achieve key company goals. This section also projects the company's five-year plan with achieving its goals and improving the organization's position in the marketplace.

Process/Systems

The Process/Systems sequence includes many items in the Baldrige system. Process/Systems recognize the workforce as the most important resource in a company. Key processes of management and control of the system are defined along the way by keeping happy customers so that they remain steady customers. For example, Item 4.0—Human Resource Development and Management—addresses the company's performance. This category also shows how the company's performance relates to employees and future data trends. Item 4.1—Human Resource Planning and Evaluation—describes how the company's human resource planning and evaluation are aligned with its strategic and business plans, and addresses the development and well-being of the entire workforce. A company's complete business goal considers human resource goals in head count, training employees, and redesigning work organizations. The challenge in this area is to keep the human resource plan aligned with the company strategic plan.

Item 4.2—High Performance Work System—shows that a company promotes and rewards individuals or teams for their contributions. Communication across units is established in this item. In this section employees are empowered to make and carry out suggestions into the system. Jobs are designed to promote employee development and to maximize the flexibility of the workforce. Functional departments and layers of management are eliminated. The design of the company system is set up to treat employees as business partners. Employees are rewarded at all levels of a company for their creativity and extra effort.

Item 4.3—Employee Education, Training, and Development—focuses on how the company's education and training policies address company plans, including building company capabilities and contributing to employee growth, progression, and development. Quality training as well as safety, job skills, and leadership is relevant in this item. A key vehicle to building a company and employee's capabilities is the company's education and training. This item is very important in the overall structure of the Baldrige Award, because education and training have indeed become part of the strategic planning of organizations. In fact, Heaphy and Gruska (1995) identify this item as a way of how "... education and training addresses: (1) key performance objectives including those related to enhancing high performance work units, (2) progression and development of all employees." Training is tailored to individual needs with just-in-time knowledge and skills that are reinforced through on the job application. A company then collects data for three to four years to prove that the training is effective.

Item 4.4—Employee Well-Being and Satisfaction—criteria are viewed as how the company maintains a work environment and a work climate that are conducive to the well-being and growth of all employees. If employees enjoy their job then they will go out of their way to delight customers. The factors that contribute to employee satisfaction are: safety, pay, benefits, job design, work environment, and workload. Special services as in counseling, recreation, day care, and flex hours within a company are mentioned in this item. Information from employee surveys would also provide complaints about the company to determine employee satisfaction.

Another category that is part of the Process/Systems in the Baldrige sequence is Item 5.0—Process Management. NIST (1995d) states that "Process Management is the focal point within the Criteria for all key work processes." In this category are the requirements for efficient and effective management, as in effective design, prevention orientation, evaluation and continuous improvement, linkage to suppliers, and total high performance.

Item 5.1—Design and Introduction of Products and Services—describes how new or modified products and services are designed and introduced. A company's key production/delivery process requirements for operational performance are designed to meet both key product and service quality requirements (NIST, 1995e,f). A systematic approach controlled by requirements of customers is needed to design new products and efficient services. The product designs and processes are tested or reviewed before being introduced into the market. A company's product or service is continuously improved in quality and in cycle time

Item 5.2—Product, Service Production, and Delivery—assesses the company's key product, service production, and delivery processes that are managed to ensure that the company's design requirements are met. Both quality and operational performance are continuously improved. This item identifies and measures key processes that are used to maintain process performance in the company. To achieve better operational performance, a company's processes are evaluated and improved by process analysis, benchmarks, new technology, and information from the customer.

Item 5.3—Support Services—describes how the company's key support service processes are designed and managed to that current requirements are met and that operational performance is continuously improved. Support Services is similar to Item 5.2, but it is concerned the support process. It sets requirements for all support functions in finance, purchasing, engineering, employees, information systems, sales/marketing, facilities, and legal. In this section all key processes, measurement plans, implemented standards, and control strategies are defined for each support area. Also in this item, the company defines ways of evaluating and improving processes to achieve better operational performance and cycle time.

Item 5.4—Management of Supplier Performance—describes how the company ensures the quality of materials, components, and services furnished by other businesses. It also describes the company's action plans to improve a supplier's performance. A company defines the supplier requirements to insure that the supplier knows exactly what is expected from them to improve their quality and service. Action plans are developed to evaluate and to improve the supplier performance. These action plans are: 1) to improve suppliers' abilities to meet requirements, 2) to improve the company's own procurement processes, including feedback sought from suppliers and from other units within the company (internal customers) and how such feedback is used, and 3) to minimize costs associated with inspection, test, audit, or other approaches in verifying supplier performance.

Item 7.2—Customer Relationship Management—describes how the company provides effective management with the use of customer responses and follow-ups to preserve and build good customer relationships. With effective management a company also increases knowledge about specific customers and customer expectations. In this item a company shows evidence on how proactively they keep their customers satisfied on an ongoing basis. Easy access into a company is supplied to a customer to seek information, to comment, and to complain. All customer complaints, services, and products are acted upon by a company for customer opinion to improve the company's processes and to build better customer relationships.

Internal Results

The next sequence in the Baldrige Criteria System is Internal Results; this is addressed by Item 6.0—Business Results. This category asks for results instead of procedures. All results are addressed except customer satisfaction results.

Item 6.1—Product and Service Quality Results—summarizes the results of key measures in current quality levels, trends, and variability for products and services. The three factors—quality levels, trends, and variability—are evaluated with results from other competitors or benchmarks. Quality level shows how a company's product or service compares to other companies or industries best. Trends are based on three or more years of data of a company's improved performance. Variability is improved performance fluctuation in product or service for five to seven years.

Item 6.2—Company Operational and Financial Results—summarizes a five-year trend of continuous improvement levels in a company's complete operational and financial performance. This item also uses comparisons from competitors and benchmarks. Operational results are measurements of a company's productivity, timeliness of delivery, environment, safety, and resources. The financial results are measured in sales, profits, and return on investment.

Item 6.3—Human Resource Results—summarizes human resource results, including employee development and indicators of employee well-being and satisfaction.

Item 6.4—Supplier Performance Results—summarizes trends and quality levels of the company's suppliers. Measurements of quality from a company's supplier are compared with other quality measurements from competitors. Supplier performance is measured on hard data as in defective merchandise or missed delivery dates, and soft data as in responsiveness and courtesy.

External Results

The last sequence in the Baldrige Criteria System is external results. This sequence focuses on the customer and customer satisfaction (7.0). It addresses the issues that deal with customer and market knowledge. Specifically, this section aims to find how the organization deals with the marketplace and the customer. In the case of the marketplace, the company's customer satisfaction results are compared with those of competitors.

In the case of the customer, the interest is in identifying the company's customer satisfaction and dissatisfaction (7.1, 7.3, 7.4) by using key measures and indicators. Customer satisfaction results focus on levels, trends, and variability for the past five years and then compares the results to those of competitors. Customer satisfaction data are segmented by market and customer type. Negative data (e.g., complaints, returns, and warrenty claims) are presented.

C. Self-Assessment Process

Before a company can use the Malcolm Baldrige Criteria for self-improvement, it should make sure that the organization's core values and concepts are consistent with the concepts and paradigms of the criteria. An organization's leadership must decide its readiness to carry out a self-improvement initiative. A very interesting study by Bemowski (1995) presents the results of a survey of people who requested the Baldrige Award criteria in order to determine their use of the criteria. While 18.4% of the respondents discarded the criteria, most of the respondents used the criteria at least once, and 70.7% of users said the criteria provided information on business excellence. More than half of the respondents used the criteria in internal communications.

D. MBNQA Evaluation Process

There are four stages in the Malcolm Baldrige National Quality Award evaluation process (NIST, 1995g). To be considered for the first stage, a company must fill out and submit a seven-page eligibility determination form with $50 fee to the NIST. The NIST reviews the form and decides if the company has met all the eligibility requirements. The categories for eligibility are currently manufacturing, service, and small business. In 1995, Pilot Evaluation teams were selected for education and health care industries to help expand the eligibility to these institutions.

The first stage is to submit an application. After the NIST has qualified an eligible company, the applicant must provide a written application package. The application package contains a four-page overview of the applicant's business and a written response to 24 examination items. The application is restricted to a maximum of 70 single-sided pages. Application fees are: manufacturing—$4000, service—$4000, small business—$1200, and supplemental sections—$1500.

During this first stage of the review process, at least five examiners are assigned to review the application sent from companies who have completed their independent assessments. Based on the ratings from this first stage, judges decide if the application should continue the review process and move up to the next stage. If an application was not selected in stage one, then an examiner consolidates a feedback report with information gathered from other assigned examiners. This feedback report is sent back to the applicant.

The second stage of review is the consensus process. All the examiners, including a senior examiner, develop a compiled score of each item and comment. The senior examiner directs the process and reconciles any scoring differences. After the consensus process, the panel of judges reviews scores and verifies that the evaluation process was properly followed. The judges then select applicants for site visits. Site visits are based on the highest scores from all applicants. If a company's application was not chosen for a site visit, then a feedback report is prepared by the consensus team and sent back to the company.

Stage three is the Site Visit Review. Site visits are conducted to obtain further information and clarification in areas listed on the original application. The site visit team verifies that the information provided on the application is correct. After the site visit the team summarizes its findings in a summary report for the judges and a feedback report for the applicant. Site visit review fees depend on the number of visits scheduled, number of examiners assigned, and the duration of their visit.

In stage four the panel of judges recommends the recipients of the Baldrige Award based on the reports from the examiners of all site-visited companies. The recommendations are given to the NIST, overseer of the award, who presents this recommendation to the Secretary of Commerce for the final award decision. The recipients are announced and presented with the award in the last quarter of the year by the President or Vice President of the United States in a special Washington, DC, ceremony. Feedback reports from all applicants are then sent out after the award recipients have been announced.

E. Scoring Process

The system for scoring applicant responses to criteria items and for developing feedback is based upon three evaluation criteria: 1) approach, 2) deployment, and 3) results. Written examination items from the award application require information relating to these three criteria.

The approach criterion refers to the methods used in addressing the item requirements in the award application. Some key factors used to evaluate approaches are techniques, integration, and improvement cycles. Evaluated and improved techniques are integrated throughout the award application to prove a prevention-based approach.

The deployment criterion addresses how broad or how narrow the approach criterion is applyied to an organizational work unit. Both approach and deployment criteria are interwoven, for without deployment the approach would only be an idea.

The results criterion refers to the outcomes and trends of data in achieving the quality levels given in an item. The factors used to evaluate results are comparing, improving, and proving performance levels.

These three evaluation criteria are critical to the scoring assessment and feedback processes; however, both approach and deployment are linked to emphasize that descriptions of approach should always convey deployment. Although approach and deployment criteria are linked, feedback to the applicant reflects strengths and/or areas for improvement in either or both criteria. For example, an approach-deployment item score of 50% represents meeting the basic objectives of the item. A results item score of 50% represents a clear indication of improvement trends or good levels of performance. An examiner will always start his scoring response in the 40–60% range of the scoring guidelines. If the item satisfies this guideline then the examiner will check for additional requirements in the next higher score range (70–90%).

On the other hand, results depend on data demonstrating performance levels and trends. The evaluation factor, breadth and importance of performance improvements, is concerned with how widespread and how significant an applicant's improvement results are. This, of course is directly related to the deployment criterion (i.e., if improvement processes are widely deployed, there should be corresponding results). A score for a results item is thus a composite based upon overall performance, taking into account the breadth and importance of improvements.

F. Feedback Report

The feedback report is a written summary of the examiners assessment of an organization. It contains the strengths and weaknesses of an organizational process based on information written in the Baldrige Application or site visit. The feedback report is considered the most valuable part of the Baldrige process because it identifies key strengths and areas for improvement when compared to the criteria.

The purpose of the feedback report is for the examiners to communicate their evaluations to a Baldrige applicant. The feedback report also is the yardstick that measures the Baldrige process. Each Baldrige applicant will receive a feedback report. The feedback report does not identify the item scores and the area to address with comments. Instead comments from the consensus are rephrased to include the intent of a specific item. The feedback report has five sections: 1) introduction, 2) background and application review process, 3) scoring: distribution of numerical scores for all applicants, 4) scoring:

summary of an applicant, and 5) strengths and areas of improvement (three to five comments per item).

The introductory section contains a few sentences explaining what the report is all about. The second section (background) explains the application review process and a summary of statistics about the number of applicants that completed each step. The third section (scoring) is a report that contains the scoring distribution of all applicants for the year. The actual scores are not revealed, only the range of where the scores fell. In the fourth section the feedback report tells the organization the range of its score. An organization that receives a consensus review score of 601–750 has an invitation for a site visit. In the last section, the examiner writes clear and concise comments of strengths and areas of improvement.

G. Barriers in Using the Baldrige Criteria

There are four common obstacles that companies must overcome to use the criteria. These barriers are: 1) time needed to prepare the written responses to the criteria's application, 2) fear of assessment results, 3) belief that everything is alright, and 4) lack of trained examiners to do assessment. Overcoming these operational drawbacks will take great leadership and management skills, determination and drive, and a performance commitment from all employees and company suppliers.

H. Future Perspective

At present, the future of the MBNQA is questionable, not because it is not delivering what it has promised, but because of political considerations. The discipline that the MBNQA offers toward the road to "best-in-class" is the most systematic and encompassing of its kind. In fact, because it offers an excellent path to improvement, many organizations that use the Baldrige criteria have no interest in applying for or winning the award. The Baldrige criteria is being used (and will continue to be used) by companies to evaluate their progress and improve their performance in becoming best-in-class. The rewards of continuing the usage of the Baldrige criteria are: 1) more satisfied customers, 2) happier employees, 3) increased sales, profits, and market share, and 4) long-term survival.

II. INTERNATIONAL AWARDS

A. The Deming Prize

The Deming prize was established in Japan by the Union of Japanese Scientists and Engineers (JUSE) in 1951. This award is named in honor of W. Edwards

Deming, who is recognized as the father of the worldwide quality movement. The Deming Prize was established to ensure that good results are achieved through successful implementation of company-wide quality control activities. It centers on process analysis and statistical methods used to reveal the causes of variation leading to continuous improvement of processes, products, and services. The Deming Prize evaluates the operations of firms against 10 criteria:

1. Company policy and planning.
2. Organization and its management.
3. Quality control, education, and dissemination.
4. Collection, transmission, and utilization of information on quality.
5. Analysis.
6. Standardization.
7. Control.
8. Quality assurance.
9. Effects.
10. Future plans.

The Deming Prize model was tailored to the emerging mass production industries of Japan and the desire to continuously improve manufacturing processes and the quality of products. As of late, it is gaining momentum outside of Japan, especially in the United States (Nakhai and Neves, 1994).

B. The European Quality Award (EQA)

Fourteen leading western European businesses took the initiative of forming the European Foundation for Quality Management (EFQM) in 1988. The European Quality Award (EQA) is fundamentally very similar to the MBNQA. The mission of the EQA organization is twofold:

- Accelerate the acceptance of quality as a strategy for global competitive advantage.
- Stimulate and assist the development of quality improvement activities.

The model contains nine criteria and are, in the order of their importance:

1. Customer satisfaction (20%).
2. Business results (15%).
3. Processes (14).
4. Leadership (10%).
5. People satisfaction (9%).
6. People management (9%).
7. Resources (9%)
8. Policy and strategy (8%)
9. Impact on society (6%)

C. Polish Quality Award

The restructuring of the Polish economy in 1989 caused rapid price liberation and a flood of western goods into Poland. This created a challenging situation for Polish industry which had to cope with many problems, including quality, to win competitive status in the market place. Theses competitive pressures created the need for quality leadership in industry and for establishing a self-assessment system for companies which are committed to continuous improvement.

There are two national quality awards in Poland. There is the "Teraz Polska" (Poland Now), which focuses on domestic products and the need to recognize leaders in Polish quality for best products, and the Polish Quality Award (PQA), which focuses on highly competitive global market conditions.

To qualify for the selection process for the Teraz Polska award, the applicant must be registered in Poland and Polish capital must exceed 50%. If these requirements are met then the selection is based on:

1. Self-assessment.
2. Expert evaluation (with a possible rejection).
3. Research on clients based on questionnaires and market feedback.
4. Secret audit.
5. Final decision by commission/collegium.

The PQA is sponsored by the National Chamber of Commerce which represents over 500,000 private businesses. The purpose of the PQA is stated in its preamble: "TQM is the only way to bring Poland out of crisis. TQM draws attention to managing people. In developed countries quality is the primary parameter for competition. The goal of the award is awareness and education of management techniques and it will represent the symbol of quality, power, unity and knowledge of strategic thinking." The structure of the award is based on nine elements with predetermined points.

1. Leadership (150 points).
2. Processes (120 points).
3. Business results (150 points).
4. Strategic policy (100 points).
5. Client satisfaction (200 points).
6. Employee satisfaction (90 points).
7. Staff management (80 points).
8. Impact on society (60 points).
9. Resources (50 points).

The assessment process is similar to the MBNQA and the Teraz Polska company award; self-assessment applications are evaluated by assessors and the best companies are visited by the examiners. To be qualified for the PQA, the applicant must be a profitable and tax paying concern, pay taxes in Poland, have been in business for more than 3 years, and Polish capital must exceed 50%.

D. India's Golden Peacock National Quality Award

The Golden Peacock National Quality Award program was introduced in February of 1991. In the words of India's President, the award "enhances the corporate commitment to quality. The institution of the award will help to bring world class quality initiatives in the Indian industry and business and encourage adoption of Total Quality strategies in all sectors: manufacturing, service, education, training, research and development, and so on."

The award is based on the MBNQA with some additional blending with the best features of the EQA as well as the British Quality Award. The criteria for the award are based on eight elements, each of which has a predetermined points.

1. Organizational leadership (100 points).
2. Customer satisfaction (300 points).
3. Strategic quality planning (100 points).
4. Human resource utilization (120 points).
5. Competitive benchmarking (120 points).
6. Product quality and service quality assurance (80 points).
7. Suppliers quality progress (80 points).
8. Impact on society (100 points).

The evaluation of the application for the award is dependent on the sound judgement of the panel of judges who are experts from industry and commerce, or who are active professionals and academicians.

E. The Singapore Quality Award (SQA)

The Singapore Government recently unveiled a National Quality Strategy to help Singapore produce more quality goods, services, and systems in order to compete successfully in the global market. The strategy includes 13 initiatives which seek to inculcate quality consciousness in Singapore companies and workers. Extending ISO 9000 certification to services, commerce, and other industry sectors, as well as helping local companies move beyond ISO 9000 towards TQM, are some of the initiatives. In addition, the National Productivity and Quality Council will give out the Singapore Quality Award every year to institutionalize a quality culture in Singapore. The award criteria are modeled on the best features of the Malcolm Baldrige National Quality Award and the Deming Prize. Table 20.2 gives the criteria for the Singapore Quality Award.

Table 20.2 Criteria for the Singapore Quality Award

Examination categories and items		points
1.0 Leadership and quality culture		150
1.1 Senior executive leadership	45	
1.2 Quality culture	50	
1.3 Management involvement	30	
1.4 Public responsibility	25	
2.0 Management of process quality		160
2.1 Design and introduction of quality products and services	50	
2.2 Process management	50	
2.3 Quality assessment	35	
2.4 Supplier quality	25	
3.0 Human resource development and management		160
3.1 Human resource planning and management	25	
3.2 Employee involvement	35	
3.3 Employee education, training and development	35	
3.4 Employee satisfaction	35	
3.5 Employee performance and recognition	30	
4.0 Strategic quality planning		70
4.1 Strategic quality planning process	45	
4.2 Quality and operational goals	25	
5.0 Use of information and analysis		60
5.1 Scope and management of information system	15	
5.2 Competitive comparison and benchmarking	25	
5.3 Analysis and effective use of data and information	20	
6.0 Customer focus and satisfaction		250
6.1 Customer expectations—current and future	40	
6.2 Customer relationship management	50	
6.3 Customer satisfaction—process	50	
6.4 Customer satisfaction—results and comparison		110
7.0 Quality and operational results		150
7.1 Product and service quality results	60	
7.2 Organization operational results	60	
7.3 Supplier quality results	30	
Total points		1,000

Table 20.3 Hierarchy of Award Criteria for MBNQA and SQA

Hierarchy	Singapore Quality Award	Malcolm Baldrige National Quality Award
Categories	7	7
Examination items	25	24
Areas to address	72	52

The SQA is less known than the MBNQA, but it does present itself as a national award with a specific focus on quality. The hierarchy of the two awards is shown in Table 20.3. The evaluation for the SQA award is also based on the MBNQA structure:

1. *Approach*: the process used to achieve quality products and services
2. *Deployment*: how widely the approach has been executed
3. *Results*: the outcome(s) of efforts in specific award categories

Based on these criteria, the assessors examine:

1. Approach
   - Is it innovative or creative?
   - Is it prevention-based or detection-based?
   - Is it focused on continuous improvement and use the Plan-Do-Study-Act cycle?
   - Is it systematic, integrated, and consistently applied?
   - Does it use quantitative, objective, and reliable data?
   - Does it make appropriate and effective use of tools, techniques, and methods?
   - Is it based upon cooperation and participation of all levels of staff?
2. Deployment
   - Is there appropriate and effective application to all product and service characteristics?
   - Is it applied to all transactions with internal and external customers?
   - Is it applied to all internal processes, activities, facilities, and employees?
   - Are the quality improvement approaches applied to suppliers?
3. Results
   - What is the rate and speed of quality improvement?
   - What is the breadth of quality improvement?
   - What are the current and past quality and performance levels?

- Are there trends showing sustained improvement over the past several years in all key result indices?
- Is there a cause-and-effect relationship between quality improvement efforts and quality results?
- What is the significance of quality improvement to the company's business?
- How do the company's quality results cpompare with world and industry leaders?

REFERENCES

Bemowski, K. (1995). How do people use the Baldrige award criteria? *Quality Progress*. May: 43–47.

Brown, M. G. (1996). *Baldrige award winning quality: how to interpret the Malcolm Baldrige award criteria*. Quality Press. Milwaukee, WI.

Brown, M. G. (1995). *The pocket guide to the Baldrige Award Criteria*. (rev. ed). Quality Resources. New York.

DeCarlo, N. J. and Sterett, W. K. (1990). History of the Malcolm Baldrige National Quality Award. *Quality Progress*. Mar.: 21–27.

Heaphy, M. S. and Gruska, G. F. (1995). *The Malcolm Baldrige National Award: a yardstick for quality growth*. Addison-Wesley. Reading, MA.

Nakhai, B. and Neves, J. S. (1994). The Deming, Baldrige, and European quality awards. *Quality Progress*. Apr.: 33–37.

National Institute of Standards and Technology (NIST) Office of Quality Programs (1995a). *Award criteria*. Malcolm Baldrige National Quality Award. Gaithersburg, MD.

National Institute of Standards and Technology (NIST) Office of Quality Programs (1995b). *Application scorebook*. Malcolm Baldrige National Quality Award. Gaithersburg, MD.

National Institute of Standards and Technology (NIST) Office of Quality Programs (1995c). *Handbook for the Board of Examiners*. Malcolm Baldrige National Quality Award. Gaithersburg, MD.

National Institute of Standards and Technology (NIST) Office of Quality Programs (1995d). *Colony Fasteners evaluation notes*. Malcolm Baldrige National Quality Award. Gaithersburg, MD.

National Institute of Standards and Technology (NIST) Office of Quality Programs (1995e). *Colony Fasteners case study*. Malcolm Baldrige National Quality Award. Gaithersburg, MD.

National Institute of Standards and Technology (NIST) Office of Quality Programs (1995f). *Colony Fasteners feedback report*. Malcolm Baldrige National Quality Award. Gaithersburg, MD.

National Institute of Standards and Technology (NIST) Office of Quality Programs (1995g). *Site visit evaluation book*. Malcolm Baldrige National Quality Award. Gaithersburg, MD.

21

International Standards of Quality and Corporate Quality Programs

The world of quality is expanding exponentially. All over the world, quality seems to be in the driver's seat; however, with this globalization many problems and opportunities have risen primarily due to inconsistent standards in the market place. What the international standards have tried to do—and seem to be very successful in doing so—is to standardize a minimum level of quality. This chapter focuses on these standards and summarizes their contribution to quality. Specifically, the chapter addresses the ISO 9000, ISO 14000 and QS-9000. Furthermore, it addresses some of the issues of the proposed TE-9000, and Z1.11. Finally, it presents a short discussion on both the Q1 and TQE programs of Ford Motor Co.

I. INTERNATIONAL STANDARDS

Quality has always been around; the problem, however, has been acceptance, consistency, and repeatability of quality throughout the world. The international standards make an attempt to harmonize these items and at the same time establish some of the basic concepts of quality as minimum requirements for everyone. The first attempt of these standardized issues came in 1987 with the introduction of ISO 9000 followed by the introduction of environmental standards (ISO 14000) and the specific industry requirements known as the QS-9000.

II. ISO 9000

The ISO 9000 series of standards provide a starting place, a foundation, for all encompassing quality efforts. The standards merely stipulate where organizations need documentation to validate processes and approaches but never dictate how much they require.

ISO 9000 is not a product registration standard and in no way measures or recognizes the quality of a company's product, nor does it mean that two companies with ISO 9000 registrations are equivalent. On the other hand, the ISO 9000 is a registration system which requires at least four items. They are:

1. Management that is committed, involved, focused, and responsive.
2. People who are organized, responsible, authorized, competent, empowered, and knowledgeable.
3. Processes that are visible, traceable, consistent, repeatable, measurable, and documentable.
4. Documents that are appropriate, relevant, simple, understandable, and consistent with processes in use.

A. Benefits of ISO 9000 to an Organization

Implementing ISO 9000 leads to improved competitiveness because participants:

- Enforce an explicit statement of declared aims and specifications.
- Enforce a system of monitoring and keeping records.
- Establish clearly documented procedures that are understood by everyone concerned.
- Provide adequate quality training for everyone that includes general comprehension of what quality means and training in the use of specific tools.
- Focus on customer needs.
- Apply a supplier/customer relationship with well-defined and mutually agreed upon requirements.
- Develop a prevention attitude throughout the company, accompanied by an early detection and correction system.
- Provide an auditable system that can be verified by external auditors.
- Define responsibilities.

B. Registration Road Map

In order for an organization to claim certification to ISO, it must go through the process of registration. The steps included in the process are:

1. Know what you are doing; if necessary hire consultants.
2. Make sure that management is committed to the implementation process and understand the concepts. Actively involve them.
3. Establish a quality steering committee with empowerment.
4. Train management and employees.
5. Communicate a registration plan.
6. Develop an implementation plan.
7. Create self-assessment questions.
8. Establish a quality manual and procedures.
9. Establish an internal auditing system.
10. Measure compliance to the procedures.
11. Establish a comprehensive corrective action system.
12. Review results and act appropriately.
13. Call independent assessors in to audit, including: preassessment by the registrar (optional), corrective actions if applicable, the official audit, corrective action to nonconformances (if applicable), and registration.
14. Surveillance

For a complete discussion on the ISO 9000, see Peach (1996), Stamatis (1995), and Lamprecht (1992, 1993).

III. ISO 14000

Continuing the barrage of international standards, the world of business (and especially manufacturing) is waiting the arrival of the ISO 14000 Environmental Management System (EMS) standard. The standard is expected to become a requirement for doing business in Europe, Asia, and other markets. In fact, it has been rumored that lack of implementing the ISO 14000 may cause nontariff trade barriers.

Proponents of the ISO 14000 series of standards say these documents provide guidance to any size company in any industry on managing, measuring, improving, and communicating the environmental aspects of their operations. These standards also can influence how a company designs and manufactures its products, chooses its raw materials, and ultimately disposes of the product (Hemenway and Hale, 1996; Hemenway, 1995).

On the other hand, implementing the EMS can boomerang, at least in the United States. The U.S. Environmental Protection Agency (EPA) promises leniency if you meet several conditions, including "... correcting the violation (nonconformance to the standard) ... implementing preventive measures ... [and] cooperating with EPA's requests for information"; however, here is the problem: an organization going through an environmental audit is required to notify the state, local, and federal officials of any nonconformance(s), and these agencies are not bound by EPA terms.

The EPA's policy also states it will not refer your case to the Department of Justice if you meet three additional criteria: the violation did not involve a company policy to conceal environmental harm, management was not involved in the violation, and serious, actual harm to human health or the environment did not occur. But even if the organization clears these hurdles, the Justice Department is not bound by what the EPA recommends and could prosecute you (National Institute of Business Management, 1995).

The ISO 14000 EMS specification is the core document and the only normative standard in the series. It is expected to be published for international use by late 1996; however, with all its shortcomings, it is expected that its adaptation in the United States will come earlier.

The series is made up of some 15 standards of various aspects of environmental management that can be classified into the categories of organization, process standards, and product-oriented standards. The EMS, auditing, and environmental performance evaluation standards make up the set of organization and process evaluation standards. Life-cycle assessment standards, labeling standards, and guidance to ISO on considering environmental aspects is in its product-oriented set of standards. There is also a harmonizing "terms and definitions" standard. Like the ISO 9000 series of quality management system standards, the ISO 14000 EMS standards are designed to guide companies toward structured management systems.

Not all 15 standards have been published. In fact, their expected completion published schedule date is between 1997 and 1999. The following have been published as of 1996:

ISO 14000: Environmental Management Systems: General guidelines on principles, systems, and supporting techniques.

ISO 14001: Environmental Management Systems: Specification with guidance for use.

ISO 14010: Guidelines for Environmental Auditing: General principles of environmental auditing.

ISO 14011: Guidelines for Environmental Auditing: Audit procedures —Part 1: Auditing of environmental management systems.

ISO 14012: Guidelines for Environmental Auditing: Qualification criteria for environmental auditors.

The ISO 14000 is a very formidable standard, but is voluntary. Nevertheless, several of the international companies have already begun to implement

ISO 14001, including AT&T, IBM, 3M, Toyota Motor Corporation, Georgia Pacific Corporation, Motorola Inc., Volvo Inc., and Texas Instrument Inc.

It seems that the implementation of the standard is a culmination of many systems for environmental management; therefore, it is very appealing across industries and countries. For example, standard organizations in Austria, Switzerland, and Turkey have already adopted the draft ISO 14001 standard as their national standard. In the mean time, some 17 certification bodies are registering companies to BS 7750 in the United Kingdom, Japan, the Netherlands, Korea, and Brazil. This standard, published in 1992, is considered the blueprint for ISO 14001. For more information on BS 7750, implementing environmental management standards and documentation, see Rothery (1993) and Stamatis (1995, 1996).

As of September 1995, more than 115 companies had registered to BS 7750. Companies justify the cost of registration—estimated to be $12,000–$100,000, depending on the size and status of the company—by citing the benefits to the company's upper management, which are:

- Reduced liability and risk.
- Profit in the market for "green" products.
- proof of good management before pollution-incident insurance coverage is issued.
- Identify areas for reduction in energy and other resource consumption.
- Improve compliance with legislative and regulatory requirements.
- Prevent pollution and reduce waste.
- Improve community goodwill.
- Display interest in creating a high quality workforce.
- Display and demonstrate the intent of the organization to be a good citizen.
- Increase response to pressure for environmentally sound policies and procedures from legislative bodies.
- Benefit from regulatory incentives that reward companies showing environmental leadership through certified compliance with an EMS.

For more details on the EMS, see Cascio (1996).

IV. QS-9000

As competition in virtually every industry becomes keener, more importance is placed on quality, cost reduction, on-time delivery, and continuous improvement. Nowhere are these business issues more important than in the automotive and truck industries. Requirements for quality products and business system improvements are being placed on tier-one and tier-two automotive suppliers. These requirements, although good, prove to be very difficult as each of the companies developed their own requirements.

A. Quality System Requirements

QS-9000 was developed by Chrysler, Ford, General Motors, and select truck manufacturers—collectively known as the Automotive Task Force—to eliminate the complexity and, more importantly, the duplication of efforts. In 1988, the Task Force began to standardize reference manuals, reporting methods/formats, and technical terminology. Five manuals have been published since that time. Previously, Chrysler, Ford, General Motors each had their own expectations and documents for supplier quality systems.

Chrysler, Ford, and General Motors require their first-tier suppliers to adopt the QS-9000, as it is generally known, which incorporates the ISO 9001 standard. In addition, it is a harmonization of Chrysler's *Supplier Quality Assurance Manual, Ford's Q-101 Quality System Standard*, and General Motors' *NAO Targets for excellence*. The QS-9000 requirement, although it is based on ISO 9000, has additional requirements pertaining to both industry and company specific quality requirements. General requirements include: process innovation, fast and accurate delivery time, and incremental cost reduction. On the other hand, specific areas that QS-9000 requires and goes beyond ISO 9000 are in the areas of Advanced Quality Planning and Production Part Approval Process.

The goal of QS-9000 is to develop fundamental quality systems that provide for continuous improvement that emphasizes defect prevention and the reduction of variation and waste in the supply chain. It applies to all internal and external suppliers of production materials, production or service parts, and heat-treating, painting, plating, or other services provided directly to Chrysler, Ford, General Motors, or other original equipment manufacturer that subscribe to the standard. (For tooling manufacturers see next section.) All the reference manuals and the QS-9000 document may be purchased from the Automotive Industry Action Group (AIAG), phone (800) 358–3003.

At first glance, it seems that this new requirement is overly demanding and, in some respects, very difficult to achieve without investing a tremendous amount of time and money. These are resources which many companies believe they cannot free up. Documentation is one of the major concerns. A lot of time can be wasted determining the appropriate depth of documentation as well as developing supporting records and writing procedures. Many companies are also concerned about the potential hours that might be needed to set up extensive calibration, preventive maintenance, tooling, and die and mold programs. To overcome these obstacles and to view this process as achievable, preparation for certification can be approached in phases:

1. System evaluation. Conduct a needs assessment to find out where the organization is and what it will take to get to certification.

2. System documentation and deployment. Review, update, and create the appropriate and applicable documentation.
3. Process implementation. Use a team approach to implement both the concept of QS-9000 and documentation.

A successful implementation is facilitated with at least the following:

1. A continuous improvement commitment by everyone in the organization.
2. A clear mission.
3. Well-defined goals.
4. Training in the requirements.
5. Consistency in reporting.
6. Two-way communication throughout the organization by everyone.

One of the biggest benefits of a company pursuing the QS-9000 is a reduction of duplicated compliance efforts on the part of the suppliers; this will relieve them of multiple on-site assessments and inconsistent reporting procedures. Full compliance to QS-9000 will be mandatory for tier-one suppliers to Chrysler, Ford, and General Motors (Chrysler, by July 31, 1997, Ford, no definite date, and General Motors, by December 31, 1997).

Since the QS-9000 is based on ISO 9000, there is a need for third-party registration and ongoing periodic surveillance similar to the procedures established by ISO 9000. Furthermore, it is very important to recognize that ISO 9000 registration does not mean QS-9000 registration. On the other hand, QS-9000 registration means that the registration is recognized also for ISO 9000. For more details on the QS-9000 see Stamatis (1996a) and Smith (1996).

V. TE-9000

The TE-9000 standard is expected to be released as a supplement to the QS-9000 requirement and it will address issues and concerns for the approximately 50,000 tooling and equipment suppliers of nonproduction parts for the automotive industry, especially Chrysler, Ford, and General Motors. TE-9000 is expected to be published in a final draft form towards the end of the 1996, and final approval is expected sometime in 1997. From what we already know, however, TE-9000 is to improve the quality, reliability, maintainability, and durability of products delivered to Chrysler, Ford, and General Motors. In addition, the TE-9000 supplement (as it is commonly called) will provide a consistent interpretation of QS-9000 and communicate additional common system requirements unique to the producers of tooling and equipment.

The following requirements have either been amplified or added to the ISO 9000 base requirements (Chrysler, Ford, and General Motors, 1995);

however, because they are in a state of flux, the areas of the pending changes are without any discussion. The letter "S" after the element number signifies that there is a potential of either adding or modifying the specific subelement. The proposed changes are:

Section I.
- Management responsibility—Element 4.1
- Quality policy—Element 4.1.1 S
- Quality system—Element 4.2
- Quality planning—Element 4.2.3 S
- Design control—Element 4.4
- Design and development planning—Element 4.4.2 S
- Design verification—Element 4.4.7 S
- Design changes—Element 4.4.9 S
- Product identification and traceability—Element 4.8
- Process control—Element 4.9
- Process monitoring and operator instructions—Element 4.9.1 S
- Preliminary process capability requirements—Element 4.9.2 S
- Verification of job setups—Element 4.9.5 S
- Process changes—Element 4.9.6 S
- Appearance items—Element 4.9.7 S
- Inspection and testing—Element 4.10
- General—Element 4.10.1 S
- Final inspection and testing—Element 4.10.4 S
- Control of nonconforming product—Element 4.13
- Control of reworked product—Element 4.13.3 S
- Engineering approved product authorization—Element 4.13.4 S
- Handling, storage, packaging, and delivery—Element 4.15
- Delivery—Element 4.15.6 S
- Training—Element 4.18
- Servicing—Element 4.19
- Statistical techniques—Element 4.20
- Procedures—Element 4.20.2 S

Section II: Sector-specific requirements
- Machinery qualification runoff requirements
  - purpose—Element 1.1 S
  - procedure—Element 1.2 S
- Continuous improvement: techniques for continuous improvement —Element 2.3 S

Section III. Customer specific requirements: as yet, no customer specific requirements have been identified.

VI. BSR/ASQC Z1.11

In the quest for standardizing quality practices in an organization, one of the latest developments is the Z1.11 standard for education and training organizations. The standard is based on the ISO 9001, but it reflects the peculiarities of education and training organizations. The standard is not completed and a final draft is expected sometime in early 1997. If the reader is interested in this international standard and its progress contact the American Society for Quality Control and ask for its status and a draft copy. (See also Russo, 1995).

VII. CORPORATE QUALITY PROGRAMS

As discussed in Chapter 20, the national awards of quality are increasing at a very fast pace. By the same token, state, industry, and corporate awards are at an all time high. They all focus on quality, but depending on the award, differ in the aspects of quality that are emphasized. Here, we addrerss two specific awards because we feel that the magnitude and scope of these two not only go beyond the basics of ISO 9000, but they build a sound foundation of product quality. Because of their exposure, other international companies are using the structure and their contents for evaluating quality. These programs are discussed to show that a company committed to quality will go on its own the extra mile necessary to accomplish "quality excellence."

A. Ford's Q1 and Total Quality Excellence Programs

Since the introduction of the QS-9000 requirement, there has been much concern about the current and future state of the two most frequently used automotive standards of quality: the Q1 and the Total Quality Excellence Programs of Ford Motor Company. Where the QS-9000 is indeed an international standard recognized in the automotive industry—at least within the Chrysler, Ford, and General Motors domains—there is a lot of confusion about the Q1 program and the Total Quality Excellence Program.

The Q1 is a very specific quality program that was developed by Ford to recognize producers that strive to improve continuously their quality and productivity. To qualify for the Q1 Award, producers must score well by the following criteria:

- Adequacy of quality system to control product quality, including the use of statistical methods for significant process parameters and product characteristics.
- Absence of product rejections, initial sample rejections, and field quality concerns.

- Commitment and active involvement of the producer's management to continuous quality and productivity improvement.
- Willingness and ability to conduct manufacturing feasibility studies in the producer's area of expertise.

Ford customer plants, product engineering, purchasing, and supplier quality must concur that a specific producer location meets the above criteria. This location will then be designated as a Q1 location.

Q1 producers have prefered position in new product development programs and in source selection. Additionally, since Q1 locations have demonstrated excellence in product quality, they are exempt from routine Ford surveys. Q1 producers are expected to take the lead in the advanced quality planning process for new products and to review and obtain Ford concurrence on FMEAs and control plans in the same manner as other producers. On an annual basis, Q1 suppliers may be asked to review their continuous improvement in an on-site meeting with Ford product engineering, purchasing, and a supplier technical assistant (STA).

The Q1 Award is permanent, subject to continuous improvement in quality and productivity. Producers must maintain their quality systems at the level that originally qualified for Q1. Should quality concerns arise, Ford product engineering, purchasing, and a STA will assist the producer in resolving such concerns. Cancellation of Q1 status will be considered only as a last resort, to be used when the producer's efforts to resolve concerns have failed. The Q1 designation does not carry over to new products added to a Q1 facility or to relocations of Q1 facilities.

When new commodities are sourced to a Q1 manufacturing location, or when control items are sourced to a manufacturing location that has not previously produced control items, a Ford STA will visit the manufacturing location to update the system survey relative to the new products.

The Q1 Award may be won by both manufacturing and service organizations. Additional information may be obtained from Ford's purchasing outside suppliers, from the corporate quality office for company plants, in Dearborn, Michigan and/or Ford Motor Co. (1989, 1990a-c, 1993).

The Total Quality Excellence (TQE) Award provides incentive and recognition for achievement of a high level of continuous improvement related to the quality of products and services provided to Ford. The TQE Award goes beyond the Q1 Award. While the Q1 Award recognizes excellence in product quality, the TQE Award recognizes excellence in all aspects of meeting Ford's needs and expectations. Specifically, the technical requirements of Ford's engineers, the commercial requirements of Ford's buyers, and the delivery requirements of Ford's plants will be considered. The TQE Award, therefore, will honor the top echelon of Ford's Q1 suppliers and will provide its recipients with further distinction in sourcing decisions and an ongoing partnership with Ford.

The TQE Award is a logical extension of Ford's mission, values, and guiding principles. It represents a real challenge and will provide producers with a strong incentive for excellence. The steps to attain the TQE Award are:

1. The TQE Award will be made on a commodity basis.
2. All producer plants manufacturing the same commodity must have held the Q1 Award for at least twelve months.
3. All such producer plants must be free of validated quality and initial sample rejections for at least six months.
4. The candidate producer will complete a self-evaluation and must exceed ratings of 90% on:
   - The quality system survey, with all questions rated at eight or better, and a score of 45 points or better on the five SPC questions.
   - The engineering criteria
   - The commercial criteria
   - The delivery criteria
5. After these criteria have been met, the producer petitions the appropriate Ford buyer for consideration for the award. With the petition, the producer must include the self-evaluation, supporting rationale for the ratings, and an explanation and evidence for the continuous improvement efforts.
6. Ford personnel from material planning and control, product engineering, purchasing, and quality will review the producer's submission.
7. If the producer meets all requirements, a cross-functional Ford team will visit the producer's plant(s) and make an on-site evaluation.
8. Upon unanimous endorsement by the team and all Ford divisions that are customers, the candidate's qualification are presented to a Ford senior management TQE Committee for approval. Any producer found not to meet TQE requirements cannot repetition for twelve months.
9. TQE Award recipients will reconfirm continuous improvement every two years by repeating the above process.

For more information on the TQE Award the reader is encouraged to come in contact with Ford's corporate quality office or the individual contact buyer for a given supplier and/or see Ford Motor Co., 1990d).

Both the Q1 and the TQE Awards are Ford's way of assuring that their supply base is quality oriented and exemplify excellence in both product and system quality. Both awards depend on second-party auditing and are totally—in fact, mutually—exclusive with the ISO 9000 standards and QS-9000 requirements. Therefore, they should not be confused and used interchangeably with any other standards.

REFERENCES

Cascio, J. (Ed.) (1996). *The ISO 14000 handbook*. CEEM & Quality Press. Fairfax, Virginia. Milwaukee, WI.

Chrysler, Ford, and General Motor (1995). *Quality system requirements: tooling and Equipment supplement*. Chrysler, Ford, and General Motors. Distributed by the Automotive Industry Action Group. Southfield, MI.

Ford Motor Co. (1989). *Supplier quality improvement guidelines for production parts*. Ford Motor Co.. NAAO/DPO Production Purchasing. Dearborn, MI.

Ford Motor Co. (1990a). *Q1 preferred quality award: for suppliers to Ford Motor Company*. Ford Motor Co.: Total Quality excellence and Systems Management. Dearborn, MI.

Ford Motor Co. (1990b). *The quality system survey and scoring quidelines*. Ford Motor Co.: Quality related publications. Corporate Quality Office. Dearborn, MI.

Ford Motor Co. (1990c). *Worldwide supplier quality rating system*. Ford Motor Co.: Purchasing. Corporate Quality Office. Dearborn, MI.

Ford Motor Co. (1990d). *Total quality excellence*. Ford Motor Co.: Supply Policy and Planning. Corporate Quality Office. Dearborn, MI.

Ford Motor Co. (1993). *Q1 preferred quality award: for service organizations*. Ford Motor Co.: Operations Support. Corporate Quality Office. Dearborn, MI.

Hemenway, C. G. and Hale, G. J. (1996). Are you ready for ISO 14000. *Quality*. Nov.: 26–28.

Hemenway, C. G. (1995). 10 things you should know about ISO 14000. *Quality Digest*. Oct.: 49–51.

Lamprecht, J. L. (1993). *Implementing the ISO 9000 series*. Marcel Dekker. New York.

Lamprecht, J. L. (1992). *ISO 9000 preparing for registration*. Marcel Dekker. New York.

National Institute of Business Management (1995). Research recommendations. *National Institute of Business Management*. 46(39): 4.

Peach, R. W. (Ed.) (1996). *The ISO 9000 handbook*. 3rd. ed. Irwin Professionals. Burr Ridge, IL.

Rothery, B. (1993). *BS 7750: Implementing the environment management standard and the EC Ecomanagement scheme*. Gower Press. Hampshire, England.

Russo, C. W. R (1995). *ISO 9000 and Malcolm Baldrige in training and education*. Charro Publishers. Lawrence, KS.

Smith, R. M. (1996). *The QS-9000 answer book*. Paton Press. Red Bluff, CA.

Stamatis, D. H. (1995). *Understanding ISO 9000 and implementing the basics to quality*. Marcel Dekker. New York.

Stamatis, D. H. (1996). *Documenting and auditing to ISO 9000 and QS-9000*. Irwin Professionals. Burr Ridge, IL.

Stamatis, D. H. (1996a). *Integrating QS-9000 with your automotive quality system*. 2nd ed. Quality Press. Milwaukee, WI.

Epilogue

As we have seen, Quality is here to stay. In order to be effective, efficient, and productive, there are many techniques, methodologies, and tools that have to be appropriately applied for optimum results. Some of these have been discussed and some have been reviewed in this book. In closing this book, however, we would like to emphasize that for quality to be implemented in any organization, it has to be internalized by both management and employees alike. In essence, quality has to be personal. It can start anywhere. The main ingredients are: enthusiasm, conviction, perseverance, determination to be the best and above all, belief in the people of the organization.

If you are a member of the management team the following recommendations are for you. If you expect people in your jurisdiction to be creative and productive, you have to provide an atmosphere and an environment that is conducive to that good performance. In addition, you must be sensitive to how these people think and know why they may act the way they do. The better you get to know the people on your staff, the easier it will be for you to guide and direct them. This kind of knowledge will also enable you to determine what it might take to motivate them individually on a day-to-day basis. Some recommendations are the following:

1. *Set an example.* Your behavior is observed all the time. Remember you have a high visibility with subordinates, so when you set a pattern, you demonstrate what you expect them to do.

2. *Teach employees and subordinates about their jobs.* It is imperative to remember that one of the cardinal functions that you have as a manager is the responsibility to educate, train, coach, and develop subordinates. If you want people to make a good showing, you will have to show them how.

3. *Set minimum standards of performance.* Let staff members know that you feel they are capable of surpassing your standards, which actually provide basic guidelines for quality and quantity of output. People on the job always need to know what is expected of them.

4. *Make known what is valued by you and the company.* Stress the need for ethics, honesty, and integrity in business operations and point out the importance of protecting the company's good name by being ethical. Always promote positiveness in behavior and in outlook for all employees.

5. *Encourage participation in setting objectives.* Make sure each person has realistic and attainable goals of his or her performance. Always admire enthusiasm and ambition. You can inspire initiative if you help them to take control of their work areas well enough to make things happen.

6. *Absolutely refuse incompetence.* Aim and encourage excellence. You know that if you put up with incompetence in just one employee, others will lose respect for you. On rare occasions, an employee's incompetence may be due to a special problem. If that problem cannot be solved, or if the employee does not show interest, then let it be known to everyone that you feel the employee does not belong in your department or with the company. To avoid this incompetence, always encourage and reward excellence. Failure to arrive at a solution is not necessarily incompetence. The failure may be due to external sources, or things beyond your/their control. What is important here is positive attitude, willingness to tackle a problem, perseverance, enthusiasm, and so on.

7. *Spur creativity and innovation.* Promote suggestions and support risk taking. Remember that one way of learning is through failures. By allowing different and new approaches to problem solving, you can make even the mundane and repetitive jobs more interesting. Develop foresight by pointing out the advantages of planning and scheduling.

Alternatively, if you consider yourself an individual within the organization, you should view quality as a personal agenda and always try to do the best you can during the day. Some of the recommendations in Table E.1 may help you define and plan quality in individual terms.

Finally, when we think of quality, we must all understand that any successful implementation quality program is an issue for everyone. After all, none of us is as smart as all of us. Good luck!

Table E-1 Quality as a Personal Agenda

Get up 5:30 AM	Preparation and commuting. Spiritual activities. Exercise. Use commuting time to listen to tapes and other general information. *Remember*: Winner is the man who thinks he can.
Start 8:00–9:00 AM	Start with a resolution such as: "I want; I deserve; I can." Take out activity log. Review today's "to do" list. Emphasize well-defined goal for results. Prioritize and allot time to each activity. Communicate and/or review with team all about past 24 hours. *Remember and practice the five Japanese words*: Seiri—sorting, Seiton—systematizing, Seisoh—sweeping, Seketsu—sanitizing, Shitsuke—self-discipline.
9:00–10:00 AM	Go through meeting agenda for effective control. Go for shop visits with sincerity and purpose. Stress housekeeping. Listen a lot, keep temper cool. Be ruthless for implementing what is agreed (SOP's, SPC, ISO, and so on). Keep internal customer satisfaction in mind. *Remember the Olympic motto*: Citius (faster), Altius (higher), Fortius (stronger).
10:00–11:00 AM	Think of specific implementation projects for improvement. Invite subordinates to discuss issues/concerns. Address chronic problem for solution. Take up next activities on the "to do" list. Catch up on paper work, filing, dictation, reports, and so on. Conduct a reality check. Compare time consumed with time allotted. *Remember*: The human mind is like a parachute, it works when it opens.
11:00 AM–12:00 noon	Keep meeting(s) on time and take care of agenda. Be prepared with facts and figures. Emphasize importance of being on track, being on time, and responsibility. Monitor meeting activity, as per responsibility. *Remember*: Good is not good enough when better is expected.
12:00 noon–1:00 PM	Finish what is pending in plan. Think of changes to improve. Chase, expedite, write. Keep an open slot for unexpected items. *Remember*: All of us are stupid, however, thank God we are stupid at different things.

Table E-1 (Continued)

1:00–2:00 PM	Lunch. Look and evaluate slogans used in your organization. Catch up on your professional reading. Identify the seven wastes in your jurisdiction: waiting time, transport, motion, process, inventory, overproduction, defects. *Remember*: When we do not do right things right, right there is a loss.
2:00–3:00 PM	Take next activity. Review progress of shift. Communicate shortfall seriously. Visit shops, expedite projects. Keep up with routine functions of your job. Keep time slot for unforeseen events. *Remember*: Minds like hearts, go where they are appreciated.
3:00–4:00 PM	Go by the win-win philosophy on your dealings with everyone. Be in touch with everything that is important to the success of quality in your organization. Do creative thinking. Take the next activity. Review procurement and future plans. Keep a slot open for visitors, employees, and so on. *Remember*: What is rewarded gets done.
4:00–5:00 PM	Take up leftover activities. Keep nothing pending, if possible; overstay if you must. Always take stock of the day and communicate findings. Prepare the activities for the next day. Prepare a "do list log." *Remember*: The three disciplines for progress: self-discipline, technical discipline, organizational discipline
5:00–6:00 PM	Social activity. Networking, family visits. Reading for knowledge and skill. Set aside some time for personal projects. Set aside some time for family time. *Remember*: A prayer before going to retire. God! Give me the courage to change the things I can. Give me the patience to bear with those I can't. Give me the wisdom to distinguish the two.

Appendix A: Selection of Statistical Techniques Based on Three Types of Data

This appendix provides a selection process for specific statistical techniques using the three types of data: ordinal, interval, and nominal (see next page).

I. SELECTION OF STATISTICAL TECHNIQUES BASED ON ORDINAL DATA

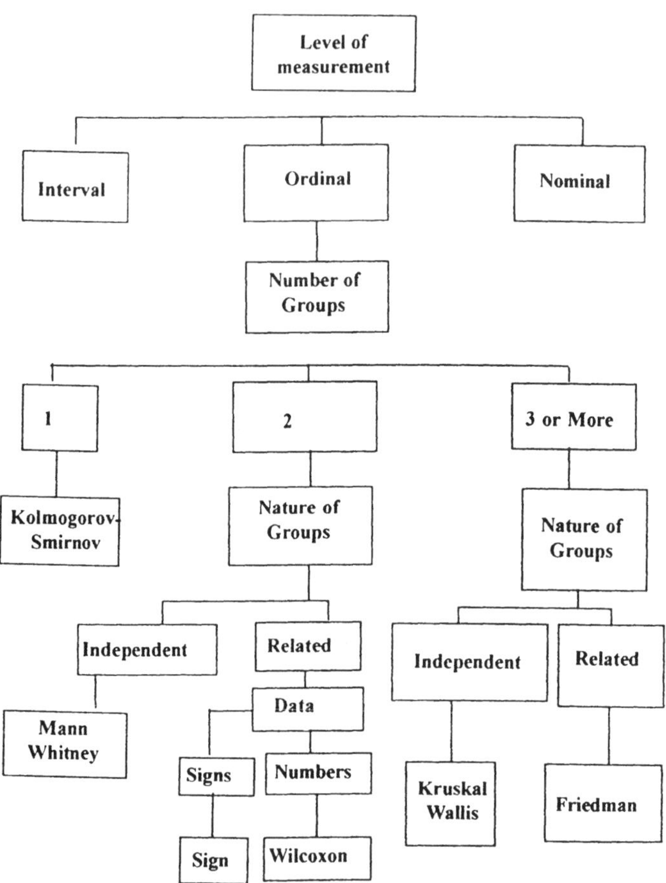

II. SELECTION OF STATISTICAL TECHNIQUES BASED ON INTERVAL DATA

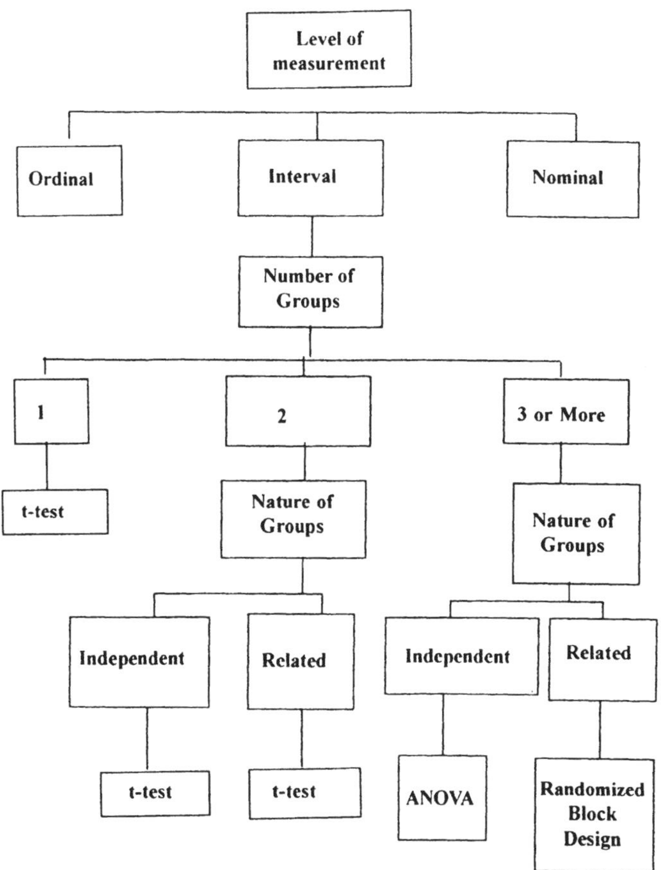

III. SELECTION OF STATISTICAL TECHNIQUES BASED ON NOMINAL DATA

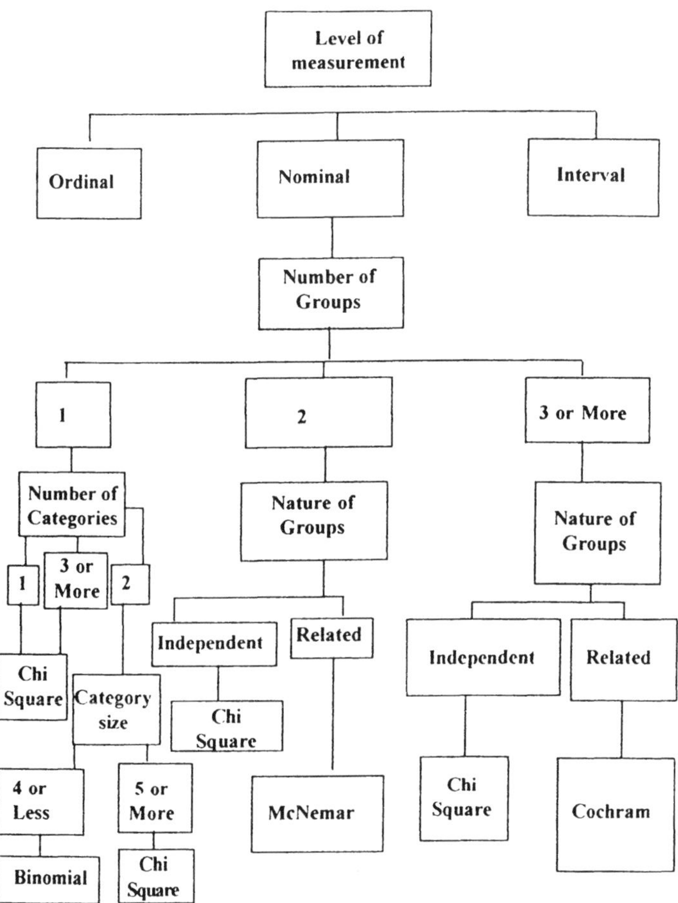

Appendix B: DISC Leadership Model

To understand the leader and the leadership style we must also understand the types of personality. One of the many available models of leadership is the DISC model, which is presented here in a summary format. Basically, the DISC model is based on the platinum rule—do unto others as they want to be done unto. Furthermore, the model is also based on the four types of personality styles:

1. *D* Dominant director.
2. *I* Interacting socializer.
3. *S* Steady relater.
4. *C* Cautious thinker.

What is important about these basic personality styles is the fact that there is no best type of personality. From a leadership perspective, however, we must remember that leaders for specific tasks and/or orientation of a project must be identified through one or a combination of the following categories with the respective personality traits:

1. Task oriented.

- Cautious thinker (C)
- Dominant directors (D)

2. People oriented.
   - Steady relaters (S)
   - Interactive socializers (I)
3. Fast paced.
   - Dominant directors (D)
   - Interactive socializers (I)
4. Slower paced.
   - Steady relaters (S)
   - Cautious thinkers (C)

In addition to recognizing the traits necessary for the task(s), we must also recognize "Who seeks what?" For example:

D Seeks power and control.
I Seeks popularity and prestige.
S Seeks sincerity and appreciation.
C Seeks accuracy and precision.

From a decision perspective, the leader will also display certain propensities for decision making, based on his or her own personality traits. They are:

D Decisive
S Conferring
I Spontaneous
C Deliberate

I. ELEMENTS OF THE CLASSICAL PATTERNS

1. *Emotional tendencies*: a specific set of emotional responses characteristic of each classical pattern.
2. *Goals*: personal goals that drive us and that help to direct us toward achieving maturity.
3. *Judging others*: criteria we use in judging others; recognizing, accepting, and controlling how we think about others can prevent misunderstandings and unhealthy biases.
4. *Influencing others*: characteristic techniques we use in influencing others; knowing the means we use to influence others helps us to measure if we are encouraging growth and understanding within our relationships.
5. *Value to an organization*: realizing our unique contributions to a team effort creates an objective measurement for an organizational environment.

6. *Overuses*: the way the specific strengths of a particular classical pattern can be overextended to the point that they become a weakness that dulls the person's ability to live and work effectively with others.
7. *Response under pressure*: deals with the characteristic way a particular classical pattern reacts to pressure; pressure situations place us under stress that can positively or negatively influence our response to life events.
8. *Fears*: recognizing emotional tendencies to protect ourselves can be useful when we are faced with a real or perceived challenge.
9. *Would increase effectiveness by*: the individual may need to learn new ways of responding to challenges or may need to learn to consciously gather around him people who posses the skills that compliment his behavioral style.

II. HOW TO RECOGNIZE A BEHAVIORAL TYPE

Two important dimensions for recognizing another person's behavioral type are directness/indirectness and the degree to which they are supporting or controlling. We all exhibit a range of these characteristics in our expressed, observable behaviors, but we need to focus on how people act in order to determine their core type.

A. Directness

This is the tendency to move forward or act outwardly by expressing thoughts, feelings, or expectations. For example:

1. Points fingers.
2. Is fast paced.
3. Will not take no for an answer.
4. Uses open arm movements.
5. Hugs and shoves.
6. Is verbally intense.
7. Uses assertive body movements.
8. Emphasizes their points of view with confident and strong vocal intonations.
9. Is fast talking.
10. Is brutally blunt.
11. Is impatient.
12. Seeks willing listeners (usually the indirect variety).
13. Motto: it is easier to beg forgiveness than seek permission, so when in doubt do it anyway; you can apologize later.

B. Indirectness

This is the tendency to move around and about through implication (in some cases). The movement may be through expressing thoughts, feelings, and/or expectations. For example:

1. Is more quiet and reserved.
2. Is more easygoing or at least more than they talk.
3. Asks questions and listens before they talk.
4. Is less confronting, less demanding, less assertive, and socially competitive.
5. Is slower paced and more security-conscious.
6. Tends to take set backs personally.
7. Predictability is important.
8. Seeks to meet their needs by accommodating the requirements of their environments.

C. Supporting

This is the tendency to support through expressing thoughts, feelings, and/or expectations a specific item. For example:

1. Uses more body language.
2. Uses more vocal inflection.
3. Uses continual eye contact.
4. Speaks in terms of feelings.
5. Uses animated facial expressions.
6. Has a flexible time perspective.
7. Likes to tell stories and anecdotes and make personal contact.
8. People needs are more important than tasks.

D. Controlling

This is the tendency to control through expressing thoughts, feelings, and/or expectations a specific item. For example:

1. Does not use facial expressions.
2. Keeps distance both physically and mentally.
3. Does not touch you and you do not touch them.
4. Tends to stand farther away even when shaking hands.
5. Has a strong sense of personal space.
6. Is time oriented.
7. Pushes for facts and details.
8. Is task oriented.

9. Puts a higher priority on getting thing done.
10. Likes structure.
11. Is more disciplined.
12. Views the planning and supervision processes as ways of reaching goals.

III. IDENTIFYING A PERSON'S BEHAVIORAL STYLE

Simply use the process of elimination. If a person is more direct than indirect, then you can eliminate the steady relater and cautious thinker types. If a person is also more supporting than controlling, then eliminate the dominant director. Now you arrive at the remaining style, in this case, an interacting socializer.

Behavior adaptability is something applied more to yourself (to your patterns, attitude, and habits) than to others. As a general rule, adaptable people try to meet the expectations of others by practicing tact. Remember, the willingness to try behaviors that are not necessarily characteristic of your type is called behavioral adaptability.

Another way of looking at this whole matter is from the perspective of maturity. Mature persons know who they are. They understand their basic behavioral type and freely express their core patters; however, when problems or opportunities arise, they readily and deliberately make whatever adjustments are necessary in their core patterns to meet the need.

- Usually your off-the-job behavior is the closest to your core behavioral type.
- Acquired style includes those additional tendencies which most of us consciously add to our core behavioral type in order to function better in our social, personal, and work environments.
- Core plus acquired styles represent 95% of the person.
- By contrast our work style is usually a learned one, where we have adapted to perform many roles that meet not only our individual expectations, but also those of others.

From a quality perspective, since we want to have winners as leaders throughout our organization, we must understand that each type has its *forte*, and therefore, depending on the task at hand, the organization should try to match the style of the leader with the task at hand.

IV. DOMINANT DIRECTOR

- They want to win.
- They fear boredom and routine.

- They may think his or her way is the only way.
- They seek new challenges.
- They want to find their own answers.
- They interpret rules.
- They accept responsibility for mistakes.
- They may manipulate people and situation.
- Under pressure, they do things alone and can become belligerent.
- They often looks for their own answers, instead of seeking them from so-called higher authorities.
- At the extreme, they may tend to manipulate people and situations.

A Dominant Director may be recognized in any organization by the following characteristics:

1. They are driven by the inner need to lead and be in personal control.
2. They take charge of people and situations, so that they can reach their goals.
3. Their key to success is achievement; they seek no-nonsense, bottom-line results.
4. Their motto is "lead, follow, or get out of the way."
5. They want to win, so they may challenge people or rules.
6. They fear falling into a routine, being taken advantage of, and looking soft.
7. They may act impatient, but they make things happen.
8. They are natural renegades; they want to satisfy their need for autonomy.
9. They want things done their way or no way at all.
10. They prefer strong directive management and work quickly and impressively by themselves.
11. They try to shape their environments to overcome obstacles in route to their accomplishments.
12. They demand maximum freedom.
13. They often have a good administration and delegation skills.
14. They appear cool, independent, and results oriented
15. They opt for measurable results.
16. They like to initiate changes (the most of all types of personalities).
17. They naturally prefer to take control of others, and inadequacies of coworkers, subordinates, friends, families, and romantic interests.
18. They process data conceptually by deductive reasoning from general to specific information.
19. They are more comfortable using the left brain than the right.
20. They do one task at a time.
21. They are able to block out distractions.

22. They channel all their energies into specific jobs.
23. They rid themselves of anger by ranting, raving, or challenging others.
24. Their bark is usually worse than their bite.
25. They soon forget what upset them in the first place.

A. Action Plan for Working with a Dominant Director

1. Show them how to win and/or new opportunities.
2. Display reasoning.
3. Provide concise data.
4. Agree on a goal and boundaries; then give support or get out of their way.
5. Allow them to do their thing within limits.
6. Vary routines.
7. Look for opportunities to modify their work load focus.
8. Compliment them on what they have done.
9. Let them lead but within limits.
10. Do not argue on a personality basis; stick to the facts.

B. How to Recognize a Dominant Director

1. Office
   a. Their tone suggests authority and control.
   b. Their desks may be covered with projects and papers stacked in neat piles.
   c. Their in-and-out baskets are bulging.
   d. They surround themselves with trophies.
   e. There is a large chair behind a massive desk.
   f. The Desk keeps them at a distance.
2. Walls
   a. There are diplomas and other evidence of success.
   b. There is a planning calendar.
   c. Their family photos hang behind them and are not readily available for them to see.
3. They have constant phone calls and interruptions (crisis room management)
4. They walk fast.
5. They dress comfortably and typically pay less attention to their appearance than the other types.
6. They usually wear navy blue or dark gray suits.

7. They like people to know they have made it without having to tell anyone about it; therefore, they often prefer possessions that emit success and authority messages. (Someone once suggested that they would buy a Sherman tank if they could.)
8. They often call someone on the phone without saying hello or a greeting, and leave a short concise message.
9. Their letters and memos tend to be brief and to the point.
10. They tend to opt for brevity, perhaps in their efforts to get as many things done as possible.

C. Distinguishing Characteristics of a Dominant Director

- They are direct and controlling.
- They have a faster, decisive pace.
- Their priority is the task and results.
- Their fears are being taken advantage of.
- They gain security through control and leadership.
- They measure personal worth by quality or impact of results, track record, and process.
- Their internal motivation is the win.
- Their appearance is businesslike and functional.
- Their workplace is efficient, busy, and structured.
- They seek control.
- Their strengths are administration, leadership, and pioneering.
- Their weaknesses are impatience, insensitivity to others, and being a poor listener.
- Their irritations are inefficiency and indecision.
- Under stress, they are dictatorial and critical.
- Decision making pattern is decisive.

A typical Dominant Director point of view consists of seeing oneself as a problem-oriented manager who enjoys a challenge just because its there.

- They are independent, strong-willed, restless, and goal-oriented.
- They are pragmatic, reactive, decisive, competitive, and harsh.
- As pressure mounts, they try to minimize the quantity of tasks or the quality of the results.
- Limitations are selective listeners, need to have ego fed constantly
- They need a coworker who draws them into the group. Dominant Directors often take themselves too seriously and can benefit from gentle reminders to take life less seriously and to laugh at themselves.
- They see themselves in a perpetual contest with others.

- They need to control, so one-upmanship can become a favorite game.
- They seek personal control, prefer time frames, express high ego strength, prefer to down play feelings and relationships, implement changes in the workplace, and tend to freely delegate duties (enabling them to take on more projects).
- They may become harsh and stubborn if confined to a system.
- They do not like detail work.
- They are motivated to manage, lead, organize, and rule.
- They often picks friends from their work pool.
- They may enjoy making fun of others.
- They tend to take themselves too seriously.
- They love to take the credit.
- They have a basic need to feel (and to be) in personal control.
- They have the dubious distinction of being the least natural listeners among the four types.
- They want the role of initiating the action and then taking charge.
- They set the tone for moving out to reach a specific goal.
- They tend to take an autocratic approach with other people. In an organization with a Dominant Director at the top, there is an emphasis on defining responsibilities, implementing action, and managing trouble; a hierarchy of leadership is usually installed so that there is a direct line of authority.
- They have a tendency to be insensitive to the feelings and goals of others. Rarely is this deliberate neglect, but the intensity with which they strive to meet objectives can cause them to consider emotional expressions as obstacles. Prone to see life as a battle during which any walls in their way must be torn down. Unfortunately, this approach is likely to result in emotional casualties along the way.
- They will seek a physical stress release. Once emotions are released they will begin to respond better to the people around them. Unfortunately, they have selected a stress release that looks to others like a personal attack and results in alienation.
- They usually get involved with physical activity as a way of working out the build-up-up of stress (e.g., golf, handball, cutting down trees).

D. Understanding the Dominant Director

- They are driven to overcome trouble and solve problems.
- They drive to overcome opposition in order to accomplish results.
- They possess an unusual ability to thrive in negative environments.
- Challenges give them the chance to create something better than was there before.

- They have confidence that they can accomplish just about anything they desire.
- They enjoy talking about their accomplishments, but they seldom mention mistakes or failures.
- Their task oriented leadership requires confidence
- They can easily become self-centered.
- They sometimes view the opinions of others as obstacles rather than helpful insights.
- They want results before all other things
- They prefer that information be given to them in bites of ten words or less, or at least they want people to get to the point as quickly as possible.
- You can usually tell if they are interested in what you are saying by the directness with which they approach you and request more affirmation.
- They fear being taken advantage of.
- They quickly seize opportunities that allow them to assume control over their destiny.
- They fear losing control; this often leads them to take offense at the aggressive behavior of others.
- They see confrontation as the best method of dealing with challenges to their goals.
- They tend to be insensitive to others.
- They need to learn to listen to others and be sensitive to their needs.
- Their failure to listen can prove costly.

E. Responding to the Dominant Director

- Their fear of losing out of hand usually means that a personal strength has gotten out of control and has turned into weakness.
- They need to learn some type of structure and accept it.
- They often become bored with routine work but operate best in an environment that offers challenging goals.
- They like being in charge and usually resist constraints.
- Because of their strong drive to reach their goals, confrontation may be necessary in order to get their attention.
- Show them how their actions affect you and prevent the fulfillment of mutual goals.
- Be brief and to the point in your explanations; expect them to disagree, but feel free to leave if the volume gets too loud and direct.
- Allow a cooling down period to reflect on their options before wanting a decision.

- Be prepared for a quick change once a commitment is made.
- Limit your expectations by understanding their greatest struggle.
- They are results oriented with an "I can do it attitude" (whether they have done it before or not).
- They do not willingly give up (persistent)
- They stand up for their rights.
- They prefer that others get things done quickly.
- They are naturally independent and dominant.
- They are self-confident that they can accomplish and make a difference.
- They are achievement-oriented and show annoyance when goals are thwarted.
- They fear others taking advantage of them.
- Under pressure, they may become critical and resistant.

Dominant Director/Interacting Socializer

- They tend to be wary and distrustful of others who show these same characteristics.
- They like who they are.
- They frequently make judgements based on how quickly others accomplish tasks, sometimes at the expense of quality.
- They want to see outputs (quantity).
- They often like confrontation and may challenge authority figures by making demands and imposing their wills.

Dominant DIrector/Inspirational

- They accept competition and even aggression from themselves and others.
- They appear to reject affection (on the surface).
- They want to control the environment.
- They judge others by how effectively they project strength and power.
- They may leap before they look.
- They tend to overcome pragmatism (the ends justify the means).
- They fear seeming to soft and dependent.
- Under pressure, they tend to argue and become belligerent.

The inspirational style is a dominant director type with nearly equal interacting socializer characteristics.

Dominant Director/Inspirational Results-Oriented Pattern

- They are driven mainly by dominance and secondarily by influencing traits.
- They display a gift for taking command of situations.

- They are quick and decisive and give direct and forceful instructions.
- When they speak few people challenge their plans.
- They command respect and other rely upon their strength and persistent character.
- They have determination and are willing to move forward no matter what the barriers.
- They see obstacles as challenges waiting to be conquered.
- Under stress, they have a tendency to be impatient and to use confrontation and intimidation as a means of gaining control.
- They have a very strong ego.

F. Inspirational

- They are quick to identify the best way to motivate those around them.
- They have a keen sense of knowing what other people need or want.
- They do not mind confrontation and have the skills to dominate any debate through force or character.
- They have equal measures of Dominant Director and Interactive Socializer traits.
- They have the ability to be both confrontational and entertaining.
- They are skilled with words and have no peers in their ability to persuade others to their point of view.
- In addition to being persuasive, they are usually tactful in getting their ideas across.
- They have the ability to manipulate the rules to work their personal advantages.
- They often forgive themselves for their shortcomings and when failure occurs, they shift personal guilt to someone else.

G. Dominant Director/Cautious Thinker Creative Pattern

- They have the ability to pick out the flaw in a position or philosophy, change it, and give it a new level of meaning and understanding.
- They have the ability to organize their thoughts in writing so that a clear, corrective plan of action is communicated.
- They are excellent debaters. (Never debate this creative pattern on their home turf—unless, of course you are working on humility.)
- The negative tendency of this profile is to be extremely critical and condescending toward others when they do not measure up to the creative pattern's personal standards.

- They have a gift for analyzing a system and determining its shortcomings.
- They have the ability to see the big picture but also see the details of the picture.
- They influence others by setting the pace for communicating new ideas.
- They impact relations by their ability to initiate activates that bring about change.
- They fear not having sufficient authority to change a course of events or the lives of others.
- They want to change the way things are done.
- They answer to their own personal ideas and standards.
- They want to influence others, and may sulk.
- They may control outward emotions and expressions.
- They strive to accomplish the unusual.
- They need tangible recognition.
- They are critical of themselves in route to becoming stars.
- They fear they will not meet their standards.
- Under pressure, they can become bored and blue.
- The creative person is a dominant director with a somewhat higher or lower cautious thinker pattern.
- They are quite controlling, though somewhat both direct and indirect (depending upon the situation).
- Their quick-thinking inventive mind may be restrained by their inner need to analyze all options before making decisions.
- This type both asks questions and supplies the answers (often internally).
- When positively motivated, creative people focus more on the future than they do on either the present or the past.
- They are change agents who seldom leave or continue things as they found them,

V. INTERACTING SOCIALIZER (I)

- They seek approval from others and are motivated by their acceptance.
- They are naturally direct and supporting.
- They fear loss of social acceptance and resulting loss of self-esteem.
- They give both time and emotions free rein.
- They like to jump from one activity to another.
- They operate emotionally.
- They prefer to start many activities, but do not necessarily finish them.

- Under pressure, they can become sentimental and careless.
- They are natural communicators.
- They are natural talkers and speak freely to let others know where they stand.
- They are overcomplimentary of others as a way to achieve goal of attaining recognition.
- They are naturally more accepting than many other styles and can often overlook physical, emotional, political, and philosophical differences that others may not.
- They like to be where the action is.
- They are outwardly energetic or fast-paced.
- Their relationships tend to take priority over tasks.
- They try to influence others in an optimistic, friendly way that is focused on positive outcomes
- They are motivated by recognition and approval.
- They want admiration and thrive on acknowledgment, compliments, and applause.
- Their favorite subject is themself.
- Their biggest fear is public humiliation.
- Their strength is enthusiasm, persuasiveness, and friendliness.
- They build alliances to accomplish results.
- Their weaknesses are too much involvement, impatience, being alone, and short attention spans.
- They are easily bored.
- They tend to make sweeping generalizations on little data.
- To an extreme, their behaviors can be seen as superficial, haphazard, erratic, and overly emotional.

A. Action Plan for the Interacting Socializer

1. Show them that you admire and like them.
2. Behave optimistically and provide upbeat setting.
3. Avoid involved details; focus on the big picture.
4. Vary the routine.
5. Do it together.
6. Compliment them personally and often.
7. Act nonaggressively and avoid arguing directly on a personal basis.
8. Support their ideas and do not poke holes in their dreams.
9. Mention their accomplishments and progress and your appreciation.

B. How to Recognize an Interacting Socializer

1. Office.

   a. Paperwork is strewn across their desks and sometimes on the floor.
   b. They react to visual stimuli, so they like to have everything where they can see it.
   c. Their desks often appear cluttered and disorganized
2. Walls.
   a. Motivational slogans, generalized personal comments or stimulating posters are prominant.
   b. Notes are taped all over the place.
3. Furniture arrangement often indicates warmth, openness, and contact.
4. They seldom sit behind a desk, but prefer to sit next to people at a table.
5. They dress in the latest style and like bright colors and unusual clothes that prompt others to compliment them.
6. They speak rapidly and emotionally on the phone.
7. Their letters reveal overuse of exclamation points, underlining, and bold highlighting.

C. Distinguishing Characteristics of the Interacting Socializer

- They are direct and supporting.
- They have a faster, spontaneous pace.
- Tehir priority is the relationship/interaction.
- They fear loss of social recognition.
- They gain security through playfulness and approval of others.
- They measure personal worth by acknowledgments, applause, compliments.
- Thier internal motivator is the show.
- Their appearance is fashionable and stylish.
- Their workplace is interacting, busy, and personal.
- They seek recognition.
- Their strengths are persuading, enthusiasm, entertaining.
- Their weaknesses are inattention to detail, short attention span, low follow-through.
- Under stress, they are sarcastic and/or superficial.
- They have a spontaneous decision-making pattern
- Personal characteristics:
  - They are supporting, trusting, instinctive, exploring, appeasing.
  - People are their business.
  - They prefer to think out loud.
  - They prefer a casual, relaxed environment.

- They need inclusion by others, popularity, social recognition, and probably freedom from a lot of detail.
- They are dreamers.
- They Like spontaneous expressive actions for noticeable results.
- They crave companionship and social recognition.
- They are true extroverts and give enthusiasm and energy to the group.
- They tend to talk more on average.
- They like to brainstorm and interact with colleagues.
- They want freedom from control, details, or complexity.
- They like to have the chance to influence or motivate others.
- They like the feeling of being a key part of an exciting team.
- They are easily bored with routine and repetition.
- They may trust others without reservation.
- Typically, they have short attention spans so they do well with many short breaks.
- They need immediate feedback to get or stay on course.
- Do not give them too much at once or they will become overwhelmed.
- They find that sorting out priorities and solutions becomes very difficult when too many opportunities bombard them.
- They see mental pictures first, then convert those pictures into words.
- Remember that with this type emotions rule.
- They are people who love people.
- They specialize in socializing.
- They hate isolation.
- They may stand with their arms around each other, hug, or show some other outward expression of their support for others.
- They tend to exhibit the most emotionally sensitive qualities.
- Their feelings tend to get hurt rather easily.
- They are lovers of gadgets and bells and whistles.
- They show "flight before fight."
- Typically, they diffuse conflict with a joke or funny observation.
- They favor fun with family members
- They tend to experience wide mood swings ranging from feelings of ecstasy to agony
- Of all the types, they are the most likely to forget about things.

D. The Interacting Socializer Influencing Style

- They initiate with poise and persuasion.

- They have the ability to provide practical insight to assist and encourage others to carry on, especially when they are discouraged.
- They generally have a positive outlook on life.
- They are expressive, enthusiastic. and the life of any party.
- They seldom have difficulty in meeting people.
- They have a strong capacity for trusting and accepting others.
- The counselor type has the greatest capacity for accepting someone perceived as an outcast.
- They are socially oriented.
- Tensions can occur when a highly Interactive Socializer must share attention with another, for the need for recognition is a strong force in their life.
- The need for recognition is one of the most important measurements of their self-acceptance.
- They fear of social rejection; if this fear is not controlled, it can lead to failure.
- They have a tendency toward disorganization.
- Their excuses and good intentions often keep them from being productive and achieving their potential.
- They never have sufficient time to finish the projects.

Interactive Socializer/Promoter

- Of all the classical patterns, they possesses the greatest natural ability to express verbal approval and acceptance.
- They have a gift for expressing themselves.
- They are second to none in conveying encouragement and acceptance.
- They are extremely skilled in getting others to support a project or causes.
- When they make mistakes, they frequently use the praise of others in an effort to cover up their errors.

Interactive Socializer/Persuader

- They collect status symbols.
- They admire how well others express themselves.
- They fear routine and the same old thing.
- They cultivate friendliness and openness.
- They look at life positively and enthusiastically.
- They persuade others to pursue a common vision with their optimism.
- They have inherent trust of others.
- Under pressure, they can become soft and persuadable.
- They are interacting socializers with many dominant director traits.

- They are basically supporting, but can also be controlling.
- Their primary worry or fear in life is a fixed environment including complex relationships which can limit both retention of and continued group strength in attaining social approval from new experiences.
- They are eternal optimists.

E. Dominant Director/Interactive Socializer: Persuader

- They are natural salesmen.
- They have the ability to communicate with illustrations, pictures, and colorful stories so that a group of people personally relate to whatever the persuader is selling.
- They have the gift of dominating a group, to the point of making a decision based on the information the persuader has provided.
- They are likely to give into social pressure
- When they must make a decision between an agreed-upon principle and social rejection, the persuaders will generally lean toward compromising the principle in order to maintain relational acceptance.
- In any negative environment, they will use their verbal skills to deny responsibility.
- They have the special ability to reach out to strangers with openness and a friendly spirit.
- They naturally speak to anyone, quickly discern needs, and warmly extend a helping hand to others.
- Physical contact is also a natural trait for them; they are quick to shake hands, give hugs, pat shoulders, and kiss cheeks.
- They are gifted communicators and are particularly comfortable in making impromptu speeches that are personalized to the needs of any group.
- They tend to listen selectively.
- They often react to the first few words of a speaker and then inappropriately offer comments out of context.

F. Interactive Socializer: Counselor

- They have a tendency to empathize and understand; this can be overused so that others take advantage
- They can become overly attached to those they care about.
- They are dependable, caring, and responsible.
- They are naturally good listeners.
- They have a tendency to trust others.

- They prefer indirect actions.
- They like to keep everybody happy.
- Under pressure, they can become too intimate and possibly gullible.
- Basically, they are an interacting socializer with many steady relater characteristics.
- They are a supportive type who is primarily direct, although can become somewhat indirect under less favorable circumstances.
- Their personal goal is friendship, especially as it results in acceptance by others.
- Their fear is being perceived as harming others.
- They have a natural dislike for pressuring or telling people what to do.
- They do not want to inconvenience anyone.
- Counselors are there when the going is tough.

G. Counselor Interactive Socializer/Steady Relater

- They are approachable, affectionate, and understanding.
- They have good listening skills and are sensitive to feelings.
- They have the ability to see the good in others, rather than just look for their faults.
- They project stability and dependability as well as considerable interpersonal skills.
- They are often key participants in an informal communication system and generally exert positive influence because of their optimistic attitudes.
- Their goal is to maintain friendships and keep everybody happy and satisfied.
- They have a tendency to use an indirect approach and tolerance in dealing with problems.
- This caring nature can become a weakness if overused in situations where establishing firm boundaries would lead to responsible behavior of others.

H. Interactive Socializer/Cautious Thinker: Appraiser

- They want to win and get results with flair.
- They judge others by their ability to start activities and make things happen.
- They may overuse authority and positions.
- They fear losing.
- They want to do things the best way.

- They are naturally enthusiastic.
- Under pressure, they may become restless, aggressive, and impatient (i.e., an interacting socializer with many cautious thinker traits).
- They will not step on others.
- They fear looking bad to others.
- They are driven to win, but prefer to achieve their victories by working with and through people
- They usually have the charisma to persuade team members to join them in achieving their personal goals and objectives.
- They are committed to the win-win philosophy.
- They strive to succeed with flair and style, but truly desire that others be a part of any success achieved.
- They have a gift of making others feel like they are important valued members of the team.
- They are committed to communicating the point that the efforts of others really count toward achieving a team's goals.
- They have two emotions that continually pull at them: one positive and one negative. They are driven both by a need for close relationships and a need to accomplish tasks. Under pressure these two drives produce a person who acts like a coiled spring. They can become restless, impatient, and aggressive.
- They need to learn the art of relaxation in order to guard against being driven to burn out.

I. Responding to Interactive Socializer

- Since they are first and foremost relational creatures, they desire a positive social environment and are particularly sensitive to maintaining a positive relationship with their peers.
- The fear of rejection is a real and dynamic force in their lives.
- They need support in dealing with the issue of stress, not criticism.
- Give clear instructions.
- They tend to cave-in to social pressure when under stress.
- Their basic needs are social recognition, popularity, people to talk to, freedom of speech, freedom from control and detail, recognition of abilities, and opportunities to help and motivate others.
- They avoid events that might disrupt their personal harmony.
- If a choice had to be made as to whether to achieve goals or preserve relationships, they will choose to maintain relationships.
- Recognize that they have a need to be leaders in groups and to express themselves verbally.
- Watch out for their tendency to test boundaries to their limit.

- Realize that they have high levels of confidence in themselves and others.
- Do not be surprised if they have difficulty following through on commitments.
- Seek to create a fun and friendly environment for them.
- Allow them the opportunity to express their own thoughts and opinions.
- Provide ideas for transferring talk to action.
- Recognize they have a need for positive social recognition.
- Use confrontation only when necessary.

VI. STEADY RELATER

- They want and are motivated by acceptance, respect, and organization.
- They are naturally reserved and supporting.
- They fear loss of security.
- They follow established procedures and traditions.
- They try to stick to limited activities at a time and follow through until completion.
- They think logically and realistically.
- They prefer tangible, identifiable tasks.
- Under pressure, they comply
- They want stability at almost any price.
- They look before they leap; more accurately, they look before they step.
- Their common drive is toward personal stability.
- They naturally want to provide services and favors for people.
- They like to see people getting along regardless of the environment.
- They are especially tolerant and accommodating of the behavior of others.
- They tend to give people the benefit of the doubt.
- They look at the bright side and seek the best of people.
- They may cover up for another's unacceptable actions, choosing to overlook the possibility that the person may be taking advantage of them; any behavior including viewing others as potential friends and treating them accordingly can be taken to an extreme.
- Steadiness and follow-through actions are their trademark.
- They prefer a slower, easier pace of life.
- They focus on building trust and getting acquainted because they aim for long-standing personal relationships.
- They are very irritated by pushy, aggressive behavior.

- They strive for security.
- Their goal is to maintain the stability they prefer in a more constant environment.
- They do not appreciate the unknown in anything.
- For them, risk is an ugly word.
- They fear change and disorganization, especially sudden changes as they are very concerned with what may happen
- Any disruption in their routine patters can cause distress.
- They need to think and plan for changes.
- They are low keyed.
- They act very sincerely.
- They are an easy type to get along with.
- They prefer stable relationships which do not jeopardize anyone, especially themselves.
- They will plan and follow through.
- They have difficulties in speaking up and seem to go along or conditions while inwardly they may or may not agree.
- They are reluctant to express themselves if the result can be hurt feelings.
- They lack assertiveness, which can take a toll on their health and well being.
- They yearn for more tranquillity and security in their lives than the other types.
- They often act pleasant and cooperative, but seldom incorporate emotional extremes.
- They have a natural need for composure, stability and balance.

A. Action Plan for the Steady Relater

1. Show how your idea minimizes risk
2. Show reasoning.
3. Provide data/proof.
4. Demonstrate your interest in them.
5. Provide outline or 1–2–3 instructions as you personally walk them through.
6. Compliment for their steady follow-through.
7. Act nonaggressively, focus on common interest.
8. Allow them to provide service or support for others.
9. Provide relaxing friendly atmosphere.
10. Provide them with a cooperative group.
11. Acknowledge their easy going manner and helpful efforts.

B. How to Recognize a Steady Relater

Steady relaters have persistence and people-to-people strengths—patience, follow-through, and responsiveness. They are good listeners and empathizers, and react to our feelings.

1. Office
   a. They have personal slogans and group photos along with a serene landscape pictures.
   b. They have family pictures and mementos, usually turned so they can view them from their desk chair.
   c. They favor nostalgic memories of stabilizing experiences.
   d. They prefer to arrange seating in a side-by-side, more congenial cooperative manner; no big desks for them.
2. They tend to wear subdued colors and conservatively cut clothing of conventional styles.
3. They often like beige or light blue.
4. They are cheerful on the phone.
5. In their written correspondence, they may send letter just to keep in touch or to let you know they are thinking of you. Of the four personality types, this one is likely to send thank you notes for almost anything. They may even send a thank you note to acknowledge your thank you note. Again, they are likely to organize their letters, writing as they do their other to do tasks lists. Since they tend to write in a slower, more methodically paced manner, their work tends to follow a systematic outline pattern.

C. Distinguishing Characteristics of the Steady Relater

- They are indirect and supporting.
- They are slower, with a relaxed pace.
- They put priority on the relationship/communication.
- They fear sudden changes, instability.
- They gain security through friendship and cooperation.
- They measure personal worth by compatibility with others and the depth of their contribution.
- Their internal motivator is participation.
- Their appearance is casual and conforming.
- Their workplace is friendly, functional, and personal.
- They seek acceptance.
- Their strengths are listening, teamwork, and follow-through.

- Their weaknesses are being oversensitive, slow to begin actions, and lack of global perspective.
- Their irritations are insensitivity and impatience.
- Under stress, they are submissive, and indecisive.
- Their decision-making pattern is conferring.
- They are observing, reflecting, implementing, applying, and avoiding.
- They take one day at a time and avoid gambles.
- They respect traditions.
- They are most compatible of all working relationships as they have patience, staying power, and commitment.
- They are very uncomfortable with conflict.
- Typically, they will say nothing negative about their observations.
- Their adherence to following instructions and maintaining the status quo can limit their actions.
- So when problems bombard them, they try to solve them by helping or working with others, following tried and true procedures, or a combination of the two. If these tactics fail, they may quietly do nothing. Doing nothing may include higher absenteeism, for when when conflict and stress increase, their tolerance may decrease.
- They are inherently accommodating.
- They have difficulties in taking tough stands.
- They seek security and inclusion with the group.
- They are optimistic realists; why change?
- They need to know the order of procedures.
- They operate well as members of a work group.
- They focus on how and when to do things.
- They work in a steady and predictable manner
- They like to perform the same kinds of duties day to day.
- They often think that if the boss does not see what is going on, they do not want to be the one to confront others about an unpleasant reality.
- As naturally interested listeners, they appreciate this same from others.
- They tend to speak indirectly, and seldom come out and say what is on their minds.
- When complimenting them, mention their teamwork and dependability.
- They express their feelings less directly, so draw them out through questioning and listening responses.
- They tend to take things personally, so remove the something-is-wrong-with-you barrier as quickly as possible.
- Be ready to do more talking than listening with them since they do not naturally feel the limelight is focused on them.
- They are quiet, evenly paced, and inwardly focused individuals.
- They tend to distrust the intangible.
- At a party they may converse with the same person all night.

- They prefer having others approach them.
- They often choose friends by the test of time method.
- They depend on ways of keeping their lives stable and secure.
- They may respond over carefully to people so as to not hurt their feelings. Consequently, they find it hard to muster up an occasional no, often allowing more assertive types to take advantage of them.
- When it comes to venting, they can be virtual shock absorbers.
- They exhibit a rather dry, straightforward, seemingly uncomplicated sense of humor.
- Their emotions bruise easily; sometimes they are rather easily confused or hurt by others and try even harder to improve matters.
- They usually do not say anything when something bothers them. They do not want anyone to dislike them.
- They avoid confrontation—their biggest fear—at almost any cost.
- Their primary need is to maintain stability in their own setting.
- They do not like being in a crowd.
- They like to do things which are planned or at least known in advance.
- Part of their appeal is that they tend to naturally accept and tolerate others, warts and all.
- They often tend to be collector types; they like to accumulate personal treasures, photographs, and souvenirs that aid them in reflecting about the way things were.
- They can overdo their affiliation with family routine and become possessive.
- They like moderation in all things.
- They prefer that significant others notice their thoughtfulness without prompting.
- They can have problems communicating directly with loved ones about what may be bothering them. Since they often make sacrifices and act as peacemakers in the family unit, they may inwardly expect acknowledgment and thanks, but fail to say anything about it. Taken to an extreme, this reticence about voicing their concerns coupled with their modest, sometimes places them in the tenuous position of martyrs or victims.
- They can benefit by realizing that those close to them may need reminders to throughly understand and be responsive to their needs, wants, and expectations.

D. Understanding the Steady Relater

- They are team players with a desire to please and maintain peace and stability in a group, even if it means sacrificing their own personal goals.

- They usually do not stand out in a crowd.
- They have consistent and steady work habits over time.
- They are reluctant to aggressively step forward as a leader.
- As a leader, they usually bring consistency, support, and stability to the position

E. General Tendencies of the Steady Relater

- Their opinions usually consist of practical advice.
- They will generally reach a solution to a problem after everyone else involved has spoken and after they have had an opportunity to process his thoughts. You will be amazed at the insights if they are allowed time to review the alternatives. If time is short for a decision, they will defer to others to keep the peace.
- They need security, peace, and support.
- They place tremendous importance on stability within their family.
- Should conflict exist among family members, they tend to become distressed and prone to worry and anxiety. This often leads to avoiding the conflict and letting others take the lead in creating solutions.
- They tend to prefer the status quo and honor traditions.
- In a work environment, they provide a stabilizing influence by the consistent manner with which they carry out their duties.
- They usually do well at handling routine matters and will usually make sure that everything is in it's proper place
- Change is usually unwelcome; they tend to cling to the present and battle the forces of change.
- When initiating change with the high Steady Relater, recognize first that a slowed down performance is a normal response to change for this personality type.
- They need to be allowed the opportunity to communicate with others who also will be affected by change

F. The Steady Relater/Specialist

- They have consistent work performances day in and day out.
- They prefer working quietly behind the scenes.
- They have the ability to maintain the pace that others have started.
- Under pressure, they adapt to those in authority and, in the case of a sudden, forced change, they usually make accommodations to the demand place upon them.

G. The Steady Relater: Agent

- They display loyalty and expect the same from others.
- They like displays of affections/approval and dislike aggression.
- They are naturally giving and friendly.
- They work toward mutual harmony.
- They may overdo hopefulness and empathy.
- They want acceptance.
- Under stress, they can stand up for their rights if they backed up with appropriate information.
- This is a Steady Relater with many Interacting Socializer characteristics.
- They are basically a very supportive person.
- They need to be a person who is wanted and needed by others.
- They fear dissension or conflict, both within themselves and with others.
- They sincerely appreciate thoughtfulness, words of encouragement, and expressions of friendship.
- They are good listeners, the best of all styles.
- They work to keep the peace and maintain harmony, even if it means taking less for themself.
- They are excellent listeners.
- They are good teachers because they mix empathy and concern with the capacity for information and instruction.
- Their negative side is that they want to keep the peace at any cost.
- They avoid conflict and fear dissension.
- They influence others by their special gift for caring.
- They have a willingness to share both the highs and the lows of a close personal relationship.
- They reach out to meet the needs of people in distress.
- They often use their friendly support styles to persuade others to be forgiving in difficult situations.
- They create environments in which people feel comfortable sharing their concerns and problems.
- Unconditional acceptance seems to come quite naturally for them.
- They are a model of what it means to be a true friend.
- Their greatest fear is the fear of aggression. They will run from it or do anything from facing it head on.

H. Steady Relater/Dominant Director: Achiever

- They have a busy and industrious work style.
- They are responsible for their tasks and pull their own weight.

- They break jobs down into small sections and complete them one by one.
- They may do too much and have difficulty delegating.
- They may show impatience and frustration with themself and others.
- They want concrete results.
- They fear looking bad because somebody else did something wrong.
- Under stress, they become frustrated and impatient.
- Their goal becomes personal accomplishments.
- Their primary fear is other individuals with competing or inferior standards; this can become a barrier in the process of achieving desired results.
- They prethink their steps.
- They demonstrate a large measure of self-reliance.
- They are industrious and diligent.
- Thsy have a tendency always to be busy accomplishing something while displaying an intensity during their involvement.
- They have a high degree of organizational ability and the rare combination of being both process and product oriented.
- They make excellent administrators.
- They have the ability to look at a goal and devise a logistical plan to achieve it in a realistic time frame.
- They have the special talent for setting up procedures and schedules with an accountability structure so that everyone knows what they are to do.
- Under pressure, they become frustrated and impatient with others.

I. The Steady Relater/Cautious Thinker/Dominant Director: Investigator

- They prefer a logical approach.
- They hang on until they he get the answer (tenacious).
- They like one-on-one, and small group settings.
- They can become blunt and seem unfeeling.
- They carry grudges, but often do not verbalize them.
- They dislike and fear large groups and untenable ideas.
- They keep emotions on an even keel and do not outwardly express euphoria or rage.
- Under stress, they may turn inward and think of wrongs done to them.
- A steady relater with many cautious thinker and dominant director characteristics.
- They are very subtle.

- Their unique goal is the power provided by formal role positions.
- They are basically self-disciplined people who prefers a logical, unemotional approach.
- They can be counted upon to maintain composure in emergencies and to factor out feelings to get to the facts.
- They often show an unorganized, prestructured quality that other styles may lack.
- They are fully capable of listening unemotionally to a barrage of complaints and even some verbal abuse without acting ruffled, since they naturally factor out the feeling elements in life; there is little that seems to outwardly disquiet them.
- They tolerate negative conditions from others because of their almost uncanny persistence and their belief that they are right. As frustrating as this trait can sometimes be for their styles, they are often correct which in turn can arouse negative feelings in others.
- They possess reason, logic, and self-discipline.
- They influence others by their determination and tenacity.
- Once they set their sights on a goal, they never waver until the task is complete.
- They prefer to work alone and, year after year, will quietly go about getting things done.
- When in a disagreement with others, they tend to internalize their feelings.
- Instead of expressing their anger and forgetting about it, they store up a list of grievances, which in turn leads them to be suspicious of others.

J. Responding to the Needs of the Steady Relator

- Under stress, the Interacting Socializer will talk, the Dominant Director becomes intense and physical, the Cautious Thinker escapes, and the Steady Relater sleeps.
- They need a stable, predictable environment if they are to feel comfortable.
- Of particular importance is a home life free of conflict. If the home is in constant disarray, it is not unusual for the high Steady Relater to begin to experience various physical problems.
- They can develop a sense of security with routine.
- They will generally not make decisions.
- They have found it is easier and safer to go along with someone else's plan than to defend their own.

- They often will react to change by compliance to keep the peace, but will not aggressively support the new direction.
- Depending on how strong the confrontation is, their normal reaction is to not get involved or to retreat into a shell of silence.
- They usually seek acceptance and try to maintain stability.
- They prefer environments that maintain the status quo unless specific reasons are given for change.
- They seek to avoid conflict and dissension within relationships.
- They demonstrate loyalty and support for those they respect.
- They strive for security for their family.
- They need to be given time to adjust to an opportunity and to visualize requests made of them.
- They need personal assurances of support.
- They seek out close personal friendships.
- They need the persons dealing with them to modify their expectations of them by understanding their greatest struggles.

VII. CAUTIOUS THINKER (OBJECTIVE THINKER)

- They are concerned with appearances.
- They are naturally process-oriented.
- They seek quality control.
- They fear irrationality and loss of emotional control.
- They need to be right.
- They tend to act and think logically.
- They are naturally geared toward reasoning and intellectualism.
- They like to acquire data,
- Under stress, they need to worry.
- They are both indirect and controlling.
- They have a concern for correctness.
- Their core fear is irrational actions.
- They are right much of the time!
- They are considered by many to be the most complex and difficult to understand among the four basic behavior types.
- They are driven by controlling force in the form of self-direction.
- They strive to please themselves and satisfy their own standards of quality.
- They reject and avoid conflict and aggressions, often seeking a quiet private place to regain a sense of calmness.
- They are more concern with content than with congratulations.
- Their concern is accuracy.
- Human emotions take a back seat for them.

- They prefer involvement with the performance of products and services under specific and preferably controlled conditions so the process and the results can be correct.
- Their biggest fears of uncontrolled emotions and irrational acts relate to their goals.
- They fear emotionality and irrationality of others.
- They strive to avoid embarrassment so they attempt to control both themselves and their emotions.
- Their strengths are dependability, independence, clarification, and testing skills, follow-through, and organization.
- They focus on expectations and outcomes.
- They want to know how things work so they can evaluate how correctly they function.
- They are resourceful and careful.
- Because they need to be right, they often check processes themselves.
- They have a tendency toward perfectionism; taken to an extreme, this can result in paralysis by overanalysis.
- They worry that the process is not progressing right, which further promotes their tendency to behave in a more critical, detached way.
- They prefer tasks over people.
- They prefer clearly defined priorities and a known pace which is agreeable to them especially where task time lines are deadlines are involved.
- They have a complex mental make-up as they constantly are relating processes to the past, present, and future.
- They tend to see the serious, more complicated sides of situations as well as the lighter or even bizarre side—which accounts for their natural mental wit.
- They are the most cerebral of the four types.
- Their motto is: It is not whether you win or lose, it's how you play the game—the more technically perfect the better.
- They comply to their own personal standards, they demand a lot of themselves and others, and they may succumb to overly critical tendencies.
- They often keep their criticisms to themselves, hesitating to tell people they think they are deficient.
- They quietly hold their ground, they do so as a direct result of their proven knowledge of facts and details, or their evaluation that others will tend to react less assertively.
- They can be assertive when they perceive they are in control of a relationship or their environment.
- Having determined the specific risks, margins of error, and other variables that significantly influence the desired results, they act.

- They are introverted and reflective, they ponder both the why and how elements of situations.

A. Action Plan for the Cautious Thinker

1. Approach them in an indirect, nonthreatening way.
2. Show your reasoning.
3. Give it to them in writing.
4. Provide explanations and rationale.
5. Allow them to think, inquire, and check-out-out before they make decisions.
6. Let them check on others' progress and performance.
7. Compliment them on their thoroughness and correctness.
8. Let them assess and be involved in the process, when possible.
9. Tactfully ask for any clarification and assistance you may need.
10. Allow them time to find the best or correct answer with available limits.
11. Tell them why and how.

Cautious thinkers typically are motivated by their planning and organizational tendencies. If we want a task done precisely, find a Cautious Thinker; of the four types, they are the ones most motivated to be correct (i.e., the quality control experts).

B. How to Recognize a Cautious Thinker

1. Office
   a. They carry their organizational tendencies into their work environments.
   b. They have highly organized desks with cleared tops.
   c. Charts, graphs, exhibits, models, job-related pictures are often placed neatly on their office walls or shelves.
   d. They favor a functional decor that enables them to work more efficiently.
   e. They tend to keep most objects within reach and readily available when needed.
   f. They have state-of-the-art technology to further enhance their efficiency.
2. They are usually people of few words.
3. They are reluctant to reveal personal feelings; they often use thinking words as opposed to feelings words.
4. They are noncontact people.

5. They are generally not fond of huggers and touchers; they prefer a cool handshake or a brief phone call.
6. They usually walk slowly and methodically toward a known destination.
7. They use formal greetings on the phone.
8. They talk in structured, careful speech patterns, as if weighing their words as they say them.
9. They typically send letters or memos that are filled with data to clarify or explain positions.

C. Distinguishing Characteristics of the Cautious Thinker

- They are indirect and controlling.
- They have a slower, systematic pace.
- Their priority is the task or process.
- They fear personal criticism of their work efforts.
- They gain security through preparation and thoroughness.
- They measure their personal worth by precision, accuracy, and quality of results.
- Their internal motivator is the process.
- Their appearance is formal and conservative.
- Their workplace is formal, functional, and structured.

D. Understanding the Cautious Thinker

- They prefer to set the standard for a group, and they expect everyone to comply always with the objective of insuring quality, accuracy, and order.
- They prefer the status quo.
- Sudden change can often produce an unsettling response in them.
- When approached with new ideas, they will usually ask numerous questions for clarification.
- They take a cautious approach toward change and can uncover errors in a planning process before it is to late.
- As they fulfill their need for quality control and accuracy, such people often save organizations much time and money.
- They can pay too much attention to detail (a weakness).
- They are task rather then people related.
- Their attention is on the quality of the product.
- They have a strong desire for a structured environment.
- They prefer a step-by-step and inch-by-inch approach to a task.

- Their preoccupation with detail lead them to prefer an environment where everything has a name.
- Because of their high standards of performance, they can be extremely critical of their own work and have a tendency to internalize criticism from others.
- They have difficulty with criticism because they see the criticism as dissatisfaction with their standard performance.
- They can be extremely harsh in evaluating their own work, and their tendency to internalize criticism usually leads them to experience feelings of insecurity and depression and, in the most extreme cases, even thoughts of suicide.
- They are demanding of others as well, and expect commitment to goals and a follow through.
- They seek accuracy.
- Their strengths are planning, systematizing, and orchestration.
- Their weakness are being perfectionists, critical, and unresponsive.
- Their irritations are disorganization and impropriety.
- Under stress, they are withdrawn and headstrong
- Their decision-making pattern is deliberate.
- They are analyzing, evaluating, planning, investigating, and criticizing.
- They see themselves as problem-solvers who like structure.
- They want to know why something works, since such insight allows them to determine for themselves the most logical way to achieve the expected results from themselves and others.
- They are practical and realistic.
- They may become mired down with data collection.
- They tend to generate the most native creativity of the four types and often find new ways of viewing old questions, concerns, and opportunities.
- They are basically more introverted individuals who seek solace and answers by turning inward.
- Because they want peace and tranquillity, they avoid and reject hostility and outward expressions of aggression.
- They are concerned with process.
- They are more interested in quality than quantity.
- Since this is type is characterized by the most complex thinking pattern, they base their decisions on the proven information and track records.
- Appeal to their need for accuracy and logic.
- Keep praise simple and concise.
- Since this type does not communicate well, persist in your attempts to get them to talk.

- Allow them to save face as they fear being wrong.
- They tend to start with what they have to work with, then personalize it (almost from the beginning), so that it works better as they see it.
- As highly intuitive, astute observers of their surroundings, they are like electronic sponges, taking in and processing information about people and things.
- They tend to process many of the complexities of life that escape the other types.
- They value privacy, individual space, and discretion in their relationships with others
- They are tactful, serious, and organized.
- Since they do not readily discuss their feeling or even their thoughts, their initial responses can speak volumes about how they really feel.
- They have memories akin to the size of an elephant; they seldom forget, especially when they have been wronged.
- Their humor typically shows a down-home witty perspective, often from an unexpected third angle.
- They usually opt for small numbers of people with whom they are comfortable socially.
- They have a basic need for accuracy.
- They are the most private of the four basic behavioral patterns, so they require more time and space in their dealings with people.
- They keep their involvement down to few people.
- They are extremely diplomatic and accommodating to others, at least on the surface, because they often mask their own inner thoughts and feelings.
- They fix mistakes and downplay emotions.
- They gravitate toward practical hobbies and interests. They often enjoy computers, novels, or other educational subjects, collections, and other pursuits that involve more private individual involvement.
- Their children hold their emotions within themselves.
- They tend to lose themselves in activities and are often are seen, but not heard. In fact, they can become so involved in what they are doing they may not even hear others speaking to them.
- They intuitively invent their own structure, method, or model for understanding and processing what they need to know or do.
- They have the need to be correct.
- Because they tend to be more comfortable thinking about their feelings rather than expressing them to others, they may expect people to know what they think, especially those who know them well.
- They are often perceived by others as too picky as they ask lots of questions.

E. Objective Thinker (High Cautious Thinker)

- They prefer to do things correctly and in order.
- They are respected for who they are; they tend to make extremely loyal friends, employers, and partners.
- They desire to fit in quietly with a team and will tend to reject interpersonal aggression.
- In a peaceful environment, they will faithfully carry out the instructions given to them; however, if conflict arises, they will tend to avoid personal relationships.
- They are valuable to organizations because of their ability to define, clarify, and follow systematic procedures.
- They usually enjoy researching information and testing results for accuracy.
- They have an obsession with detail that can result in a failure to see the big picture (i.e., paralysis by analysis).
- Developing the ability to be more flexible is a mark of maturity for the objective thinker.

F. Cautious Thinker/Steady Relater: Perfectionist

- They want to be in charge of quality control.
- They go by the book and follow established rules (if they were formed by people they respect).
- They want to get things done right.
- They prethink procedures and attend to details accordingly.
- They are masters of diplomacy and tact.
- They dislike opposition and hostility.
- They want stability and security (no surprises).
- Under pressure, they become tactful.
- They are Cautious Thinkers with many Steady Relater characteristics.
- Their essential fear is antagonism.
- They prefer to follow proven practices.
- They may behave quietly and privately around others, waiting for them to initiate relationships.

G. Cautious Thinker/Dominant Director: Perfectionist

- They pay meticulous attention to detail and have a strong concern for quality control and the maintenance of standards.

- They are unique in ability to work alone for long periods of time and still enjoy it.
- Under stress, they have a tendency to be extremely cautious, ask a lot of questions, and refuse to accept change quickly.
- They are models of quality control.
- They respond well to those in authority and follow written and verbal instructions in a methodical orderly manner.
- They are valuable to an organization because of their conscientiousness and their efforts to maintain high standards.
- Under pressure, they become tactful and diplomatic, particularly when their physical and psychological security is threatened.

H. Cautious Thinker/Interacting Socializer/Steady Relater: Practitioner

- They think they can do better than others.
- They have high aspirations and ambitions.
- They expect a high level of expertise from others as well as themselves.
- They dislike criticism, especially regarding their expert ideas.
- They cultivate self-control.
- They are good problem solvers.
- They want others to value them for being exciting, unique, and proficient.
- Under pressure, they become restrained and sensitive to criticism.
- They fit into the cautious thinkers style with some steady relater and interacting socializer traits.
- They are fascinating studies as social chameleons.
- Their goal is relatively high ambition for personal growth.
- They are motivated to advance themselves and want others to recognize them for their accomplishments and knowledge.
- They often know the rules and how to get things done and can explain these procedures to others.
- They judge others by how self-disciplined they are.
- They are people who are motivated by a preference for compliance to the standards they have established for themselves and others.
- To understand their actions is to realize that they are driven by the interaction of these separate needs: control, influence, and cooperation.
- They are skilled at breaking down complex ideas into practical steps and systematic procedures.
- They are able to persuade people that what they have developed will be useful.

- They often give the impression of knowing something about many areas.
- Others will benefit from their determination and ability to solve problems successfully.
- They have high ambitions for themselves and others; this causes them to stand out in a crowd.
- They strive to be innovative and unconventional.
- They can come across as somewhat arrogant in their attitude and can be harsh in their criticism of others under pressure.
- They possess good verbal skills.

I. Responding to the Needs of the Cautious Thinker

- An important part of their life is obeying laws, following the rules, and fulfilling commitments.
- They fear antagonism, aggression, and conflict.
- They can look at society and see the flaws.
- They are often labeled a pessimist.
- They need reassurance of support.
- They typically choose independent resources to assimilate information.
- They have a need for order and defined responsibility.
- The major tip-off that a Cautious Thinker is moving into the stress mode is to listen to the kinds of questions they ask. If their inquires take on a personal tone, it strongly indicates that they sense a problem.
- Under stress, it is common for Cautious Thinkers to feel that they are the only people in history of the world to face the problems they are experiencing.
- They have a tendency to have "pity parties" if not aware of the problem.
- They prefer working alone.
- They have a preference for intellectual pursuits.
- When someone requests their participation in a project, be prepared to give specific details.
- They will usually exercise caution and restraint.
- They need assurances of support for their efforts.
- Be prepared to answer their questions in a patient and persistent manner.
- Be willing to provide them reassurances that no surprises will occur.
- Support your plans with accurate data and specific information.
- Strongly reject any "poor me" statements that they may give.
- Disagree with the facts and not the person.
- Be willing to offer assistance if their fears persist.

- Recognize that they are conscientious, maintain high standards, and complete assignments given to them.
- Modify your expectations by understanding their greatest struggles.
- They enjoy exercising their reasoning abilities.
- Life of the mind is of prime importance and takes precedence over emotions.
- Spending hours alone devolving new thoughts can be fun and pleasurable for them.
- They usually find instruction manuals intellectually stimulating.
- They are sensitive people; if necessary, disagree with the facts but avoid personal attacks.
- They are driven by a commitment to excellence.
- One of their greatest struggles is with unrealistic expectations.
- They struggle with the fear of criticism and may keep good ideas to themselves.

VIII. HOW EACH PATTERN CAN INCREASE THEIR EFFECTIVENESS

1. Developers (Dominant Directors)
   - Show more patience.
   - Collaborate with others.
   - Demonstrate empathy.
2. Results-oriented
   - Demonstrate genuine concern for others.
   - Verbalize reasons for their views; be nonjudgmental.
   - Show more humility and empathy.
3. Inspirational (Dominant Director/Interactive Socializer)
   - Show genuine sensitivity to others.
   - Facilitate and promote the success of others.
   - Use a quieter, more systematic thoughtful approach.
4. Creative (Dominant Director/Cautious Thinker)
   - Increase genuine warmth and caring for others.
   - Remain tactful when under pressure and stress.
   - Cooperate.
5. Promoters (Interactive Socializer)
   - Increase emotional self-control.
   - Clarify expectations, reactions, events, and outcomes more objectively.
   - Follow through on tasks.
6. Persuaders (Interactive Socializer/Dominant Director)
   - Objectively analyze situations, including facts and feelings.
   - Respond more democratically to other's ideas.

- Provide personal attention to follow-up.

7. Counselors (Interactive Socializer/Cautious Thinker)
   - Cultivate a sense of priority, thus avoiding procrastination.
   - Monitor the tendency for overtolerance.
   - Attend and commit to realistic deadlines and constraints.
8. Appraisers (Interactive Socializer/Cautious Thinker)
   - Show empathy, even when feeling disapproval and frustration.
   - Develop added patience with themselves and others.
   - Pace themselves, prioritize and recognize degrees of achievement.
9. Specialists (Steady Relater)
   - Share ideas, experiences, and insights with others.
   - Develop long-term self-confidence by focusing on their strengths.
   - Focus on critical essential task priorities and sometimes on shortcuts.
10. Agents (Steady Relater/Interactive Socializer)
    - Act more self-assertive and less indirect in communications.
    - Be firmer when it is appropriate to say no or take a stand.
    - Learn to respond to conflict as a opportunity for greater stability.
11. Achievers (Steady Relater/Dominant Director)
    - Look toward longer-term results in mutual collaboration with people.
    - Respond to feelings and needs of others on each persons terms.
    - Clarify priorities for long range personal development.
12. Investigators (Steady Relater/Cautious Thinker/Dominant Director)
    - Develop tolerance and sensitivity to the needs and difficulties of others.
    - Learn to communicate in a warmer, more open shared manner.
    - Develop more open exploratory approaches to their views.
13. Objective thinkers (Cautious Thinker)
    - Learn to share insights and views with others.
    - Accept imperfections and mistakes.
    - Open themselves up more to facilitate collaborative relationships.
14. Perfectionists (Cautious Thinker/Steady Relater)
    - Develop more open, accepting attitude toward different approaches.
    - Learn interdependence through collaboration with others.
    - Develop genuine unconditional acceptance of themselves as human.
15. Practitioners (Cautious Thinker/Interactive Socializer/Steady Relater)
    - Learn to genuinely collaborate with others for the common benefit.
    - Share discoveries and procedures with others.
    - Learn to more appropriately delegate key terms.
    - Greater independence through collaboration with others.
    - Develop genuine unconditional acceptance of themselves as human.

Appendix C: Constants for Constructing Some of the Most Common Control Charts

This appendix provides the reader with some of the most common constants used in control charting (see next page).

Appendix D: Basic Formulas Used in the Construction of Control Charts

This appendix provides some typical formulae used in the control charting process. The list by no means is an exhaustive one, but it provides the quality professional with some basic approaches to statistical understanding for the everyday application of some common charts. The first section provides the most common descriptive statistical formulae, and the second section provides the formulae for calculating control chart limits and capability.

I. GENERAL FORMULAE

1. Arithmetic mean (average) from ungrouped data

For a population: $\mu = \frac{\sum X}{N}$

For a sample: $\bar{x} = \frac{\sum x}{n}$

For the average of the average: $\overline{\overline{X}} = \frac{\sum \overline{x}}{n}$

where

N	=	number of observations in the population
n	=	number of observations in the sample
X	=	observed population value
x	=	observed sample value
$\sum X, \sum x$	=	sum of observed population (sample) values
$\sum \overline{X}, \sum \overline{x}$	=	average of population or sample values

2. Arithmetic mean (average) from grouped data

For a population: $\mu = \frac{\sum fX}{N}$

For a sample: $\overline{X} = \frac{\sum fX}{n}$

where

f	=	class frequency
X	=	class midpoint

3. Median from ungrouped data

For a population: $M = X_{(N+1)/2}$ in an ascending ordered array

For a sample: $m = X_{(n+1)/2}$ in an ascending ordered array

4. Median from grouped data

For a population: $M = L + \frac{(N/2) - F}{f} w$

For a sample: $m = L + \frac{(n/2) - F}{f} w$

where

L	=	lower limit of the median class
f	=	absolute frequency of the median class
w	=	width of median class
F	=	sum of frequencies up to (but not including) the median class

5. Mode from grouped data

For a population or a sample: $\text{MO or mo} = L + \dfrac{d_1}{d_1 + d_2} w$

where

L	=	lower limit of the modal class
w	=	width of modal class
d_1, d_2	=	differences between the modal class frequency density and that of the preceding or following class, respectively.

6. Weighted mean from ungrouped data

For a population: $\mu_w = \dfrac{\sum wX}{\sum w}$

For a sample: $\bar{x}_w = \dfrac{\sum wX}{\sum w}$

where

w	=	weight
X	=	observed value in population
x	=	observed value in sample
$\sum w$	=	N or n

7. Mean absolute deviation

From ungrouped data

For a population: $\text{MAD} = \dfrac{\sum \lvert X - \mu \rvert}{N}$

For a sample: $\text{MAD} = \dfrac{\sum \lvert x - \bar{x} \rvert}{n}$

where

MAD	=	mean absolute deviation
$\bar{X}$	=	population mean
$\bar{x}$	=	sample mean
μ	=	population mean
$\lvert X - \mu \rvert$	=	absolute differences between X and μ
$\lvert X - \bar{X} \rvert$	=	absolute differences between X and $\bar{X}$

From grouped data

For a population: $\text{MAD} = \dfrac{\sum f|X-\mu|}{N}$

For a sample: $\text{MAD} = \dfrac{\sum f|X-\overline{X}|}{n}$

where

f = absolute class frequency
X = class midpoints

Note: occasionally, absolute deviations from the median rather than from the mean are calculated; in which case μ is replaced by M, and $\overline{X}$ is replaced by m.

8. Variance from ungrouped data

For a population: $\sigma^2 = \dfrac{\sum (X-\mu)^2}{N}$

For a sample: $s^2 = \dfrac{\sum (x-\overline{x})^2}{n}$

where

X = observed population value
x = observed sample value
μ = population mean
$\overline{x}$ = sample mean
$\sum (X-\mu)^2$ = sum of squared deviations between X and μ
$\sum (x-\overline{x})^2$ = sum of squared deviations between x and $\overline{x}$
σ^2 = variance of the population
s^2 = variance of sample

9. Variance from grouped data

For a population: $\sigma^2 = \dfrac{\sum f(X-\mu)^2}{N}$

For a sample: $\sigma^2 = \dfrac{\sum f(x-\overline{x})^2}{n}$

where

f	=	absolute class frequency
X	=	class midpoint of population
x	=	class midpoint of sample
σ	=	population standard deviation
s	=	sample standard deviation

10. Standard deviation from ungrouped data

For a population: $\sigma = \sqrt{\dfrac{\sum (X-\mu)^2}{N}}$

For a sample: $s = \sqrt{\dfrac{\sum (x-\bar{x})^2}{n-1}}$

where

$\sum (X-\mu)^2 =$ sum of squared differences between X and μ

$\sum (x-\bar{x})^2 =$ sum of squared differences between x and $\bar{x}$

11. Standard deviation from grouped data

For a population: $\sigma = \sqrt{\dfrac{\sum f(X-\mu)^2}{N}}$

For a sample: $s = \sqrt{\dfrac{\sum f(x-\bar{x})^2}{n}}$

where

f	=	absolute class frequencies
X	=	grouped population class midpoints
x	=	grouped sample class midpoints

12. Variance from ungrouped data (shortcut method)

For a population: $\sigma^2 = \dfrac{\sum X^2 - N\mu^2}{N}$

For a sample: $s^2 = \dfrac{\sum x^2 - N\bar{x}^2}{n}$

where

$\sum X^2$ = sum of squared population values

$\sum x^2$ = sum of squared sample values

μ^2 = squared population mean

$\bar{x}^2$ = squared sample mean

13. Variance from grouped data (shortcut method)

For a population: $\sigma^2 = \dfrac{\sum fX^2 - N\mu^2}{N}$

For a sample: $s^2 = \dfrac{\sum fx^2 - n\bar{x}^2}{n-1}$

where

f = absolute-class-frequency
X = population squared-class-midpoint

x = sample squared-class-midpoint
μ^2 = squared population mean
$\bar{x}^2$ = squared sample mean

II. FORMULAE FOR CALCULATING CONTROL CHART LIMITS

A. Control Chart Limits

1. $\bar{X}$ and R chart
Process average:

$$\text{UCL} = \bar{\bar{X}} + A_2\bar{R}$$

$$\text{Centerline} = \bar{\bar{X}}$$

$$\text{LCL} = \bar{\bar{X}} - A_2\bar{R}$$

Process variation

$$\text{UCL} = D_4\bar{R}$$

$$\text{Centerline} = \bar{R}$$

$$\text{LCL} = D_3\bar{R}$$

where

UCL, LCL	=	upper and lower control limits of the process
A_3, B_3, and B_4	=	constants
$\bar{X}$	=	process average
$\bar{R}$	=	average range

2. Individual (X) and moving range chart
   Process average:

   $$UCL = \bar{\bar{X}} + E_2\bar{R}$$

   Note that in this case the $\bar{X}$ and $\bar{\bar{X}}$ are the same.

   $$\text{Centerline} = \bar{X}$$
   $$LCL = \bar{\bar{X}} - E_2\bar{R}$$

   Process variation:

   $$UCL = D_4\bar{R}$$
   $$\text{Centerline} = \bar{R}$$
   $$LCL = D_3\bar{R}$$

   where E_2, D_3, andD_4 are constants.

   To calculate the range, the experimenter must couple the individual data points in groups. The most sensitive (and recommended) grouping is a sample of two observations. More may be grouped together, but much of the sensitivity will be lost.

3. $\bar{X}$ and s chart
   Process average:

   $$UCL = \bar{\bar{X}} + A_3\bar{S}$$
   $$\text{Centerline} = \bar{X}$$
   $$LCL = \bar{\bar{X}} - A_3\bar{S}$$

   Process variation:

   $$UCL = B_4\bar{S}$$
   $$\text{Centerline} = \bar{S}$$
   $$LCL = B_3\bar{S}$$

   where

$A_3, B_3,$ and B_4 = constants
$\overline{S}$ = average standard deviation

4. Median chart
Process average:

$$\text{UCL} = \tilde{\tilde{X}} + \tilde{A}_2 \overline{R}$$

$$\text{Centerline} = \tilde{\tilde{X}}$$

$$\text{LCL} = \tilde{\tilde{X}} - \tilde{A}_2 \overline{R}$$

Process variation:

$$\text{UCL} = D_4 \overline{R}$$

$$\text{Centerline} = \overline{R}$$

$$\text{LCL} = D_3 \overline{R}$$

where

$\tilde{A}_2, D_3,$ and D_4 = constants

$\tilde{\tilde{X}}$ = process median (median of sample medians)

5. p chart

$$\text{UCL} = \overline{p} + 3\sqrt{\frac{\overline{p}(1-\overline{p})}{\overline{n}}}$$

$$\text{Centerline} = \overline{p}$$

$$\text{LCL} = \overline{p} - 3\sqrt{\frac{\overline{p}(1-\overline{p})}{\overline{n}}}$$

where
$\overline{n}$ =
$\overline{p}$ = average proportion defective

6. np chart

$$\text{UCL} = n\overline{p} + 3\sqrt{\frac{n\overline{p}(1-\overline{p})}{n}}$$

$$\text{Centerline} = \overline{p}$$

$$\text{LCL} = n\overline{p} - 3\sqrt{\frac{n\overline{p}(1-\overline{p})}{n}}$$

where $n\bar{p}$ is the number of average proportion defective

7. Standardized-value p chart

$$p_s = \frac{(p-\bar{p})}{\sqrt{(\bar{p}\bar{q})/n}}$$

where

p_s	=	standard p value
p	=	observed sample proportion defective
q	=	$(1-p)$ yield
n	=	sample size
$\bar{p}$	=	process average proportion defective

8. c chart

$$\text{UCL} = \bar{c} + 3\sqrt{\bar{c}}$$

$$\text{Centerline} = \bar{c}$$

$$\text{LCL} = \bar{c} - 3\sqrt{\bar{c}}$$

where $\bar{c}$ is the average defect

9. u chart

$$\text{UCL} = \bar{u} + 3\sqrt{\frac{\bar{u}}{n}}$$

$$\text{Centerline} = \bar{u}$$

$$\text{LCL} = \bar{u} - 3\sqrt{\frac{\bar{u}}{n}}$$

where $\bar{u}$ is the average of the number defects

B. Capability

1. Process capability

$$C_p = \frac{\text{USL} - \text{LSL}}{6\sigma_x}$$

where σ_x is the standard deviation.

2. Capability ratio

$$C_p = \frac{6\sigma_x}{\text{USL} - \text{LSL}}$$

3. Capability index

$$C_{pk} = \frac{Z_{\min}}{3}$$

where $Z_{\min}$ is the lesser value of $(USL - \bar{X})/3\sigma$ and $(\bar{X} - LSL)/3\sigma$

4. Target ratio percent

$$TR_p = \frac{3\sigma}{Z_{\min}} \times 100$$

Appendix E: General Statistical Formulas Used in the Pursuit of Quality

This appendix provides the reader with some common statistical formulae that are used in the pursuit of quality.

1. Coefficient of variation

   For a population: $\nu = \frac{\sigma}{\mu}$

   For a sample: $\nu = \frac{s}{X}$

   where

   μ = the population mean
   X = the sample mean
   σ = population standard deviation
   s = sample standard deviation

2. Pearson's coefficient of skewness

   For a population: $\text{sk} = \frac{\mu - \text{MO}}{\sigma}$

For a sample: $sk = \frac{\overline{X} - mo}{s}$

where

sk	=	skewness
MO	=	population mode
mo	=	sample mode

3. Coefficient of kurtosis
   From ungrouped data:

For a population: $K = \frac{\left[\sum (X - \mu)^4\right] / N}{\sigma^4}$

For a sample: $K = \frac{\left[\sum (x - \overline{X})^4\right] / n}{s^4}$

where

K	=	kurtosis
X	=	observed population values
x	=	observed sample values
N	=	population size
n	=	sample size
σ^4	=	squared population variance
s^4	=	squared sample variance

From grouped data:

For a population: $K = \frac{\left[\sum f(X - \mu)^4\right] / \sum f}{\sigma^4}$

For a sample: $K = \frac{\left[\sum f(x - \overline{X})^4\right] / \sum f}{s^4}$

where

X	=	class midpoints
f	=	class frequencies

4. The probability of event A: classical approach

$$p(A) = \frac{P(A)_{\text{fav}}}{p(A)_{\text{pos}}} = \frac{n}{N}$$

where

$p(A)_{fav}$ = number of equally likely basic outcomes favorable to the occurrence of event A

$p(A)_{pos}$ = number of equally likely basic outcomes possible

5. The probability of event A: empirical approach

$$p(A) = \frac{p(A)_{past}}{p(A)_{max}} = \frac{k}{M}$$

where

$p(A)_{past}$ = number of times event A occurred in the past during a large number of experiments

$p(A)_{max}$ = maximum number of times that event A could have occurred during these experiments

6. The factorial

$$n! = n \times (n-1) \times (n-2) \times \ldots \times 3 \times 2 \times 1$$

where 0! and 1! are equal to 1 by definition.

7. Permutations for x at a time out of n distinct items (with no repetition among the x items)

$$P_x^n = \frac{n!}{(n-x)!}$$

where $x \leq n$

8. Permutations for x out of n items, given k distinct types and $k < n$

$$P_{x=1,\, x=2,\, \ldots\, x=k}^n = \frac{n!}{x_1!\, x_2!\, \ldots\, x_k!}$$

where x_1 items are of one kind, x_2 items are of a second kind, ... and x_k items are of a kth kind and where $x_1 + x_2 + \ldots + x_k = n$.

9. Combinations for x at a time out of n distinct items (with no repetitions allowed among the x items)

$$C_x^n = \frac{n!}{x!\,(n-x)!}$$

where $x \leq n$. Note that this formula collapses to 1 for the limiting case of $x = n$:

$$C_n^n = \frac{n!}{n!(n-n)!} = \frac{n!}{n!0!} = \frac{n!}{n!(1)} = 1$$

10. General addition law

$$p(A \text{ or } B) = p(A) + p(B) - p(A \text{ and } B)$$

11. Special addition law for mutually exclusive events

$$p(A \text{ or } B) = p(A) + p(B)$$

12. General multiplication law
    a. $p(A \text{ and } B) = p(A) \times p(B|A)$
    b. $p(A \text{ and } B) = p(B) \times p(A|B)$

13. Special multiplication law for independent events

$$p(A \text{ and } B) = p(A) \times p(B)$$

14. Bayes's theorem

$$p(E|R) = \frac{p(E)\, p(R|E)}{p(E) \times p(R|E) + p(\overline{E}) \times p(R|\overline{E})}$$

15. Bayes's theorem rewritten

$$p(E|R) = \frac{p(E \text{ and } R)}{p(R)}$$

16. Summary measures for the probability distribution of a random variable, R

$$\mu_r = E_R = \sum p(R = x) \times x$$

$$\sigma_R^2 = \sum p(R = x) \times (x - \mu_R)^2$$

$$\sigma_R = \sqrt{\sigma_R^2}$$

where

μ_r	=	arithmetic mean
σ_R^2	=	variance
σ_R	=	standard deviation of the probability distribution of random variable, R
$p(R = x)$	=	probability of the random variable = x
x	=	an observed value of that variable

17. The binomial formula

$$p\left(R = x|n,\Pi\right) = c_x^n \times \Pi^{\alpha} \times \left(1 - \Pi\right)^{n-x}$$

where

x	=	0, 1, 2, ...
n	=	number of successes in n trials
C_x^n	=	$n!/[x!(n-x)!]$
Π	=	probability of success in any one trial

18. Summary measures for the binomial probability distribution of random variable, R

$$\mu_R = E_R = n \times \Pi$$

$$\sigma_R^2 = n \times \Pi\left(1 - \Pi\right)$$

$$\sigma_R = \sqrt{\sigma_R^2}$$

Skewness: zero for $\Pi = 0.5$, positive for $\Pi < 0.5$, and negative for $\Pi > 0.5$

19. The hypergeometric formula

$$p(R = x|n,N,S) = \frac{C_x^S\, C_{n-x}^{N} - S}{C_n^N}$$

where

x	=	number of "successes" in a sample
n	=	sample size
N	=	population size
S	=	number of population with success characteristic

Clearly, $x = 0, 1, 2, ..., n$ or N (whichever is smaller); $n < N$; $S < N$.

20. Summary measures for the hypergeometric probability distribution of random variable, R

$$\mu_R = E_R = n \times \Pi = n(S|N)$$

$$\sigma_R^2 = n\Pi\left(1 - \Pi\right)\left(\frac{N-n}{N-1}\right)$$

$$\sigma_R = \sqrt{\sigma_R^2}$$

where $\Pi = S/N$

21. The Poisson formula

$$p(R = x|\mu) = \frac{e^{-\mu}\,\mu^{x}}{x!}$$

where

x	=	number of occurrences ($x = 0, 1, 2, \ldots, \infty$)
μ	=	mean number of occurrences within the examined units of time or space
e	=	2.71828.

22. Summary measures for the Poisson probability distribution

$$\mu_R = E_R = \lambda t$$

$$\sigma_R^2 = \lambda t$$

$$\sigma_R = \sqrt{\sigma_R^2}$$

where

σ_R	=	standard deviation of Poisson random variable, R
λ	=	Poisson process rate
t	=	number of units of time or space examined

Skewness: always positive.

23. The normal probability density function

$$f(x) = \frac{1}{\sigma_R\sqrt{2\pi}}\, e^{-1/2}\left(\frac{x - \mu R}{\sigma_R}\right)$$

where

$f(x)$	=	function of observed value of x
π	=	3.14159

24. The standard normal deviate

$$Z = \frac{x - \mu_R}{\sigma_R}$$

where x = observed value of random variable, R

25. The standard normal probability density function

$$f(z) = \frac{1}{\sqrt{2\pi}\, e^{-z^2/2}}$$

where $z = (x - \mu_R) / \sigma_R$ and $-\infty \leq z \leq \infty$

26. The exponential probability density function

$$f(x) = \lambda e^{-\lambda x}$$

where

$$x > 0$$
$$\lambda > 0$$

27. Greater-than-cumulative exponential probabilities

$$p(R > x) = e^{-\lambda x} = e^{-x / \mu_R}$$

28. Less-than cumulative exponential probabilities

$$p(R < x) = 1 - e^{-\lambda x} = 1 - e^{-x / \mu_R}$$

29. Combination formula for exponential probabilities

$$p(x_1 < R < x_2 = 1 - \left[e^{-\lambda x_2} + (1 - e^{-\lambda x_1})\right] = e^{-\lambda x_1} - e^{-\lambda x_2}$$

where $1 / \mu_R$ can be substituted for λ

30. The uniform probability density function

$$f(x) = \frac{1}{b - a} \text{ if } a \leq x \leq b \text{ otherwise } f(x) = 0$$

31. Less-than-cumulative uniform probabilities

$$p(R \leq x) = \frac{x - a}{b - a} \text{ if } a \leq x \leq b \text{ otherwise } p(R \leq x) = 0$$

32. Summary measures for the uniform probability density

$$\mu_R = \frac{a + b}{2}$$

$$\sigma_R^2 = \frac{(b - a)^2}{12}$$

$$\sigma_R = \frac{\sqrt{(b - a)^2}}{12}$$

33. Summary measures of the sampling distribution of $\overline{X}$
When selections of sample elements are statistically independent events; typically refered to as "the large-population case," because $n < 0.05N$.

$$\mu_{\bar{x}} = \mu$$

$$\sigma^2_{\bar{X}} = \frac{\sigma^2}{n}$$

$$\sigma_{\bar{X}} = \frac{\sigma}{\sqrt{n}}$$

When selections of sample elements are statistically dependent events; typically referred to as "the small-population case," because $n \geq 0.05N$.

$$\sigma^2_{\bar{X}} = \frac{\sigma^2}{n}\left(\frac{N-n}{N-1}\right)$$

$$\sigma_{\bar{X}} = \frac{\sigma}{\sqrt{n}}\sqrt{\frac{N-n}{N-1}}$$

where

$\mu_{\bar{X}}$ = mean of the sampling distribution of the sample mean
$\sigma^2_{\bar{x}}$ = variance of sampling distribution of the sample mean
$\sigma_{\bar{x}}$ = standard deviation of the sampling distribution of the sample mean
$\bar{X}$ = sampling distribution of the sample mean
μ = population mean
σ^2 = population variance
σ = population standard deviation

34. Summary measures of the sampling distribution of P
When selections of sample elements are statistically independent events; typically referred to as "the large-population case," because $n < 0.05N$.

$$\mu_p = \Pi$$

$$\sigma^2_p = \frac{\Pi(1-\Pi)}{n}$$

$$\sigma_p = \sqrt{\frac{\Pi(1-\Pi)}{n}}$$

When selections of sample elements are statistically dependent events; typically referred to as "the small-population case," because $n \geq 0.05N$.

$$\sigma_p^2 = \frac{\Pi(1-\Pi)}{n}\left(\frac{N-n}{N-1}\right)$$

$$\sigma_p = \sqrt{\left[\Pi\frac{(1-\Pi)}{n}\right]}\sqrt{\frac{N-n}{N-1}}$$

where

μ_p	=	mean of the sampling distribution of the sample proportion
σ_p^2	=	variance of the sampling distribution of the sample proportion
σ_p	=	standard deviation of the sampling distribution of the sample proportion, p
Π	=	population proportion

35. Upper and lower limits of confidence intervals for the population mean, μ and the proportion, Π (using the normal distribution).

$$\mu = \overline{X} \pm (z\sigma_{\overline{X}})$$

$$\Pi = p \pm (z\sigma_p)$$

where

$\overline{X}$	=	sample mean
p	=	sample proportion
$\sigma_{\overline{x}}$	=	standard deviation of sample
σ_p	=	standard deviation of the sampling distribution of the sample proportion, p

Assumptions: sampling distribution of $\overline{X}$ or p is normal because: a) the population distribution is normal, or b) the central-limit theorem applies [for the mean, $n \geq 30$, $n < 0.05N$; for the proportion, $n\Pi \geq 5$ and $n(1-\Pi) \geq 5$].

36. Upper and lower limits of confidence interval for the population mean (using the t distribution)

$$\mu = \overline{X} \pm \left(t\frac{s}{\sqrt{n}}\right)$$

Assumption: $n < 30$ and the population distribution is normal.

37. Upper and lower limits of confidence interval for the difference between two population means (large and independent samples from large or normal populations)

$$\mu_A - \mu_B > \overline{X}_A - \overline{X}_B \pm \left[\sqrt{\frac{S_A^2}{n_A} + \frac{s_B^2}{N_b}} \right]$$

Assumptions: $n_A \geq 30$ and $n_B \geq 30$ and sampling distribution of $\overline{X}_A - \overline{X}_B$ is normal because $n_A < 0.05N_A$ and $n_B < 0.05N_B$ or because population distributions are normal. Note: for small and independent samples, the same formula applies except that t replaces z. In that case, $n_A < 30$ and $n_B < 30$, but the two populations are assumed to be normally distributed and to have equal variances.

38. Upper and lower limits of confidence interval for the difference between two population means (large and matched-pairs samples from large or normal populations)

$$\mu_A - \mu_B = \overline{D} \pm \left(z \frac{s_D}{\sqrt{n}} \right)$$

where

$\overline{D}$	=	mean of sample matched-pairs differences
s_D	=	standard deviation of sample matched-pairs differences
n	=	number of matched sample pairs

Assumptions: $n > 30$ and the sampling distribution of $\overline{D}$ is normal because $n < 0.05N$ or because the populations are normally distributed. Note: for a small and matched-pairs sample, the same formula applies except that t replaces z. In that case, $n < 30$, but the populations are assumed to be normally distributed and to have equal variances.

39. Upper and lower limits of confidence interval for the difference between two population proportions (large and independent samples from large populations)

$$\Pi_A - \Pi_B > (p_a - p_b) \pm \left(\sqrt{\frac{p_A(1-p_A)}{n_A} + \frac{p_B(1-p_b}{n_b}} \right)$$

Assumptions: $n_A \geq 30$ and $n_B \geq 30$ and the sampling distribution of $(p_A - p_B)$ is normal because $n\Pi \geq 5$ and $n(1 - \Pi) \geq 5$ for both samples.

40. Required sample size for specified tolerable error and confidence levels when estimating the population mean

$$n = \left(\frac{z\,\sigma}{e}\right)^2$$

where

z	=	standard normal deviate appropriate for the desired confidence level
σ	=	population standard deviation
e	=	tolerable error level

41. Required sample size for specified tolerable error and confidence levels when estimating the population proportion

$$n = \frac{z^2\,p(1-p)}{e^2}$$

where p is an estimate of the population proportion.

42. A chi-square statistic

$$\chi^2 = \sum \frac{(f_o - f_e)^2}{f_e}$$

where

f_o	=	observed frequency
f_e	=	expected frequency

43. Alternative chi-square statistic

$$\chi^2 = \frac{s^2\,(n-1)}{\sigma^2}$$

44. Expected value and standard deviation of the sampling distribution of W (given: H_0: the sampled populations are identical)

$$\mu_W = \frac{n_A\,(n_A + n_B + 1)}{2}$$

$$\sigma_W = \sqrt{\frac{n_A\,n_B\,(n_A + n_B + 1)}{12}}$$

Note: if W were difined as the rank sum of sample B, one would have to interchange subscripts A and B in the μ_w formula.

45. Normal deviate for the Wilcoxon rank-sum test

$$z = \frac{w - \mu_W}{\sigma_W}$$

Assumption: $n_A \geq 10$ and $n_B \geq 10$

46. Mann-Whitney test statistic

$$U = \left[(n_A\, n_B) + \frac{n_A\,(n_A + 1)}{2} \right] - W$$

Note: W is defined as the rank sum of sample A. If it were defined as the rank sum of sample B, one would have to interchange subscripts A and B in this formula.

47. Normal deviate for Mann-Whitney test

$$s = \frac{U - \mu_U}{\sigma_u}$$

Assumption: $n_A \geq 10$ and $n_B \geq 10$

48. Expected value and standard deviation of the sampling distribution of U (given: H_0: the sampled populations are identical)

$$\mu_U = \frac{n_A\, n_B}{2}$$

$$\sigma_U = \sqrt{\frac{n_A\, n_B\,(n_A + n_B + 1) susp}{12}} = \sigma_W$$

49. The Wilcoxon signed-rank test statistic

T = sum of signed ranks among the n matched pairs with nonzero differences

50. Expected value and standard deviation of the sampling distribution of T (given: H_0: there are no populations differences)

$$\mu_r = 0$$

$$\sigma_r = \sqrt{\frac{n\,(n+1)\,(2n+1)}{6}}$$

51. Normal deviate for Wilcoxon signed-rank test

$$z = \frac{T - \mu_T}{\sigma_T} = \frac{T}{\sigma_T}$$

Assumption: $n \geq 10$.

Appendix F: Cost of Quality

I. IDENTIFICATION OF MANUFACTURING PROCESS IMPROVEMENT OPPORTUNITIES

1. Identify the process to be evaluated.
2. Become acquainted with the process by reviewing process sheets and through discussion with line supervision.
3. Visit each operation to review for type of cost incurred (e.g., appraisal, internal failure).
4. Talk to individual operators to define further what goes wrong at each operation (e.g., misassembly, wrong tools, poor setup); note machine numbers, part numbers, and shift.
5. Identify and quantify failures at each operation (e.g., scrap, damage, rework) by shift.
6. Use the existing financial system to assign costs (e.g., direct/indirect labor, benefits, material), to each operation within the process.
7. For each operation calculate the cost of scrap, rework, testing, inspection, production checks, sorting, and audits. Also calculate the costs associated with return sales, warranty, and customer loyalty.

8. Sum these costs to obtain the total cost of quality within the process.
9. State this cost as a fraction of the total cost of the process or as a dollar amount that represents the opportunity for improvement in the process.
10. Assure continuous improvement through ongoing process analysis (The Deming cycle: Plan—Do—Study—Act).

II. IDENTIFICATION OF ADMINISTRATIVE PROCESS IMPROVEMENT OPPORTUNITIES

1. Identify the process or procedure to be evaluated.
2. Become acquainted with the process or procedure by reviewing instruction sheets and procedure manuals, and by generating a unique process flow diagram; discuss with local supervision.
3. Review each operation for the type of cost incurred: appraisal (e.g., checks, reviews) and internal failure (e.g., blueprint errors, incomplete forms).
4. Talk to individual employees to define further what goes wrong at each operation (e.g., redundant operations, misfiling, improper direction, delays).
5. Identify and quantify failures at each operation and their effect on subsequent operations.
6. Use the existing financial system to assign the cost of labor and material to each operation.
7. Calculate the cost of losses associated with items identified in Nos. 4 and 5.
8. Sum these costs to obtain the total cost of quality within the process.
9. State this cost as a fraction of the total cost of the process or as a dollar amount that represents the opportunity for improvement in the process.
10. Assure continuous improvement through ongoing process analysis (The Deming cycle: Plan—Do—Study—Act).

III. PROCEDURE FOR QUALITY IMPROVEMENT

1. Organize a team.
2. Describe the problem (opportunity).
    - Estimate the magnitude of quality costs.
    - Identify the key business processes that have the greatest impact on the costs.
3. Define the root causes; identify and prioritize the root causes of process problems.

4. Implement interim corrective action; establish control of the business process.
5. Implement permanent corrective action; improve the capability of the business process.
6. Verify effectiveness of actions; measure effect of actions identified in Nos. 4 and 5.
7. Prevent recurrence; modify management and operating systems, practices, procedures and processes.
8. Give credit and congratulate team.

IV. MATRIX OF TYPICAL COST ELEMENTS IN A COST OF QUALITY PROGRAM

See Table F.1.

V. TOTAL COST OF QUALITY (COQ) SUMMARY: FINANCIAL PERFORMANCE STATEMENT

See Table F.2.

Table F.1 Typical Cost Elements in a Cost of Quality Program

Category	Cost element	Description	Where to obtain
Prevention	1. Quality planning by Quality Department Quality engineers SQA engineers Reliability engineers Statisticians Other planning by other departments Manufacturing engineers Controller's office Systems Administrative Purchasing Other	All costs (salary and administrative) related to the planning of an effective quality system that translate customer requirements into the manufacturing process. Test and inspection planning costs are reported separately (see No. 2). Allocated costs for time spent in quality planning by personnel not reporting to the Quality Department.	Salary budget reports Expense budget reports Estimates Department budget reports (allocated) Time sheets Purchase orders Estimates
	2. Test and inspection planning	Costs of planning, procuring, and developing test and inspection equipment (excluding actual costs part of appraisal costs). Development costs for test and inspection processes are also included.	Department budget reports (allocated) Purchase orders Estimates
	3. Qualification of new products/processes/equipment	Costs for qualifying new products, processes, and equipment (including of test and inspection) to meet customer requirements.	Department budgets (allocated) Launch budget (allocated) Purchase orders Estimates

Category	No.	Element	Description	Sources
	4.	Quality training	All costs for developing, implementing, operating, and maintaining formal quality training (including statistical training).	Training budget Purchase orders Estimates
	5.	Other prevention expenses	All other costs associated with planning, implementing, and maintaining a quality system not specifically included elsewhere.	Estimates Adjustments (including negative costs)
Appraisal	1.	Incoming and receiving inspection and test	All costs of inspectors, supervision, lab, and clerical personnel working on incoming material. Includes costs to visit or station personnel at supplier locations.	Department budgets (allocated) Process sheet standards Inspection sheet standards Estimates
	2.	In-process inspection and test	Salaries and associated costs of all staff performing in-process inspection and testing either 100% or sampling. Includes materials consumed during tests. Same as No. 1	
	3.	Test and inspection equipment	Costs of tests, inspection, and lab equipment. Equipment maintenance and purchased services are also included.	Department budgets (allocated) Purchase orders/maintenance contracts Estimates
	4.	Product quality reviews	Personnel expenses for performing quality reviews on in-process or finished products.	Department budgets (allocated) Estimated
	5.	Field performance evaluations	Costs incurred in field testing for product acceptance at a customer's site, prior to releasing the product.	Field inspection reports Estimates

Table F.1 (Continued)

Category	Cost element	Description	Where to obtain
	6. Other appraisal costs	All other appraisal costs not specifically covered elsewhere.	Estimates Adjustments (including negative costs)
Internal failure costs	1. Rework and repair Internal fault Supplier fault	Costs of reworking defective product, including costs associated with the review and dispositioning of nonconforming purchased products and products.	Costs accounting reports Defective material reports Department budgets (allocated) Estimates
	2. Scrap	All scrap losses incurred resulting from defective purchased materials/products and incorrectly performed manufacturing operations. Costs charged to suppliers are not included. Scrap value, less handling charges, may be included as an offset.	Salvage reports Defective material reports Estimates
	3. Trouble shooting and failure analysis	Costs incurred in analyzing nonconforming product to determine causes.	Department budgets (allocated) Problem reports Estimates
	4. Reinspect and retest	Costs to reinspect or retest products that had previously failed.	Department budgets (allocated) Estimates
	5. Excess inventory	Inventory costs resulting from producing defective products. Include storage of defective product, and added inventory of good product to cover production shortfalls.	Cost accounting reports Department budgets (allocated) Estimates

Category	No.	Element	Description	Source
	6.	Design and process changes	Costs to revise a product or process due to production of defective product.	Estimates
	7.	Other internal failure costs/offsets	All other costs related to the production of defective product not specifically included elsewhere.	Estimates Adjustments (including negative costs)
External failure costs	1.	Warranty	All warranty costs that can be allocated to a manufacturing location due to the production of defective product or incoming material. Includes internal processing and investigation of warranty.	Warranty reports (allocated) Department budgets (allocated) Estimates
	2.	Recalls and product liability claims	All costs associated with manufacturing location fault for recall campaigns or liability claims.	Recall reports Corporate liability settlement reports Department budgets (allocated)
	3.	Products returned or rejected	Costs of handling and accounting for defective product returned or rejected by the consuming plant or customer.	Department budgets (allocated) Returned material reports Sales and service reports Estimates
	4.	Reinspection and retest	Costs to reinspect or retest defective product at the customer's site.	Department budgets (allocated) Estimates
	5.	Customer and field contacts	Salary and administrative costs to handle meetings, visits, and so on, with customer personnel resulting from the receipt of defective product.	Department budgets (allocated) Estimates

Table F.1 (Continued)

Category	Cost element	Description	Where to obtain
	6. Design and process changes	Costs to revise the product or process to satisfy the customer who received defective product.	Department budgets (allocated) Estimates
	7. Customer goodwill	Extraordinary costs that result from attempting to satisfy a customer whose expectations were not met with previously received defective product.	Travel and expense reports Department budgets (allocated) Estimates
	8. Other external failure costs	All other costs related to defective product reaching the customer not specifically covered elsewhere.	Estimates Adjustments (including negative costs)

Table F.2 COQ Financial Performanace Statement

	Operating budget ($000)	COQ ($000)	COQ (%) of operating budget	Remarks
Volume				
Sales revenue				
Variable costs				
Material				
Freight				
Direct labor				
Overhead				
Scrap				
Other				
Total variable costs				
Fixed costs				
Specific				
Indirect hourly				
Salary				
Tax				
Insurance				
Utilities				
Fixed costs (cont)				
Maintenance/sundry				
Depreciation				
Launch				
Project expense				
Tooling				
Subtotal				
Allocated				
Indirect hourly				
Salary				
Utilities				
Maintenance/sundry				
Depreciation				
Engineering				
Administration				
Tax				
Insurance				
Warranty				
Subtotal				
Total fixed costs				
Total costs				

Appendix G: A Comparison of Process Improvement

This appendix compares some common process improvement methods.

Method	Strengths	Limitations
Process performance check		
A spot check of the process based on collection of a small period (one week) data.	Will indicate the centering of the process and the variation at the time of check.	Does not give indication of process capability.
Very few data are required, hence study is simple and the analysis is done using Pareto diagrams or frequency distribution charts.	Extracts knowledge of the percent detectives and the breakdown of defectives at the time of check in case of attribute data.	Will solve only very simple process problems (one output quality characteristic and directly related causes.
Usually carried out by one person.	Can lead to quick corrective action.	

Method	Strengths	Limitations
Process performance evaluation		
Data should be collected for about one month.	Gives a broader view to estimate process capability.	Useful mainly in ascertaining a single output characteristic and a few related causes.
Analysis done using Pareto or p and np charts in case of attribute data, or $\bar{X}$ and R charts in case of variables data.	It is a fast and reliable way of determining what has been going on in a process.	Based on what has been done and not what can be done.
Usually (but not always) carried out by one person.	Many processes can be optimized from this study.	
Process capability study		
This is a study made on current production and could continue for several months.	Determines the best performance that can be expected from the existing process running under normal control condition, without any major expenses for process improvement.	Can be useful only to study a few output characteristics at a time.
Analysis methods are p and np charts for attributes data and $\bar{X}$ and R charts for variables data.	Significant process improvements.	Might take considerable time, hence might cause loss of opportunity.
Requires a skillful team.		
Process improvement program		
Usually undertaken when, even running at full capability, the process cannot meet the required standards.	Will improve a process to run at an acceptable rate, and is in control and will produce a product that is within specifications.	Consumes a long period of time.
Might include major expenses for equipment, better raw materials, improvement to material flow, and so on.		Is useful only when optimizing on a few unrelated quality characteristics.

Method	Strengths	Limitations
Process improvement program (continued)		
Can be undertaken only after process capability is fully established.		
Needs commitment of the various groups of an organization, including top management.		
Evolutionary operations (EVOP)		
Select independent process variables that are likely to influence quality characteristics.	A reliable method of improving a process to a desired level at a low cost.	Best applied when the desired output characteristic is measurable.
Change these variables according to a plan.	Can be applied efficiently to situations such as the chemical process industry.	
Calculate the effect on the quality characteristic.		
When the effect of a variable is significant, change the variable mid-point in the favorable direction.		
If no variable affects the said quality, then change the range or select new variables.		
When a maximum is achieved, incorporate the new variable in the process.		
Simulation experiments		
Define the input and output variables and the relationship between them.	Mostly used for quality problems such as reliability, interacting tolerances, and circuit designs.	Can handle situations only when a complete relationship of input and output characteristics can be defined.

Method	Strengths	Limitations
Simulation experiments (continued)		
Collect data on the distribution of the input variables.	Can accommodate several output characteristics at a time.	
Use computer software to simulate values of output characteristics for different relationships between input variables.		

Glossary

This glossary provides some of the terms associated with the Total Quality Management philosophy and the process of implementing and sustaining it. This glossary is not an exhaustive one, but rather it presents frequent terms used with total quality management.

Appearance items. Parts designated on the engineering drawings as requiring customer approval for appearance characteristics, including: color, grain, texture, and so on.

Attributes. Qualitative data that have only two categories (e.g., pass/fail) and can be counted for recording and analysis.

Benefit. See outcome.

Bilateral specifications. Those specifications which state both a minimum and a maximum value.

Black box (proprietary assembly). An assembly purchased by the customer, the components of which are designed by the supplier and for which only a performance specification has been developed by the customer.

Boundary. The beginning or end point in the portion of a process from a supplier to a customer that will be the focus of the process improvement effort.

Brainstorming. A group decision-making technique designed to generate a large number of creative ideas through an interactive process. Brainstorming is used to generate alternative ideas to be considered in making decisions.

Capability. When the process average and $\pm 3\sigma$ spread of the distribution of individuals is contained within the specification tolerance (variables data), or when at least 99.73% of individuals are within specification (attribute data), a process is said to be capable. This can be determined only after the process is in statistical control. Efforts to improve capability must continue, however, consistent with the operational philosophy of never-ending improvement in quality and productivity.

Capability indices. A convenient way to refer to the capabilities of a process, after the process has been verified to be in a state of statistical control. These indices are ratios that indicate the ability of a process to produce products that conform to a given specification. Examples of capability indices are C_p, C_{pk}, C_r, and so on. While it is appropriate to have interim targets for these indices, the goal must be continual improvement of capability for all processes affecting significant characteristics.

Common cause. A source of variation that is always present. It is an inherent variation of the process itself. Its cause can usually be traced to elements of the system which only management can correct.

Cause-and-effect chart. A graphic tool used to explore and display options about those components of a process that effect every occurrence. A cause-and-effect chart is used to clearly illustrate the various causes affecting a given key quality characteristic (KQC) by sorting and relating the causes to the effect and to create a starting point for determining the key process variables (KPV). It is also called a fishbone diagram as well as a Ishikawa diagram.

Center line. The line on a control chart that represents the average (mean or median) value of the items being plotted.

Check sheet. A data collection form consisting of multiple categories. Each category has an operational definition and can be checked off as it occurs. Properly designed, the check sheet helps to summarize the data, which is often displayed in a Pareto chart.

Coach. A key resource person from within the organization who will support the CEO's leadership of the quality improvement process. A respected peer from the work force who is enthusiastic and knowledgeable about quality improvement and eager to learn and eager to help others learn.

Common cause variation. The inherent variation of the process. It is the result of interactions within the process that affect every occurrence.

Control chart. A display of data in the order that they occur with statistically determined upper and lower limits of expected common cause variation.

It is used to indicate special causes of process variation, to monitor a process for maintenance, and to determine if process changes have had the desired effect.

Control items. The products identified on drawings and specifications with a special notation (such as an inverted delta) preceding the part and/or the specification number and which contain one or more critical characteristics.

Control limits. Expected limits of common cause variation. Sometimes they are referred to as upper and lower control limits. They are not specification or tolerance limits.

Control plans. Written descriptions of the system for controlling processes producing products for a given customer. Control plans must be established for all new products by the producer prior to issuance of the tooling order. As a general rule to have an effective control plan, an FMEA must have been conducted.

C_p. A capability index which is a ratio of the 6σ process spread to part specification tolerance without regard to the location of the data.

C_{pk}. A capability index which considers both process spread and relation of that spread to the target value or specification. In order for the C_{pk} to be appropriate for a process, that process must be consistent, repeatable and predictable (i.e., in statistical control).

C_p. A capability index similar to the C_p; in fact, the relationship between the two is an opposite one.

Critical characteristics. Those product features or process parameters which can affect compliance with government regulations or safe process/product function. Usually, critical characteristics are determined based on the product design and will be identified on the engineering drawing, specification, and/or installation manual with a special symbol (e.g., the inverted delta, shield, and so on) next to the characteristic and the part number. During the preparation of the FMEA and control plan, the producer will identify additional critical characteristics as determined by the producer's product and/or process knowledge and expertise.

Cross-functional teams. Consist of members from different functional areas (e.g., team members from design, manufacturing, quality, sales, purchasing, and maintenance.

Customer. The receiver of an output of a process, either internal or external to an organization or corporate unit. A customer could be a person, a department, a company, and so on.

Data collection. Gathering facts on how a process works and/or how a process is working from the customer's point of view. All data collection is driven by knowledge of the process and guided by statistical principles.

Deming cycle for continuous improvement. A visualization of the quality improvement process usually consisting of four points—Plan, Do, Check (study), Act—linked by quarter circles.

Deming's 14 principles. The foundation upon which the organization-wide quality improvement process is built. The points are a blend of leadership, management theory, and statistical concepts which highlight the responsibilities of management while enhancing the capacities of employees.

Detection. A past-oriented strategy that attempts to identify unacceptable output after it has been produced and separate it from the good output. (*See also* prevention.)

Deviation. A document issued by the customer authorizing a temporary departure from engineering specifications. A deviation is normally restricted to a specified number of pieces or period of time.

Failure mode and effect analysis (FMEA). A methodology that allows for identification of both current and potential problems to be addressed, based on a formal priority.

Fishbone chart. *See* cause-and-effect chart.

Flowchart. A graphical representation of the flow of a process. A useful way to examine how various steps in a process relate to each other, to define the boundaries of the process, to identify customer/supplier relationships in a process, to verify or form the appropriate team, to create common understanding of the process flow, to determine the current "best method" of performing the process, and to identify redundancy, unnecessary complexity, and inefficiency in a process.

FOCUS-PDCA. A strategy that provides a roadmap for continuous process improvement when linked to a quality definition. It is an acronym: **F**ind a process to improve, **O**rganize a team that knows the process, **C**larify current knowledge of the process, **U**nderstand causes of process variation, **S**elect the process improvement, **P**lan the improvement and continued data collection, **D**o the improvement, data collection, and analysis, **C**heck and study the results, **A**ct to hold the gain and to continue to improve the process.

Forced field analysis. A systematic method for understanding competing forces that increase or decrease the likelihood of successfully implementing change. It provides a framework for developing change strategies aimed at decreasing restraining forces and increasing driving forces.

Functional check. The evaluation performed on initial samples by the customer to ensure they assemble properly, conform to operational requirements, meet engineering specifications for assembly, and are adaptable to usage.

Future state. A description of an organization that has gone through a quality improvement process transformation. In this state, explicit messages from the leadership clearly and consistently support and reinforce the new way. People work more efficiently together. Common language evolves related to mission, customers, and daily organizational life. Standards are never enough, and the status quo never good enough. Listening to customers drives the organization.

Gray box. An assembly purchased by the customer for which the producer has design, development, and detail drawing responsibility, but for which the customer provides design or material specification inputs.

In process tests. Functional or durability tests required by product engineering to monitor a particular design requirement on a continuing basis during production. Sampling and reaction plans for these tests must be included in the control plan.

Lot. A homogeneous quantity of a product.

Lot traceability. A system for tracking and identifying a batch of raw material through all steps in the process and identifying the final product when it is shipped to the customer(s).

Key process variable (KPV). A component of the process that has a cause-and-effect relationship of sufficient magnitude with the key quality characteristic such that manipulation and control of the KPV will reduce variation of the KQC and/or change its level.

Key quality characteristics (KQC). The most important quality characteristics. Key quality characteristics must be operational defined by combining knowledge of the customer with knowledge of the process.

Mentor. A highly skilled quality improvement process (QIP) professional with extensive training and experience in the initiation and operation of an organization wide quality improvement process. An external resource person who visits periodically to counsel the CEO, and other leaders in the initiation of the process.

Multiple voting. A group decision-making technique designed to reduce a long list to a few ideas.

New quality technology tools. A group of techniques and charts used to collect, organize, display, and evaluate knowledge about a process (i.e., affinity chart, correlation chart, distribution diagram, matrix chart, PERT, matrix data analysis, and process decision program chart).

Nonconforming units. Units which do not conform to a specification.

Nonconformities. Specific occurrences of a condition which does not conform to specifications or other standards; sometimes called discrepancies.

Old quality tools. A group of techniques and charts used to collect, organize, display, and evaluate knowledge about a process (i.e., brainstorming, flow chart, cause-and-effect chart, check sheet, Pareto chart, histogram, run chart, control chart).

Operational definition. A description in quantifiable terms of what to measure and the steps to follow to consistently measure it. The purpose of this measurement is to determine the actual performance of the process. An operational definition is developed for each key quality characteristics before data are collected.

Opportunity statement. A concise description of a process in need of improvement, its boundaries, and the general area of concern where a quality improvement team should begin its efforts.

Outcome (benefit). The degree to which outputs meet the needs and expectations of the customer.

Output. What is produced by the actions of the process.

Owner. The person who has or is given the responsibility and authority to lead the continuing improvement of a process. Process ownership is driven by the boundaries of the process.

Paradigm shift. A point in time when the knowledge or structure which underlies a science or discipline change in such a fundamental way that the beliefs and behavior of the people involved in the science or discipline are changed.

Pareto chart. A bar graph used to arrange information in such away that priorities for process improvement can be established. It displays the relative importance of data and is used to direct efforts to the biggest improvement opportunity by highlighting the vital few in contrast to the many others.

Present state. A description of an organization as it currently exists. It includes what is happening now, both formally and informally in the organization (i.e., the status quo). A condition in which managers at all levels and employees throughout the organization are pulling in different directions because of diverse demands, short-term requirements, and unclear values, goals, and roles.

Prevention. A future-oriented strategy that improves quality by directing analysis and action towards correcting the production process. Prevention is consistent with a philosophy of never ending improvement. (*See also* detection.)

Process. A series of actions which repeatedly come together to transform inputs provided by a supplier into outputs received by a customer. All work is a process.

Process change. Any change in processing concept which could alter the capability of the process to meet the design requirements or durability of the product. This includes new, different, relocated, or rehabilitated production machinery/equipment, any change in subcontracted products or services including the use of engineering-approved alternate materials, changes in manpower, new processing methods/concepts (including major changes in the sequence of operations), and in environment.

Process control. The gathering of data about a process, the use of control charts and the establishment of a feedback loop to prevent the manufacture of nonconforming products.

Process failure mode and effects analysis (process FMEA). An analytical technique which uses the potential failure modes of a process and the resulting effects to prioritize corrective actions. The FMEA should be treated as a living document that is updated as necessary whenever the process changes.

Process improvement. The continuous endeavor to learn about all aspects of a process, and to use this knowledge to change the process to reduce variation and complexity and to improve customer judgments of quality. Process improvement begins by understanding how customers judge quality, how processes work, and how understanding the variation in those processes can lead to wise management action.

Process owner. *See* owner.

Process parameters. Those variables that are part of the process and affect its output (e.g., speeds feeds, temperatures, chemical concentrations, pressures, and voltages).

Producer. A manufacturing or assembly plant or outside supplier that provides product or service to a customer.

Product. Any part, component, assembly, or material provided to a customer's location and covered by the standard. In some cases, this definition includes additional items such as perishable tools and nonproduction materials.

Process variation. The spread of process output over time. There is variation in every process, and all variation is caused. The causes are of two types: 1) normal or common or inherent, and 2) special or assignable. A process can have both types of variation at the same time or only common cause variation. The management action necessary to improve the process is very different in each situation.

Quality assurance. Designing a product or service so well that quality is inevitable.

Quality characteristics. Characteristics of the output of a process that are important to the customer. The identification of quality characteristics requires knowledge of the customer needs and expectations.

Quality control. A traditional outdated way of thinking about quality. The focus of quality control is always on appraising systems after the product or service has been delivered. It is a sorting mechanism.

Quality function deployment (QFD). A management system to assist in translating the "voice of the customer" into operational definitions that can be used to produce and deliver products desired by the customers. Specifically, QFD is the development and use of an array to compare customer requirements with the design elements of a product. It highlights conflicting customer requirements so that they can be reconciled.

Quality improvement council (QIC). A group composed of the coach and a senior leadership of an organization which is primarily responsible for planning, developing strategy, deploying, monitoring, educating, and promoting the quality improvement process.

Quality improvement process (QIP). The application of the new quality technology tools in the day-to-day operation and management of an organization.

Quality improvement storytelling. A major accelerator of the process of organization-wide quality improvement that uses quality improvement storybooks to follow steps in the FOCUS-PDCA strategy. Storybooks and storyboards help teams organize their work and their presentations so that others can more readily learn from them. Use of storyboards and storybooks reduces variation in the process of quality improvement storytelling so that the focus of learning is on content, not the method of telling. Storybooks form a permanent record of a team's actions and achievements and all the data generated; storyboards can function as the working minutes of a team.

Quality improvement team (QIT). A specially constituted group, usually five to eight people, chosen to address a specific opportunity for improvement. It consists of those people who have regular contact with the process.

Special cause variation. Variation in the process that does not affect every occurrence but arises because of special circumstances.

Quality inspection. Usually consists of three stages: sampling, measuring, and sorting. While many organizations rely on inspection to improve quality, the better way is to design quality into the product or service in order to improve the process. This may include some inspection as a means of data gathering.

Quality management process (QMP). The application of the new quality technology tools in the day-to-day operation and management of a given organization.

Quality performance records. Documents which show the results of inspections and tests performed on materials, parts, and assemblies.

Quality systems records. Documents such as control plans, inspection instructions, laboratory test instructions, and gage and test equipment verifications and calibrations. These documents define the operation of the supplier's quality system.

Red bead experiment. A simple exercise to demonstrate that many managers hold workers to standards beyond their control, that variation is part of any process, and that workers work within a system beyond their control. The game also shows that some workers will always be above average, some average and some below average, that the system—not the skills of individual workers—determines to a large extent how workers in repeating processes perform, and that only management can change the system or empower others to change it.

Refreezing. Recognizing, reinforcing, and rewarding new organizational attitudes and behaviors so they become the norm. Making processes, systems, and methods throughout the organization support the quality improvement process.

Rework. The act of doing something again because it was not done right the first time. It can occur for a variety of reasons, including insufficient planning, failure of a customer to specify the needed input, and failure of a supplier to provide a consistently high quality output.

Run. A point or a consecutive number of points that are above or below the central line in a run chart. Too long a run or too many or two few runs can be evidence of the existence of special causes of variation.

Run chart. A display of data in the order that they occur. A run chart is used to indicate the presence of special causes of process variation in the form of trends, shifts, or other nonrandom patterns in a key quality characteristic.

Sample. One or more individual events or measurements selected from the output of a process.

Self-certifying suppliers. Suppliers whose performance has justified waiving the on-site customer evaluation of initial samples and first production shipments.

Setup. A change in the adjustments and/or fixturing to change over a process from the production of one product to another.

Shewhart cycle. *See* Deming cycle for continuous improvement.

Sigma (σ). The Greek letter used to designate a standard deviation.

Significant characteristics. Those characteristics that are normally established from a review of the design and/or process FMEAs and for which quality planning actions must be summarized on a control plan. Significant characteristics will include:

- Critical characteristics with a special symbol as identification. (All critical characteristics are automatically significant characteristics).
- Characteristics identified by the customer and the producer on the basis of process/product knowledge, criticality for customer satisfaction, function, fit, durability, and/or appearance.

Special causes. Sources of variation that are intermittent, unpredictable, and unstable (e.g., broken tools, interruptions in power, changes in raw materials); sometimes they are called, assignable causes. They are signaled by a point beyond the control limits or a run or other nonrandom pattern of points within the control limits.

Special samples. Advance samples produced from less than complete production tooling/processes. Special samples are required to maintain the functional check/build schedule and are requested in advance of and in addition to initial samples. Special samples must be made from specified materials and must have received dimensional and laboratory verification by the producer prior to shipment.

Specification. The engineering requirement for judging acceptability of a particular characteristic. Chosen with respect to functional or customer requirements for the product, a specification may or may not be consistent with the demonstrated capability of the process (if it is not, out-of-specification parts are certain to be made). A specification is never to be confused with a control limit.

Specification requirements. Drawings, engineering standards (ESs), or other media issued to detail the product requirements established by the customer and manufacturing requirements issued by the customer's manufacturing engineering.

Sponsor. The person or group of people responsible for making the business decision that improving the process is important enough to provide team members and a coach/facilitator adequate time and resources to work on the improvement. The sponsor is most often a department manager or the quality improvement council.

Stable process. A processes in which variation in output arise only from common causes. A process that is in statistical control and therefore predictable.

Stability. The absence of special causes of variation, the property of being in statistical control (i.e., predictable).

Standard deviation. A measure of the spread of the process output or the spread of a sampling statistics from the process (e.g., subgroup averages); denoted by the Greek letter σ (sigma).

Statistic. A value calculated from or based upon sample data (e.g., a subgroup average or range), that is used to make inferences about the process that produced the output from which the sample came.

Statistical control. The condition describing a process from which all special causes of variation have been eliminated and only common causes remain. Statistical control is evidenced on a control chart by the absence of points beyond the control limits and by the absence of any nonrandom patterns of trends.

Statistical process control. The use of statistical techniques such as control charts to analyze a process or its output so as to take appropriate actions to achieve and maintain a state of statistical control and to improve the capability of the process.

Statistical thinking for process improvement. A data driven method for decision making based primarily on an understanding of process variation.

It results in wise management actions which contribute to the continuous improvement of quality.

Storyboard. *See* quality improvement storytelling.

Storybook. *See* quality improvement storytelling.

Storytelling. *See* quality improvement storytelling.

Supplier. The party or entity responsible for an input to a process. A supplier could be a person, department, a company, and so on. A supplier differs from a vendor, in that the supplier relationship is always win-win. A vendor relationship is that of a commodity/price and therefore win-lose.

Supplier quality assistance (SQA). The customer's activity primarily responsible for contacts with outside suppliers on quality. The primary objective of SQA is to assist suppliers in developing systems for the prevention (rather than detection) of defects and the continuing reduction of process variability.

Supplier quality engineering (SQE). In some organizations it is used to designates the activity that combines the supplier quality assistance, supplier feasibility analysis, and tooling liaison functions.

Supplier quality improvement (SQI). In some organizations it is used to designate the activities that work intensively on quality improvement with selected outside suppliers and company plants during the prototype and launch phases of new products.

Tampering. Taking action without taking into account the difference between special and common cause variation.

Team leader. A person designated to lead the quality improvement team. An individual who has team leadership skills and basic quality improvement skills.

Teams (cross-functional and multidisciplinary). A group of usually five to nine people from two or more areas of an organization who are addressing an issue which impacts the operations of each area. A *functional team* is a group of five to eight people addressing an issue where any recommended changes would not be likely to affect people outside the specific area.

Tools. *See* new and old quality technology tools.

Transformation. A major organizational change from the present state to a new/preferred state in which the quality improvement process flourishes. The primary steps involved in moving an organization through a transformation are present state, unfreezing, transition period, refreezing, and new/preferred state.

Transition period. A description of the time when an organization is visibly moving away from the old way toward the new way. During this time, employee attitudes and behaviors range from being excited and busy to being confused and resistant. The support for change is building, new leaders emerge, champions of the change come forward, and confusion over roles begins to clear.

Unfreezing. Reassessing old values and behaviors and becoming open to the acceptance of a new culture.

Unilateral specifications. Specifications which state either a maximum value only or a minimum value only. Examples are concentricity (diameters A and B concentric within 0.5 mm maximum) and flatness (flat within 0.1 mm maximum).

Variables. Characteristics of a part or process which can be measured, as opposed to be classified. (*See* attributes.)

Variation. The inevitable differences among individual outputs of a process. The sources of variation can be grouped into two major classes: common causes and special causes.

Vendor. *See* supplier.

Waiver of change. *See* deviation.

Selected Bibliography

Adam, P. and Vandre Water, R. (1995). Benchmarking and the bottom line: Translating business reengineering into bottom-line results. *Industrial Engineering*. Feb.: 24–26.

Akao, Y. (1991). *Hoshin kanri policy deployment for successful TQM*. Productivity Press. Portland, OR.

Akiyama, K. (1991). *Function analysis: systematic improvement of quality and performance*. Productivity Press. Portland, OR.

American Productivity and Quality Center. (1993). *The benchmarking management guide*. Productivity Press. Portland, OR.

Amsden, R. T., Butler, H. E., and Amsden D. M. (1986). *SPC simplified practical steps to quality*. Quality Resources. New York.

Arbuckle, H. J. (1996). Partnership personified. *Metal Forming*. Mar.: 39–45.

Arbuckle, H. J. (1996). Education and empowerment. *Metal Forming*. Jan.: 50, 54.

Asaka, T. and Ozeki, K. (1990). *Handbook of quality tools: the Japanese approach*. Productivity Press. Portland, OR.

Atkinson, D., Haas, R., and Holden, J. (1995). Quality and achievement. *Technology*. Oct.-Nov.: 6–20.

Aznavorian, M. et al. (1995). Personnel: a crisis facing manufacturing in the 90s. *Metal Forming*. Dec.: 38–42.

Baumgardner, M. and Tatham, R. (1988). 'No preference' in paired-preference testing. *Quirk's Marketing Research Review*. June/July: 18–20.

Band, W. (1991). Instill Kaizen in your organization. *Marketing News*. June 24: 30–31.

Bandyopadhyay, J. K. (1995). *QS-9000 handbook*. St. Lucie Press. Delray Beach, FL.

Baker, A. (1994). Virtual reality moves into design. *Design News*. Feb. 7: 40.

Barker, J. (1996). The service ideal. *Successful Meetings*. Mar.: 45–52.

Barnthal, P. R. (1995). Evaluation that goes the distance. *Training and Development*. Sept.: 41–48.

Barnett, A. J. and Andrews, R. W. (1994). Are you getting the most out of your control charts? *Quality Progress*. Nov.: 75–80.

Barrett, D. (1994). *Fast focus on TQM: A concise guide to company wide learning*. Productivity Press. Portland, OR.

Barrett, J. G. (1996). Designing for control reliability in safety critical machine controls. *Metal Forming*. July: 35–44.

Barry, D., (1991). Managing the bossless team: Lessons in distributed leadership. *Organizational Dynamics*. 20(1): 31–47.

Beck, J. D. W. and Yeager, N. M. (1996). How to prevent teams from failing. *Quality Progress*. Mar.: 27–32.

Bergstrom, R. Y. (1996). The Baldrige: It's about more than quality. *Automotive production*. June: 50–53.

Bergstrom, R. Y. (1996). Voices of the customer and supplier. *Automotive Production*. Mar.: 53–55.

Bertalanffy, L. (1968). *General systems: foundations, development, applications*. Braziller. New York.

Berry, L. L. (1995). *On great service: a framework for action*. The Free Press. New York.

Bianchi, A. (1994). Virtual reality gets real—commercial and industrial use. *Inc*. July: 31.

Bittel, L. R. and Ramsey, J. E. (1983). New dimensions for supervisory training and development. *Training and Development Journal*. Mar.: 13–20.

Bohan, G. P. (1995). Focus the strategy to achieve results. *Quality Progress*. July: 89–94.

Bounds, G. (1994). *Beyond total quality management*. McGraw-Hill. New York.

Bowles, J. and Hammond, J. (1991). *Beyond quality: how 50 winning companies use continuous improvement*. G. P. Putnam's Sons. New York.

Bracey, G. (1990). Education still not looking at the big picture. *Electronic Learning*. May-June: 20.

Brache, A. (1983). Seven prevailing myths about leadership. *Training and Development Journal*. June: 120–126.

Bradford, C. L. and Bradford, R. W. (1993). *Simplified strategic planning*. Center for Simplified Strategic Planning, Inc. Vero Beach, FL.

Braithwalte, T. (1996). *The power of it*. Quality Press. Milwaukee, WI.

Brennan, M. (1996). Virtual design: real results. *Detroit Free Press: Business Monday*. Feb. 26: 6F-7F.

Breyfogle, F. W. (1992). *Statistical methods for testing, development and manufacturing*. Wiley and Sons. New York.

Broydrick S. C. (1996) *The 7 universal laws of customer value*. Irwin Professional. Burr Ridge, IL.

Brown, M. G. (1995). *Baldrige award winning quality*. Quality Resources. New York.

Brown, M. G., Hitchcock, D. E., and Willard, M. L. (1995). *Why TQM fails and what to do about it*. Irwin Publishing. Burr Ridge, IL.

Brumm, E. K. (1995). *Managing records for ISO 9000 compliance*. Quality Press. Milwaukee, WI.

Bryk, A. S. and Raudenbush, S. W. (1992). *Hierarchical linear models*. Sage Publications. Thousand Oaks, CA.

Bryson, J. M. (1995). *Strategic planning for public and nonprofit organizations*. Jossey-Bass. San Francisco.

Bryson, J. M. and Alston, F. K. (1995). *Creating and implementing your strategic plan*. Jossey-Bass. San Francisco.

Burns, G. (1995). The secrets of team facilitation. *Training and Development*. June: 46–54.

Bushe, G. R., Havlovic, S. J., and Goetzer, G. (1996). Exploring empowerment from the inside out. *Journal for Quality and Participation*. Mar.: 36–45.

Butz, H. E. (1995). Strategic planning: the missing link in TQM. *Quality Progress*. May: 105–108.

Cameron, J. and Pierce. W. David. (1994). Reinforcement, reward, and intrinsic motivation: a meta analysis. *Review of Educational Research*. Fall: 363–424.

Cannie, J. K. and Caplin, D. (1991). *Keeping customers for life*. The American Management Association. New York.

Cartin, T. J. (1993). *Principles and practices of TQM*. Quality Press. Milwaukee, WI.

Chase, R. and Acuilano, N. (1992). *Production and operations management*. Irwin. Homewood, IL.

Chawala, S. and Renesch, J. (Eds). (1995). *Learning organizations: Developing cultures for tomorrow's workplace*. Productivity Press. Portland, OR.

Checkland, P. (1981). *Systems thinking, system practice*. J. Wiley and Sons, Inc. Chichester, England.

Chesbrough, H. (1996). When is virtual virtuous? *Harvard Business Review*. Jan.: 65–73.

Clark, J. and Koonce, R. (1995). Meetings go high-tech. *Training and Development*. Nov.: 32–39.

Clausing D. and Simpson, B. H. (1990). Quality by design. *Quality Progress*. Jan.: 41–44.

Clemants, R., Sidor, S. M., and Winters, R. E. (1995). *Preparing your company for QS-9000: A guide for the automotive industry*. Quality Press. Milwaukee, WI.

Clogg, C. C. (1994). *Statistical models for ordinal variables*. Sage Publications. Thousand Oaks, CA.

Cokins G. (1996). *Activity-based cost management making it work*. Irwin Professional. Burr Ridge, IL.

Cole, R. E. (1993). Learning from learning theory: implications for quality. *Quality Management Journal*. Oct.: 9.

Cole, R. E. (1994). Reengineering the corporation: a review essay. *Quality Management Journal*. July: 77.

Collins, R. D. (1988). Essentials of delivery analysis. *Quality Progress*. Aug.: 55–57.

Constantineau, L. A. (1995). Reengineering the marketing research function. *Quirk's Marketing Research*. Oct.: 26, 63–64.

Cooper, D. R. and Emery, C. W. (1995). *Business Research methods*. 5th ed. Irwin. Chicago.

Corrigan, J. P. (1995). The art of TQM. *Quality Progress*. July: 61–66.

Crom, S. and France, H. (1996). Teamwork brings breakthrough improvements in quality and climate. *Quality Progress*. Mar.: 39–44.

Crosby, P. B. (1979). *Quality is free*. McGraw-Hill Book Co., Inc. New York.

Cusimano, J. M. (1995). Turning blue collar workers into knowledge workers. *Training and Development*. Aug.: 47–49.

Damelio, R. (1995). *The basics of benchmarking*. Quality Resources. New York.

Damelio, R. and Englehaupt, W. (1995). *An action guide to making quality happen*. Quality Resources. New York.

Davidson, F. (1996). *Principles of statistical data handling*. Sage Publications. Thousand Oaks, CA.

Davis, A. W. (1996). Putting teeth into total customer service. *Computer and Communications OEM*. May: 82.

Day Overmyer, L. E. (1995). Benchmarking training. *Training and Development*. Nov.: 26–31.

Dean, J. W. and Evans, J. R. (1994). *Total quality*. West Publishing, Minneapolis, MN.

Dedhia, N. S. (1995). The basics of ISO 9000. *Quality Digest*. Oct.: 52–54.

Debra, C. (1993). Teaching is believing. *Technology Review*. Aug.-Sept.: 9.

D'Egidio, F. (1990). *The service era: Leadership in a global environment*. Productivity Press. Portland, OR.

DeRose, G. J. and McLaughlin, J. (1995). Outsourcing through partnerships. *Training and Development*. Oct.: 51–55.

Desai, M. P. (1996). Implementing a supplier scorecard program. *Quality Progress*. Feb.: 73–80.

Destefani, J. D. (1996). Taking quality to the next level. *Quality in Manufacturing*. Jan.-Feb.: 14–15.

Didis, S. (1990). Kaizen. *Internal Auditor*. Aug.: 66–69.

Dreikorn, M. J. (1995). *Aviation industry quality systems: ISO 9000 and the federal aviation regulations*. Quality Press. Milwaukee, WI.

Einsiedel, A. A. (1995). Case studies: indispensable tools for trainers. *Training and Development*. Aug.: 50–53.

Enrick, N. L. (1988). The exponential reliability curve. *Quality*. Nov.: 63.

Entinghe, R. (1994). Using radar charts to present customer satisfaction data. *Quirk's Marketing Research Review*. Oct.: 10, 54.

Ettlie, J. E. (1995). Quality is back. *Production*. Dec.: 14–15.

Farkas, C. M. and De Backer, P. (1995). *Maximum leadership: The world's leading CEOs share their five strategies for success*. Henry Holt. New York.

Feinberg, S. (1995). Overcoming the real issues of implementation. *Quality Progress*. July: 79–82.

Fellers, G. (1995). *Creativity for leaders*. Pelican Publishing Co. Gretna, LA.

Fernandez, J. E. and Jacobs, R. M. (1995). Trying out teams: reorganizing for quality, Chrysler style. *Technology*. Oct.-Nov.: 40–44.

Ford Motor Co. (1990). *Planning for quality*. Ford Motor Co. Corporate Quality Office. Dearborn, MI.

Fraker, A. T. and Spears, L. C. (Eds.). *Seeker and servant*. Jossey-Bass. San Francisco.

Frame, J. D. (1995). *Managing projects in organizations: how to make the best use of time, techniques, and people*. rev. ed. Jossey-Bass. San Francisco.

Frame, J. D. (1994). *The new project management*. Jossey-Bass. San Francisco.

Frick, D. M. and Spears, L. C. (Eds.). *On becoming a servant-leader: the private writings of R. K. Greenleaf*. Jossey-Bass. San Francisco.

Fuchs, E. (1993). Total quality management from the future: practices paradigm. *Quality Management Journal*. Oct.: 26.

Fukuda, R. (1990). *CEDAC: a tool for continuous systematic improvement*. Productivity Press. Portland, OR.

Fukuda, R. (1986). *Managerial engineering: techniques for improving quality and productivity in the workplace*. Productivity Press. Portland, OR.

Fulcher, J. (1996). Working out the kinks: partnering replaces competition in work with suppliers. *Manufacturing Systems*. May: 22.

Galagan, P. (1996). Sign of the times. *Training and Development*. Feb.: 32–36.

Galbrath, J. R. (1995). *Designing organizations*. Jossey-Bass. San Francisco.

Gale, B. T. (1994). *Managing customer value*. The Free Press. New York.

Galgano, A. (1994). *Company wide quality management*. Productivity Press. Portland, OR.

Garvin, D. A. (1988). *Managing quality*. Free Press. New York.

Gibbons, A. S. and Rogers, D. H. (1991). The practical concept of an evaluator and its use in the design of training systems. *Educational Technology*. Nov.: 7–15.

Glenn, C. F. (1994). Virtual or Real? The mind in Cyberspace. *The Futurist*. Mar.-Apr.: 22.

Godfrey, A. B. (1996). Integrating quality and strategic planning. *Quality Digest*. Mar.: 15.

Goldratt, E. M. and Cox, J. (1992). *The goal*. 2nd rev. ed. North River Press. Great Barrington, MA.

Golstein, J. (1994). *The unshackled organization: facing the challenge of unpredictability through spontaneous reorganization*. Productivity Press. Portland, OR.

Goodden, R. (1995). Quality and product liability. *Quality Digest*. Oct.: 35–41.

Gottschalk, M. A. (1994). Engineering enters the virtual reality world; VR technology steps out of the arcade and into the engineering tool box. *Design News*. May 9: 23–24.

Graves, S. B. (1995). Common principles of quality management and development. *Quality Management Journal*. Winter: 65.

Greene, A. (1996). Looking beyond ERP for a supply chain advantage. *Managing Automation*. Jan.: 6–14.

Griffin, J. (1995). *Customer loyalty*. Lexington Books. New York.

Grossman, S. R. (1994). Why TQM does not work ... and what you can do about it. *Industry Week*. Jan. 3: 57–62.

Gunther, B. H. (1989). The use and abuse of C_{pk}. (Three part series.) *Quality Progress*. Jan.-May: 72–76, 79–80.

Hammer, M. and Champy, J. (1993). *Reengineering the corporation*. Harper Business. New York.

Hammock, G. (1995). Enter IBM's virtual world. *Manufacturing Engineering*. Mar.: 50.

Handy, C. (1995). Trust and the virtual corporation. *Harvard Business Review*. May/June: 40–50.

Heckman, F. (1996). The participative design approach. *Journal for Quality and Participation*. Mar.: 48–51.

Hemenway, C. G. and Hale, G. L. (1996). Leveraging ISO 9000 for environmental management system documentation. *Quality Digest*. Mar.: 29–34.

Herzberg, F., Mausner, B., and Snyderman, B. (1959). *The motivation to work*. John Wiley and Sons. New York.

Herzberg, F. (1968). One more time: how do you motivate employees? *Harvard Business Review*. Jan.-Feb.: 53–62.

Heyes, G. B. (1989). Interaction box plots. *Quality*. Apr.: 71–73.

Hirano, H. (1995). *5 pillars of the visual workplace: The sourcebook for 5S implementation*. Productivity Press. Portland, OR.

Holp, L. (1992). Masking choices: self-directed work teams or total quality management? *Training*. May: 69–76.

Hopkins, E. J. (1995). Strategies for defusing customer defections. *Quality Digest*. Dec.: 26–29.

Huber, P. (1996). The quantum consumer. *Forbes*. May 20: 291.

Hurley, H. (1996). Cycle-time reduction: your key to a better bottom line. *Quality Digest*. Apr.: 28–34.

Information Mapping, Inc.(1994). *Demystifying ISO 9000*. 2nd ed. Quality Press. Milwaukee, WI.

Janson, R., Attenello, D., and Uzzi, J. A. (1995). *Reengineering for results: a step by step guide*. Quality Resources. New York.

Jaycox, M. (1996). How to get nonbelievers to participate in teams. *Quality Progress*. Mar.: 45–50.

Jensen, M. (1996). Give'em a jolt. *Performance Strategies*. Mar.: 22–27.

Johnson, N. M. (1995). Achieving consensus with the group writing technique. *Quality Digest*. Dec.: 30–33.

Johnson C. (1996). Before you apply SPC, identify your problems. *Test and Measurement World*. Apr.: 43–48.

Joiner, B. L. (1996). Quality, innovation, and spontaneous democracy. *Quality Progress*. Mar.: 51–58.

Joines, S. and Ayoub, M. A. (1995). Design for assembly: an ergonomic approach. *Industrial Engineering*. Jan.: 42–44.

Juran, J. M. (1962). *Quality control handbook*. McGraw-Hill Book Co. Inc. New York.

JUSE Problem Splving Research Group (Ed.). *TQC solutions: the 14–step process*. Productivity Press. Portland, OR.

Kaeter, M. (1995). An outsourcing primer, *Training and Development*. Nov.:

Kanji, G. K. (1993). *100 statistical tests*. Sage Publications. Thousand Oaks, CA.

Katzenback, J. and Smith, D. (1993). *The wisdom of teams: creating the high performance organization*. Harper Collins. New York.

Kaydos, W. (1991). *Measuring, managing and maximizing performance*. Productivity Press. Portland, OR.

Keenan, T. (1995). Flip flop: automakers, suppliers exchange views on demand. *Ward's Auto World*. July: 62–63.

Keenan, T. (1995). Quest for QS-9000. *Ward's Auto World*. Nov.: 41–44.

Keeney, K. A. (1995). *The audit kit*. Quality Press. Milwaukee, WI.

Keeney, K. A. (1995). *The ISO 9000 Auditor's companion*. Quality Press. Milwaukee, WI.

Kessels, J. P. A. M, and Korthagen, F. A. J. (1996). The relationship between theory and practice: back to the classics. *Educational Researcher*. Apr.: 17–22.

Kezsbom, D. S. (1995). Making a team work: techniques for building successful cross-functional teams. *Industrial Engineering*. Jan.: 39–41.

King, E. (1996). Virtual prototype finds design flaws. *Integrated System Design*. May: 19–28.

Kobayashi, I. (1995). *20 keys to workplace improvement*. rev. Productivity Press. Portland, OR.

Kolka, J. W. (1995). ISO/DIS 14000 and liability exposure—General guidelines and principles contained in ISO/DIS 14004 and the liability relationship to ISO/DIS 14001. *The Complete European Trade Digest*. Nov.: 15–19.

Kolka, J. W. (1995). ISO/DIS 14000 and liability exposure—A symbiotic relationship: Part I. *The Complete European Trade Digest*. Oct.: 14–17.

Koonce, R. (1996). The importance of new learning. *Training and Development*. Mar.: 17.

Kovich, S. (1993). What employees want. *Quality Digest*. Mar.: 32.

Kuei, C-H. (1995). Manager's perceptions of factors associated with quality. *Quality Management Journal*. Spring: 67–80.

Kula, T. R. (1995). Raising customer satisfaction through expert system analysis. *Quirk's Marketing Research*. Oct.: 26, 51–53.

Kouzes, J. M. and Posner, B. Z. (1995). *The leadership challenge*. Jossey-Bass. San Francisco.

Lamprecht, J. L. (1996). *ISO 9000 implementation for small business*. Quality press. Milwaukee, WI.

Lamprecht, J. L. (1994). *ISO 9000 and the service sector: a critical interpretation of the 1994 revisions*. Quality Press. Milwaukee, WI.

Langemo, M. (1995). The rules of records retention. *Office Systems 95*. Oct.: 41–44.

Larson, M. (1996). QS-9000: Small suppliers get behind the wheel. *Quality*. Mar.: 34–37.

Lavinski, D. (1995). In customer satisfaction measurement you do not always get what you expect. *Quirk's Marketing Research Review*. Apr.: 16, 42–43.

Layden, J. E. (1996). A rapidly changing landscape: MES, ERP. *Manufacturing Systems*. Mar.: 10A-18A.

Leach, L. P. (1996). TQM, reengineering, and the edge of chaos. *Quality Progress*. Feb.: 85–96.

Leonard, R. K., Wismer, R. D., and Bosserman, S. (1994). Building an integrated engineering organization. *Research Technology Management*. Nov.-Dec.: 14–20.

Lincoln, S. and Price, A. (1996). What benchmarking books do not tell you. *Quality Progress*. Mar.: 33–44.

Locke, D. (1996). *Global supply*. Irwin Professional. Burr Ridge, IL.

Lovitt, M. (1996). Continuous improvement through the QS-9000 road map. *Quality Progress*. Feb.: 39–44.

Lucas, H. C. (1995). *T-Form organization*. Jossey-Bass. Sanfrancisco.

Lawler, E. E., Mohrman, S. A., and Ledford, G. E. (1995). *Creating high performance organizations*. Jossey-Bass. San Francisco.

MacGregor, J. F (1996). Using on-line process data to improve quality. *ASQC Statistics Division Newsletter*. Winter: 6–13.

MacCann, H. (1996). Straight talk in teamland. *Ward's Auto World*. Feb.: 27–33.

MacLean, G. E. (1993). *Documenting quality for ISO 9000 and other industry standards*. Quality Press. Milwaukee, WI.

Madansky, A. (1995). The significance of significance. *Quirk's Marketing Research Review*. June/July: 12–14.

Majumdar, N. (1996). Preparing for battle. *The Strategist*. New Delhi, India. Feb. 20: 1.

Manji, J. F. (1996). Hughes revisited: MRP II makes a positive impact. *Managing Automation*. Apr.: 62–64.

Martin, T. (1996). Seeking a registrar for QS-9000. *Automotive Excellence*. Winter: 8–9.

Martin, S. and Spaulding, S. (1994). Ensuring satisfaction. *Quirk's Marketing Research Review*. Oct.: 16–17, 39.

Maskell, B. H. (1991). *Performance measurement for world class manufacturing: A model for American companies*. Productivity Press. Portland, OR.

Maslow, A. H. (1943). A theory of human motivation. *Psychological Review*. (50). pp 374–396.

McArthur, C. D. and Womack, L. (1995). *Outcome management: Redesigning your business systems to achieve your vision*. Kraus Organization Limited. New York.

McDermott, L. (1995). Reengineering middle management. *Training and Development*. Sept.: 36–40.

McIlvaine B. (1995). What is design for assembly—and what is not? *Managing Automation*. Nov.: 53–54.

McIlvaine, B. (1996). Cycling through the supply chain logistics maze. *Managing Automation*. Apr.: 28–31.

McIlvaine B. (1996). What price virtual reality. *Managing Automation*. Jan.: 95

McNealy, R. M. (1993). *Making quality happen: A step-by-step guide to winning the quality revolution*. and Hall. London.

Mestel, R. (1994). Is cyber shopping in the bag? *New Scientist*. June 25: 42.

Meyer, R. S. (1994). Total quality management and the National Labor Relations Act. *Labor Law Journal*. Nov.: 718–721.

Michael, A. and Stover, D. (1993). Living in a virtual world. *Popular Science*. June: 82.

Miltenburg, J. (1995). *Manufacturing strategy: how to formulate and implement a winning plan*. Productivity Press. Portland, OR.

Mitskavich, D. (1996). A joyful ride in uncertain times. *Manufacturing Systems*. Apr.: 88–89.

Mizuno, S. (Ed.). (1988). *Management for quality improvement: The seven new QC tools*. Productivity Press. Portland, OR.

Mohanty, R. P. and Dahanayka, N. (1989). Process improvement: evaluation of methods. *Quality Progress*. Sept.: 45–47.

Mohrman, S. A., Cohen, S. G., and Mohrman, A. M. (1995). *Designing team-based organizations*. Jossey-Bass. San Francisco, CA.

Monagham, R. (1995). Customer management teams are here to stay. *Marketing News*. Nov. 6: 4.

Monden, Y. (1995). *Cost reduction systems: target costing and Kaizen costing*. Productivity Press. Portland, OR.

Morecroft, J. D. W. and Sterman, J. D. (1994). *Modeling for learning organizations*. Productivity Press. Portland, OR.

Morgan, R. B. and Smith, J. E. (1996). *Staffing the new workplace*. Quality Press. Milwaukee, WI.

Mosley, D. C., Moore, C. C. (1994.) TQM and partnering: an assessment of two major change strategies. *PM Network*. Sept.: 22–26.

Mullet, G. M. (1994). Regression. *Quirk's Marketing Research Review*. Oct.: 12–15.

Nadkarni, R. A. (1995). A not-so-secret recipe for successful TQM. *Quality Progress*. Nov.: 91–96.

Noer, D. M. (1995). *Healing the wounds: overcoming the trauma of layoffs and revitalizing downsized organizations*. Jossey-Bass. San Francisco.

Nuese, C. J. (1995). *Building the right things right: a new model for product and technology*. Quality Resources. New York.

Odenwald, S. (1996). Global work teams. *Training and Development*. Feb.: 54–57.

Odenwald, S. B. and Matheny, W. G. (1996). *Global impact*. Irwin Professional. Burr Ridge, IL.

Orme, B. (1996). Helping managers understand the value of conjoint. *Quirk's Marketing Research Review*. Mar.: 14–17, 44–45.

Pace, P. G. (1996). Safety versus productivity it's not an economic issue. *Metal Forming*. Mar.: 55– 58.

Parker, G. M. (1995). *Team players and teamwork*. Jossey-Bass. San Francisco.

Parry, S. B. (1994). *From managing to empowering: an action guide to developing winning facilitation skills*. Quality Resources. New York.

Patterson, J. (1995). Adding value by managing supply chain activities. *Marketing News*. July 17: 6.

Paul, D. S. (1988). Reliability growth monitoring. *Quality*. Feb.: 87–88.

Pennucci, N. J. (1988). MTBF and reliability. *Quality*. Mar.: 73.

Perigord, M. (1991). *Achieving total quality management: a program for action*. Productivity Press. Portland, OR.

Poirier, C. C. and Tokarz, S. J. (1996). *Avoiding the pitfalls of total quality*. Quality Press. Milwaukee, WI.

Prince, J. B. (1994). Performance and reward practices for total quality. *Quality Management Journal*. Jan.: 36.

Pyzdek, T. (1993). Process control for short and small runs. *Quality Progress*. Apr.: 51–60.

Rado, L. G. (1989). Enhance product development by using capability indexes. *Quality Progress*. Apr.: 38–41.

Raia, E. (1993). Quality: a journey without end. *Purchasing*. Jan. 14: 55.

Rhinesmith, S. H. (1996). *Manager's guide to globalization*. Irwin Professional. Burr Ridge, IL.

Risher, H. and Fay, C. (Eds). (1995). *The performance imperative*. Jossey-Bass. San Francisco.

Ristelhueber, R. (1995). What is behind management pay: competition for talent and changes in responsibilities reshape compensation packages in the 1990s. *Electronic Business Today*. Dec.: 42–48.

Robinson, A. (Ed.) (1991). *Continuous improvement: a systematic approach to waste reduction*. Productivity Press. Portland, OR.

Rogers, E. M. (1995). *Diffusion of innovations*. The Free Pres. New York.

Roy, R. (1996). Getting to the core. *The Strategist*. New Delhi, India. Feb. 20: 2.

Rummler, G. A. and Brache, A. P. (1995). *Improving performance: how to manage the white space on the organization chart*. 2nd ed. Jossey-Bass. San Francisco.

Russell, J. P. (1995). *Quality management benchmark assessment*. 2nd. ed. Quality Press. Milwaukee, WI.

Russell, J. W. (1996). Safety programs and audits. *Metal Forming*. July: 22–32.

Rust, R. T., Zahorik, A. J., and Keiningham, T. L. (1995). *Return on quality*. IRWIN. Burr Ridge, IL.

Rydholm, J. (1994). Dwelling on satisfaction. *Quirk's Marketing Research Review*. Oct.: 6–7, 32–33.

Sakurai, M. (1995). *Integrated cost management: a company wide prescription for higher profits and lower costs*. Productivity Press. Portland, OR.

Salegna, G. and Fazel, F. (1996). An integrative framework for developing and evaluating a TQM implementation plan. *Quality Management Journal*. 3(1): 73–84.

Schmauch, C. H. (1995). *ISO 9000 for software developers*. rev. ed. Quality Press. Milwaukee, WI.

Schmidt, J. (1994). Partnering with your client. *PM Network*. Sept.: 27–30.

Schnaars, S. P. (1995). *Managing imitation strategies*. The Free Press. New York.

Semich, S. (1994). Information replaces inventory at the virtual corporation. *Datamation*. July 15: 37–43.

Shiba, S., Graham, A., and Walden, D. (1993). *A new American TQM: four practical revolutions in management*. Productivity Press. Portland, OR.

Scicchitano, P. (1995). Managing the environment with ISO 14000. *Quality Digest*. Nov.: 43–49.

Shaw, D. S. (1992). Computer-aided instruction for adult professionals: a research report. *Journal of Computer-Based Instruction*. Spring: 54–57.

Shertz, R. S. (1988). Effecting a quality plan of action. *Quality*. Jan.: 59.

Shreve, G. S. (1996). Evolution of wastewater treatment technologies in manufacturing industries. *The Journal of Environmental Engineering and Management*. Winter: 4–10.

Slurzberg, L. (1996). Let's put survey errors in perspective. *Quirk's Marketing Research Review*. Feb.: 22–23.

Smith, G. P. (1996). *The new leader: bringing creativity and innovation to the workplace*. St. Lucie Press. Delray Beach, FL.

Smith, J. and Oliver, M. (1992). 6σ: realistic goal or PR ploy? *Machine Design*. Sept. 10: 71–74.

Smith, R. (1996). *The QS-9000 answer book*. Paton Press. Red Bluff, CA.

Spector, P. E. (1993). *SAS programming for researchers and social scientists*. Sage Publications. Thousand Oaks, CA.

Stamatis, D. H. (1996). *TQM in healthcare*. Irwin. Burr Ridge, IL.

Stamatis, D. H. (1997). *150 ways to improve teamwork in your organization*. Paton Press. Red Bluff, CA.

Stamatis, D. H. (1995). QS-9000 revisions: Not far enough? *Quality Digest*. Dec.: 46–49.

Stanowski, A. and Danish, A. (1995). Vendor selection criteria for a health needs assessment. *Quirk's Marketing Research Review*. June/July: 10, 26–28.

Stephenson, E. W. (1995). In process using measurement to drive solutions, not just derive data. *Technology*. Oct.-Nov.: 31–35.

Stevens, N. (1996). The challenge of change. *Manufacturing Systems*. Apr.: 84–86.

Stevens, K. S. (1994). ISO 9000 and total quality. *Quality Management Journal*. Fall: 57.

Stinson, W. A. (1996). *The robust organization*. Irwin Professional. Burr Ridge, IL.

Sudman, S., Bradburn, N. M., and Scharz, N. (1995). *Thinking about answers: the application of cognitive processes to survey methodology*. Jossey-Bass. San Francisco.

Sujansky, J. (1991). *The power of partnering*. Pfeiffer and Co. San Diego.

Swartz, J. (1994). *The hunters and hunted: A nonlinear solution for reengineering the workplace*. Productivity Press. Portland, OR.

Swift, J. A. (1995). *Introduction to modern statistical quality control and management*. St. Lucie Press. Delray Beach, FL.

Takei, T. (1993). Fujitsu's neuro computer and virtual reality development projects. *Japan 21st*. Mar.: 20–22.

Tadikamalla, P. R. (1994). The confusion over 6σ quality. *Quality Progress*. Nov.: 83–85.

Tamini, N. (1995). Assessing the psychometric properties of Deming's 14 points. *Quality Management Journal*. Spring: 38–52.

Taylor, C. M. (1995). Benchmarking: the sincerest form of flattery. *Quality in Manufacturing*. Nov./Dec.: 18–19.

Taylor, D. and Ramsey, R. (1993). Empowering employees to 'just do it.' *Training and Development Journal*. May: 71–76.

Tibor, T. (1995). *ISO 14000: a guide to the new environmental management standards*. Irwin. Burr Ridge, IL.

Tibor, T. and Feldman I. (1995). ISO/DIS 14000 and the eco-management and audit scheme. *The Complete European Trade Digest*. Sept.: 6–8.

Tibor, T. and Feldman, I. (1996). *Implementing ISO 14000*. Irwin Professional. Burr Ridge, IL.

Tomakich, T. (1995). Toward 'best practices' in manufacturing too, handsome is as handsome does. *Technology*. Oct.-Nov.: 36– 39.

Index

Printed and bound by CPI Group (UK) Ltd, Croydon, CR0 4YY
23/10/2024
01778237-0008